Ultrafast Infrared and Raman Spectroscopy

PRACTICAL SPECTROSCOPY

A SERIES

ADDITIONAL VOLUMES IN PREPARATION

Ultrafast Infrared and Raman Spectroscopy

edited by

M. D. Fayer
Stanford University
Stanford, California

CRC Press
Taylor & Francis Group
Boca Raton London New York

CRC Press is an imprint of the
Taylor & Francis Group, an **informa** business

First published 2001 by Marcel Dekker, Inc.

Published 2019 by CRC Press
Taylor & Francis Group
6000 Broken Sound Parkway NW, Suite 300
Boca Raton, FL 33487-2742

© 2001 by Taylor & Francis Group, LLC
CRC Press is an imprint of Taylor & Francis Group, an Informa business

First issued in paperback 2019

No claim to original U.S. Government works

ISBN 13: 978-0-367-44728-1 (pbk)
ISBN 13: 978-0-8247-0451-3 (hbk)

Visit the Taylor & Francis Web site at
http://www.taylorandfrancis.com

and the CRC Press Web site at
http://www.crcpress.com

Preface

The field of ultrafast infrared and Raman spectroscopy is advancing at a remarkable rate. New techniques and laser sources are making it possible to investigate a wide range of problems in chemistry, physics, and biology, using ultrafast time domain vibrational spectroscopy. Although the first infrared measurements were made by Isaac Newton in the early 1700s, it is only recently that an explosion of activity using ultrafast pulsed techniques has moved vibrational spectroscopy along the path that magnetic resonance spectroscopy followed almost from its inception.

Vibrational spectroscopy examines the internal mechanical degrees of freedom of molecules and the external mechanical degrees of freedom of condensed matter systems. It is the direct connection among vibrational spectra, molecular structure, and intermolecular interactions that has made vibrational spectroscopy an indispensable tool in the study of molecular matter. In addition, most chemical, physical, and biological processes are thermal. Such processes involve the time evolution of the mechanical degrees of freedom of molecules on their ground electronic state potential surfaces. This is the purview of vibrational spectroscopy. The advent of ultrafast pulsed vibrational spectroscopy, using both resonant infrared and Raman methods, is fundamentally changing the nature of the information that can be obtained about condensed matter molecular materials. It is now possible to examine the structural evolution of systems on the time scales on which the important events are occurring.

All the powerful methods of magnetic resonance, from solid-state nuclear magnetic resonance (NMR) to medical magnetic resonance imaging, depend on measuring the time evolution of a spin system following the application of one or more radio frequency pulses. In the visible and ultraviolet, ultrafast optical pulse sequences have been used for many years to measure both population dynamics and coherence phenomena. At low temperatures, electronic transitions of complex molecules can have narrow, homogeneous line widths even if the absorption spectra display broad, inhomogeneous lines. In low-temperature crystals and glasses, optical coherence methods, such as photon echoes and stimulated photon echoes, have been highly successful at extracting a great deal of information about dynamics and intermolecular interactions. As visible pulse durations became increasingly short, photon echoes and related sequences have been applied to molecules in room-temperature liquids. Many elegant experiments have begun to extract some information from such systems. However, there is an intrinsic problem: Because of the exceedingly short electronic dephasing times of complex molecules at high temperatures, ultrashort pulses (tens of femtoseconds or less) are required to perform the experiments. Ultrashort pulses have very large bandwidths, resulting in the excitation of a vast number of vibronic transitions in complex molecules. Experiments of this type cannot be described properly in terms of two states coupled to a medium. The complex multistate superposition that is initially prepared by the broad bandwidth radiation field has a time evolution that depends on the nature and magnitude of the many states that comprise the superposition as well as the system's interactions with the medium. It is difficult to develop a detailed understanding of such experiments except when they are performed on simple molecules (e.g., diatomics).

The electronic absorption spectra of complex molecules at elevated temperatures in condensed matter are generally very broad and virtually featureless. In contrast, vibrational spectra of complex molecules, even in room-temperature liquids, can display sharp, well-defined peaks, many of which can be assigned to specific vibrational modes. The inverse of the line width sets a time scale for the dynamics associated with a transition. The relatively narrow line widths associated with many vibrational transitions make it possible to use pulse durations with correspondingly narrow bandwidths to extract information. For a vibration with sufficiently large anharmonicity or a sufficiently narrow absorption line, the system behaves as a two-level transition coupled to its environment. In this respect, time domain vibrational spectroscopy of internal molecular modes is more akin to NMR than to electronic spectroscopy. The potential has already been demonstrated, as described in some of the chapters in this book, to perform pulse sequences that are, in many respects, analogous to those used in NMR. Commercial equipment is available that can produce the necessary infrared (IR) pulses for such experiments, and the equipment is rapidly becoming less expensive, more compact, and more reliable. It is possible, even likely, that coherent IR pulse-sequence vibrational spectrometers will become available for general use, much as NMR spectrometers have gone from home-built, specialized machines to instruments widely used in many areas of science.

While the internal vibrational modes of molecules can display sharp spectral features, the vibrational spectra of modes of bulk matter are broad and relatively featureless. Nonetheless, Raman and infrared methods can be used to study the bulk, the intermolecular degrees of freedom of condensed matter systems. A great deal of information on bulk degrees of freedom has been extracted from electronic spectroscopy, particularly at low temperatures. Such experiments, however, rely on the influence of the medium on an electronic transition. Using ultrafast Raman techniques, including multidimensional methods, and emerging far-IR methods, it is possible to examine the bulk properties of matter directly.

A remarkable collection of individuals has been assembled to contribute to the book — experimentalists and theorists who are at the forefront of the advances in ultrafast infrared and Raman spectroscopy. They discuss a diverse set of important chemical, physical, and biological problems and a broad range of experimental and theoretical methods. While the experimentalists all use theory to understand their results, the inclusion of top theorists adds to the comprehensive nature of the book. The theorists are developing descriptions of the new techniques and methods for interpreting the results. The wealth of data that has emerged from the application of new methods has spawned a great deal of theoretical effort. In turn, new theoretical methods drive the experiments by placing them in proper context and indicating lines for new experimental endeavors.

The experiments discussed in this book are diverse, but they break down into two broad categories: (1) resonant infrared methods in which ultrafast IR pulses are tuned to the wavelength of the vibrational transition and (2) Raman methods (in some instances referred to as *impulsive stimulated scattering*), in which two visible wavelengths have a difference in frequency equal to the vibrational frequency. In some experiments, infrared and Raman techniques are combined in a single measurement.

There is another manner in which the experiments can be separated into two broad categories. In some of the experiments, the time evolution of vibrational populations are studied. For example, a particular vibration may be excited with an infrared pulse of light, and then the time evolution of the population is followed with either infrared or Raman probe techniques. In other experiments, a chemical reaction is begun with an ultrafast visible pulse, and the time evolution of the chemical reaction is followed with ultrafast infrared pulses that monitor the time dependence of the vibrational spectrum. In another class of experiments, vibrational coherence experiments are performed. Experiments such as the infrared vibrational echo or Raman vibrational echo are closely analogous to NMR spin echo. Such experiments, in one- and two-dimensional incarnations, examine the time evolution of the phase relationship among vibrations.

Both population and coherence experiments provide information on the dynamics and interactions of condensed matter systems. In addition, time domain vibrational experiments can extract spectroscopic information that is hidden in a conventional measurement of the infrared or Raman spectra. This book will provide the reader with a picture of the state of the art and a perspective on future developments in the field of ultrafast infrared and Raman spectroscopy.

M. D. Fayer

Contents

Contributors

Philip A. Anfinrud, Ph.D. Laboratory of Chemical Physics, National Institute of Diabetes and Digestive and Kidney Diseases National Institutes of Health, Bethesda, Maryland

Mark A. Berg, Ph.D. Department of Chemistry and Biochemistry, University of South Carolina, Columbia, South Carolina

David A. Blank, Ph.D.* Department of Chemistry, University of California at Berkeley, and Lawrence Berkeley National Laboratory, Berkeley, California

Binny J. Cherayil, Ph.D. Department of Inorganic and Physical Chemistry, Indian Institute of Science, Bangalore, India

Vladimir Chernyak, Ph.D. Department of Chemistry, University of Rochester, Rochester, New York

Minhaeng Cho, Ph.D. Department of Chemistry, Korea University, Seoul, South Korea

Timothy F. Crimmins, Ph.D. Department of Chemistry, Massachusetts Institute of Technology, Cambridge, Massachusetts

John C. Deàk, Ph.D.† Department of Chemistry, University of Illinois at Urbana-Champaign, Urbana, Illinois

---

* *Current affiliation*: University of Minnesota, Minneapolis, Minnesota
† *Current affiliation*: Procter & Gamble Company, Ross, Ohio

James L. Skinner, Ph.D. Department of Chemistry, University of Wisconsin–Madison, Madison, Wisconsin

Richard M. Stratt, Ph.D. Department of Chemistry, Brown University, Providence, Rhode Island

Andrei Tokmakoff, Ph.D. Department of Chemistry, Massachusetts Institute of Technology, Cambridge, Massachusetts

Haw Yang, Ph.D.* Department of Chemistry, University of California at Berkeley, Berkeley, California

----

* *Current affiliation*: Harvard University, Cambridge, Massachusetts

1

Ultrafast Coherent Raman and Infrared Spectroscopy of Liquid Systems

Alfred Laubereau and Robert Laenen
Technische Universität München, Garching, Germany

Vibrational and structural dynamics in condensed molecular systems are of special interest because they provide a basis for the understanding and manipulation of important material properties and processes in various fields in material science, chemistry, and biology. Simple examples include heat transport, shear viscosity, and ultrasonic absorption, which originate from complex intermolecular interactions. A special role is played here by hydrogen bonding, which is abundant in nature and has profound effects on microscopic structures. While there is a wealth of information on the structure and certain dynamical properties of H-bonded systems obtained using various techniques such as neutron/x-ray scattering, nuclear magnetic resonance (NMR), or dielectric relaxation, (1,2) direct time-resolved observations of structural relaxation are still missing. Some first steps in this direction involve time-resolved spectral holburning observations. These studies benefit from novel laser sources emitting pulses in the infrared (IR) spectral region with picosecond to femtosecond duration and exploit the OH stretching vibration as a local probe for the hydrogen-bonding environment.

The first time-resolved investigations on vibrational dephasing and vibrational lifetimes of molecules in the liquid phase were reported in 1971 and 1972 by Kaiser et al. utilizing nonlinear Raman scattering (3,4). A combination of infrared excitation with spontaneous Raman probing was applied a few years later by the same group (5). Infrared pump and probe measurements were first conducted by Chesnoy and Ricard (6) and Heilweil and coworkers (7), the latter one in liquids, but with only one tunable IR pulse. The more elaborate two-color versions of such experiments representing time-resolved infrared spectroscopy were demonstrated by Laubereau and coworkers (8). With the help of this powerful spectroscopic method, detailed information on intra- and intermolecular energy relaxation processes of molecules was obtained. The technique was first applied to smaller polyatomic molecules like $CHBr_3$ in different solvents (9,10). Implementing polarization resolution for the measured probe absorption, molecular reorientation times were also measured (11).

While the first experiments of time-resolved IR spectroscopy were conducted with pulse durations exceeding 10 ps, the improved performance of laser systems now offers subpicosecond (12) to femtosecond (13–15) pulses in the infrared spectral region. In addition, the pump-probe techniques have been supplemented by applications of higher-order methods, e.g., IR photon echo observations (16).

In this chapter we will first discuss coherent anti-Stokes Raman scattering (CARS) of simple liquids and binary mixtures for the determination of vibrational dephasing and correlation times. The time constants represent detailed information on the intermolecular interactions in the liquid phase. In the second section we consider strongly associated liquids and summarize the results of time-resolved IR spectroscopy (see, e.g., Ref. 17) on the dynamics of monomeric and associated alcohols as well as isotopic water mixtures.

I. COHERENT ANTI-STOKES RAMAN SPECTROSCOPY OF SIMPLE LIQUIDS

A. Introduction

Understanding the mechanism governing the shape and width of spectroscopic lines has challenged a great number of spectroscopists (18–20). For molecular vibrations in condensed matter, the role of dephasing processes in addition to energy relaxation and molecular reorientation was recognized more than 30 years ago, providing a qualitative description. From the observed linewidth in liquids at room temperature with values of the order of magnitude of a few cm^{-1}, the time scale of 10^{-12} s was readily estimated (20). With the advent of ultrashort laser pulses, direct time-resolved techniques became experimentally accessible in addition to conventional infrared and Raman spectroscopy. To study the dephasing properties of vibrational transitions, nonlinear Raman spectroscopies have been developed, representing special versions of the pump-probe technique, e.g., coherent anti-Stokes Raman scattering (CARS) and coherent Stokes Raman scattering (CSRS) (21). The first liquid examples were stretching vibrations of carbon tetrachloride, ethanol at room temperature (3), and the fundamental mode of liquid nitrogen (22), while phonon modes of calcite (23,24) and diamond (25) were addressed in the early solid-state investigations. Over the past decades a variety of gases and liquid and solid state systems have been studied using time-domain CARS (26–28). Higher-order Raman techniques were also demonstrated (29,30).

B. General Considerations

The CARS and CSRS processes are generally described as four-wave mixing (31,32); in the time domain spectroscopy with delayed pump and probe fields the elementary scattering mechanism is split into a two-step two-wave interaction (21). For excitation two laser pulses are applied, i.e., two coherent electromagnetic waves with appropriate frequency difference interact with the molecular ensemble and drive a specific vibrational mode with transition frequency ω_0 resonantly (or close to resonance); "Raman" is used here as a synonym for "frequency difference resonance." The same interaction is involved in the stimulated Raman effect, so that the latter process was applied in early measurements for the excitation process. The probing process is coherent scattering of the additional interrogation pulse off the phase-correlated vibrational excitation, i.e., classical scattering involving the induced polarization of the molecular ensemble and producing side bands $\omega_P \pm \omega_0$ (Stokes and anti-Stokes) of the probe frequency ω_P. The process is the optical phonon analog for light scattering of coherent acoustic phonons in ultrasonics (e.g., Debye-Sears effect). The two-step interaction is illustrated in Fig. 1. The pumping process is represented by the simple energy level scheme of Fig. 1a with the ground and the first excited levels of the considered vibration; the vertical arrows represent the involved pump photons with frequencies ω_L ("laser") and ω_S ("Stokes"), respectively. The wave vector diagram is also depicted in Fig. 1a; an off-axis beam geometry is assumed for the input fields represented by wave vectors k_L and k_S. The resulting vector k_v represents the spatial phase relation of the vibrational excitation imposed on the molecular ensemble.

The coherent anti-Stokes scattering of a probing pulse generating radiation with frequency $\omega_A = \omega_P + \omega_0$ and wave vector k_A is depicted in

relative to the coincident excitation components. It is important to note that the relative contributions of P depend on the orientations of the electric field vectors, i.e., chosen polarization geometry. The latter effect is described in Equations (2)–(4) by the time-independent prefactors F that are explicitly known (see below). The F's also contain the different coupling elements of $\partial\alpha/\partial q$ and χ_{nr} [Equation (1)]. Φ_{vib} and Φ_{or}, respectively, represent the vibrational and orientational autocorrelation functions of individual molecules and enter Equations (2)–(4) in various ways; the resulting differences in temporal behavior of the scattering parts are significant. The equations above refer to moderate pulse intensities so that stimulated amplification of the Stokes pulse and depletion of the laser pulse can be ignored.

The measured CARS signal S^{coh} is proportional to the time integral over the absolute value squared of the total third-order polarization, $P = P_{iso} + P_{aniso} + P_{nr}$, because of the slow intensity response of the detector:

$$S^{coh}(t_D) = \text{const} \times \int_{-\infty}^{\infty} |P(t, t_D)|^2 \, dt \tag{5}$$

The signal S^{coh} represents a convolution integral of the intensity of the probing pulse $\propto |E_P(t - t_D)|^2$ with the molecular response; the latter is governed by the autocorrelation functions Φ_{vib} and Φ_{or}. Numerical solutions of Equations (2)–(5) are readily computed and will be discussed in the context of experimental results.

Φ_{vib} and Φ_{or} also show up in the theory of spontaneous Raman spectroscopy describing fluctuations of the molecular system. The functions enter the CARS interaction involving vibrational excitation with subsequent dissipation as a consequence of the dissipation-fluctuation theorem and further approximations (21). Equations (2)–(5) refer to a simplified picture; a collective, delocalized character of the vibrational mode is not included in the theoretical treatment. It is also assumed that vibrational and reorientational relaxation are statistically independent. On the other hand, any specific assumption as to the time evolution of Φ_{vib} (or Φ_{or}), e.g., if exponential or nonexponential, is made unnecessary by the present approach. Homogeneous or inhomogeneous dephasing are included as special cases. It is the primary goal of time-domain CARS to determine the autocorrelation functions directly from experimental data.

Regarding the relationship between CARS and conventional Raman spectroscopy, as is evident from the equations above, the scattered anti-Stokes field amplitude (proportional to P) depends linearly on the autocorrelation functions. With respect to molecular dynamics and disregarding the minor point that the field amplitude is not directly measured, CARS is a linear spectroscopy and cannot provide more information than is available from conventional Raman spectroscopy. On the level of present theoretical approaches, both methods are simply related by Fourier transformation and deliver the same information. This is of course only true in principle, not in practice for real measurements, because of the different role of experimental accuracy in the two techniques. For example, the asymptotic exponential decay of Φ_{vib} was observed over more than three orders of magnitude, while the Raman bandshape could not be measured with similar precision because of the contributions of neighboring lines, especially in congested parts of the spectrum. In short, coherent experiments can provide dephasing data of superior accuracy. On the other hand, conventional Raman spectroscopy is well suited for measuring frequency positions or shifts. The time- and frequency-domain versions of vibrational spectroscopy are complementary, and the combination of the respective results is particularly rewarding.

As far as CARS distinguishing between homogeneous and inhomogeneous broadening mechanisms, some investigators supported the idea that CARS as a linear technique with respect to molecular response does not do this (36). The present authors question that opinion; in fact, examples will be discussed below in which dephasing in the homogeneous, intermediate, or inhomogeneous case was distinguished on the basis of femtosecond CARS data. On the other hand, it is generally accept that higher-order techniques like infrared echo or Raman echo measurements can more directly differentiate between homogeneous and inhomogeneous dephasing mechanisms (37).

Two important improvements in time-domain CARS spectroscopy have been made in recent years and will be briefly discussed in the following areas:

> High-precision CARS (38)
> CARS with magic polarization geometry (35,39)

C. Experimental Aspects

In the early days of time-resolved CARS it was often convenient to use laser and probing pulses at the same frequency position, leading to two-color CARS ($\omega_P = \omega_L$). The approach has the disadvantage that secondary interaction processes of the excitation pulses also generate emission at the anti-Stokes frequency position $\omega_L + \omega_0$, representing an undesirable background (not depending on delay time) for the detection of the coherent probe scattering at $\omega_P + \omega_0$. In more advanced approaches, therefore, the frequency coincidence is avoided (22,38). The latter version, three-color

CARS, can provide more accurate data because of its higher sensitivity and lower intensity level of the excitation pulses. The preferred frequency position of the probing pulse, in general, is between the laser and Stokes components, $\omega_L > \omega_P > \omega_S$. We mention here that phase matching arguments for anti-Stokes scattering (21) would suggest a frequency position close to the Stokes frequency, but the finite bandwidth of ultrashort pulses makes a significant frequency shift necessary between the (intense) laser pump and (weak) anti-Stokes scattering at $\omega_P + \omega_0$.

As an example the experimental apparatus used by the authors' group is briefly discussed. The system is based on femtosecond dye laser technology and depicted schematically in Fig. 2b (38,40). Using an amplified and frequency-doubled, modelocked Nd-YLF laser with repetition rate 50 Hz for synchronous pumping, a hybrid modelocked dye-laser oscillator is operated. After multipass dye amplification of a single pulse, part of the laser radiation is directed to a quartz plate for continuum generation. Out of the produced spectral broadening, two frequency bands are selected by pairs of interference filters and amplified in two additional dye amplifiers for the generation of the Stokes and probe pulses. Together with the second part of the laser pulse that also passes narrow-band filters, three different input pulses of approximately 250 fs duration and $50-70$ cm^{-1} width are accomplished. For a given set of three pairs of interference filters and amplifier dyes, tuning ranges of the three pulses are accomplished by angle variation of the filters ($565-571$ nm, $675-689$ nm, and $605-619$ nm for L, S, and P, respectively). A nonlinear absorber cell (NA) in the probe beam in front of the sample improves the pulse contrast and helps to increase the dynamical range of the CARS scattering signal.

Applying $\lambda/2$ plates and a Glan polarizer (Pol1), parallel linear polarization of the input laser and Stokes pulses is adjusted. For reasons discussed below the polarization plane of the probe pulse (Pol2) is inclined by an angle $\theta_P = 60°$ with respect to the pump polarization, while in earlier work an angle of 90° was used. High-quality polarization optics including a 2 mm sample cell practically free of stress birefringence are used. An off-axis beam geometry is adopted providing phasematching for the anti-Stokes scattering of the probe pulse, as calculated from refractive index data.

The coherent Raman scattering is measured behind an analyzing polarizer (Pol3) transmitting radiation with the polarization plane oriented at angle θ_A relative to the vertical pump polarization. A small aperture (AP) defines the solid angle of acceptance ($\approx 10^{-5}$ sr) along the phasematching direction. The scattering is detected at the proper anti-Stokes frequency position, using dielectric filters (IF) with a bandwidth of 80 cm^{-1}, variable

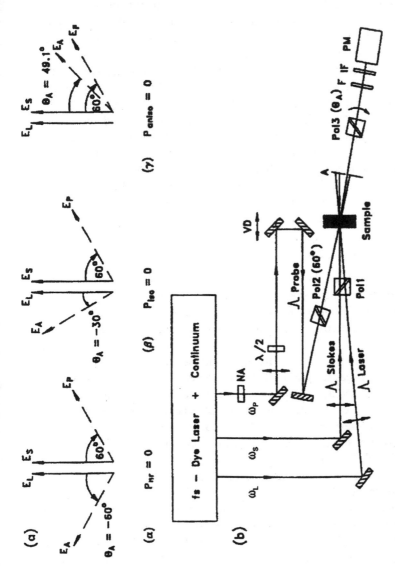

Figure 2 (a) Polarization geometries for the suppression of the nonresonant (α), resonant-isotropic (β), and resonant-anisotropic (γ) CARS components. Constant polarization of the input fields E_L, E_S, and E_P; magic angles θ_A for the orientation of the detected anti-Stokes field E_A. (b) Schematic diagram of the experimental system for three-color CARS with magic polarization conditions. NA, nonlinear absorber; VD, variable delay; Pol1-Pol3, polarizers; A, aperture; F, calibrated neutral filters; IF, interference filters; PM, photomultiplier.

neutral filters (F), and a photomultiplier (PM). The input pulse energies are also monitored and used to correct the signal amplitude for the single shot fluctuations (<20%) of the input pulses. The instrumental response function, determined by a measurement of the nonresonant CARS signal of carbon tetrachloride [compare Equation (4)] decays exponentially over an accessible dynamical range of 10^6, suggesting exponential wings of the input pulses. From the decay of the curve with a slope of $1/60$ fs^{-1}, the available experimental time resolution is deduced. In earlier applications of the experimental setup a slightly different time resolution of 80 fs was achieved. An example is shown in Fig. 3a (open circles, dashed curve). For the adjusted frequency difference in wavenumber units of $(\omega_L - \omega_S)/2\pi c = 2925$ cm^{-1} in CCl_4, off-resonance CARS via the nonresonant part χ_{nr} of the third-order nonlinear susceptibility is measured and plotted in the figure on a logarithmic scale. The signal maximum is normalized to 1, while its abscissa position defines zero delay. The observed steep signal decay by a factor of 10^6 within 1 ps is noteworthy.

1. High Precision fs-CARS

For a demonstration of the performance of the instrumental system, some results for neat acetone at room temperature are depicted in Fig. 3a (38). The symmetrical CH_3 stretching mode at 2925 cm^{-1} is resonantly excited. The anti-Stokes scattering signal of the probing pulse with perpendicular polarization plane relative to the pump beams is plotted versus delay time (full points, logarithmic scale). The maximum scattering signal (exceeding the off-resonance scattering of CCl_4 by two orders of magnitude) is normalized to unity and displays a small delay relative to the instrumental response function. For $t_D > 0.5$ ps the signal transient decreases exponentially over a factor $>10^6$ corresponding to a linear dependence in the semi-log plot. From the slope of the decay curve the time constant $T_2/2 = 304 \pm 3$ fs is directly deduced. For long delays a weak background signal shows up. The solid curve in Fig. 3a is calculated from Equations (2)–(4). The relevant fitting parameter for the resonant CARS signal is the dephasing time T_2.

The accuracy of the data is illustrated by Fig. 3b. The ratio of the signal amplitudes of the experimental points to that of the calculated signal curve of Fig. 3a is plotted. It is interesting to see the minor scatter of the data with approximately constant experimental error ($\leq 10\%$) in spite of the signal variation over many orders of magnitude. Each experimental point represents the average of approximately 400 individual measurements. The reproducibility of the slope of the signal decay is better than $\pm 3 \times 10^{-3}$.

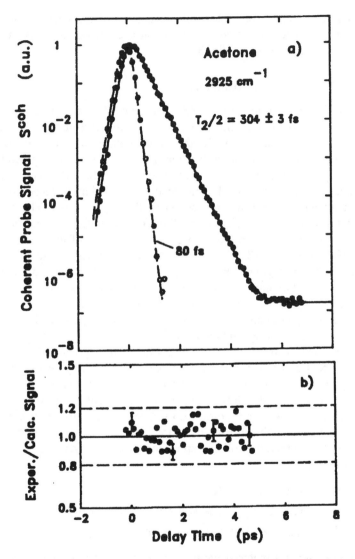

Figure 3 Femtosecond nondegenerate CARS in liquids: (a) Coherent probe scattering signal versus delay time; open circles, dashed curve: nonresonant scattering of CCl_4 yielding the instrumental response function and the experimental time resolution of 80 fs; full points, solid line: resonant CARS signal from the CH_3-mode of acetone at 2925 cm^{-1}, obtaining $T_2/2 = 304 \pm 3$ fs. (b) Ratio of experimental and calculated scattered data of (a) for acetone versus delay time; the small experimental error of the data points extending over 6 orders of magnitude is noteworthy.

Taking into account a possible calibration error of the neutral filters used to detect the CARS signal, an experimental accuracy of $\pm 1\%$ is estimated for the T_2 measurement of Fig. 3.

2. Magic Polarization Conditions

Early work on time-domain CARS was devoted to the measurement of the vibrational dephasing time T_2, i.e., the time constant accounting for the asymptotic signal decay. In the general case (not fully depolarized vibrational transition, sufficiently short pulses), the latter originates from the isotropic component of the nonlinear polarization P, since the other parts decrease more rapidly. The nonresonant contribution responds almost instantaneously and follows the wings of the input pulses. The decay of the anisotropic part is accelerated by the additional effect of reorientational motion compared to the purely vibrational relaxation of the isotropic scattering [Equations (2), (3)]. The remaining problem for the spectroscopist, of course, is to recognize when the signal transient has reached the asymptotic behavior. For more information on molecular dynamics, it is highly desirable to separate the three scattering contributions.

A remedy obviously should be available using polarization tricks. In conventional Raman spectroscopy, the isotropic and anisotropic components are deduced from linear combinations of the "polarized" and "depolarized" spectra, while a nonresonant part is not clearly recognized (41). In frequency-domain CARS it is known how to suppress the nonresonant contribution and solely measure resonant scattering (isotropic plus anisotropic part) (42). In time-domain CARS, polarization interference can do an even better job with three "magic" cases (derived in Refs. 35,39). These authors derived explicit expressions for the coupling factors F in Equations (2)–(4):

$$F_{iso} = i\kappa\alpha^2 \cos(\theta_P - \theta_A) \tag{6}$$

$$F_{aniso} = i\kappa 2/45 \times \gamma^2[2\cos\theta_P\cos\theta_A - \sin\theta_P\sin\theta_A] \tag{7}$$

$$F_{nr} = \chi_{nr}/2 \times [3\cos\theta_P\cos\theta_A + \sin\theta_P\sin\theta_A] \tag{8}$$

κ combines several material parameters. α and γ denote the isotropic and anisotropic parts of the Raman polarizability tensor $\partial\alpha/\partial q$. χ_{nr} represents here the χ_{xxxx} element of the nonresonant third-order susceptibility. The above equations refer to the parallel pump polarization depicted in Fig. 2b.

The above expressions show that for the polarization geometry often adopted in earlier investigations with $\theta_P = \theta_A = 90°$, the isotropic contribution is maximal but the two other components are also present. It is more attractive to choose $\theta_P \neq \theta_A$ and consider situations with constant $\theta_P = 60°$ and variable θ_A. Three magic values are found, where one of the coupling factors alternatingly vanishes, $F_i(\theta_A) = 0$:

$$\theta_A = \tan(2/\sqrt{3}) \simeq 49.1° \qquad \text{(no anisotropic contribution)}$$
$$\theta_A = -30° \qquad \text{(no isotropic component)}$$
$$\theta_A = -60° \qquad \text{(no nonresonant contribution)}$$

Simply adjusting these values for the analyzer orientation, different signal transients are measured where the CARS signal contains only two contributions. The magic polarization geometries are depicted in Fig. 2a. The theoretical results were verified experimentally (35,39). Reduction of the suppressed components by several orders of magnitude was accomplished.

A set of measurements with the three magic angles allows one to determine the three scattering components with different time dependencies separately. Examples are presented in the next section. The following pieces of information become accessible in this way:

Isotropic scattering: In addition to the dephasing time T_2, the correlation time τ_c of the purely vibrational relaxation process can be measured, providing quantitative information on the question of homogeneous/inhomogeneous line broadening.

Anisotropic part: The reorientational relaxation of the vibrating molecular subgroup becomes directly experimentally accessible.

Nonresonant part: Instrumental response function and zero setting of delay time scale are provided.

Peak amplitudes: The relative magnitudes of the coupling parameters α, γ, and χ_{nr} can be determined.

The mechanism selecting two scattering components out of three is polarization interference. The polarization of each scattering contribution (for sufficiently weak, linearly polarized input fields) is linear but with tilted polarization planes. The isotropic scattering, for example, occurs in the plane of the incident probing field. Blocking of this component simply requires a crossed analyzer with $\theta_A = \theta_P - 90°$.

The polarization dependence of the individual contributions can be measured in special cases when the presence of the other two can be excluded. Figure 4a presents results for the nonresonant CARS of neat carbon tetrachloride excited for $(\omega_L - \omega_S)/2\pi c = 2925$ cm^{-1} while a resonant vibrational mode does not exist; i.e., resonant scattering is absent. The time evolution of the signal curve was presented in Fig. 3a (open circles).

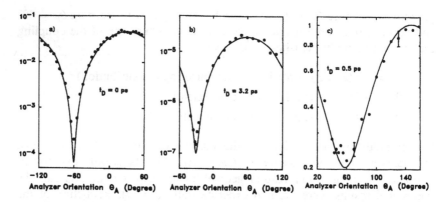

Figure 4 Suppression of scattering components of three-color CARS for magic polarization conditions. The anti-Stokes scattering signal is plotted on a logarithmic scale versus analyzer orientation (for polarization of the input pulses, see Fig. 2). (a) Nonresonant scattering of neat CCl_4 at $t_D = 0$ with a sharp signal minimum at the magic angle $-60°$; (b) resonant isotropic scattering of the ν_2 mode of neat acetone (2925 cm^{-1}) at $t_D = 3.2$ ps with a minimum at $-30°$; (c) dominant resonant anisotropic scattering of neat DMSO excited at 3000 cm^{-1} for $t_D = 0.5$ ps that disappears for the magic angle 49.1°.

The peak scattering signal at $t_D = 0$ is plotted in Fig. 4a versus analyzer orientation θ_A. For the magic angle $-60°$, a sharp signal minimum occurs as predicted by Equations (4) and (8). For the resonant isotropic scattering of the ν_2-mode of acetone, some data are depicted in Fig. 4b. For the chosen delay time of $t_D = 3.2$ ps, the nonresonant scattering has already disappeared and anisotropic scattering is negligible because of the smallness of anisotropy γ. The isotropic scattering, on the other hand, has decayed only by approximately four orders of magnitude, as shown by the signal transient of Fig. 3a (full points). The angle dependence of the isotropic component is shown in Fig. 4b. As predicted, the minimum signal occurs in the figure for the magic angle $\theta_A = -30°$.

Experimental evidence for the elimination of the resonant anisotropic scattering at the magic angle $\theta_A = 49.1°$ is presented in Fig. 4c, where the CARS signal off neat DMSO at room temperature is plotted ($t_D = 0.5$ ps). For excitation at $(\omega_L + \omega_S)/2\pi c \simeq 3000$ cm^{-1} an approximately fully depolarized CH stretching mode at 2996 cm^{-1} is preferentially excited (depolarization factor $\rho \simeq 0.74$, negligible isotropic component). In addition, an adjacent strongly polarized vibration is also weakly pumped that decays more slowly. It can be shown that the anisotropic scattering dominates at 0.5 ps delay by a factor of $\simeq 5$. The corresponding angle dependence of the signal amplitude S^{coh} is depicted in Fig. 4c with the expected minimum at $\simeq 50°$.

Elimination of intense scattering contributions requires high-quality optical components. The finite contrast factor of polarization optics (10^6-10^7) and possible higher-order processes (e.g., six-wave mixing) to the nonlinear scattering limit the suppression potential (39).

An experimental example for the three signal transients under magic polarization conditions is depicted in Fig. 5. Neat bromochloromethane was studied at room temperature. A frequency difference of $(\omega_L - \omega_S)/2\pi c \simeq$ 2987 cm^{-1} was adjusted for the resonant excitation of the symmetrical methylene stretching mode ν_1. The measurements were carried out for the different analyzer orientations during the same experimental runs while the other experimental parameters were kept constant. The measured anti-Stokes scattering signal was plotted on a semi-logarithmic scale versus probe delay (experimental points). The curves were calculated from the theory of coherent Raman scattering [Equations (2)–(8)].

Fig. 5a shows the signal transient for $\theta_A = -30°$ with elimination of resonant-isotropic scattering, so that the signal amplitude represents

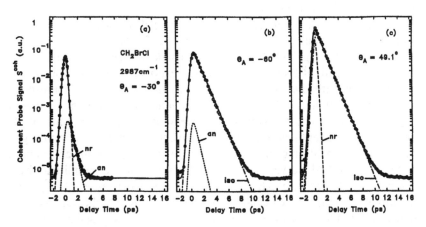

Figure 5 Coherent probe scattering of three-color CARS for the symmetrical CH$_2$ stretching vibration of neat CH$_2$BrCl at 2987 cm^{-1} vs. delay time t_D: (a) with elimination of the isotropic component ($\theta_A = -30°$); the nonresonant and anisotropic contributions are measured (dashed and dotted lines, respectively); (b) with elimination of the nonresonant part ($\theta_A = -60°$) observing a dominant isotropic contribution (dot-dashed curve); (c) with elimination of the anisotropic component ($\theta_A = 49.1°$); experimental points, calculated curves.

the superposition of the nonresonant and resonant-anisotropic components. The former contribution clearly dominates around $t_D = 0$ and provides the instrumental response, i.e., information on the input pulse shapes (broken line in Fig. 5a). The zero-setting of the abscissa scales of Fig. 5a–c is obtained with high accuracy from a comparison of computed data of the nonresonant part and the experimental points around the signal maximum. The finite lifetime of the anisotropic part gives rise to a trailing wing of the signal curve (dotted line in Fig. 5a) from which the reorientational time constant can be determined (see below). The small amplitude of the anisotropic component in comparison with the isotropic part (Fig. 5b,c) is due to the small anisotropy γ as indicated by the depolarization factor $\rho \simeq 0.05$ of the ν_1 Raman band.

A different situation is found for $\theta_A = -60°$, depicted in Fig. 5b. Now the nonresonant component is suppressed and the resonant-isotropic part dominates (dash-dotted line). The anisotropic contribution is negligible, as indicated by the calculated dotted line; the latter curve is obtained from the analysis of the data of Fig. 5a and the known angle dependence of the amplitude factor F_{aniso} [Equation (5)]. The asymptotic exponential decay of the dominant isotropic scattering is verified over approximately four orders of magnitude. The dephasing rate $2/T_2$ is directly determined from the curve yielding $T_2 = 1.65 \pm 0.03$ ps.

The time evolution of the superimposed resonant-isotropic and nonresonant components is depicted in Fig. 5c for $\theta_A = 49.1°$, where the anisotropic scattering is eliminated. After a sudden increase to a maximum at $t_D \simeq 150$ fs, the signal curve decreases exponentially over four orders of magnitude with the same decay time as in Fig. 5b. The nonresonant contribution around $t_D = 0$ is noticeable from the data of Fig. 5c by careful inspection, e.g., from a small signal overshoot and minor shift of the maximum to smaller t_D as compared to the calculated purely isotropic signal (dash-dotted line). Using the data of Fig. 5a the nonresonant contribution of Fig. 5c (dashed curve) is computed with enlarged amplitude by a factor of 3.57 that originates from the angle dependence of F_{nr}^2 [note Equations (7), (8)]. The calculated solid curve represents the superposition of the two scattering components including the cross product term. The latter notably enhances the nonresonant contribution. It is readily seen that around $t_D = 0$ the nonresonant component modifies considerably the total scattering amplitude. Details of the vibrational dynamics for short times, $t_D < 1$ ps, can be inferred from the data of Fig. 5c only if the nonresonant part is properly taken into account.

D. Results and Discussion

The separation of the individual scattering components in time-domain CARS provides a wealth of experimental information not accessible in earlier work from the spectroscopic method. As a result, different aspects of molecular dynamics in condensed matter can be investigated.

1. Reorientational Motion of Liquid Molecules in Time-Domain CARS

In analogy to conventional Raman spectroscopy and using the same approximation of statistically independent relaxation channels, the reorientational dynamics can be simply deduced from a comparison of isotropic and anisotropic signal transients. First measurements of this kind on the picosecond time scale and using different polarization conditions were reported in Ref. 43. An example with superior time resolution is presented in Fig. 6 (35). The CH_2-stretching vibration ν_1 of dichloromethane at 2986 cm^{-1} is studied for resonant excitation at room temperature. The measured signal transients with dominant isotropic contribution are depicted in Fig. 6a for two analyzer orientations. The full points refer to the magic angle 49.1°, where the anisotropic part is suppressed. The open circles, on the other hand, were measured with magic angle $\theta_A = -60°$, eliminating the nonresonant scattering. Simple exponential dephasing is observed over many orders of magnitude providing the dephasing time $T_2/2 = 875 \pm 8$ fs. The signal overshoot at $t_D \simeq 0$ for $\theta_A = 49.1°$ (full points) originates from nonresonant scattering that is absent for $\theta_A = -60°$ (open circles). A contribution of anisotropic scattering cannot be seen for the latter case with the available measuring accuracy.

Of special interest are the data for the magic analyzer orientation $-30°$ suppressing the isotropic scattering, measured in the same experimental runs and for identical conditions as the data of Fig. 6a. The results are presented in Fig. 6b. The ordinate scale refers to the same units as in Fig. 6a, while the time scale of the abscissa is stretched. The measured transient consists of two features: a pronounced signal overshoot around $t_D = 0$ due to nonresonant scattering and an exponential tail, obviously representing the anisotropic contribution. The data represent novel evidence for the exponential time dependence of the reorientational autocorrelation function Φ_{or} [see Equation (3)]:

$$S_{aniso}^{coh}(t_D) \propto (\Phi_{vib}\Phi_{or})^2 \rightarrow \exp(-2t_D/\tau_{an}) \qquad (9)$$

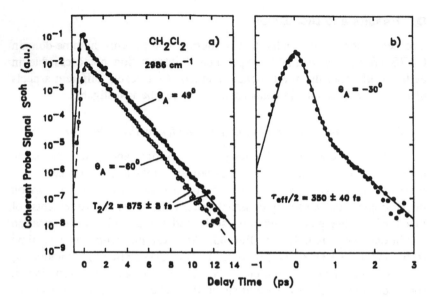

Figure 6 Coherent anti-Stokes scattering $S^{coh}(t_D)$ of the CH stretching mode (2986 cm^{-1}) of neat CH_2Cl_2 vs. delay time of the probe pulse for $\theta_P = 60°$ (295 K): (a) for $\theta_A = 49.1°$ suppressing the anisotropic component but with nonresonant contribution; similar data for $\theta_A = -60°$ with elimination of the coherence peak (open circles); $T_2/2 = 875 \pm 8$ fs; (b) for $\theta_A = -30°$ with suppression of the resonant-isotropic component. A coherence peak around $t_D = 0$ and resonant-anisotropic scattering ($t_D > 0.5$ ps) are measured allowing one to determine the anisotropic relaxation time τ_{an}, $\tau_{an}/2 = 350 \pm 40$ fs. Experimental points, theoretical curves.

where ($t_D > 0.5$ ps). The observations do not support a more complex temporal behavior sometimes inferred from spontaneous Raman band shapes (20). τ_{an} denotes the decay time of the anisotropic scattering component. Since Φ_{vib} decays exponentially (Fig. 6a), the same behavior results for Φ_{or}. From the slope of the signal curve a time constant $\tau_{an}/2 = 350 \pm 40$ fs is directly deduced, in contrast to the value of $T_2/2$ of Fig. 6a for purely vibrational dephasing mentioned above. The result for the reorientation time is $\tau_{or} = (1/\tau_{an} - 1/T_2)^{-1} = 1.2 \pm 0.2$ ps.

Comparing the signal amplitudes of the data in Fig. 6a and b, the peak level of the nonresonant scattering is fully consistent with the prediction of Equation (8). The ratio of coupling constants of the resonant scattering is determined to $\gamma^2/\alpha^2 = 0.8 \pm 0.2$ equivalent to the depolarization factor $\rho = 3\gamma^2/(45\alpha^2 + 4\gamma^2) = 0.05 \pm 0.01$. The number nicely agrees with the

Table 1 Summary of Dephasing Parameters

Parameter	CH$_2$BrCl (2987 cm^{-1})	(CH$_3$)$_2$SO (2913 cm^{-1})	CH$_2$Cl$_2$ (2986 cm^{-1})
T$_2$ (ps)	1.65 ± 0.03	0.982 ± 0.01	1.75 ± 0.02
τ_{or} (ps)	2.2 ± 0.2	2.0 ± 0.2	1.2 ± 0.2
τ_c (fs)	330 ± 70	240 ± 50	—
$\delta\nu_{exp}$ (cm^{-1})	6.3 ± 0.2	10.4 ± 0.2	6.2 ± 0.2
$\delta\nu_{calc}$ (cm^{-1})	6.32 ± 0.12	10.37 ± 0.17	—
$\delta\nu_L$ (cm^{-1})	6.43 ± 0.12	10.81 ± 0.11	6.10 ± 0.07

reported value of $\rho = 0.045 \pm 0.003$ (44) from Raman spectroscopy and strongly supports our interpretation of the data of Fig. 6b. The values of T$_2$ and τ_{an} are consistent with the measured linewidths of the isotropic and anisotropic Raman bands. Similar results for two other liquids are listed in Table 1.

2. Time Scale of the Dominant Dephasing Mechanism

The increased time resolution of fs pulses makes it possible to study rapid features of molecular dynamics, e.g., the non-Markovian behavior of vibrational relaxation at short times (45,46). The analysis of experimental data close to the maximum of the signal transient is, however, made difficult by the additional factors discussed above that obscure the resonant-isotropic scattering around t$_D$ = 0. The experimental problem is solved applying the magic polarization geometries.

In the following we consider pure dephasing, i.e., rapid frequency changes by the fluctuating molecular environment, as the dominant source for the linewidth of the isotropic Raman line and for the time evolution of the corresponding CARS signal. Using the Abragam-Kubo theory an explicit expression can be derived for the vibrational autocorrelation function (20,47):

$$\Phi_{vib}(t) = \exp\{-t/T_2 - [\exp(-t/\tau_c) - 1]\tau_c/T_2\} \qquad (10)$$

for t > 0. The oscillatory part of the autocorrelation function is omitted in Equation (10). Two temporal parameters, τ_c and T$_2$, govern the vibrational dephasing. Equation (10) has model character because of the assumed exponential shape of the underlying memory function (20). The modulation time

τ_c describes the time scale of the dephasing mechanism, while the asymptotic decay time T_2 (dephasing time) is also governed by the strength of the phase-changing events (mean amplitude of the frequency fluctuations).

Some numerical examples are shown in Fig. 7a. The nonexponential part of $\Phi_{vib}(t)$ for short times is readily seen and leads to a delayed onset of the exponential decay. The latter represents the Markovian limit (straight line in the semi-log plot). A time interval of approximately $2 \times \tau_c$ is required to establish the asymptotic dephasing rate $1/T_2$. The purely exponential case, $\tau_c = 0$, is fictitious and shown in the figure only for comparison.

The corresponding shape of the isotropic part of the spontaneous Raman band, computed by Fourier transformation of $\Phi_{vib}(t)$, is depicted in Fig. 7b. For variable τ_c and constant T_2 distinct changes of the bandshape are noticed. $\tau_c = 0$ corresponds to an exact Lorentzian shape, while the line wings decay faster for a finite time scale of the dephasing mechanism. For the halfwidth of the band, on the other hand, only minor changes occur for variable τ_c.

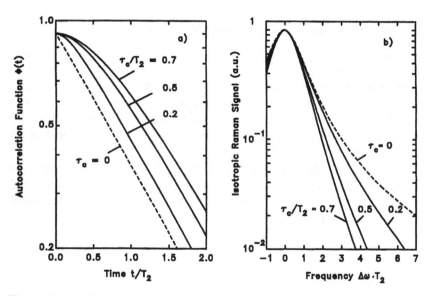

Figure 7 (a) Vibrational autocorrelation function in the Abragam-Kubo formalism (logarithmic scale) vs. normalized time in units of the asymptotic decay time T_2 for the several values of the correlation time τ_c, also in units of T_2. (b) Corresponding bandshape of the isotropic Raman line, plotted on a logarithmic scale vs. normalized wavenumbers in units of $1/T_2$; $\Delta\omega = 0$ denotes the band center.

Since T_2 is readily determined from time-domain CARS with high accuracy ($\leq 2\%$), a combined analysis of frequency- and time-domain data was proposed and demonstrated (45), plotting the spontaneous Raman data in normalized frequency units, $\Delta\omega \times T_2$ (note abscissa scale of Fig. 7b). In this way the bandshape only depends on the ratio τ_c/T_2, and only this ratio has to be deduced from the wings of the Raman band. With respect to the experimental uncertainties (ordinate value of the baseline, overlap with neighboring combination tones), the approach is more reliable than the determination of two quantities, τ_c and T_2, from the spectroscopic data.

An experimental demonstration is depicted in Fig. 8 for the ν_1 mode of neat bromochloromethane (45). Part of the data of Fig. 5b are replotted in Fig. 8a with an enlarged ordinate scale. We recall that the anisotropic component (dotted line in Fig. 5b) is negligible for the special case here,

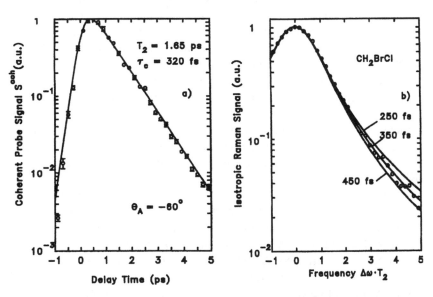

Figure 8 (a) Same as Fig. 5b for the ν_1 mode of neat CH_2BrCl, but with enlarged coordinate scales, representing isotropic scattering; a value of 320 ± 100 fs is inferred for the correlation time τ_c from the delayed onset of the signal decay after the scattering maximum; (b) isotropic component of the spontaneous Raman scattering of the same vibration versus circular frequency in units of $1/T_2$ suggesting a value of 350 ± 100 fs for τ_c. Experimental points, calculated curves including a convolution with the approximately Gaussian instrumental response function of the conventional Raman spectrometer.

so that only the isotropic scattering remains. The finite τ_c modifies the peak of the isotropic signal transient with a shift of the decaying part to larger t_D. Correct determination of the delay zero-setting is detrimental for an accurate determination of τ_c. Fitting theoretical curves for the special Φ_{vib} of Equation (9) to the data of Fig. 8a yields a value of 320 ± 100 fs for τ_c. The same value is also derived from the data of Fig. 5c after proper deconvolution of the nonresonant contribution (data not shown).

Corresponding lineshape data of the isotropic component of spontaneous Raman scattering are presented in Fig. 8b. Knowing T_2 within ± 30 fs (Fig. 5), the measured lineshape (experimental points) is readily plotted versus normalized frequency units $\Delta\omega \times T_2$. Comparison with the computed lineshapes [Fourier transformation of Equation (9)] leads to 350 ± 100 fs for τ_c, in good agreement with the time-domain results. Combining the measurements, one arrives at $\tau_c = 330 \pm 70$ fs (45).

Since more than one mechanism may notably contribute to the dephasing dynamics in general, the result is a mean value characterizing the dominant relaxation path. Only a few examples (neat liquid) have been studied by time-domain CARS (Table 1). For the symmetrical mode at 2913 cm^{-1} of neat DMSO, a similar value of $\tau_c = 240 \pm 50$ fs was measured. The measured values lie in the expected sub-ps range for liquids under normal conditions; numbers as small as 60 fs [acetonitrile (48)] are reported in the literature. Larger values ($\tau_c > 1$ ps) may occur close to the critical point, e.g., $\simeq 10$ ps, for supercritical liquid nitrogen (49). The result on CH_2BrCl agrees within experimental error with the τ_c value ($\simeq 400$ fs) of Ref. 50, determined from a different (more indirect) time-domain technique using picosecond pulses.

The time constants τ_c and T_2 provide information on the question of inhomogeneous broadening. The short values of τ_c in Table 1 indicate rapid modulation, $\tau_c/T_2 < 0.3$. In other words, (quasi-)homogeneous linebroadening is observed, as expected for dephasing via the repulsive (short-range) intermolecular interaction. A contribution from the (long-range) attractive part of the intermolecular potential that may lead to inhomogeneous broadening (51) is not apparent in the neat liquid at room temperature.

The spectral shape of the vibrational band differs from a Lorentzian. The deviation is evident for the wings, a factor of $\simeq 20$ below the maximum, and necessitates an accurate determination of neighboring bands and the baseline level. The isotropic Raman linewidth $\delta\nu$ (FWHM) is predicted to be smaller by a few percent than the Lorentzian width $\delta\nu_L = (\pi c T_2)^{-1}$ (FWHM in wavenumber units) for the same T_2 (45). The numbers for the example DMSO in Table 1 (last three lines) give some experimental support to this finding. In the inhomogeneous limit, $\tau_c/T_2 \gg 1$, a Gaussian shape is predicted by the model function [Equation (9)].

3. Measurements of Specific Dephasing Channels

The decay of coherent vibrational excitation in simple liquids is caused by the interaction of the considered molecule, with its fluctuating environment affecting the amplitude and/or phase of the vibration. Energy dissipation and/or loss of phase correlation result. In simple cases the various kinds of processes may be statistically independent, leading to a total (asymptotic) dephasing rate of

$$1/T_2 \simeq 1/(2T_1) + 1/T_{2,a} + 1/T_{2,b} + 1/T_{2,c} \tag{11}$$

Here T_1 is the energy (or population) lifetime that can be directly measured with other time-domain techniques (4,21). The constants $T_{2,a}-T_{2,c}$ represent pure dephasing contributions not changing the vibrational population (52,53). The following interactions have to be considered (20,53):

1. Dephasing induced by the translational motion (VT coupling) via the repulsive (a1) and attractive (a2) parts of the intermolecular potential
2. Dephasing induced by rotational motion in combination with the repulsive (b1) and attractive (b2) intermolecular interaction
3. Resonance energy transfer via the short-range (repulsive, case c1) and long-range (attractive, case c2) parts of the potential among the identical vibrational transition of different molecules
4. Dephasing via concentration fluctuations in the solvation layer of the vibrating molecule in liquid mixtures

Repulsive interaction channels are obviously restricted to the nearest neighbor shell of the vibrating molecule, while long-range attractive forces of other mechanisms also involve more distant neighbors. The relative role of the various relaxation processes a–c is not known in general. The effect of concentration fluctuations in mixtures will be discussed in the following subsection. Early theoretical estimates based on an isolated binary collision (IBC) model by Fischer and Laubereau (52) and a hydrodynamic model by Oxtoby et al. (54) gave only semiquantitative agreement with experimental data for a series of liquids. A more advanced theoretical treatment of the dephasing processes in the liquid state is highly desirable.

As a rule it may be surmised that the attractive contributions a2, b2 are insignificant compared to the repulsive counterparts a1 and b1, whereas the relation of a1 to b1 is unclear. For resonance transfer, on the other hand, (electric) transition dipole–transition dipole coupling contributing to c2 was shown to play a dominant role (relative to c1 and a, b) for the Raman linewidth of vibrations with a large transition dipole element.

For a comparison of theoretical predictions with experimental data, the isolation of specific relaxation channels is necessary. Standard techniques for this purpose developed by Raman spectroscopists and adopted for CARS studies are the variation of concentration (e.g., isotopic dilution), temperature, and pressure (40,53).

4. Resonant Vibrational Dephasing

The high precision of CARS investigations makes it possible to study a relaxation mechanism even when it contributes little to the total dynamics. An example is the CH-stretching mode of acetone at 2925 cm^{-1} (investigated in Fig. 3), which was measured in the binary mixture with deuteroacetone and carbon tetrachloride (40). The relaxation rate $2/T_2$ is plotted in Fig. 9. A small decrease of $2/T_2$ of approximately 10% is found for isotopic dilution (full points); assuming intermolecular T_1 processes to be negligible, the change is obviously due to the elimination of resonant energy transfer of vibrational quanta in the isotopic solution. The dominant nonresonant dephasing processes are practically unchanged by the replacement of isotopic molecules. The observed linear concentration dependence (see Fig. 9) supports the theoretical estimation that mechanism d is negligible for the isotopic mixture; the latter is expected to be nonlinear with a maximum for mole fraction $x = 0.5$ and vanishing contributions for $x = 0.1$. Comparison with theoretical estimates suggests that resonance transfer c1 via the repulsive intermolecular interaction is observed. The ratio of resonant to nonresonant dephasing rate for the neat liquid deduced from the isotopic dilution data of Fig. 9 is 0.12 ± 0.01, consistent with predictions of an isolated binary collision model (40).

For the mixture with CCl_4 (see Fig. 9) concentration fluctuations d are again found to be insignificant. Comparison of the larger decrease of the dephasing rate for CCl_4 relative to isotopic dilution gives information on the nonresonant mechanisms a and b. In terms of IBC models the difference is related to the changes of mass and intermolecular collision frequency when neighboring molecules of the vibrating acetone species are replaced by CCl_4 (52). Considering mechanism a1 and using the Enskog hard sphere model to estimate the collision frequency (51), the slightly nonlinear concentration dependence of Fig. 9 (dashed line) is calculated for consistent, reasonable

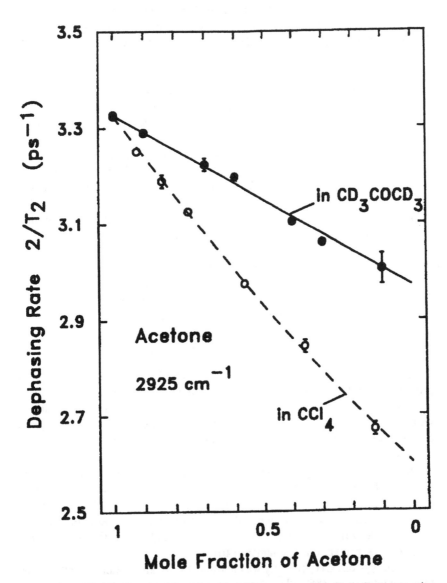

Figure 9 Dephasing rate of the ν_2 mode of acetone as a function of concentration for the solvents acetone-d_6 (full points) and carbon tetrachloride (open circles); experimental points, calculated curves; see text.

parameter values, e.g., hard sphere radii. The evidence from the figure in favor of relaxation channel a1 compared to b1 is not convincing. A study of the temperature and pressure dependence of T_2 would clarify the situation. IBC models predict opposite temperature gradients for the dephasing rates via processes a1 (negative) and b1 (positive). For pressure dependence the gradients are also opposite, but with inverted signs (53).

In this context the τ_c values are also of interest. The correlation times of mechanisms a1 and b1 are different and related to the time between the translational and rotational collisions, τ_v and τ_J, respectively. Little information is available on the latter constants from experiments or simulations. Using the Enskog hard sphere model with collision time τ_E (55), we have $\tau_c(VT) = \tau_v \simeq \tau_E$ for VT coupling (a1). For the VR interaction (b1) the rough hard sphere model is applicable, where $\tau_c(VR) \simeq \tau_J \simeq 2.3\tau_E$ (51). In other words the time scale of VR dephasing is slower. Assuming reasonable values for the effective hard core diameter for the examples of Table 1 and Fig. 9, Enskog times in the range of $\tau_E \simeq 60–160$ fs are estimated. Comparison with the experimental numbers for BCM and DMSO in the table indicates that the order of magnitude is estimated correctly, while the quantitative disagreement between τ_c and τ_E points towards a dominant VR relaxation at room temperature for the two examples.

5. Vibrational Dephasing in the Intermediate Case

Time domain CARS is well suited for detailed studies of relaxation processes. Under carefully chosen experimental conditions an individual dephasing channel may be selected, enabling a quantitative comparison with theory. The special case considered in this subsection is dephasing via concentration fluctuations (δ) in the liquid mixture $CH_3I:CDCl_3$.

Measurements of the concentration dependence allow one to uniquely separate the dephasing contribution via concentration fluctuations in the solvation layer of the vibrating CH_3I molecule, the time scale of which is determined to be 3.2–1.1 ps in the temperature range 242–374 K. Striking deviations from simple exponential relaxation dynamics are reported for the ν_1 mode with correlation times $\tau_c > 1$ ps, indicating the intermediate case for this relaxation mechanism (46).

The system $CH_3I:CDCl_3$ was studied by IR and Raman spectroscopy two decades ago (56–58). The ν_1 frequency in wavenumber units is $\omega_0/2\pi c = 2950$ cm^{-1} in the neat liquid and varies approximately linearly with dilution. The linewidth (isotropic scattering component) in the neat liquid is 4.8 cm^{-1} with approximately the same value for infinite dilution. For the 1:1 mixture a considerably larger linewidth of $\simeq 8.6$ cm^{-1} has been measured. The peculiar concentration dependence was first explained in terms of quasistatic concentration fluctuations around the reference molecules (56). Knapp and Fischer (59) included the dynamical aspect in the theory, i.e., time-dependent concentration fluctuations with exchange rate R of molecules in the solvation layer. The available IR and Raman data, however, did not help determine the time scale of the concentration fluctuations. Muller et al. (60) recently reported evidence for spectral diffusion in Raman echo observations of the 50% mixture at room temperature. From a comparison with CARS and conventional Raman data, an inhomogeneous broadening contribution was deduced by these authors with a lifetime of 4–7 ps at room temperature assigned to concentration fluctuations.

Examples of our time-resolved CARS data are presented in Fig. 10 for a molar fraction of x = 0.515 and two temperature values of 271 and 374 K. Since the mixture has a boiling point of $\simeq 325$ K at ambient pressure, a minor pressure increase up to 5×10^5 Pa of the sample was used. The small increase of several bars is expected to have a negligible effect on the dephasing dynamics (53). The data refer to the magic angle $\theta_A = 49.1°$ suppressing the anisotropic scattering contribution that would perturb the determination of purely vibrational dephasing. The nonresonant part (dotted line) produces the signal overshoot around zero delay and was also separately measured for $\theta_A = -60°$ (data not shown). For larger delay, $t_D > 0.5$ ps, the isotropic scattering dominates. It is interesting to see the pronounced deviation from the asymptotic exponential slope in Fig. 10a over a time interval of several picoseconds, which indicates a rather long correlation time around 1 ps or, correspondingly, a relatively slow exchange rate R. At 374 K (Fig. 10b) τ_c is obviously notably shorter, leading to a more exponential decay.

The signal transients for $\theta_A = -60°$ and for other concentrations were also measured (data not shown). For the neat liquid CH_3I no temperature dependence of T_2 was found within experimental accuracy and a short correlation time, $\tau_c < 300$ fs.

The time-resolved data are analyzed in terms of two (statistically independent) mechanisms with autocorrelation functions Φ_1 and Φ_2 making up the total autocorrelation function Φ_{vib} [Equations (2),(3)]:

$$\Phi_{vib}(t) = \Phi_1(t) * \Phi_2(t) \tag{12}$$

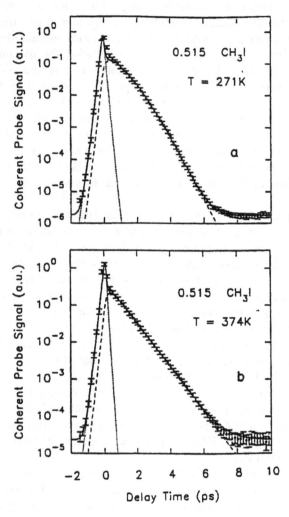

Figure 10 Coherent probe scattering of the symmetrical stretching vibration ν_1 of CH_3I in the mixture with $CDCl_3$ vs. delay time for the magic angle $\theta_A = 49.1°$, mole fraction x = 0.515 of CH_3I and temperatures of 271 K (a) and 374 K (b). Experimental points; calculated curves: isotropic scattering component (dashed), nonresonant part (dotted), and total CARS signal (solid line).

The Φ_i values are governed by individual parameter sets $T_{2,i}$ and $\tau_{c,i}$, analogous to Equation (10). The first relaxation channel represents the effect of concentration fluctuations of neighboring molecules A or B in the solvation layer around the probed molecule A_{pr} because of different frequency shifts exerted by the neighbors A, B on the sample molecule. The second mechanism accounts for the dephasing in the neat liquid A that may originate from the repulsive part of the intermolecular interaction including energy relaxation (52). For the special case under investigation, the corresponding dephasing interaction of A_{pr} with solvent molecules B is also incorporated in Φ_2 (e.g., infinite dilution) since it can be taken to be numerically equal. Knapp and Fischer (59) developed an explicit expression for Φ_1 in terms of the exchange rate R and number N of interacting molecules in the solvation layer and the concentration-dependent frequency shift $\Delta\Omega$ of the vibration known from Raman spectroscopy.

Measurements at 298 K for various concentrations in the range 0.076–1 yield $R = 0.55 \pm 0.1$ ps^{-1} and support the constant value of R anticipated in the theoretical model. For the number of interacting neighbors in the solvation layer, one finds $N = 5.0 \pm 0.9$, a reasonable value because of the large size of the iodine atom so that only some nearest neighbors can couple to the methyl group. The corresponding residence time of a molecule in the solvation layer is $1/R \simeq 1.8$ ps at 298 K.

The temperature dependence of the dephasing process for the liquid mixture with $x = 0.515$ allows an additional test of the Knapp-Fischer model (59). For the exchange rate R an increase from 0.31 ps^{-1} (at 242 K) to 0.9 ps^{-1} (374 K) is found, while $N \simeq 5$ remains constant. The results on R are plotted in Fig. 11 as a function of temperature (experimental points). It is interesting to compare the experimental numbers with estimates of R from jump diffusion (61). Using viscosity data available in the temperature range 0–40°C, the exchange rate was estimated in good agreement with the experimental results (solid curve in Fig. 11).

The Anderson-Kubo model [Equation (10)] and the Knapp-Fischer theory both account for the observed dephasing via concentration fluctuations. The average τ_c of the total dephasing displays a shortening with rising temperature by a factor of approximately 4 in the range 240–380 K, which is qualitatively reasonable because of the faster thermal motion. The opposite effect for the effective dephasing time T_2 is explained by motional narrowing. The numbers at room temperature yield $\tau_c/T_2 \simeq 0.9$, i.e., relaxation in the intermediate regime.

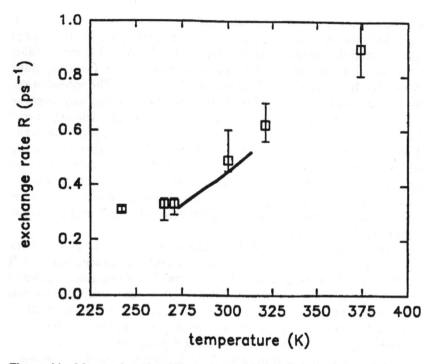

Figure 11 Measured exchange rate R (experimental points) as a function of temperature for the solvation layer containing $N \simeq 5$ molecules that contribute to the dephasing of the probing molecule CH_3I; the data are deduced from the mixture with molar fraction $x = 0.515$ using the Knapp-Fischer model. The solid line is estimated for jump diffusion from available viscosity data. Rapid concentration fluctuations are found leading to dephasing in the intermediate regime.

II. TIME-RESOLVED IR SPECTROSCOPY OF STRONGLY ASSOCIATED LIQUIDS

A. Introduction

The vibrational spectra of strongly associated liquids display broad bands that are difficult to analyze. The structural and dynamical information contained in these spectral features provided a field of speculation over the last decades. Obviously additional experimental evidence is urgently needed for a reliable interpretation. In this chapter, nonlinear pump-probe spectroscopy with intense ultrashort tunable pulses is considered a powerful spectroscopic tool that may provide an improved basis of understanding. This is because of its photo-physical hole-burning potential to reveal inhomogeneous broadening mechanisms, on the one hand, and its experimental time resolution, on the other hand, so that vibrational and structural dynamics can be directly followed as a function of time. One of our main interests in this context is the investigation of hydrogen bonds in condensed matter that play a major role in nature. The H-bridge bond vibrations themselves show up in FIR at frequencies below ≈ 400 cm^{-1}, which is difficult to access for time-resolved pump-probe spectroscopy. On the other hand, the OH-stretching mode of the bonded hydroxylic group in the 3000^{-1} region is known from conventional spectroscopy to be a sensitive indicator of the hydrogen-bonding situation, which is more readily accessible. In fact, it was demonstrated that the vibration can serve as a spectral probe with ultrashort time resolution (62). Qualitatively speaking, the OH-stretching vibration distinctly shifts towards lower frequencies for stronger H bonds, while the IR absorption cross section correspondingly increases by a factor up to an order of magnitude (1). This particular feature makes IR studies of the OH stretch particularly sensitive to structural changes as compared to Raman scattering or other vibrational modes, where the amplitude effect is practically missing.

B. General Considerations

In conventional spectroscopy, inhomogeneous broadening of a vibrational band is difficult to verify. Spectral hole burning can provide direct evidence for a band substructure and reveal subcomponents hidden by a featureless band contour. Exploiting the spectral and temporal resolution of pump-probe experiments in the OH-stretching region, structural and dynamical processes of H-bonded systems can be studied. The inherent problem of relating measured spectral shifts to a local structural picture may be tackled by computer simulations of the liquid ensemble. The method is, of course, limited by the uncertainty principle; a compromise between spectral and time resolution requirements has to be found for proper experimental investigations.

The principle of the experiment is illustrated by Fig. 12. On the left-hand side the absorption band of an inhomogeneously broadened transition, e.g., the OH-stretching band in different H-bonded local environments, is depicted schematically. Structures with different OH-O bond angles and/or O-O bond lengths show up in the spectrum with different OH frequencies; linear bonds and shorter bond length correspond to larger red shifts

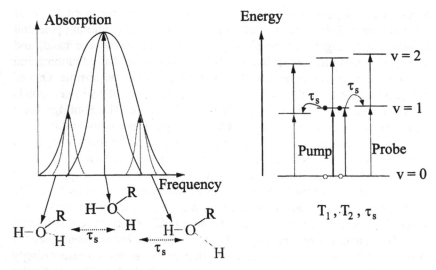

Figure 12 The principle of a pump-probe experiment performed on an inhomogeneously broadened OH-stretching band is depicted. On the left-hand side three structural components with decreasing H-bond strength and corresponding increasing OH frequency are shown. The corresponding energy level scheme describing the pump-probe experiment on the illustrated OH band is seen on the right-hand side. The pump pulse excites predominantly one structure. Determination of the sample transmission with an independently tunable probe pulse yields the temporal evolution of the structural components via spectral hole burning and structural relaxation with time constant τ_s. In addition, the vibrational lifetime T_1 and in special cases phase relaxation with time constant T_2 can be determined.

of the OH frequency. Three spectral subcomponents with a certain quasi-homogeneous width via dephasing processes are explicitly shown in the figure (thin lines). The total band contour (thick line) results from the superposition of the subspecies. A second mechanism that also effects the OH frequency position via the local H-bonding situation will be discussed below. Structural relaxation of the molecular environment corresponds to fluctuations of the individual OH frequencies (spectral relaxation) and is described by a time constant τ_s. With respect to the experimental situation the process is assumed to be sufficiently slow compared to the inverse quasi-homogeneous width of a subcomponent. The IR pump pulse with a spectral width notably smaller than the total band resonantly interacts only with a fraction of the OH groups. In this way a subensemble is selected according to its frequency position and promoted from the ground state to an excited vibrational level. A situation is depicted where the structural components located at the peak of the OH band are predominantly excited.

The corresponding energy level scheme is shown on the right-hand side of Fig. 12. Arrows denote the pump process (thick arrow) and possible probing transitions (thin arrows) between the $v = 0 \rightarrow v = 1$ transitions (ground state bleaching) and $v = 1 \rightarrow v = 2$ transitions (excited state absorption). The induced changes of the sample transmission are measured by the help of a weak, independently tunable probe pulse. Probing transitions are indicated by thin vertical arrows. Spectral relaxation is indicated in the level scheme and appears as an energy transfer process, although the mechanism is a shift of excited individual OH groups to a different frequency position. The process has to be distinguished from energy migration among neighboring groups. The time constant of spectral relaxation may be determined from the temporal evolution of the ground state population, e.g., the missing molecules in the vibrational ground level leading to a transient sample bleaching.

The population and reorientational dynamics are not indicated in the figure, but may be also derived from the pump-probe measurements. The population lifetime T_1 in the first excited level can be inferred from the excited state absorption monitoring the $v = 1 \rightarrow v = 2$ transition. The molecular reorientation becomes experimentally accessible, introducing polarization resolution in the probing step. For known structural and population dynamics the temporal evolution of the width of the observed spectral hole also provides information on the dephasing time T_2 of the vibrational transition (63).

For a quantitative treatment the molecular ensemble is described by an n-level system with random orientational distribution and a corresponding set of rate equations that implies short lifetimes of the induced transition dipole moments compared to the experimental pulse duration ($T_2 \ll t_p$). Each level represents a spectral subemsemble of molecules in the vibrational ground state ($v = 0$) or the first excited state ($v = 1$) of the OH-stretching vibration and possible further intermediate states involved in the relaxation pathway, depending on the respective situation. Additional levels, i.e., the terminating states of the probing transitions, need not be explicitly taken into account because of the negligible populations generated by weak probing pulses. The coherent coupling artefact for pump and probe pulses at the same frequency position is included by additional terms. The coupled rate equations and the propagation equations for the incident pulses are solved numerically. The polarization dependence of the probe transmission is computed by integration over the orientational distributions, while the reorientational motion is treated in the limit of rotational diffusion (64). In a more general approach coherent effects can be also included; the finite T_2 plays an important role for the time evolution of spectral holes (63).

As a simple example a three-level system with slow reorientation is considered. States (0) and (1) are directly coupled to the excitation pulse, while the intermediate level (2) is populated in the relaxation path of the excess population N_1 of the upper level on its way back to the ground state. The temporal evolution of the population numbers and the pump intensity I_{Pu} is given by:

$$\frac{\partial N_0}{\partial t} = -\frac{nc\varepsilon_0}{2} I_{Pu} \frac{\sigma_{01}}{\hbar\omega_{01}} (N_0 - N_1) + \frac{N_2}{T_{20}} \tag{13}$$

$$\frac{\partial N_1}{\partial t} = \frac{nc\varepsilon_0}{2} I_{Pu} \frac{\sigma_{01}}{\hbar\omega_{01}} (N_0 - N_1) - \frac{N_1}{T_{12}} \tag{14}$$

$$\frac{\partial N_2}{\partial t} = \frac{N_1}{T_{12}} - \frac{N_2}{T_{20}} \tag{15}$$

$$\left(\frac{\partial}{\partial y} + \frac{n}{c} \frac{\partial}{\partial t} \right) I_{Pu} = -I_{Pu} \sigma_{01} (N_0 - N_1) \tag{16}$$

with σ_{01} being the absorption cross section, T_{ij} the relaxation time constants (i, j = 0–2), and n the refractive index.

Numerical solutions of equations like the ones shown above for the probe transmission may be fitted to the measured signal transients yielding the relevant relaxation rates. Measured transient spectra are interpreted by comparison with calculated spectra for a set of probing transitions with assumed Lorentzian (spectral holes) and Gaussian lineshape (other spectral components), consistent with an n-level model describing the time evolution, and proper convolution with the spectral profile of the probe pulses of approximately Gaussian shape. In this way the finite instrumental resolution is taken into account and true spectral parameters of the physical system deduced from the experiment. Fitting to the experimental data is performed using a Levenberg-Marquardt algorithm (65).

C. Experimental Aspects

In order to obtain new information on the structural and vibrational relaxation of molecular systems with subpicosecond time resolution, we have developed a suitable laser system for two-color IR pump-probe spectroscopy with pulses close to the Fourier limitation and adjustable in the range 0.4–2.6 ps, derived from synchronously pumped optical parametric oscillators. For the measurement of spectral holes with widths below ≈ 30 cm^{-1}, a frequency width of 20 cm^{-1} of the pulses is available in our experiments. The instrumental parameters may be compared with other setups based on commercially available Ti:sapphire lasers with intense IR pulses (13–15) of shorter duration, down to 100 fs, but at the expense of spectral resolution due to a pulse width around 100 cm^{-1} of optical parametric amplifier devices. The latter value is not appropriate for the investigated spectral holes of widths below 50 cm^{-1}.

We start with a specially designed, Kerr-lens mode-locked Nd:YLF laser (12,66) for the synchronous pumping of two optical parametric oscillators (OPOs) in parallel (repetition rate 50 Hz). The OPOs are operated at variable pump intensities slightly above threshold up to the strong saturation regime. This allows for a tunable pulse duration of the fed-back idler component in the OPO in the range of 2.6 ps to 260 fs. Single-pulse selection, frequency down-conversion, and amplification of the OPO outputs are carried out in subsequent optical parametric amplifier stages equipped with AgGaS$_2$ crystals for an extended range in the IR. Two synchronized pulses with independent, automatic tuning ranges are accomplished. Frequency setting using computer-controlled stepping motors requires a fraction of a second. This makes the setup a real spectrometer for time-resolved investigations.

Working in the short-pulse regime, for example, excitation and probing pulses of ≥ 450 fs duration and spectral width ≤ 35 cm^{-1} are generated in the range 1600–3700 cm^{-1} (12). The pulse energy amounts to ≤ 10 μJ (pump) and ≈ 10 nJ (probe). For long-pulse conditions, on the other hand, pump (probe) pulses of 2 ps (1 ps) of spectral width 8 cm^{-1} (16 cm^{-1}) are available, providing superior spectral resolution. The data discussed in the following mostly refer to these longer pulses if use of the sub-ps pulses is not explicitly stated.

The noncollinear pump-probe experiment is depicted schematically in Fig. 13. The linearly polarized (P3) pump pulse is focused (L1) into the sample producing induced transmission changes. The polarization of the probe beam is adjusted to 45° relative to the pump with a half-wave plate ($\lambda/2$) and a Glan polarizer (P1). By the help of an analyzer (P2) simultaneous detection of the parallel (∥) and perpendicular (⊥) components of the energy transmission $T(\nu, t_D)$ of the probe through the sample is installed. For blocked excitation (chopper, Ch) the sample transmission

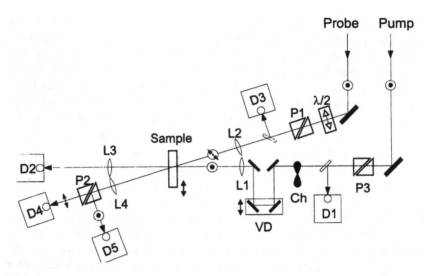

Figure 13 Schematic of the setup of the pump-probe experiment with polarization resolution for the probing of the induced change in sample transmission. $\lambda/2$: half-wave plate; P1–P3: polarizers; L1–L4: lenses; D1–D5: detectors; Ch: chopper; VD: optical delay line. The sample is permanently moved in a plane perpendicular to the beams in order to avoid accumulative thermal effects.

$T_0(\nu)$ is measured. The resulting relative transmission changes $\ln(T/T_0)_{\|,\perp}$ for variable probe frequency ν and probe delay time t_D (VD) are used in the following as the relevant signal quantities, from which the following quantities are derived, as demonstrated by Graener et al. (11):

The isotropic signal amplitude,
$$\ln(T/T_0)_{is} = (\ln(T/T_0)_{\|} + 2\ln(T/T_0)_{\perp})/3$$
The anisotropic signal $\ln(T/T_0)_{anis} = \ln(T/T_0)_{\|} - \ln(T/T_0)_{\perp}$
The induced dichroism $\ln(T/T_0)_{anis}/2\ln(T/T_0)_{is}$

The isotropic signal delivers (rotation-free) information on the temporal evolution of the population numbers of the investigated vibrational transition(s). The induced dichroism is governed by the time constant τ_{or} (second-order reorientational correlation time, $1 = 2$) and possibly population redistribution that may contribute to the loss of induced optical anisotropy. The zero-setting of the delay time scale (maximum overlap between pump and probing pulses) is determined by a two-photon absorption technique in independent measurements with an accuracy of better than ± 0.2 ps (67).

Two-color pump-probe absorption spectroscopy is carried out with moderate pump energies producing small depletions of the vibrational ground state of only a few percent in order to avoid secondary excitation steps and minimize the temperature increase of the sample due to the deposited pump energy.

D. Alcohols in Solutions

In this section we will discuss results of transient IR spectroscopy of different alcohols in solution in a wide concentration range from almost monomeric alcohol samples to strongly associated oligomers. In order to investigate the influence of hydrogen bonds on the dynamical properties of the molecules, we present first a discussion of the data on the vibrational and reorientational dynamics of the OH mode of isolated molecules in the solvent.

1. Monomers in an Apolar Solution

Early time-resolved experiments in the infrared by Heilweil and coworkers were performed on an alcohol in apolar solvents (68). Since the laser system utilized in these investigations delivered pulses of several tens of picosecond duration, a problem was to deconvolute the instrumental response function from the measured transients. A further complication for the interpretation of the data arose from the fact that one-color measurements were carried out; parts of the same pulse were used for excitation and probing. In such experiments it is difficult to separate the coherent coupling artefact and to distinguish the recovery of the ground state from the population decay of the first excited state. The latter point is important if longer-lived intermediates are involved in the relaxation pathway. It is more advantageous to conduct two-color measurements where the vibrational population lifetime can be deduced from the transient excited state absorption monitoring the $v = 1 \rightarrow v = 2$ transition. In addition, the effect of reorientational dynamics has to be removed from the data in general by the help of polarization resolution, measuring the isotropic signal transient (11,69). The technique dwells on the common anharmonic frequency shift of vibrational modes that is quite pronounced for the OH stretch (≈ 200 cm^{-1}), comparing the $v = 1 \rightarrow v = 2$ transition with the fundamental mode frequency. Time-resolved spectroscopy on binary systems in the liquid phase have recently been reported by the group of Heilweil (70) demonstrating conservation of vibrational excitation during chemical reactions.

In the following we present data on the simple alcohols methanol (Me), ethanol (Eth), decanol (De), and 2,2-dimethyl-3-ethyl-3-pentanol (DMEP) in dilute solutions of CCl_4. Methanol is additionally investigated at low concentrations in the solvents C_2Cl_4, C_4Cl_6, and C_5Cl_6 providing environments with differently shaped molecules (71). Here we are interested in the lifetime and the relaxation channels of the excited OH-stretching mode of monomers as well as the reorientational dynamics of the OH group. A concentration of 0.25 M for the DMEP and of 0.05 M for the other alcohols in the respective solvents was adjusted corresponding to a sample transmission of $\approx 50\%$ at the respective peak position of the OH stretch.

Some typical results are presented in Fig. 14 for ethanol in CCl_4. On the left-hand side the conventional (a) and time-resolved (b) spectra

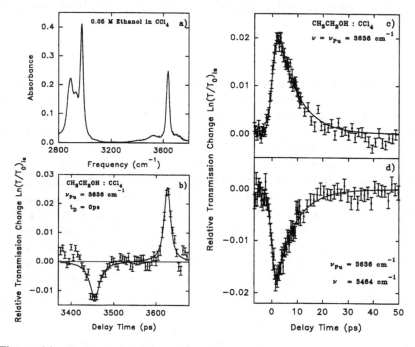

Figure 14 Conventional (a) and isotropic transient (b) spectra of a 0.05 M mixture of ethanol and CCl_4. The bleaching of the time-resolved spectrum shown at a delay time of 0 ps (b) is related to ground state depopulation while the corresponding excited-state absorption is red-shifted by 170 cm^{-1}. Time-dependent measurements taken within the maximum of the respective components are shown in (c) and (d), as denoted in the figure.

are shown; the latter represents the rotation-free isotropic component at zero-delay time (optimum overlap between pump and probing pulse) and pumping in the maximum of the OH vibration. The different abscissa scales should be noted. The conventional absorption (a) is dominated by the OH mode with a maximum at 3636 cm^{-1} and the CH$_{2,3}$-stretching modes including CH-bending overtones in the range of 2800–3100 cm^{-1}. In the blue wing of the former a weak band is assigned to a combination tone between a CH stretch and a CO- or COH-stretching mode (72) that plays a role in the relaxation of the OH vibration. The extended red wing of the OH band suggests the presence of a small number of ethanol dimers positioned around 3500 cm^{-1}.

The time-resolved spectrum (Fig. 14b) consists of a bleaching feature of the sample at the OH frequency resulting from the depopulation of the vibrational ground state and induced (excited-state) absorption (ESA), which is red-shifted because of the vibrational anharmonicity. From the smaller spectral width of the sample bleaching (17 cm^{-1}) with respect to the conventional absorption band (width 29 cm^{-1}), one concludes that spectral hole burning occurs in the OH band of monomeric ethanol, since the reorientational motion is too slow to account for the difference. The inhomogeneous character is due to solvent-solute interaction and is probably related to different preferred constellations of the OH group relative to the ethyl group.

Some time-dependent data are depicted in Fig. 14c,d. From a comparison of model computations with the measured temporal evolution of the induced bleaching (isotropic component, Fig. 14c) as well as the ESA (Fig. 14d) directly related to excited state population, we infer a lifetime of the OH-stretching mode of $T_1 = 8 \pm 1$ ps for the dissolved ethanol monomers. From the anisotropic signal component measured in the same experimental runs we determine the reorientation time to be $\tau_{or} = 2 \pm 0.5$ ps at room temperature.

The results for various alcohol solutions are compiled in Table 2. The data refer to room temperature except for methanol in C_5Cl_6, which was measured at 333 K. A few spectroscopic parameters are shown in the first four lines. There is evidence for inhomogeneous broadening of the monomeric OH-stretching vibration for the various investigated alcohols with a width of the observed spectral holes as small as 50% of the conventional absorption band (DMEP). The anharmonic shift of the OH mode is indicated in the table to amount to 170 cm^{-1}, independent of solute or solvent within experimental accuracy. The large anharmonicity of the monomeric alcohols is noteworthy. The population lifetime T_1 of the mode displays little dependence on the number of CH groups of the alcohol and on the apolar solvents leading to numbers of 8–10 ps. The data suggest that intramolecular energy redistribution is dominant; in fact, the solvent molecules do not not offer suitable accepting modes for vibrational energy transfer of the OH vibration. The relaxation pathway was studied in detail in the case of monomeric ethanol (73). From time-delayed bleaching at CH-stretching frequencies intramolecular relaxation via these vibrations was inferred. Theoretical arguments based on Fermi resonance coupling suggest energy transfer to the adjacent combination tone of CH-, CO-stretching modes that decays, in turn (in part), to the CH stretch. Considering the frequency spacing and amplitude of the combination band relative to the OH absorption, an OH lifetime of 8 ps is estimated (74) in perfect agreement with the measured value. The slightly longer values of the ground state recovery time T_{10} in Table 2 are consistent with the multiple step energy decay via CH to the ground state.

The reorientational motion of the OH group is found to proceed in the mixture with CCl_4 at room temperature with time constants of 2–4 ps (see last line of Table 2). With increasing size of the solute molecule, the reorientation seems to slow down. For the different mixtures, the largest $\tau_{or} \approx 7$ ps occurs for the most viscous liquid C_5Cl_6.

2. Ethanol Oligomers in Solution: Spectral Holes and Vibrational Lifetime Shortening

Equipped with information on the dynamical properties of the OH-stretching mode of monomers, we are now in a position to tackle the role of H-bridge bonds for the structure and dynamics of alcohols. The first time-resolved investigations of this kind were performed by Laubereau and coworkers (62,75). A mixture of ethanol and CCl_4 was investigated by two-color IR spectroscopy with 11 ps pulses. From the time evolution of the spectra clear evidence was obtained that after excitation of the OH mode of ethanol molecules with internal position in oligomer chains, ultrafast breaking of H bonds occurs, followed by partial reassociation of the broken oligomers. The measurements showed a shift of the chemical equilibrium towards higher temperatures after OH excitation, while the available time resolution was not sufficient to study several details of the dynamics including the possibility of spectral hole burning of the OH band in the liquid. The latter was found in these early studies for a hydrogen-bonded polymer film (76).

With improved laser systems (12–15) further experiments were conducted several years later. Woutersen et al. (77) reported one-color investigations with 200 fs pulses of a 0.4 M mixture of ethanol and

Table 2 Experimental Results on Transient Spectroscopy of Monomeric, Simple Alcohols in Different Solvents

	$Me:CCl_4$	$Me:C_2Cl_4$	$Me:C_4Cl_6$	$Me:C_5Cl_6$	$De:CCl_4$	$DMEP:CCl_4$	$Eth:CCl_4$
ν_c (± 1 cm^{-1})	3644	3643	3639	3639	3638	3625	3636
$\Delta\nu_c$ (± 1 cm^{-1})	22	21	22	24	17	20	29
$\Delta\nu_h$ (cm^{-1})	15 ± 3	15 ± 3	19 ± 3	15 ± 2	16 ± 3	11 ± 2	17 ± 2
$\Delta\nu_{12}$ (cm^{-1})	20 ± 3	20 ± 3	19 ± 3	30 ± 3	30 ± 3	15 ± 2	29 ± 4
$\nu_c - \nu_{12}$ (cm^{-1})	168 ± 3	170 ± 3	170 ± 3	168 ± 5	171 ± 3	170 ± 3	168 ± 5
T_{10} (ps)	12 ± 2	10 ± 1	13 ± 2	10 ± 2	9 ± 1	8 ± 1	8 ± 1
T_1 (ps)	9 ± 1	10 ± 1	8 ± 1	8 ± 3	9 ± 1	8 ± 1	8 ± 2
τ_{or} (ps)	2 ± 0.5	2 ± 0.5	2.5 ± 0.5	7 ± 3	4 ± 1	4 ± 1	2 ± 0.5

All data are taken at room temperature besides the one shown for methanol diluted in C_5Cl_6 (T = 333 K). ν_c: peak position of the OH stretch; $\Delta\nu_h$, $\Delta\nu_{12}$: spectral width of the transient hole and the excited-state absorption, respectively; anharmonic shift $\nu_c - \nu_{12}$; T_1, T_{10}: time constants for the vibrational lifetime of the OH, ground state filling and reorientation, respectively.

CCl$_4$. The authors found a faster predissociation process with time constants of 250 fs ($\nu_{Pu} = 3225$ cm^{-1}) to 900 fs ($\nu_{Pu} = 3450$ cm^{-1}). For the subsequent reassociation a time constant of 15 ps was measured. From subsequent investigations of the probe transmission with perpendicular polarization, the authors inferred a fast delocalization of the deposited vibrationally energy along the oligomer chain confirming the findings of Ref. 78.

A more detailed picture was derived from two-color spectroscopy with 2 ps pump and 1 ps probing pulses (78,79). The probing of the sample transmission in the whole frequency range of the fundamental v $= 0 \rightarrow$ v $= 1$ and excited state v $= 1 \rightarrow$ v $= 2$ OH transition provides much more spectroscopic insight and is discussed in the following.

The transient spectra for a 0.17 mixture of ethanol in CCl$_4$ at room temperature are presented in Fig. 15 for three different delay times of -2 ps (top), 0 ps (middle), and 11 ps (bottom) (78). The conventional absorption spectrum depicted in Fig. 15f (dash-dotted curve) exhibits two peaks in the investigated spectral region: one related to monomeric ethanol molecules and/or species with proton acceptor function in open oligomer chains (3636 cm^{-1}), and another involving OH groups in internal oligomer positions with proton donor and acceptor function. The band maximum occurs at 3340 cm^{-1}, where the excitation pulse is positioned (vertical arrows). The slight asymmetry of the band with a shoulder in the blue wing may be assigned to hydroxilic end groups with proton donor function (80) positioned at $\simeq 3500$ cm^{-1} (width ≈ 110 cm^{-1}). In the extended red wing of the oligomeric OH band a further spectral component around 3150 cm^{-1} (width ≈ 150 cm^{-1}) may be assumed and related to internal OH groups in long ethanol chains as seen in solid Ar matrices (81).

In order to compare primary dynamics with secondary relaxation steps, we depict on the left-hand side of Fig. 15 the "anisotropic" spectra (a–c), which consist mainly of spectral components with the same linear polarization as directly induced by the pump pulse. On the right-hand side of the figure the corresponding "isotropic" spectra (d–f) are shown. In the latter spectral components can notably contribute that result from a relaxation process, where the initially orientation of the OH transition dipole is (partially) lost.

Transient hole burning is clearly indicated by the data for the anisotropic component of the probe transmission change (a, b) with a halfwidth (true FWHM) of 45 ± 5 cm^{-1} of the Lorentzian shaped hole at early delay times, t$_D \leq 0$ (a). For the isotropic component of the probing signal, the narrow hole (dotted lines in d, e) is superimposed on a broader

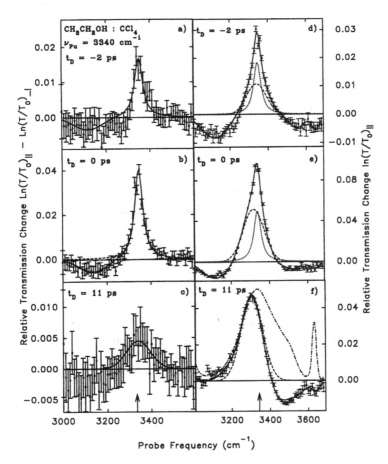

Figure 15 The evolution of the transient spectra is shown here for a 0.17 M mixture of ethanol and CCl_4 and the anisotropic (a–c) and isotropic (d–f) components. With increasing delay time from −2 ps (a,d) to 0 ps (b,e) and finally 11 ps (c,f), clearly a short-living spectral hole (a, dotted line) as well as a completely isotropic Gaussian component (d, dashed line) is noticed.

component (dashed lines) of approximately Gaussian shape as suggested by a numerical decomposition of the measured transmission changes and very similar to the findings at higher concentration (79). The latter feature obviously is of almost isotropic character, independent of the polarization of the excitation process, and thus a secondary phenomenon involving some relaxation process. The same conclusion holds for the induced absorption in the blue part of the probe spectra above 3400 cm^{-1} of Fig. 15d and e. Some dynamics obviously proceed on a time scale comparable to or shorter than the duration of the pump pulse of 2 ps. The induced absorption in the red part of the oligomer absorption, on the other hand, displays some polarization dependence, suggesting it to be (at least in part) a primary feature of the population changes of the pump process. At $t_D = 0$ (b and e) the broad absorption increase at 3140 cm^{-1} with width of 160 cm^{-1} is readily observed and attributed to excited-state absorption from the first excited level of the OH groups in internal positions. At a later delay time of 11 ps the narrow bleaching feature (spectral hole) has disappeared while the broad bleaching component at 3320 cm^{-1} of the oligomer band with approximately Gaussian shape (width 130 ± 10 cm^{-1}) has survived and even can be seen in the anisotropic spectrum (Fig. 15c and f, respectively). Part of the induced absorption at higher frequencies is also still existing.

The dynamics derived from the evolution of different spectral components is interpreted by the help of the five-level scheme shown in Fig. 16 (78). Starting from the ground state (0), the excited vibrational level (1) of an oligomeric subensemble is populated by the pumping process with fast spectral redistribution among the neighboring OH groups of the same chain represented by level 2. In addition to population decay, breaking of hydrogen bonds follows decribed by level 3. The final slow rearrangement of the chemical equilibrium is described by a long-lived modified ground

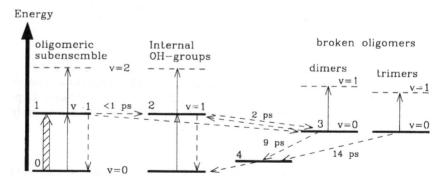

Figure 16   Simplified energy level scheme to account for the dynamics after vibrational excitation of an oligomeric subensemble of H-bonded ethanol molecules. While the thick arrow marks excitation, the thin arrows refer to the probing positions and dashed lines denote the different possibilities of relaxation. For details see text.

state (level 4). The five-level scheme is based on various pieces of experimental observations and can account for the measured spectral dynamics in the investigated frequency range.

First, transient hole burning in the ≈ 220 cm^{-1} broad ethanol band is observed and accounted for by the excited molecular subensemble (1). The measured hole width of 45 ± 5 cm^{-1} may be compared with the transient hole width of the OH dimer band of the amorphous polymer matrix PVB, where a value of approximately 100 cm^{-1} was found at room temperature and smaller values at lower temperature (76). The smaller width determined here for the ethanol sample may be explained by the more pronounced motional narrowing in the liquid phase. Furthermore, this hole width is comparable to the recently published value for higher ethanol concentration (2.4 M) (82). The lifetime of the spectral hole of $\tau_{hole} = 1.2 \pm 0.2$ ps agrees with the number found in previous investigations for a concentration of 1.2 M (79). τ_{hole} may be governed by three processes: population decay to the vibrational ground state, energy migration to neighboring ethanol molecules with shifted frequency positions, and structural relaxation of the excited OH groups also generating frequency shifts (79).

Subsequently fast spectral redistribution to level 2 is indicated by the short lifetime of the spectral hole and the early observation of the 130 cm^{-1} broad oligomer bleaching in the isotropic part of the transient spectrum. The bleaching component located at 3320 cm^{-1} is present already at a delay time of -2 ps. From time-dependent measurements taken at frequencies >3400 cm^{-1} (data not shown), a time constant of $\tau_{mig} < 1$ ps is inferred for this process. A lower limit of the proposed energy migration time is set by the transient hole width of about 0.2 ps, assuming other dephasing processes to be negligible.

On the other hand, structural relaxation involving reorientation of the OH groups with corresponding changes of the bond angles of the H bridges (and of the OH frequencies) is found to be slow with a time constant of 8 ps as inferred from the anisotropic signal component of time-dependent measurements taken at $\nu = \nu_{Pu} = 3340$ cm^{-1}. The reorientational motion cannot account for the observed rapid spectral changes; the fast hole relaxation involves obviously molecules in the v = 1 state. It is concluded that energy migration, i.e., near-resonant transfer of vibrational quanta of the OH stretching mode, explains the spectral features. The larger bandwidth of the excited state absorption (ESA) of 160 ± 15 cm^{-1} relative to the hole width may be also explained by the migration along the oligomer chains. This conclusion is supported by the fact that the band area, which is comprised by the ESA in the isotropic spectrum, exceeds the one of the hole by a factor of almost 3 (see Fig. 15d–f). A substantial part of the isotropic ESA has to be related to the broad oligomer bleaching of similar bandwidth. It is proposed that the same process accounts for the rapid loss of orientational information, i.e., the isotropic character of the oligomer bleaching and fragments produced by bond breaking. As a consequence the broad bleaching component at early delay times is interpreted in part as a manifestation of vibrationally excited ethanol molecules at different positions in the hydrogen-bonded chain, populated by energy migration. The remaining part of the transient oligomer bleaching is attributed to a depletion of the vibrational ground state, as will be discussed in the following.

The excitation of the internal OH vibrations is accompanied by a red-shifted ESA around 3120 cm^{-1} representing a direct measure of the population of the $v = 1$ level. The ESA disappears with a time constant of $T_1 = 1.7 \pm 0.3$ ps. The same value is measured for the isotropic and anisotropic signal transients, suggesting a minor contribution of reorientational motion only. Because of the large width of the ESA band, structural relaxation and migration of the vibrational quanta are not resolved and T_1 obviously represents the effective population lifetime of the oligomeric OH groups. As compared to the monomeric alcohol discussed above, a significant lifetime shortening is noticed for internal OH groups in oligomer chains; a decrease from 8 ps (see above) to 1.7 ps for a 0.17 M ethanol:CCl$_4$ mixture and 1.4 ps for a 1.2 M ethanol:CCl$_4$ sample is found (83).

The next step following spectral hole relaxation (i.e., energy migration along the oligomer chain) is breaking of a hydrogen bond producing thereby dimers (84) positioned at 3500 cm^{-1} or trimers (81) at 3450 cm^{-1} (level 3) in the OH ground state. The mechanism contributes to the population decay of the excited vibrational state. The generation of the smaller species is indicated by the time-delayed appearance of induced absorption at the respective frequencies. The delayed bond breaking is also inferred from the induced absorption at 3633 cm^{-1}, the frequency position of monomers or OH groups with proton acceptor function. It is further proposed that the delayed weak bleaching at 3140 cm^{-1} is directly related to the bond breaking (Fig. 15f): the frequency position refers to internal OH groups in long chains (N > 5). Pumping a suitable hydroxylic group of one such long chain at 3340 cm^{-1}, with subsequent bond breaking, a shorter chain results lacking absorption at 3140 cm^{-1}, as observed experimentally. According to this mechanism the delayed bleaching at 3140 cm^{-1} should build up as rapid as the induced absorption of the fragments, consistent with the observations. The equal values (within measuring accuracy) for the subsequent decay times of approximately 20 ps for the dynamics at 3140, 3340 (isotropic component), and 3633 cm^{-1} also gives some support to this physical picture.

Finally, reassociation of the broken oligomers occurs with time constants between 9 ps (dimers) and 14 ps (trimers) to a new thermal quasi-equilibrium (level 4) with lifetimes in the nanosecond region. The system does not return to the initial state (0) as a direct consequence of the deposited energy of the excitation process. The resulting temperature increase of the sample is small but produces measurable effects and is estimated from the long-lived amplitude of the signal transients to be below 1 K.

Similar experiments have also been performed on a higher concentrated mixture of 1.2 M ethanol and CCl_4. Again spectral holes could be identified from the transient spectra with a width of 25 cm^{-1} and lifetime of ≈ 1 ps. From transient spectra taken for different excitation frequencies in the OH band, evidence for a faster hole relaxation with increasing red shift (bond strength) is inferred, which is accompanied by the differences in the temporal evolution of the isotropic Gaussian component related to level 2.

Recalling the situation for monomeric ethanol, the question arises in what respect the CH-stretching modes involved in the OH relaxation (see above) may influence the structural dynamics due to H-bond breaking. Transient spectra tackling this question are shown in Fig. 17. Data are taken for a 1.2 M mixture of ethanol with CCl_4 at room temperature and excitation in the CH-stretching region at 2974 cm^{-1} (note vertical arrow) (85).

The conventional infrared spectrum of the sample is depicted in Fig. 17c (dash-dotted line, righthand ordinate scale). The absorption in the region between 2800 and 3000 cm^{-1} originates from CH_2- and CH_3-stretching vibrations mixed by Fermi resonance to overtones of CH-bending modes (86). The most intense band at 2974 cm^{-1} with a width of 21 cm^{-1} is assigned to the asymmetrical CH_3-stretching vibration, which will be referred to here as the CH vibration. The absorption at higher frequencies above 3050 cm^{-1} is attributed to the OH-stretching vibration.

The parallel component of the probe transmission change is plotted in Fig. 17 in the range 2850–3650 cm^{-1} for different delay times (lefthand ordinate scales, experimental points, calculated solid curves). The transient spectrum during the excitation process, $t_D = 0$ ps, is depicted in Fig. 17a. A bleaching at the frequency position of 2974 cm^{-1} is shown because of the excitation of the CH vibration. The excess population of the upper level $v = 1$ can be directly monitored from the induced absorption around 2952 cm^{-1} and is attributed to excited-state absorption (width of 17 ± 2 cm^{-1}). The bleaching signal at lower frequencies indicates

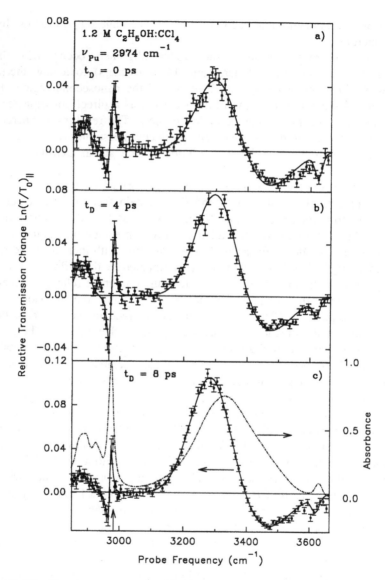

Figure 17 Transient spectra of a 1.2 M ethanol and CCl_4 mixture taken at room temperature and excitation within the CH-stretching modes at 2974 cm^{-1}. The data are shown at three different delay times of 0 ps (a), 4 ps (b), and 8 ps (c). The conventional absorption of the sample is shown for comparison in (c), right-hand ordinate scale and dash-dotted line. Measured data of the parallel signal; calculated lines.

population changes of other CH transitions that are not directly excited by the pump pulse but are obviously due to population redistribution processes. Above 3100 cm^{-1} the measured OH-absorption changes provide evidence for a rearrangement of the H-bonding system. Since excited state absorption around 3100 cm^{-1} is not found, population of the excited OH level is of minor importance. Cleavage of ethanol associates to shorter pieces is suggested by the bleaching feature peaking at 3300 cm^{-1} (lack of oligomeric absorption), while the induced absorption above 3400 cm^{-1} is to be assigned to an excess of shorter oligomers, increasing the number of OH groups with proton donor or acceptor function. Similar transient spectra are presented in Figs. 17b and c for $t_D = 4$ ps and 8 ps, respectively (note different abscissa scales). The CH-amplitude increase in Fig. 17b compared to Fig. 17a results from the completion of the excitation process, followed by a minor decay of the CH amplitude in Fig. 17c, while the transient OH amplitudes are still growing until 8 ps delay time. Evidence for rapid energy redistribution in the CH-stretching region with a time constant <0.5 ps and an effective population decay of the CH modes with time constant $T_{10}(CH) = 12 \pm 2$ ps are observed from time-dependent measurements taken with probing and pump pulse resonant to the respective absorptions (data not shown).

Three different contributions to the dynamics in Fig. 17 can be identified (85):

1. Population transfer from CH to OH
2. Anharmonic coupling between CH- and OH-stretching modes so that the fundamental transition frequency of CH $(0, 0 \rightarrow 0, 1)$ differs from the corresponding combination tone $(1, 0 \rightarrow 1, 1)$
3. Thermalization processes

The three interaction channels are responsible for the bleaching and induced absorption features in the range 2950–2990 cm^{-1} in Fig. 17a, with mechanism 2 obviously producing a small red shift of the monitored CH-transition. The population lifetime of the excited OH vibration is known to be 1.4 ± 0.3 ps (79) so that energy transfer from CH$_3$ to OH cannot populate the upper OH level notably. There is evidence for energy transfer between the OH and CH vibrations. For OH excitation the measured CH-amplitude changes suggest a lower limit of 4 ps for the transfer time constant. Assuming detailed balance, a lower limit of 24 ps is estimated for the reverse process CH $\rightarrow$ OH in Fig. 17. From previous spectroscopic experiments on associated ethanol an energy transfer time of 12 ps was reported for OH $\rightarrow$ CH and 60 ps for the reverse process (75).

It is interesting in this context to compare the dynamics of a nonbonded ethanol molecule: for ethanol monomers in CCl_4 (0.05 M) an effective population lifetime of 10 ± 2 ps was measured for the CH_3-stretching vibration. Furthermore, energy transfer from the OH- to the asymmetrical CH_3-stretching mode was seen with $\tau_{trans} = 15 \pm 10$ ps. The inverse process was below the detection limit, consistent with detailed balance arguments for the frequency difference of 660 cm^{-1} of the two vibrations in the monomer case (73).

Almost simultaneously dissociation of hydrogen bonds occurs, with a quantum yield of $70 \pm 20\%$ and time constant 2 ± 0.5 ps, as indicated by the bleaching and induced absorption features in the OH range in Fig. 17. A thermal mechanism may account for the rapid bond breaking, involving the energy redistribution process of the CH vibrations and implying large rate constants for the rearrangement of the chemical equilibrium of the H-bonding system. The transient changes after CH pumping are similar but not identical to those after the OH excitation in the center of the oligomer band, alluding to differences of the dissociation mechanism.

To account for the measured conventional and transient spectra, we have developed a structural model for associated ethanol at room temperature starting from the following asumptions (87):

1. In the liquid phase at medium to high concentrations, we expect ethanol to arrange to open chains with the number of H-bonded molecules denoted by L.

2. For each associate of length L the absorption frequency of the OH-stretching mode depends on the position n in the chain, i.e., on the H-bond strength. In accordance with the literature, we assume end groups with proton acceptor function to show up in the spectrum like ethanol monomers with absorption at 3633 cm^{-1} independent on the chain length. The frequency position of the proton donor end group of 3500 cm^{-1} is kept constant in the same way.

3. For the shortest and longest chains that occur in the computation, we choose the OH frequency of the internal groups in the middle position and interpolate the frequency position for the other Ls.

4. The absorption frequencies of internal molecules $\nu_L(n)$ are derived from assumptions 2 and 3 above by cubic interpolation with three fixed values (end groups of the chain and middle position); alternatively equal frequencies for all internal molecules are considered.

5. The proton acceptor end group exhibits a Lorentzian line shape with width 25 cm^{-1}, while all H-bonded molecules are assumed to experience local disorder (different bond angles and O-O lengths) leading to inhomogeneous broadening with Gaussian shape and width of 100 cm^{-1}. The inhomogeneous character of the OH band of associated ethanol was verified by transient hole burning (79).

6. The value of the absorption cross section is fixed at three positions in the chain (end positions and in the middle), with a parabolic dependency on position for the other groups within each associate.

7. We take a Gaussian distribution of chain lengths with a full-width half maximum ΔL and peaking at L_p to account for the expected disorder in the liquid phase at room temperature.

The conventional OH-absorption spectra of ethanol:CCl$_4$ in the range 0.05 M up to the neat liquid can be perfectly fitted using the model assumptions above. The ratio of absorption cross sections is determined to be $\sigma_{3633}:\sigma_{3500}:\sigma_{3330} = 1:2:9$. From the experimental concentration dependence the most probable length of the oligomer chains is inferred to be proportional to $(\log c)^2$, where c denotes the ethanol concentration, yielding numbers of 1, 2-3, 5, and 11–16, respectively, for c = 0.05 M, 0.17 M, 1.2 M, and the neat liquid, respectively. The model also accounts for the time-resolved spectra if some additional assumption on the breaking of the oligomer after vibrational excitation is made. A semiquantitative understanding of the formation of hydrogen bonds is provided, emphasizing the importance of open chains for tetramers and larger oligomers, while for trimers the cyclic configuration seems to prevail (lack of OH end groups in the transient spectrum after bond breaking). For details the reader is referred to the original work (87).

3. Fully Associated Ethanol in Isotopic Mixtures

It is interesting to study the structural properties of the neat hydrogen-bonded liquid without the perturbing influence of apolar molecules in mixtures. As an example in this direction the dynamics of ethanol in the isotopic mixture are investigated. Data will be presented on ethanol-d$_6$ samples containing 1 vol% (diluted) or 50 vol% (concentrated) of protonic ethanol (88). In contrast to the apolar CCl$_4$ environment, additional H bonds between the ethanol molecules and their environment can be formed representing new species with modified properties.

In Fig. 18 transient spectra of the diluted isotopic mixture are demonstrated; the measured relative transmission change (isotropic component) is plotted versus probing frequency for three different delay times of -1 ps (a), 1 ps (b), and 6 ps (c) and excitation at the peak position of the OH-stretching band of 3330 cm^{-1}. The conventional absorption band is similar to the oligomeric band of Fig. 17c (dash-dotted curve). Similar to the data of Fig. 15, spectral hole burning is seen in the transient spectra at early times (Figs. 18a,b). A hole width of 35 ± 5 cm^{-1} is inferred from the data assuming a Lorentzian shape (dotted curves). The spectral hole is accompanied by a broad bleaching component at the pump frequency with a width of 140 cm^{-1} (approximately Gaussian shape) and induced absorption at 3100 cm^{-1} (dashed lines). The spectral components are obviously related to ground state depletion and corresponding ESA of the excited OH-stretching mode. Several ps later (Fig. 18c) the spectral hole has disappeared and a blue-shifted induced absorption at 3500 cm^{-1} built up, while the broad bleaching and ESA components decreased in amplitude (note different ordinate scales). In analogy to the results for ethanol in solution, breaking of oligomer chains in the liquid is inferred from the spectra on the time scale of a few picoseconds.

The situation in the concentrated isotopic mixture (50 vol% protonated ethanol) is illustrated by Fig. 19. An extended frequency range including the OH and OD stretching vibrations is covered by the transient spectra generated again by pumping at the peak position of the OH band. For $t_D = 1$ ps (Fig. 19a) maximum bleaching is observed at the excitation frequency 3330 cm^{-1}. The feature can again be deconvolved into a Lorentzian-shaped transient hole with width of 25 cm^{-1} (dotted curve) as well as a 140 cm^{-1} wide Gaussian component (dashed line). Furthermore, the corresponding ESA can be seen at 3100 cm^{-1} (short dashed line) and a smaller induced absorption centered at 3500 cm^{-1}, the latter indicating the breaking of H bonds. Most interestingly, bleaching and induced absorption around the frequency position of the OD band is also measured. Increasing the delay time to 2 ps (Fig. 19b) the spectral hole has decreased while the induced absorption below 3400 cm^{-1} is enlarged. At a later time, $t_D = 11$ ps (c), the peak position of the induced absorption in the blue part of the spectrum has shifted towards 3430 cm^{-1}, while a bleaching in the region of the CH-stretching vibrations between 2800 cm^{-1} and 3000 cm^{-1} is also evident. The spectral features of the OD absorption of deuterated molecules have gained in amplitude.

Changing the excitation frequency to 3450 cm^{-1}, the bleaching as well as the ESA related to vibrational excitation of the OH mode are also

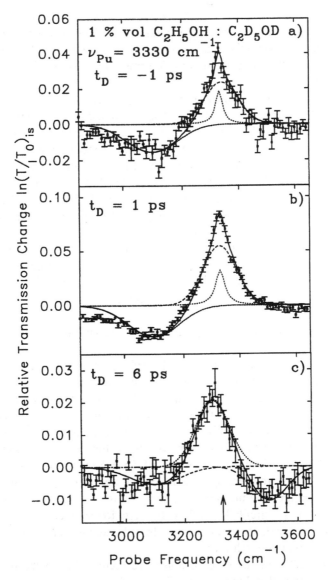

Figure 18 Isotropic time-resolved spectra of a solution of 1 vol% ethanol and fully deuterated ethanol are shown for delay times of −1 ps (a), 1 ps (b), and 6 ps (c) while excitation is adjusted to 3330 cm^{-1}. Measured data, calculated lines. Excitation is marked with an arrow. A spectral hole at the frequency position of excitation is clearly seen from the data.

blue-shifted by 80 cm^{-1}, while the width of the spectral hole is unchanged within the measuring accuracy. However, the peak position of the bleaching component with Gaussian shape moves with increasing delay time down to 3300 cm^{-1}, the same position as determined for $\nu_{Pu} = 3330$ cm^{-1} (data not shown).

The strong OD band also makes it possible to examine the influence of OH excitation on the neighboring deuterated species. From the time-resolved spectra shown in Fig. 19, we are able to infer two different dynamics at frequencies within the OD band. Together with the OH excitation a shift of the OD band is observed toward higher frequencies, i.e., weaker H bonds. This manifests itself in a bleaching of the sample peaking at 2430 cm^{-1} and an increased absorption around 2520 cm^{-1}. The finding refers to molecules in the vibrational ground state and is attributed to a weakening of the hydrogen bonds of deuterated molecules. Subsequently a second dynamical feature is seen, leading to an amplitude increase of the mentioned bleaching and absorption feature. The process builds up simultaneously with the induced OH absorption around 3500 cm^{-1} and may be related to the dissipation of vibrational energy outlined in the following.

The measured transient spectra for the isotopic mixtures are outlined in the following sections.

a. Diluted Isotopic Mixture (1 vol% protonated ethanol). Transient spectral hole burning within the OH band is observed as already known for several ethanol solutions (78,79). From the spectra of Fig. 18 and additional data taken at different excitation frequencies (not shown here), we deduce a width of the spectral hole of 35 ± 5 cm^{-1} after deconvolution with the instrumental response curve for excitation at 3210 cm^{-1} and 3330 cm^{-1}, while a larger width of 55 ± 5 cm^{-1} is found for $\nu_{Pu} = 3450$ cm^{-1}. The transient spectral holes are again interpreted in terms of specific structures within the H-bonded chains of ethanol molecules. The increase in the spectral hole width with excitation frequency may be rationalized by a wider distribution of possible H-bond angles and distances with decreasing bond strength (88).

The widths of the spectral holes are comparable to the ones determined for ethanol solutions (78). For the latter samples we infer from model computations a size of the oligomers directly excited by the pump pulse of 4–6 (87), while in the isotopic mixture even longer associates may interact with the pump. Depending on the excitation frequency, a protonated species within an associate of 5–10 d_6-ethanol molecules is estimated to be

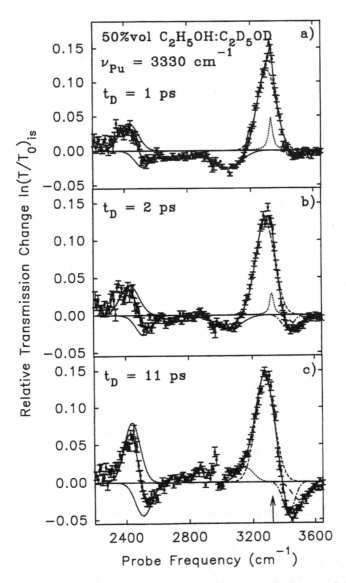

Figure 19 The isotropic signal of transient spectra of a solution of 50 vol% ethanol and fully deuterated ethanol is depicted. Data are shown at room temperature and delay times of 1 ps (a), 2 ps (b), and 11 ps (c), while excitation is performed at the peak position of the OH band at 3330 cm^{-1}. Note the induced absorption/bleaching at frequencies related to the OD-stretching band. Measured data, calculated lines.

excited (87). In this case a statistical distribution is expected for the position of the ethanol in the associate. Similar results were reported recently in Ref. 82. In this investigation a spectral hole width of ≈ 100 cm^{-1} was found at room temperature. The larger value may be explained by saturation broadening for the approximately 10-fold higher excitation level with longer pulses of 4-5 ps, leading to a pronounced heating of the sample. From the temporal evolution of the ESA we infer an average lifetime of the OH-stretching mode of associated ethanol of $T_1 < 1$ ps, $T_1 = 1.5 \pm 0.5$ ps, and $T_1 = 1.2 \pm 0.5$ ps for excitation at 3210, 3330, and 3450 cm^{-1}, respectively. It is emphasized that the T_1 values represent averages over a variety of structural components since, for example, associates of different lengths may be simultaneously excited. Ground-state recovery proceeds with a time constant similar to T_1: For excitation at band maximum we find a relaxation time of $T_{10} = 2.2 \pm 0.3$ ps for ground state repopulation. Intermediate states are apparently of minor importance to the dynamics since ground-state refilling proceeds only slightly slower than T_1. We mention that the long-lived components of the signal transients contribute little to the total signal amplitude, suggesting a minor local heating of the sample.

Because of the small concentration the protonated molecules are essentially isolated from one another in a surrounding of deuterated ethanol. The OH- and OD-stretching frequencies are largely separated by almost 900 cm^{-1}. As a consequence the deposited vibrational energy cannot be efficiently dissipated by near-resonant intermolecular energy transfer among the nearest neighbors. Other decay channels must account for the energy relaxation, e.g., mechanisms involving hydrogen bridge bond vibrations, since the OH lifetime is known to be considerably shortened by the presence of H bonds. Energy transfer to H-bond vibrations could readily explain the breaking of associates and the corresponding shift of the chemical equilibrium, indicated by the time delayed build-up of blue-shifted absorption in the transient spectra (see Fig. 18c).

b. Concentrated Isotopic Mixture (50 vol% protonated ethanol). Excitation of the OH-stretching mode of associated molecules in the concentrated mixture results in more complicated dynamics. In the time-resolved spectra similar to Fig. 19a (data not shown), spectral holes are again observed in the OH band with width 25 ± 5 cm^{-1} after deconvolution of the instrumental resolution, independent of the excitation frequency within the halfwidth of the OH band. A lifetime of the spectral holes of 1 ± 0.5 ps is suggested by the measurements. The numbers agree well with those discussed above for ethanol in CCl$_4$ (1.2 M). The appearance of a spectral hole is again interpreted as evidence for spectral subcomponents of the inhomogeneously broadened OH band related to ethanol molecules in different positions in the H-bonded oligomers. The average size of the excited associates in the case of the 1.2 M solution and the isotopic mixture is similar and estimated from a structural model (87) to be 5–10, while larger oligomers may also be pumped at the excitation frequency.

The lifetime of the OH vibration of associated ethanol depends on the OH-stretching frequency, i.e., on the strength of the OH bond. We determine T_1 to be 1.6 ± 0.3 ps, 1.5 ± 0.3 ps, and 1.3 ± 0.3 ps, respectively, for probing of the ESA peak at 3190 cm^{-1}, 3100 cm^{-1}, and 3100 cm^{-1}, while the corresponding excitation is carried out at 3450 cm^{-1}, 3330 cm^{-1}, and 3240 cm^{-1}.

Two components with different time dependence are inferred from the induced absorption in the OH spectrum for $\nu > 3400$ cm^{-1}. Consistently, the deconvolution of the time-resolved spectra strongly suggests two spectral components peaking at 3430 cm^{-1} and 3500 cm^{-1}, respectively. The latter builds up time-delayed in comparison to the ESA, with a time constant of 2.5 ± 0.5 ps. The component is assigned to predissociation of H bonds in correspondence to the findings for ethanol solutions discussed above (78,79). The former component at 3430 cm^{-1} develops more quickly with a time delay of ≈ 1 ps, quite similar to the population decay of the first excited state of the OH vibration. The faster component seems to be related to a rearrangement of the chemical equilibrium due to the thermalization of pump energy. The resulting heating of the sample leads to a weakening of the hydrogen bonds and a corresponding blue-shift of the OH absorption. While the induced absorption discussed here represents additional absorbing molecules via breaking of H bonds in the blue part of the transient spectrum, the missing larger oligomers are also seen as induced bleaching peaked at ≈ 3300 cm^{-1}. The delayed spectral red-shift of the bleaching peak and the slow amplitude increase support the physical picture.

A direct vibrational energy transfer between neighboring ethanol and d_6-ethanol molecules is not indicated by our data because a corresponding red-shifted ESA from excited OD modes below 2400 cm^{-1} is not observed. The finding is reasonable as the energy mismatch between the two stretching vibrations corresponds to approximately 850 cm^{-1}, making this relaxation channel unlikely on the basis of Fermi-resonance arguments.

4. H-Bonded Dimers: Librational Substructure of the OH Band of Proton Donors

The mechanisms for the broadening of the OH-stretching band via hydrogen bonds are still controversially discussed. Two possibilities are considered that could contribute simultaneously: one factor is local inhomogeneity via different bond angles and/or lengths considered above. The other mechanism is energetic inhomogeneity of the band via combination tones between the OH stretch and low-lying bridge bond vibrations. In associated alcohols like ethanol or in water (see below) there is evidence for the first mechanism from time-resolved spectroscopy (78,79,89–91) and from numerical investigations (92–94). Intimations of the second process were obtained from theoretical investigations (95–97). For hydrogen-bonded dimers in a polymer film the temperature dependence of transient spectral hole burning strongly suggested that variation of the local environments accounts for the large broadening of the proton-donor OH band (98). To tackle this important question in more detail in the liquid phase, the special alcohol 2,2-dimethyl-3-ethyl-3-pentanol (DMEP) was investigated by conventional and time-resolved IR spectroscopy. Quantum statistical simulations of molecular clusters show that even in the neat liquid DMEP only monomers and open dimers are present (99). The simple explanation is steric hindrance by the CH_n groups; obviously for the same reason the H bond of the open dimer of DMEP is weaker as compared to ethanol dimers. The absence of larger oligomers and the more well-defined local structure of DMEP dimers make the system well suited for detailed study.

Examples of transient spectra taken with excitation adjusted to the frequency position of proton acceptors/monomers at $T = 295$ K are shown in Fig. 20 (isotropic component) (100). At early delay times of -1 ps (Fig. 20a) a sample bleaching at the frequency of the proton acceptors/monomers (3625 cm^{-1}) and, most important, at the proton donor position (3520 cm^{-1}) is readily seen. Simultaneously an induced absorption builds up below 3480 cm^{-1}, representing red-shifted ESA of species in the $v = 1$ level. Soon thereafter at 2 ps (Fig. 20b) the spectral components have increased in amplitude, while at a later delay time of 10 ps (Fig. 20c — note different ordinate scale) the induced bleaching of the sample dominates the transient spectrum at a reduced amplitude level. To interpret the fast dynamics related to the bleaching of the donor band that proceeds with a time constant <1 ps below our temporal resolution, it is important to know the reorientational motion of the donor-acceptor H bond. The information is derived from the measured induced dichroism with probing at the excitation frequency. Reorientational relaxation proceeds with time constants of 4 and 8 ps for the proton acceptor and donor molecule, respectively (101). The (weak) H bond between two DMEP molecules significantly slows down the reorientation of the bonded OH group by a factor of 2 relative to the nonbonded species. It is concluded

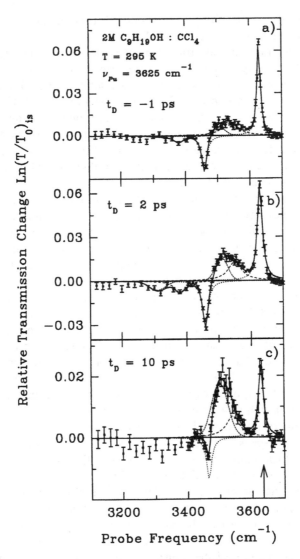

Figure 20 Isotropic transient spectra taken at 295 K and excitation of the OH-stretching vibration of proton acceptors/monomers are shown in this figure. Data are measured at −1 ps (a), 2 ps (b), and 10 ps (c), while excitation is adjusted to 3625 cm⁻¹, marked by an arrow. Migration of the vibrational quantum from the directly excited proton acceptor to the proton donor can be clearly seen from the spectra. Measured filled squares, solid line is the sum of the various spectral components indicated by different line styles.

that the initial fast bleaching of the donor absorption cannot be explained by structural changes, i.e., bond breaking.

Further transient spectra taken with an excitation frequency of 3430 cm^{-1} (see Fig. 21) show no measurable absorption change at the proton acceptor frequency 3625 cm^{-1}, i.e., some cooperative effect does not account for the fast spectral dynamics shown in Fig. 20. The bleaching of the proton donor absorption while exciting proton acceptor OH groups is explained by vibrational energy transfer, strongly supported by the observation of further red-shifted ESA in Fig. 20. The finding corresponds to the notion of energy migration in ethanol oligomers after excitation of internal OH groups (see above).

The analysis of the bleaching of the proton donor absorption in Fig. 20b suggests two spectral subcomponents (see calculated curves). To investigate the effect in detail, transient spectroscopy at four different excitation frequencies within the proton donor OH-stretching band was performed at 260 K (102). One example is shown in Fig. 21. The isotropic signal component is plotted, i.e., population dynamics without a reorientational contribution is measured; excitation is carried out at 3430 cm^{-1} for OH groups in the red wing of the dimer band. The conventional absorption spectrum of the sample is also presented in Fig. 21c (dash-dotted line, right-hand ordinate scale). It consists of a narrow line at 3625 cm^{-1}, attributed to absorption of monomers and proton acceptor groups, and a broader band at 3505 cm^{-1} of asymmetrical shape. The latter represents the OH groups with proton-donor function in dimers, as shown by comparison with other diluted alcohols (80).

It is interesting to see in the transient spectrum of Fig. 21a at early times the double-peak of a bleaching structure. The band has already relaxed to a single maximum with shoulder for $t_D = 0$ (Fig. 21b), followed by a shift of the transient band towards 3500 cm^{-1} (Fig. 21c). Again, the bleaching of the sample for $v > 3350$ cm^{-1} reflects the depletion of the vibrational ground state $v = 0$ upon excitation. Correspondingly, induced absorption of molecules in the first excited state is observed at frequencies of <3350 cm^{-1}. The time-resolved spectrum is readily attributed to a discrete substructure of the transient band, as suggested by the multiple maxima of the bleaching and ESA. The bleaching feature can be well described by a time-dependent spectral hole at the excitation frequency with approximately Lorentzian shape and width of 45 ± 5 cm^{-1} as well as additional components assigned to satellite holes with different time evolutions to account for the complex, rapidly changing band shape. The data suggest an approximately constant frequency spacing of 35 ± 5 cm^{-1}

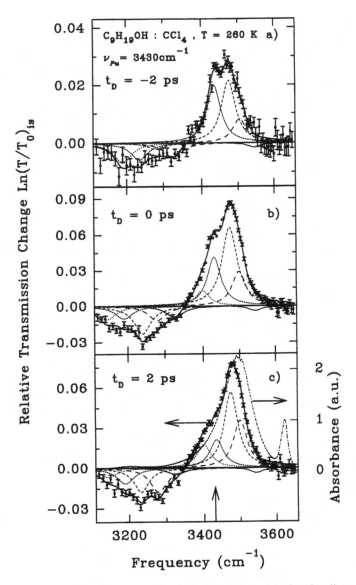

Figure 21 Conventional spectrum (c; dash-dotted line; right-hand ordinate scale) and transient spectra of a 2 M DMEP and CCl₄ mixture (isotropic signal) in the OH-stretching region taken at 260 K for different delay times: −2 ps (a), 0 ps (b), and 2 ps (c); excitation adjusted to 3430 cm⁻¹ (marked by a vertical arrow); experimental points, calculated curves.

between the subcomponents. Satellite maxima are also seen in the induced absorption and assigned to inverse spectral holes, but with a different frequency distance of 45 ± 5 cm^{-1}. The decomposition of the spectra is indicated in Fig. 21 by calculated curves. Corresponding spectral holes (inverse holes) are plotted with the same line style.

We have investigated the transient spectra of DMEP solutions in the delay interval -2 to 10 ps for further excitation frequencies in the range 3430–3555 cm^{-1}. The data (not shown) are consistently fitted with a set of Lorentzian-shaped spectral lines (transient holes and inverse holes) by determining the minimum deviation numerically (65). Some results are summarized in Table 3. The transition frequency of holes ($v = 0 \rightarrow v = 1$) is denoted by ν_{01}, while ν_{12} is that of inverse holes ($v = 1 \rightarrow v = 2$). Spectral widths of 45 ± 5 cm^{-1} and 55 ± 5 cm^{-1} for the holes and inverse holes, respectively, account for the experimental data. In order to interpret the spectral holes and satellite holes in the sample bleaching, it is important to recall that reorientational and similarly structural relaxation proceeds slowly with a time constant of 8 ps and cannot explain the fast dynamics in Fig. 21. Consequently the spectral components are interpreted in terms of an energetic substructure, i.e., combination tones between the proton donor OH stretch and a low-frequency hydrogen bridge bond vibration. The constant frequency spacing of the measured spectral holes suggests that one mode is dominant and is considered in our simplified physical picture in the following. A Franck-Condon–like situation, illustrated by Fig. 22, is assumed with significant anharmonic coupling between the high-frequency OH-stretching vibration (quantum number v) of the donor molecule and the low-frequency mode of the bridge (quantum numbers n, n′; frequency $\nu_{nn'}$). Anharmonic frequency changes of $\nu_{nn'}$, on the other hand, with quantum number n are considered to be small

Table 3 Frequency Positions of the Spectral Holes and Inverse Holes as Determined by Fitting the Time-Resolved Spectra Taken at Various Delay Times, Excitation Frequencies, and Temperatures of 260, 298, and 343 K

	$\Delta n = 2$	$\Delta n = 1$	$\Delta n = 0$	$\Delta n = -1$	$\Delta n = -2$	$\Delta n = -3$
ν_{01} (cm^{-1})	3575	3545	3505	3475	3435	3400
ν_{12} (cm^{-1})	3360	3325	3280	3240	3190	3145

The spectral width of the components derived from fitting of the sample bleaching and the excited-state absorption amounts to 45 and 55 cm^{-1}, respectively. All numbers with an accuracy of ± 5 cm^{-1}.

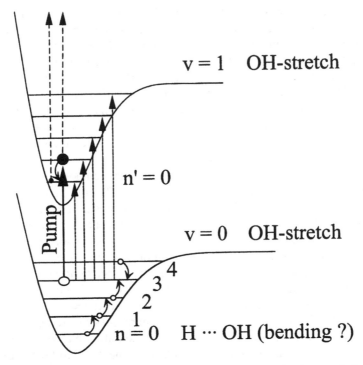

Figure 22 Simplified energy level scheme that accounts for the spectral observations with librational substructure (horizontal lines, n, n') of the OH-stretching levels (solid curves, v); the thick vertical arrow represents an example for the excitation process with $\Delta n = -2$; thin solid and dashed vertical arrows denote possible probing transitions for different Δn for ν_{01} and ν_{12} transitions, respectively.

compared to the measured holewidths and are omitted. In addition to the selection rules of the harmonic case, $\Delta v = 1$, $\Delta n = n' - n = 0$, transitions $\Delta n = \pm 1, \pm 2, \ldots$ must be included (see Fig. 22). Tentatively we assign the maximum of the conventional absorption band to a superposition of $\Delta n = 0$ transitions, suggesting the value $\nu_{OH} = 3505$ cm^{-1}. In the time-resolved measurements, the pump pulse promotes a subensemble of molecules from thermally populated $(0, n)$ levels to a modified set $(1, n')$, the population changes depending on the individual cross sections. In Fig. 22 the situation for pumping at 3430 cm^{-1} is considered corresponding to $\Delta n = -2$ transitions (thick vertical arrow). Since the n's are lowered in the excitation step, excess population of the lower quantum numbers is generated compared to the Boltzmann distribution. The perturbed occupation of the low-frequency mode gives rise to subsequent relaxation and transient bandshape changes, i.e., spectral holes and, correspondingly, inverse holes for $v = 1 \rightarrow v = 2$ probing transitions. In this picture the frequency spacing of the components is equal to the mode transition frequencies in the $v = 1$ state (v_{01} transitions, bleaching) and $v = 2$ level (v_{12} transitions, ESA), respectively. The depicted probing transitions in Fig. 22 are to be taken as examples only. The measured induced transmission change at 3430 cm^{-1} is equal to the sum over all v_{01}-probing transitions with $\Delta n = -2$ for the case assumed in the figure.

This interpretation is supported by the temporal evolution of the spectral components at, for example, 3475 cm^{-1}($\Delta n = -1$) and 3435 cm^{-1}($\Delta n = -2$) as derived from a decomposition of the measured transients. The decay times of the two components are determined to 4.5 ± 0.5 ps and 3 ± 0.5 ps, respectively. Two mechanisms obviously contribute to the time constants: population decay of the OH-stretching mode and population redistribution among the bridge bond vibration. Since the excess population of the lower n levels is transferred to larger n, probing with $\Delta n = -2$ yields a faster decay compared to $\Delta n = -1$, as indicated by the data. For larger delay times the system does not return exactly to its initial situation, since the deposited OH-excitation energy thermalizes. From the temporal evolution of the ESA component at 3240 cm^{-1} derived from fitting of the transient spectra, we determine a lifetime of the proton donor OH stretch of 3 ± 0.5 ps.

Tentatively we assign the low-frequency mode to the bending vibration of the hydrogen bridge bond, the frequency value $v_{nn'} \simeq 35$ cm^{-1} referring to the OH level $v = 1$. In the low-frequency Raman spectrum of neat DMEP a band around 27 cm^{-1} with width of approximately 50 cm^{-1} shows up that may be attributed to the OH$\cdots$H-bending mode in the $v = 0$ state (G. E. Walrafen, unpublished). The difference between the two numbers may be related to the different OH levels, $v = 0$, and 1. For comparison, in water a band located at 50 cm^{-1} was proposed to represent the bridge-bending mode [103]. With regard to the notably weaker H bond of the present dimers, our interpretation appears reasonable.

The physical situation is supported by a comparison of computed bandshapes with the measured conventional dimer absorption, the temperature dependence of which is well described. Figure 23 illustrates the situation with three spectra of a 2 M DMEP solution in CCl$_4$ taken at 260 K (a), 298 K (b), and 343 K (c). The measured data points can be fitted well with the already determined spectral components listed in Table 3. Again, the

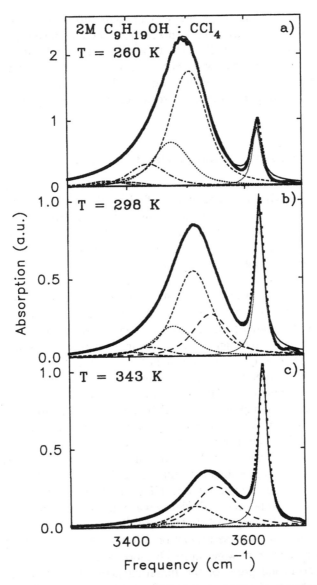

Figure 23 Here we show conventional absorption spectra of the 2 M DMEP and CCl_4 mixture in the OH-stretching region at the three indicated temperatures of 260 K (a), 298 K (b), and 343 K (c). The spectral components derived from time-resolved spectroscopy and denoted by different line styles account well for the measured data.

different spectral components are depicted with different line styles. A shift of the peak of the distribution is noted from the component at 3505 cm^{-1} (260 K) to the one at 3545 cm^{-1} (343 K). This is related to an overall weakening of the H bond with increasing temperature, which results from excitation of the low-frequency bridge bond vibrations to higher quantum numbers in correspondence to the result of the fitting of the proton donor OH band.

E. Investigations of Isotopic Water Mixtures

In the previous sections evidence for spectral holes in the absorption band related to the OH-stretching mode of H-bonded alcohols was discussed and interpreted in terms of different H-bonded structures (associated ethanol) or combination tones between the OH-stretch and an H-bridge bond vibration (DMEP). It is possible to use the same techniques for experiments on water, the most abundant liquid in nature. Water has been investigated by a variety of experimental techniques during recent decades (1,2). Nevertheless, a clear physical picture of the structure and structural relaxation of water was not established because time-resolved experiments were lacking. Contradicting results were derived from dielectric relaxation, nuclear magnetic resonance, and Raman spectroscopy. From Rayleigh scattering, for example, time constants for structural relaxation on the 1 ps time scale were reported as derived from fitting of the linewidth (104).

The potential of the OH-stretching vibration of water as a local probe for hydrogen bonding was realized in early investigations (1). Its relation to the strength of hydrogen bonds was utilized for infrared (105,106) and Raman (107) spectroscopic investigations in order to identify different structural components from observed temperature or pressure changes. The first time-resolved investigation with two-color infrared spectroscopy was performed for the isotopic mixture HDO in D_2O with the moderate time resolution of $\approx$10 ps pulses. Evidence was reported for three major spectral components in the range 3340–3520 cm^{-1} at room temperature (89). The recent advance of IR pulse generation allows one to tackle the question of structural relaxation in water in much more detail (108).

Experimental proof of the inhomogeneous broadening of the OH-stretching band of water is straightforward, following the lines of the ethanol studies discussed above (89). Because of its simpler OH spectrum with only one stretching mode above 3000 cm^{-1} and in order to minimize local heating effects, it is advantageous to investigate highly diluted HDO in D_2O.

The low concentration of 0.8 M ensures that the OH mode is utilized as an ultrafast local probe for hydrogen bonds while quasi-resonant energy transfer to neighboring OH groups is negligible.

To focus on the existence of spectral holes, we first show transient spectra in Fig. 24 for excitation around the peak position of the OH band at $t_D = 0$ and three temperature values (91). The displayed anisotropic component emphasizes the primary features of the molecular excitation that maintain the linear polarization of the pump pulse. At T = 273 K (Fig. 24a) the spectral bleaching is mainly governed by a rather broad component II (dashed curve) while two further contributions I and III give minor contributions to the wings of the measured band around the pump frequency. A spectral hole is obviously not observed. The broad induced absorption around 3160 cm^{-1} is interpreted as excited-state absorption. Changing to room temperature (Fig. 24b) the spectrum looks distinctly different with a narrow component riding on a broad bleaching band. A spectral hole (dotted curve) is inferred from the data in addition to the components I–III already observed in Fig. 24a. At a higher temperature of 343 K (Fig. 24c) a similar band shape is measured apart from a smaller amplitude. The main contributions to the bleaching are assigned to component III (long-dashed) and the spectral hole (dotted). The latter is well described by a Lorentzian shape with width 45 ± 10 cm^{-1} and located at the frequency position of the pump pulse, as verified by additional measurements with varying excitation in the blue wing of the OH absorption.

The spectral hole is attributed to a depletion of the vibrational ground state and corresponding population of the first excited state of a molecular subensemble. The lifetime of the hole is determined from the temporal evolution of its peak amplitude derived from fitting of the respective transient spectra taken at different delay times, yielding $\tau_h = 1 \pm 0.4$ ps. The time constant represents information on spectral changes initiated by structural relaxation with component III. The width of the spectral hole and its lifetime differ from the results for weak ethanol solutions. Because the spectral holes in water are only seen within component III located in the blue wing of the OH band of HDO, the information on structural relaxation refers to species with weaker hydrogen bonds. The weaker bonds seem to allow for a broader variety of OH frequencies (bond lengths/angles) leading to inhomogeneous broadening of component III.

In order to identify different spectral components within the OH band of HDO more clearly, additional transient spectra with a variation of ν_{Pu} (109) are presented in Fig. 25, where the respective components are predominantly excited. The dependence of the OH bleaching at 275 K and

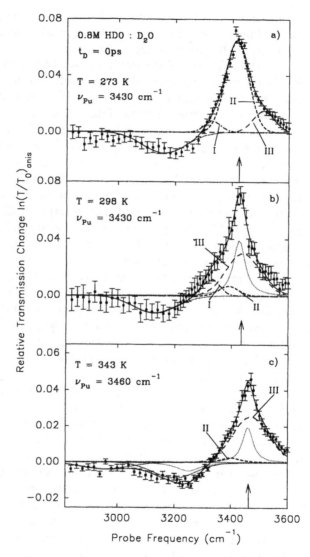

Figure 24 Transient spectra (anisotropic signal) for $t_D = 0$ ps and excitation at the peak position of the OH band at three temperature values: 273 K (a), 298 K (b), and 343 K (c). Experimental points; the calculated thin lines indicate the analysis of the transient bandshapes (thin full line) in terms of the major spectral components I–III (dash-dotted, broken, long-dashed) with Gaussian shape and a Lorentzian spectral hole contribution (dotted line, b,c); the pump frequencies are indicated in the figure by vertical arrows.

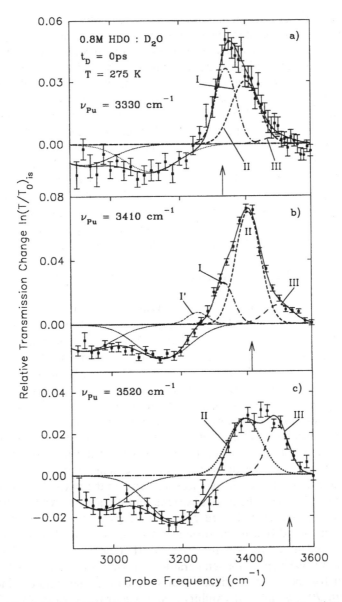

Figure 25 Same as Fig. 24, but for constant temperature T = 275 K. The isotropic signal component is this time depicted for three different frequency positions of the excitation pulse (note vertical arrows): 3330 cm^{-1} (a), 3410 cm^{-1} (b), and 3520 cm^{-1} (c). Note different ordinate scales.

zero delay time on excitation frequency is noteworthy and inconsistent with a simple bell-shaped distribution function of OH transition frequencies. With the pump pulse tuned resonantly to the position of component I (Fig. 25a) the latter (dash-dotted) dominates, but with a large amplitude of II and minor contribution of III. Alternatively exciting at the position of II (Fig. 25b) a large bleaching amplitude of II occurs with smaller contributions of I and III appearing as shoulders in the wing. Simultaneously the amplitudes of bleaching and induced absorption below 3250 cm^{-1} have gained (note different ordinate scales). Tuning the pump pulse to the blue wing of the OH absorption (Fig. 25c), we find large amplitudes of species II and III, while the contribution of I is negligible. We recall that a spectral hole is not found for the data of Fig. 25 at 275 K. This is explained by the fact that component III changes its character at lower temperatures (smaller width); the finite measuring accuracy may also explain why a spectral hole within is not identified below room temperature. The deconvolution of the transient spectra shown in Figs. 24 and 25 is obtained by self-consistent fits of theoretical models to all transient spectra taken at three different excitation frequencies in the OH band and at least 7 delay time settings at a certain temperature. Lorentzian shape of the transient holes is assumed while the spectral components are assumed to be Gaussian. The obtained spectral components are indicated in the figures by different line styles. Three major spectral species I–III are deduced from the manifold of isotropic and anisotropic transient spectra. As a fourth major contribution the spectral hole positioned at the pump frequency is introduced in the frequency band of species III and for temperatures above 290 K, since a hole was not observed at lower temperatures. A minor contribution I' at 3260 cm^{-1} has been included at lower temperatures. Only the amplitudes of the components are allowed to change with delay time and excitation frequency, while the positions and widths are kept constant.

    The results of five sets of measurements in the range 273–343 K are presented in Fig. 26. The peak positions and the spectral widths (FWHM) of the major species I–III are plotted in Figs. 26a and b, respectively. Components I and II with the strongest red-shift exhibit almost no temperature dependence. We find for species III a decrease from 3510 cm^{-1} at 273 K to 3450 cm^{-1} for temperatures above 290 K. Similiar behavior is found for the spectral widths; that of I appears temperature independent with 70 ± 5 cm^{-1}. II exhibits a slight narrowing from 95 ± 5 cm^{-1} to 80 ± 5 cm^{-1}, whereas component III with the smallest red-shift and consequently the weakest H bonds, displays significant broadening from

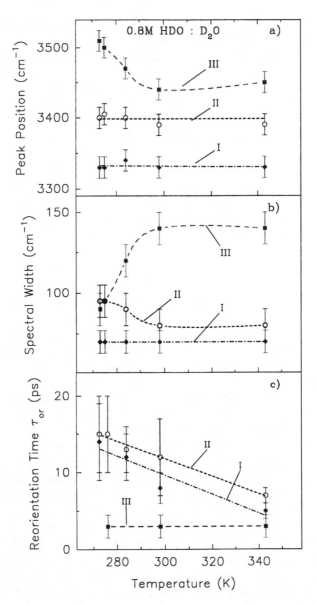

Figure 26 Results of the decomposition of the measured transient spectra: the peak positions (a), spectral widths (FWHM) (b), and reorientation times (c) of three prominent spectral components I–III, attributed to different local structures of water, as a function of temperature; experimental points; the lines are drawn as a guide for the eye.

90 ± 5 cm^{-1} at 273 K to 140 ± 7 cm^{-1} above 290 K. The increased band-width is consistent with the occurrence of spectral holes in contribution III at higher temperatures, suggesting inhomogeneous broadening of the latter. The different character of the spectral species is supported by results on the reorientation time of the OH group of HDO depicted in Fig. 26c. The time constants of I and II are similar and shorten from approximately 15 ps at 273 K to 5 ps and 7 ps, respectively, at 343 K. Species III, on the other hand, exhibits fast reorientation with $\tau_{or} = 3 \pm 1.5$ ps, independent of temperature (109).

The major spectral components I–III derived from transient spec-troscopy are assigned to distinct structural environments. The physical reason is believed to be sterical hindrance. Due to the fairly close packing of molecules in the liquid, on the one hand, the nonspherical molecular shape and the directive character of the bridge bond, on the other hand, preferen-tial arrangements of nearest neighbors for a given HDO molecule exist with preferred frequency shifts of the OH vibration. Our experimental results are in semiquantitative agreement with earlier findings (89). The spectral obser-vations are consistent with MD computations that predict three classes of sites in the water network (110). Species I is located at a frequency close to the OH frequency of HDO for the ice structure I$_h$ involving at least four strong H bonds and possibly small bond angles, as suggested by a comparison of the peak position with data of conventional infrared (111) and Raman spectroscopy (112). The ice-like situation of component I is in accordance with the slow orientational motion (see Fig. 26c). Evidence for an approximately tetrahedral local geometry in liquid water in the measuring range 273–343 K is obtained in accordance with MD simulations (113). Formation of weaker (bent) hydrogen bridges will lead to a higher OH frequency (II), a component that has been also deduced earlier from Raman investigations (112). The spectral component III obviously corresponds to an even weaker H-bonding situation and is tentatively assigned to molecules with bifurcated hydrogen bonds as concluded previously from the Raman spectrum of HDO, where a spectral feature at $\simeq 3435$ cm^{-1} was found at room temperature (114) and interpreted that way (115,116). It should be noted that a possible substructure of species III with overlapping compo-nents and different temperature dependencies cannot be excluded from the present data. Components close to the frequency positions of the species II and III determined here can be also estimated from results for the OD band of HDO (117). Our data on three major species I, II, and III do not exclude the existence of further (weaker) components. It is recalled that in principle all of the 12 possible bonding configurations of HDO (118) may contribute to the overall OH band with respective Raman and/or IR activity.

The decrease of sample transmission in Figs. 24 and 25 below 3300 cm^{-1} after excitation in the OH band is attributed to ESA from transitions $v = 1 \rightarrow v = 2$. The peak position of the approximately 180 cm^{-1} wide ESA band varies between 3120 and 3200 cm^{-1}, depending on excitation frequency and temperature. The average value of the anharmonic shift of the OH-stretching vibration is estimated to be 240 ± 20 cm^{-1}. The number is consistent with the earlier result of 270 ± 20 cm^{-1} of Ref. 89 and in satisfactory accordance with results for other hydrogen-bonded systems, e.g., 210 cm^{-1} for a dimer in a polymer matrix (76) and 230 cm^{-1} for ethanol oligomers in liquid solution (78). The anharmonic shifts seem to vary with the strength of the hydrogen bonding.

A further component shows up in induced absorption at 2940 ± 20 cm^{-1} (see Figs. 24 and 25) that is tentatively assigned to an overtone of the bending mode of HDO. A corresponding feature is found in the conventional IR spectrum at 2960 cm^{-1}. The induced absorption presents evidence that an excited level of the bending vibration is populated by energy transfer from the stretching mode. The relaxation channel was previously suggested for water vapor (119). The induced absorption at 2940 cm^{-1} displays a pronounced isotropic character in contrast to the anisotropic excited-state absorption around 3160 cm^{-1}. This notion supports the conclusion that the signal around 2940 cm^{-1} involves a secondary process with loss of orientational information, e.g., energy transfer.

Our results on the spectral substructure of the OH absorption are consistent with the bandshape in the conventional IR spectrum of HDO. For a quantitative description of the latter an additional weaker component at 3270 cm^{-1} is required below 300 K that shows up also in the transient spectrum in the same temperature range, denoted with I'. This component is located at a frequency determined for a special ice conformation from Raman spectroscopic investigations (112). A further contribution to the overall OH-absorption band of HDO is noticed at 3260 cm^{-1}; the latter is believed not to represent OH-stretching transitions and was recently assigned to a combination tone (89). Increase of temperature reduces significantly the contribution of components I and II, while species III dominates the conventional absorption band above 290 K.

The three major spectral components are converted into each other by structural relaxation. This process involves reorientational and/or translational motion of surrounding D_2O molecules relative to the considered HDO. The process is described by a structural relaxation time

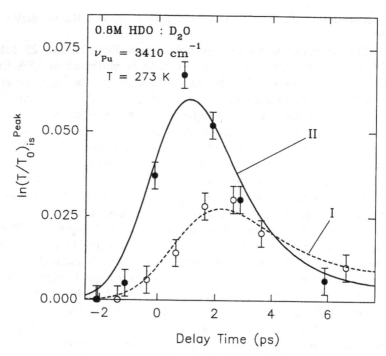

Figure 27 Structural relaxation within the OH-stretching band of HDO in D_2O at 273 K; the peak amplitudes (isotropic signal) of species I (filled circles, solid line) and II (hollow circles, dashed curve) are plotted versus probe delay, as derived from the decomposition of the measured transient spectra with excitation at 3410 cm^{-1}; experimental points, calculated lines.

τ_s and investigated by the data of Fig. 27 for 273 K and OH excitation at 3410 cm^{-1}. The relative amplitudes (peak value of the isotropic signal) of species I and II as obtained from the spectral decomposition of the transient spectra are plotted versus delay time. It is interesting to see that the resonantly pumped species II (full points, solid line) rises rapidly close to experimental time resolution, and decays via time constants τ_s and T_1, the latter being known from time-resolved experiments with fixed probing frequency performed at the excitation frequency. Component I (open circles, dashed curve), on the other hand, is pumped by the excitation pulse only to a small extent and grows notably slower because of the spectral (structural) relaxation process. The finding presents direct evidence for the time scale of the phenomenon. Comparison with the rate equation model (calculated lines in the figure) yields the time constant $\tau_s = 1.5 \pm 0.5$ ps, which is interpreted as the structural relaxation time of water at 273 K. Somewhat smaller numbers of $\tau_s = 1.0 \pm 0.4$ ps and 0.8 ± 0.4 ps are deduced from measurements at 298 K and 343 K, respectively (data not shown). Similar results for species III suffer from a smaller amplitude level and limited measuring accuracy and are omitted here.

Data from Fig. 26c on the reorientation time are at variance with the results of Woutersen et al. (90) for HDO at room temperature. These authors report a different behavior of the reorientational motion measured at three spectral positions within the OH band in a one-color experiment with pulses of 250 fs and spectral width of $70-100$ cm^{-1}. At 3320 cm^{-1} a singly exponential decay of the induced dichroism with a time constant of 13 ps was measured, whereas at 3400 and 3500 cm^{-1} biexponential relaxation with time constants of 0.7 ps and 13 ps was reported (experimental accuracy not stated). The authors concluded that with respect to the orientational dynamics, two distinct species of liquid water molecules exist. Data from Fig. 26 with superior spectral resolution indicate that three dominant species I–III have to be distinguished with different reorientational time constants of, respectively, 8 ± 2, 12 ± 5, and 3 ± 1.5 ps at 298 K. Since parts of the data are measured with pulses of 460 fs duration, a fast signal contribution cannot be overlooked in the measurement, since the induced dichroism amplitudes start at the maximum theoretical value of 0.4. The different findings of Ref. 90 may be related to the larger bandwidth of the pulses used there so that a superposition of spectral species is observed. Our data show that component III with the smallest red-shift and weakest hydrogen bonding reorients most rapidly with ≈ 3 ps in accordance with the reorientation time of water monomers in various inert solvents of approximately 2 ps (120) while the other species with stronger bonds behave more slowly.

Our findings for τ_s and τ_h may be compared with results of computer simulations for water. Values between 1 and 2 ps are stated for the average lifetime of a hydrogen bond by different authors (121–123), in satisfactory agreement with our experimental values. It is also interesting to compare with the frequency shift correlation function of the vibrational modes of water obtained from MD computations (124). Recently a slower component of this function with an exponential time constant of 0.8 ps was predicted for HDO in D_2O at 300 K and a density of 1.1 g/cm^3 (pressure ≈ 2 kbar). The existence of the slow component is a necessary prerequisite for the observation of spectral holes and the spectral relaxation time τ_s reported here. The faster component of the frequency shift correlation function with $\tau_c = 50$ fs (124) represents rapid fluctuations that contribute to the spectral bandwidths of the spectral species and of the spectral holes.

The value of τ_c suggests quasi-homogeneous broadening, $\tau_c \Gamma_i < 0.5$, of species I, II, and the spectral holes of III, introducing here the respective halfwidths Γ_i (HWHM). In fact, no evidence for hole burning was found for species I and II in our measurements.

Finally, time-resolved spectroscopy with femtosecond pulses was recently carried out by Gale and coworkers on a similar $HDO:D_2O$ sample (125). Due to the notably wider bandwidth of the applied IR pulses in the latter investigations, no details on reshaping of the transient spectra in dependence of the excitation frequency were accessible. A time-dependent position of the peak position of the induced sample bleaching was interpreted in terms of a shift within the statistical distribution of OH frequencies with a time constant of $\simeq 1$ ps. However, because only the parallel signal of the induced sample transmission was detected, the measured dynamics corresponds to a superposition of vibrational, reorientational, and structural relaxation. The data are interpreted by the help of a model of with random (bell-shaped) distribution of OH oscillators, quite different from the results of other groups.

III. CONCLUSIONS

It is the purpose of this chapter to point out the dramatic progress made during the past decade in our understanding of various dynamical processes in molecular liquids. The experimental tool is nonlinear spectroscopy with two versions — time-domain CARS and transient spectral hole burning in the infrared. We are now in the position to measure directly several significant time constants relevant to ultrafast vibrational processes of polyatomic molecules. In addition quantitative information on the rapid time evolution of the local structure of liquids is for the first time derived. The dynamics and structure of weakly associated (e.g., $CH_3I:CDCl_3$) and hydrogen-bonded liquids (e.g., alcohols and partly deuterated water) can be directly studied in time domain.

REFERENCES

1. (a) Schuster P, Zundel G, Sandorfy C. The Hydrogen Bond. Vol. I–III, Amsterdam: North-Holland, 1976; (b) Hadzi D. Hydrogen Bonding. Oxford: Pergamon Press, 1959.
2. Franks F. Water, A Comprehensive Treatise. New York: Plenum Press, 1972.
3. von der Linde D, Laubereau A, Kaiser W. Phys Rev Lett 1971; 26:954–957.

4. Laubereau A, von der Linde D, Kaiser W. Phys Rev Lett 1972; 28:1162–1165.

5. Spanner K, Laubereau A, Kaiser W. Chem Phys Lett 1976; 44:88–92.

6. Chesnoy J, Ricard D. Chem Phys Lett 1980; 73:433–437.

7. Heilweil EJ, Casassa MP, Cavanagh RR, Stephenson JC. Chem Phys Lett 1985; 117:185.

8. Graener H, Dohlus R, Laubereau A. In: Fleming GR, Siegman AE, eds. Ultrafast Phenomena V. Berlin: Springer, 1986.

9. Graener H, Dohlus R, Laubereau A. Chem Phys Lett 1987; 140:306.

10. (a) Graener H. Chem Phys Lett 1990; 165:110; (b) Graener H, Seifert G. Chem Phys Lett 1991; 185:68.

11. Graener H, Seifert G, Laubereau A. Chem Phys Lett 1990; 172:435–439.

12. Laenen R, Simeonidis K, Laubereau A. J Opt Soc Am 1998; B15:1213–1217.

13. Emmerichs U, Woutersen S, Bakker HJ. J Opt Soc Am 1997; B14:1480.

14. Gale GM, Gallot G, Hache F, Sander R. Opt Lett 1997; 22:1253.

15. Hamm P, Lim M, Hochstrasser RM. J Chem Phys 1997; 107:10523.

16. Zimdars D, Tokmakoff A, Chen S, Greenfield SR, Fayer MD. Phys Rev Lett 1993; 70:2718–2721.

17. Fleming GR, Siegman AE, eds. Ultrafast Phenomena. Vol. IV–XI. Berlin: Springer, 1986.

18. Bratos S, Pick RM, eds. Vibrational Spectroscopy of Molecular Liquids. New York: Plenum, 1979.

19. Oxtoby DW. Adv Chem Phys 1979; 40:1–42.

20. Rothschild WG. Dynamics of Molecular Liquids. New York: Wiley, 1983.

21. Laubereau A, Kaiser W. Rev Mod Phys 1978; 50:607–665.

22. Laubereau A. Chem Phys Lett 1974; 27:600–602.

23. Alfano RR, Shapiro SL. Phys Rev Lett 1971; 26:1247–1251.

24. Laubereau A, Wochner G, Kaiser W. Opt Commun 1975; 14:75–78.

25. Laubereau A, von der Linde D, Kaiser W. Phys Rev Lett 1971; 27:802–805.

26. (a) Akhmanov SA, Koroteev NI, Magnitskii SA, Morozov VB, Tarassevich AP, Tunkin VG, Shumay IL. In: Auston DH, Eisenthal KB, eds. Ultrafast Phenomena. Berlin: Springer, 1984:278–281; (b) Graener H, Laubereau A, Nibler JW. Optics Lett 1984; 9:165–169; (c) Graener H, Laubereau A. Optics Commun 1985; 54:141–145; (d) Kamalov Vf, Koroteev NI, Toleutaev BN. In: Clark RJH, Hester RE, eds. Time Resolved Spectroscopy. New York: Wiley, 1989.

27. Zinth W, Kaiser W. In: Kaiser W, ed. Ultrashort Laser Pulses and Applications. Berlin: Springer, 1988:235–277.

28. (a) Dlott DD, Schosser CL, Chronister EL. Chem Phys Lett 1982; 90:386–390; (b) Delle Valle RG, Fracassi PF, Righini R, Califano S. Chem Phys 1983; 74:179–185; (c) Geirnaert ML, Gale GM, Flytzanis C. Phys Rev Lett 1984; 52:815–818; (d) Dlott DD. Ann Rev Phys Chem 1986; 37:157–208.

29. Vanden Bout D, Muller LJ, Berg M. Phys Rev Lett 1991; 67:3700–3703.
30. Inaba R, Tominaga K, Tasumi M, Nelson KA, Yoshihara K. Chem Phys Lett 1993; 211:183–187.
31. (a) Maker PD, Terhune RW. Phys Rev 1965; 137A:801–809; (b) Shen YR, Bloembergen N. Phys Rev 1965; 137A:1786–1793.
32. Giordmaine JA, Kaiser W. Phys Rev 1966; 144:676–681.
33. D'yakov YuS, Krikunov SA, Magnitskii SA, Nikitin SY, Tunkin VG. Sov Phys JETP 1983; 57:1172–1176.
34. (a) Penzkofer A, Laubereau A, Kaiser W. Progr Quant Electron 1979; 6:55–82; (b) Zinth W, Laubereau A, Kaiser W. Optics Commun 1978; 26:457–461.
35. Li W, Purucker H-G, Laubereau A. Optics Commun 1992; 94:300–308.
36. (a) Loring RF, Mukamuel S. J Chem Phys 1985; 83:2116–2123; (b) Tanimura Y, Mukamel S. J Chem Phys 1993; 99:9496–9501.
37. Tokmakoff A, Kwok AS, Urdahl RS, Francis RS, Fayer MD. Chem Phys Lett 1995; 234:289–294.
38. Fickenscher M, Laubereau A. J Raman Spectrosc 1990; 21:857–861.
39. Purucker H-G, Tunkin V, Laubereau A. J Raman Spectrosc 1993; 24:453–458.
40. Fickenscher M, Purucker H-G, Laubereau A. Chem Phys Lett 1992; 191:182–188.
41. Bratos S, Marechal E. Phys Rev 1971; A4:1078–1085.
42. Heiman D, Hellwarth RW, Levenson MD, Martin G. Phys Rev Lett 1976; 36:189–192.
43. Kohles N, Laubereau A. Chem Phys Lett 1987; 138:365–370.
44. Fukushi K, Kimura M. J Raman Spectrosc 1979; 8:125–130.
45. Lindenberger F, Rauscher C, Purucker H-G, Laubereau A. J Raman Spectrosc 1995; 26:835–840.
46. Lindenberger F, Stöckl R, Asthana BP, Laubereau A. J Phys Chem 1999; 103:5655–5660.
47. (a) Kubo R, J Phys Soc Jap 1957; 12:570–586; (b) Rep Prog Phys 1966; 29 Pt.I:255–284; (c) Rothschild WG. J Chem Phys 1976; 65:455–462.
48. Westlund P-O, Lynden-Bell RM. Mol Phys 1987; 600:1189–1209.
49. Chesnoy J. Chem Phys Lett 1986; 125:267–270.
50. Telle HR, Laubereau A. Chem Phys Lett 1983; 94:467.
51. Schweizer KS, Chandler D. J Chem Phys 1982; 76:2296–2314.
52. (a) Fischer SF, Laubereau A. Chem Phys Lett 1975; 35:6–12; (b) Chem Phys Lett 1978; 55:189–196.
53. Aechtner P, Laubereau A. Chem Phys 1991; 419:419–425.
54. Horia M, Oxtoby DW, Freed KF. Phys Rev 1977; A15:361–371.
55. (a) Enskog D. Kungl Svenska Vetenskapsakad Handling 1922; 63; (b) Chapman S, Cowling TG. Mathematical Theory of Non-Uniform Gases. Cambridge: Cambridge University Press, 1970.
56. Bondarev AF, Mardaeva AI. Opt Spectrosc 1973; 35:167.

57. Fujiyama T, Kakimoto M, Suzuki T. Bull Chem Soc Japan 1976; 49:6.
58. (a) Döge G. Z Naturforsch 1973; 28a:919–932; (b) Döge G, Arndt R, Bühl H, Bettermann G. Z Naturforsch 1980; A35:468; (c) Döge G, Arndt R, Yarwood J. Mol Phys 1984; 52:399.
59. (a) Knapp EW, Fischer SF. J Chem Phys 1981; 74:89–95; (b) J Chem Phys 1982; 76:4730–4735.
60. Muller LJ, Vanden Bout D, Berg M. J Chem Phys 1993; 99:810–819.
61. Egelstaff PA. An Introduction to the Liquid State. New York: Academic Press, 1967.
62. Graener H, Ye TQ, Laubereau A. J Chem Phys 1989; 90:3413.
63. Laenen R, Rauscher C. Chem Phys 1998; 230:223.
64. Band YB. Phys Rev 1986; A34:326.
65. Press WH, Flannery BP, Teukolsky SA, Vetterling WT. Numerical Recipes. Cambridge: Cambridge University Press, 1986.
66. Laenen R, Simeonidis K, Rauscher C. IEEE J Sel Top Quant Electron 1996; 2:487–492.
67. Rauscher C, Laenen R. J Appl Phys 1997; 81:2818–2821.
68. Heilweil EJ, Casassa MP, Cavanagh RR, Stephenson JC. J Chem Phys 1986; 85:5004.
69. Graener H, Seifert G. J Chem Phys 1993; 98:36–45.
70. Arrivo SM, Heilweil EJ. J Phys Chem 1996; 100:11975
71. Laenen R, Simeonidis K. Chem Phys Lett 1999; 299:589–596.
72. Stuart AV, Sutherland GBBM. J Chem Phys 1956; 24:559.
73. Laenen R, Rauscher C. Chem Phys Lett 1997; 274:63–70.
74. Fendt A, Fischer SF, Kaiser W. Chem Phys 1981; 57:55.
75. Graener H, Ye TQ, Laubereau A. J Chem Phys 1989; 91:1043.
76. Graener H, Lösch T, Laubereau A. J Chem Phys 1990; 93:5365.
77. Woutersen S, Emmerichs U, Bakker HJ. J Chem Phys 1997; 107:1483.
78. Laenen R, Rauscher C. J Chem Phys 1997; 106:8974–8980.
79. Laenen R, Rauscher C, Laubereau A. J Phys Chem 1997; A101:3201–3206.
80. Smith FA, Creitz EC. J Res Natl Bur Standards 1951; 46:145.
81. Barnes AJ, Hallam HE. Trans Faraday Soc 1970; 66:1920.
82. Seifert G, Weidlich K, Hofmann M, Zürl R, Graener H. J Chim Phys 1996; 93:1763.
83. Laenen R, Rauscher C. J Mol Struct 1998; 448:115.
84. Fishman E. J Chem Phys 1961; 65:2204.
85. Laenen R, Rauscher C, Laubereau A. Chem Phys Lett 1998; 283:7–14.
86. Perchard JP, Josien ML. J Chim Phys 1968; 64:1834.
87. Laenen R, Rauscher C. J Chem Phys 1997; 107:9759–9763.
88. Laenen R, Rauscher C, Simeonidis K. J Chem Phys 1999; 110:5814–5820.
89. Graener H, Seifert G, Laubereau A. Phys Rev Lett 1991; 66:2092–2095.
90. Woutersen S, Emmerichs U, Bakker HJ. Science 1997; 278:658–660.
91. Laenen R, Rauscher C, Laubereau A. Phys Rev Lett 1998; 80:2622–2625.
92. Saiz L, Padro JA, Guardia E. J Phys Chem 1997; B101:78.

93. Jorgensen WL. J Phys Chem 1986; 90:1276.
94. Stillinger FH. Science 1980; 209:451.
95. Bratos S. J Chem Phys 1975; 63:3499.
96. Robertson GN, Yarwood. J Chem Phys 1978; 32:267.
97. Abramczyk H. Chem Phys 1990; 144:305.
98. Graener H, Laubereau A. J Phys Chem 1991; 95:3447.
99. Laenen R, Simeonidis K, Ludwig R. J Chem Phys 1999; 111:5897–5904.
100. Laenen R, Simeonidis K. Chem Phys Lett 1998; 292:631–637.
101. Laenen R, Simeonidis K. J Phys Chem 1998; A102:7207–7210.
102. Laenen R, Simeonidis K. Chem Phys Lett 1998; 290:94–98.
103. Walrafen GE, Hokmabadi MS, Yang WH. J Phys Chem 1988; 92:2433.
104. Conde O, Teixeira. J Mol Phys 1984; 53:951.
105. van Thiel M, Becker ED, Pimentel GC. J Chem Phys 1957; 27:95.
106. Marechal Y. J Chem Phys 1991; 95:5565.
107. Walrafen GE. J Chem Phys 1964; 40:3249.
108. Heilweil EJ. Science 1999; 283:1467.
109. Laenen R, Rauscher C, Laubereau A. J Phys Chem 1998; B102:9304–9311.
110. Rahman A, Stillinger FH. J Chem Phys 1971; 55:3336.
111. Ford TA, Falk M. Can J Chem 1968; 46:3579.
112. Walrafen GE. J Solut Chem 1973; 2:159.
113. Stillinger FH. Science 1980; 209:451.
114. Walrafen GE. J Chem Phys 1969; 50:560.
115. Giguere PA. J Chem Phys 1987; 87:4835.
116. Sciortino F, Geiger A, Stanley HE. Phys Rev Lett 1990; 65:3452.
117. Senior WA, Verrall RE. J Phys Chem 1969; 73:4242.
118. Clarke ECW, Glew DN. Can J Chem 1972; 50:1655.
119. Finzi J, Hovis FE, Panfilov VN, Hess P, Moore CB. J Chem Phys 1977; 67:4053.
120. Graener H, Seifert G, Laubereau A. Chem Phys 1993; 175:193.
121. Geiger A, Mausbach P, Schnitker J, Blumberg RL, Stanley HE. J Phys 1984; 45 C7:13.
122. Luzar A, Chandler D. Nature 1996; 379:55.
123. Marti J, Padro JA, Guardia E. J Chem Phys 1996; 105:639.
124. Diraison M, Guissani Y, Leicknam JCl, Bratos S. Chem Phys Lett 1996; 258:348.
125. Gale GM, Gallot G, Hache F, Lascoux N, Bratos S, Leicknam JCl. Phys Rev Lett 1999; 82:1068–1071.

2
Probing Bond Activation Reactions with Femtosecond Infrared

Haw Yang* and Charles Bonner Harris
University of California at Berkeley, Berkeley, California

I. INTRODUCTION

A great number of chemical reactions occur in liquids and at interfaces. In describing the chemical dynamics in the above environments, one usually simplifies a system to portray only the reactive site and its immediate surroundings. The thought of a localized active site forms the basis for conceptualizing the chemical reactivity. Our understanding of chemical reactivity comes from knowledge of elementary steps from such a localized viewpoint. The elementary steps include changes in the electron distribution, molecular structure, and translocation of chemical moieties. However, these processes are inevitably modulated by the surrounding media, the time scales of which are typically on the order of picoseconds or less. Therefore, a critical test of our fundamental understanding of chemical reactivity requires experimental techniques that allow resolution of those ultrafast events. Femtosecond infrared (fs-IR) offers such an opportunity.

One problem that is amenable to this technique is transition metal–mediated chemical reactions. They range from industrial processes such as petroleum refinery to biological cycles such as nitrogen fixation and photosynthesis. These reactions have been thought to take place at an unsaturated metal site, for example, the kinks or step edges on a metal surface. Ideally, one would like to observe the chemical events that evolve

---

* *Current affiliation*: Harvard University, Cambridge, Massachusetts

around such an active site as a reaction develops. This can be accomplished by monitoring the vibration of a CO ligand attached to the reactive metal center. Due to the CO-σ → metal-d_σ coordination and CO-σ^* → metal-d_π back-donation, the vibrational frequency of the CO ligand is a very sensitive measure of the chemical environment about the metal. Qualitatively, an increase (or decrease) of the metal electron density reduces (augments) the C–O bond order to result in a lower (higher) CO-stretching frequency. In experiments, the unsaturation is artificially created with an ultrafast UV pulse that dissociates the metal fragment from a photolabile ligand. A cascade of chemical events then ensued. Note that such reactions are *bimolecular* in nature. As such, a direct dissociation process is welcome since the time it requires to create an unsaturated metal center determines the inherent time resolution for studies of bimolecular reactions of this kind.

This chapter describes studies of an important class of organometallic reactions — oxidative addition reactions — utilizing femtosecond infrared spectroscopy. Since numerous excellent reviews have been published on this subject, only a brief account is presented below.

II. BACKGROUND

Transition metals capable of existing in multiple oxidation states may afford new ligands and consequently increase their formal charge (oxidation). Such translocations of atoms, molecules, or groups onto a metal center constitute a special class of reaction known as *oxidative addition* (1). These reactions are of fundamental importance not only in inorganic chemistry, but also in organic synthesis, catalysis (2), and biological sciences (3), to name just a few. The number of chemical moieties the metal received categorizes this type of reaction into two-electron or one-electron oxidative addition, (Fig. 1). One notable feature of the reaction patterns in Fig. 1 is the weakening, or "activation," of an otherwise strong chemical bond to the extent that the reactions proceed under mild conditions similar to a biological environment.

For a two-electron oxidative-addition reaction, the weakening and eventual cleavage of a chemical bond can be understood from the bonding of dihydrogen transitional metal complexes such as $W(CO)_3(PPr_3^i)_2$ (H_2) $(Pr^i = isopropyl)$ (4). Figure 2 shows a qualitative molecular orbital (MO) picture for a dihydrogen complex (5,6). Such a σ complex is stabilized by two factors. On the one hand, the hydrogen molecule interacts with the metal center in a manner similar to a Lewis acid-base pair through electron donation from the filled H_2 σ orbital to the empty metal-d_σ orbital. The

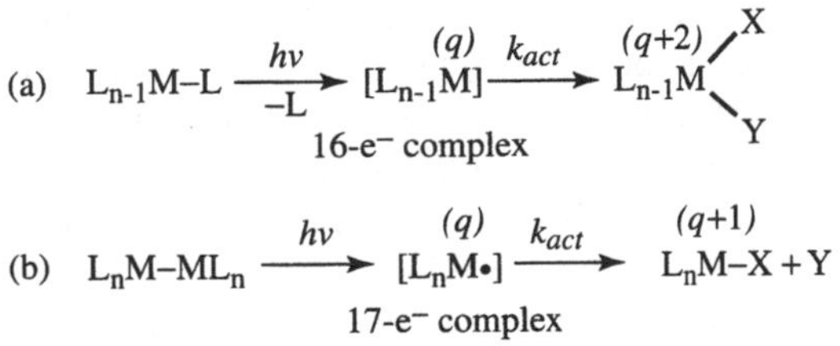

(a) $L_{n-1}M–L \xrightarrow[-L]{h\nu} \overset{(q)}{[L_{n-1}M]} \xrightarrow{k_{act}} \overset{(q+2)}{L_{n-1}M} \begin{smallmatrix} \nearrow X \\ \\ \searrow Y \end{smallmatrix}$

16-e⁻ complex

(b) $L_nM–ML_n \xrightarrow{h\nu} \overset{(q)}{[L_nM\bullet]} \xrightarrow{k_{act}} \overset{(q+1)}{L_nM–X} + Y$

17-e⁻ complex

Figure 1 An illustration of (a) two-electron and (b) one-electron oxidative addition.

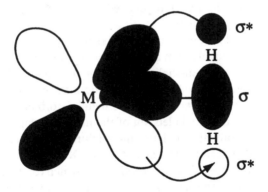

Figure 2 A qualitative molecular orbital presentation for the chemical bonding in a dihydrogen σ complex.

same interaction also weakens, but does not break, the H–H bond. On the other hand, the metal-dihydrogen complexation is strengthened by electron "back-donation" from filled metal d_π orbitals to the empty H_2 σ^* orbital, which further weakens the H–H bond. It follows that, from the MO viewpoint, the degree to which the H–H bond is activated depends on the energetic and spatial overlaps between the metal and dihydrogen MOs. The concept of a precursory σ-complex has been extended to understanding the reactivity of other two-electron oxidative addition reactions, including C–H and Si–H activation. In particular, the observed selectivity of primary > secondary > tertiary C–H bonds in C–H activation is attributed to more stable (less steric hindrance) σ-complex intermediate for a primary C–H bond in the reaction. A substantial amount of indirect evident in the literature, mostly by the primary kinetic isotope effect, has demonstrated the existence of a σ-complex intermediate in the alkane C–H bond activation (7). Theoretical calculations also provide much insight into the possible involvement of a precursory intermediate in a two-electron oxidative-addition reaction. For example, Musaev and Morokuma have computed the reaction profiles for σ-bond activation of H_2, CH_4, NH_3, H_2O, and SiH_4 by η^5-CpRh(CO) (Cp = C_5H_5). Their calculations predict a precursory complex in the activation of C–H, N–H, and O–H bonds but not in the activation of H–H or Si–H bonds (8). The marked difference along the calculated reaction coordinates for the isoelectronic C–H and Si–H bonds can be understood by the aforementioned MO picture. The Si d orbitals allow a better interaction between the Si–H bond and the metal d orbitals such that the reaction proceeds to completion without passing through a precursory complex.

Along a parallel vein, it has also been proposed that a 19-electron intermediate may be involved in atom-abstraction reactions via one-electron oxidative addition to open-shell 17-electron organometallic compounds (9–12). In a halogen-atom abstraction reaction, for example, the lone-pair electrons of the halogen overlap with the metal d_σ orbital, hence 19 formal electrons, while the unpaired electron of the metal overlap with the carbon-halogen (C–X) σ^* orbital. The latter interaction may reduce the C–X bond order and eventually lead to cleavage of the C–X bond (13). Unlike closed-shell organometallic reactions, the reactivity of which can be conceptualized by the 16- or 18-electron rule (14), a general principle has yet to be formed for the reactivity of a open-shell 17-electron species. Therefore, a detailed knowledge of the reactive intermediates in metal-centered radical reactions is crucial in shaping a general principle for the reactivity of 17-electron organometallics.

In light of the above discussion, it is evident that the putative intermediates play a pivotal role in oxidative-addition reactions. As such, detailed knowledge of the dynamics of the intermediates is crucial for our fundamental understanding both of reactions and of the nature of chemical bonds in general. Recently, the development in this area has greatly accelerated because of increasingly sophisticated and efficient quantum chemical computational methods. Theoretical calculations provide valuable insight (e.g., the electronic structure and chemical bonding) for transient intermediates, the lifetimes of which are too short for conventional characterization (15). However, they also often rely on a priori knowledge of an approximate, sometimes idealized, reaction coordinate that may not necessarily be true in actuality. Without experimental results, theoretical consideration may easily overlook, for instance, the dynamical partitioning of reaction pathways in the excited state or during solvation, such as in the case of Si–H bond activation, to be discussed later. A close interplay between experiments and theories will be necessary in order to unveil the reaction dynamics underlying a complicated reaction. The following three sections describe the recent results of studies of prototypical two-electron (C–H and Si–H bonds) and one-electron (Cl atom) oxidative addition to organometallic compounds.

III. C–H BOND ACTIVATION BY η^3-Tp*Rh(CO)$_2$

The natural abundance of alkanes makes it an ideal feedstock as the starting material for organic synthesis. The first step towards converting an alkane molecule into other synthetically advantageous compounds is to cleave or activate the strong C–H bond (bond energy ~ 100 kcal/mol). Consequently, the chemistry needed to selectively activate a C–H bond has been vigorously pursued in the past decade (16–20). Advancements in this area include the development of homogeneous intermolecular C–H bond activation by organometallic compounds (21,22). They offer the possibility to tailor a specific reaction utilizing well-developed synthetic strategies. The rational design of such a reaction relies significantly on knowledge of both the detailed reaction mechanism and the nature of the reaction intermediates. Previous mechanistic studies of photochemical C–H bond activation have focused on the η^5-Cp*M(CO)$_2$ (M = Ir, Rh) (16) compounds, which, unfortunately, have relatively low quantum yields (23). Instead, the present work employs η^3-Tp*Rh(CO)$_2$ (Tp* = HB-Pz$_3^*$, Pz* = 3, 5-dimethylpyrazolyl; Fig. 3), which has been shown to photochemically activate the alkane C–H bond to form the final product η^3-Tp*Rh(CO)(H)(R) (R = alkyl) with $\sim 30\%$ quantum yields (24–27).

A. The Dynamics of Reaction Intermediates — Vibrational Relaxation and Molecular Morphology Change

Photolysis of a CO ligand from the parent compound initiates the reaction. The resultant η^3-Tp*Rh(CO) complex has a coordinatively vacant site that is quickly occupied by a solvent molecule. The time scale for such a solvation process is generally on the order of 2 ps in room temperature liquids (28). The solvated η^3-Tp*Rh(CO)(S) (S = alkane) complex exhibits a single CO-stretching band (ν_{CO}) at 1972 cm^{-1} (the v = 0 $\rightarrow$ 1 band) as shown in Fig. 4a. For this particular system the large excess energy (~ 40–50 kcal/mol) deposited by the UV photon creates

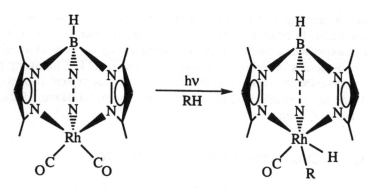

Figure 3 Alkane C–H bond activation by η^3-Tp*Rh(CO).

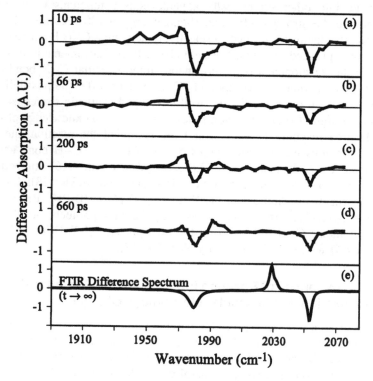

Figure 4 Transient difference spectra in the CO stretching region for η^3-Tp*Rh(CO)$_2$ in room-temperature alkane solution at various time delays following 295 nm photolysis. Panel (e) is an FTIR difference spectrum before and after 308 nm photolysis. A broad, wavelength-independent background signal from CaF$_2$ windows has been subtracted. (Adapted from Ref. 29.)

a non-Boltzmannian population distribution such that higher vibrational levels may also be occupied to the extent that they become observable. As a result, the $v = 1 \rightarrow 2$ and $v = 2 \rightarrow 3\nu_{CO}$ transitions appear at 1958 and 1945 cm^{-1}, respectively. The two hot bands gradually decay away while the $v = 0$ band (1972 cm^{-1}) rises as the system approaches thermal equilibrium. The time scale for such intramolecular relaxation (IVR) is measured by monitoring the population dynamics of the $v = 0$ state. The 1972 cm^{-1} band shows a fast rise of ~23 ps, attributed to the above IVR process, and a slower decay of ~200 ps (Fig. 5a). The ~200 ps decay

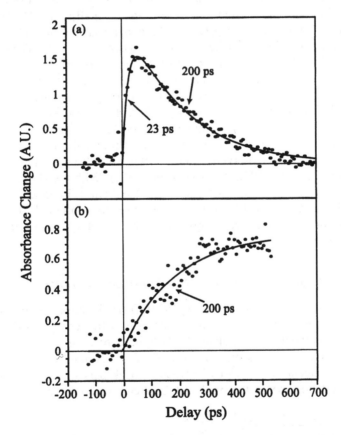

Figure 5 Ultrafast kinetics (dots) of η^3-Tp*Rh(CO)$_2$ in room-temperature alkane solution after 295 nm photolysis at (a) 1972 cm^{-1}, the CO stretch of the solvated η^3-Tp*Rh(CO) (alkane) intermediate, and (b) 1990 cm^{-1}, the CO stretch of the arm-detached η^2-Tp*Rh(CO) (alkane) intermediate. (Adapted from Ref. 29.)

indicates that the η^3-Tp*Rh(CO)(S) solvate reacts to form another species on this time scale. Indeed, another peak appears at 1990 cm^{-1} at later time delays (Fig. 4c and d). The correlation between η^3-Tp*Rh(CO)(S) and the 1990 cm^{-1} species is established by the $\sim$200 ps rise of the 1990 cm^{-1} band (Fig. 5b) (29).

The 1990 cm^{-1} band is attributed to an η^2-Tp*Rh(CO)(S) solvate where one of the three chelating pyrazolyl ligands detaches itself from the Rh center. To verify this assignment, the fs-IR spectra following UV photolysis of η^3-Tp*Rh(CO)$_2$ are compared to those of an analogous compound η^2-Bp*Rh(CO)$_2$ (Bp* = H$_2$B-Pz$_2^*$) (Fig. 6). Photolysis of η^2-Bp*Rh(CO)$_2$ in alkane solutions results in the η^2-Bp*Rh(CO)(S) solvate that also exhibits a single CO-stretching band at $\sim$1990 cm^{-1}, thereby providing experimental evidence for the assignment (30). Theoretically, Zaric and Hall computed the reaction using a CH$_4$ to model the alkane solvent (31). Their DFT calculations show that the η^2-TpRh(CO)(CH$_4$) (Tp = HB-Pz$_3$, Pz = pyrazolyl) complex is energetically more stable than the η^3-TpRh(CO)(CH$_4$) complex by 7.7 kcal/mol. In addition, the DFT frequency of the model η^2 complex is found 22 cm^{-1} higher than that of the η^3 complex. The calculated frequency shift in the model systems is consistent with the observed 18 cm^{-1} blue shift from η^3-Tp*Rh(CO)(S) (1972 cm^{-1}) to η^2-Tp*Rh(CO)(S) (1990 cm^{-1}). With the above-discussed evidence, it can be concluded that the 1990 cm^{-1} intermediate should be assigned to η^2-Tp*Rh(CO)(S). Therefore, the observed 200 ps time constant measures the free energy barrier $\Delta G^{\ddagger} \approx 4.1$ kcal/mol for the η^3-to-η^2 isomerization.*

B. The Activation Barrier — The Bond-Breaking Step

The fact that the final product η^3-Tp*Rh(CO)(H)(R) does not appear on the ultrafast time scale (<1 ns,) (Fig. 4) indicates a free energy barrier greater than 5.2 kcal/mol for the alkane C–H bond activation. Nanosecond step-scan FTIR experiments on the η^3-Tp*Rh(CO)$_2$/cyclohexane system show that the remnant of the η^2-Tp*Rh(CO)(S) peak persists for $\sim$280 ns after photoexcitation, while the product CO stretch at 2032 cm^{-1} rises with a

---

*   The free-energy barrier $\Delta G^{\ddagger}$ is estimated from the transition-state theory: the reaction rate k = 1/τ = k$_B$T/h exp($-\Delta G^{\ddagger}$/RT), where τ is the measured lifetime, k$_B$ Boltzmann constant, h Plank's constant, R the ideal gas constant, and T temperature in Kelvin.

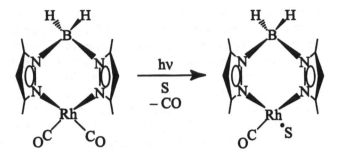

Figure 6 Photodissociation and subsequent solvation of η^2-Bp*Rh(CO)$_2$.

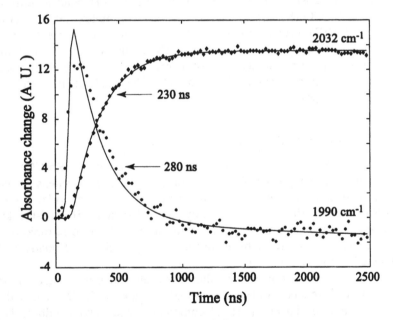

Figure 7 Nanosecond kinetics (dots) of η^3-Tp*Rh(CO)$_2$ in room-temperature alkane solution after 295 nm photolysis at 1990 cm^{-1}, the η^2 intermediate, and 2032 cm^{-1}, the final product. (Adapted from Ref. 30.)

time constant of 230 ns (Fig. 7) (30). Notice that the previously detached pyrazole ring rechelates back in the final product η^3-Tp*Rh(CO)(H)(R). Since no other transient intermediates appear on the time-resolved IR spectra prior to formation of the final product, it is concluded that the rate-limiting step consists of both the C–H bond cleavage and the pyrazole rechelation processes. At this point, the available experimental data do not allow differentiation of the bond-cleavage step from the ring-rechelation one. Nonetheless, a comparison of the energy barrier of the current system with literature values of analogous systems provides more insight into this reaction. The 230 ns time constant for the last rate-limiting step permits an estimate of the apparent free energy barrier $\Delta G^{\ddagger} \approx 8.3$ kcal/mol. This value is comparable to the ~ 7.2 kcal/mol barrier extrapolated from low-temperature studies of the $Cp^*Rh(CO)_2$ ($Cp^* = C_5Me_5$, Me $= CH_3$) system (32), the rate-limiting step of which is thought to consist of only the C–H bond-breaking process (23,33). This leads to the proposition of the following rationale for the bond-cleavage step. The η^2-$Tp^*Rh^{(I)}(CO)(RH)$ complex with a Rh(I) center reacts to cleave the C–H bond to form the $[\eta^2$-$Tp^*Rh^{(III)}(CO)(R)(H)]^*$ complex. The latter complex has a electron-deficient Rh(III) center, providing the electronic environment for the electron-donating pyrazole ring to reattach to the Rh center on a time scale much shorter than the that of barrier crossing. The ring closure drives the reaction to completion and forms the stable final product η^3-$Tp^*Rh(CO)(H)(R)$.

C. The Reaction Mechanism

The ability to follow the C–H bond activation by η^3-$Tp^*Rh(CO)$ from its initiation to completion allows one to assemble a comprehensive reaction profile delineated in Fig. 8. The proposed scheme provides a direct assessment of the time scale and energy barrier for the C–H bond-breaking step. A comparison of these results with those of the isoelectronic Si–H bond activation requires a better understanding of chemical bonds, to be discussed in the next section.

The dynamics unveiled in this study also address several important issues in related systems. In a study of alkane C–H bond activation by the $K[\eta^2$-$Tp^*Pt^{(II)}(CH_3)_2]$ compound, Wick and Goldberg have proposed a dissociative mechanism involving an $[\eta^2$-$Tp^*Pt^{(II)}(CH_3)]^*$ intermediate prior to formation of the final product η^3-$Tp^*Pt^{(IV)}(CH_3)(R)(H)$ (34). Similar η^2 intermediates have also been proposed for the vinyl C–H bond activation by $Tp^*M(L)$ (M = Rh, Ir; L = ligand) (35). For this type of reaction, theoretical works by Jiménez-Cataño, Niu, and Hall suggest that the two possible η^2-$Tp^*M(CH_2H_4)(L)$ and η^3-$Tp^*M(C_2H_4)(L)$ intermediates are connected by a small energy barrier. Both intermediates are found to transverse through the same transition state, which involves rechelation of a pyrazolyl ligand before reaching the final product η^3-$Tp^*M(H)(HCCH_2)(L)$

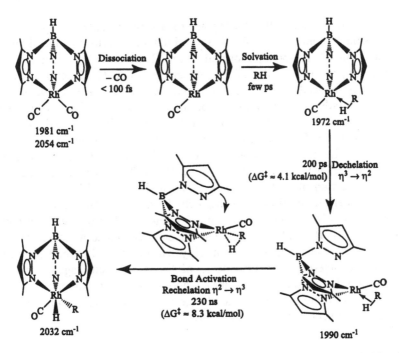

Figure 8 A proposed reaction mechanism for the alkane C–H bond activation by η^3-Tp*Rh(CO)$_2$ covering the ultrafast dynamics to nanosecond kinetics.

(36). In all of the above examples, the η^3-to-η^2 isomerization plays a significant role in the reactivity of the transition metal complex. In this regard, the reaction scheme in Fig. 8 provides not only a measure of the relevant time scales but also an understanding of the roles the reactive intermediates assume in the course of a reaction.

IV. Si–H BOND ACTIVATION BY η^5-CpM(CO)$_3$, (M = Mn, Re)

The insertion of a transition metal complex into a Si–H bond is an industrially important process in the production of substituted silane and silyl polymers (2,37). The reaction is also helpful for understanding the two-electron oxidative addition in general through comparison to the isoelectronic C–H bond activation (38). As mentioned earlier, the existence of the proposed σ-complex intermediate in the C–H bond activation has been a critical element in our general understanding of oxidative addition and reductive elimination (39). Analogous to a dihydrogen σ complex, in a C–H bond σ complex the filled d_{xy} orbital of a transition metal overlaps with the σ^* anti-bonding orbital of an alkane C–H bond (see Fig. 2). As a result, the bond order of the otherwise strong C–H bond in the σ complex is reduced to such an extent that the metal is able to cleave the C–H bond under the mild ambient conditions. In a recent theoretical study, Koga, Musaev, and Morokuma have related the activation barriers along a series of R–H (R = H, C, N, O, and Si) bonds to factors including the stability of the precursory complex, the R–H and M–R bond strengths, and the directionality of the R–H bond (8,40). They have also predicted that the insertion of a metal complex into a silane Si–H bond is a barrierless exothermic reaction, inferring a Si–H bond activation time scale on the order of solvent coordination time ($\sim$ few ps). The theoretical prediction of the Si–H bond activation barrier contrasts those measured at low temperatures, which suggest an enthalpy barrier of 7.9 kcal/mol (41–44). In order to determine the time scale of Si–H bond cleavage — thereby the magnitude of the energy barrier — and to understand the nature of the apparent rate-limiting step, a knowledge of the mechanism composed of elementary reaction steps is necessary.

A. The Reaction Intermediates — Solvation-Partitioned Pathways and Intersystem Crossing

The η^5-CpMn(CO)$_3$ and η^5-CpRe(CO)$_3$ complexes are two of the most extensively studied transition metal compounds that photochemically activate the silane Si–H bond (39). Hence, the initial fs-IR studies focus on these two compounds (45,46). Although both Mn and Re belong to the same group in the periodic table, the two η^5-CpMn(CO)$_3$ and η^5-CpRe(CO)$_3$ complexes differ appreciably in their photochemical properties. There appear to be two electronic excited states in the experimentally accessible region (290 nm < λ < 400 nm) for the Mn compound, but only one for the Re compound (Fig. 9). It turns out that excitation of the Mn compound to the higher-energy states (<290 nm) leads to η^5-CpMn(CO)$_2$ in its singlet electronic state, whereas the lower-energy state ($\sim$330 nm) leads to triplet η^5-CpMn(CO)$_2$. For the Re compound, on the other hand, the reaction that follows 295 nm excitation proceeds along the singlet electronic surface of η^5-CpRe(CO)$_2$. We first discuss the solvation dynamics of the metal carbonyls.

For both metals, photolysis of the parent η^5-CpM(CO)$_3$ results in the coordinatively unsaturated η^5-CpM(CO)$_2$, which is quickly solvated by the

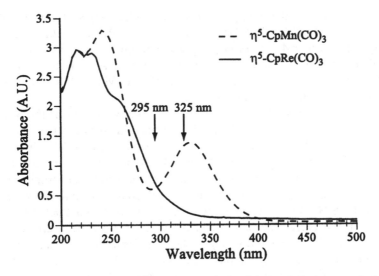

Figure 9 UV-Vis spectra of η^5-CpMn(CO)$_3$ and η^5-CpRe(CO)$_3$ in neat triethylsilane taken under experimental conditions. The excitation wavelengths are indicated by arrows. (Adapted from Ref. 45.)

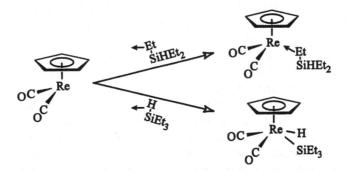

Figure 10 Solvation of the coordinatively unsaturated η^5-CpRe(CO)$_2$ intermediate by a Et$_3$SiH molecule partitions the silane Si–H bond activation reaction to two pathways.

surrounding solvent. The ensuing reaction, however, may be further partitioned by solvation of a solvent molecule such as triethylsilane (Et$_3$SiH, Et = C$_2$H$_5$) that has two chemically distinct Si – H and C$_2$H$_5$ sites. Using the Re complex as an example, Fig. 10 illustrates such a partitioning via solvation of the singlet η^5-CpRe(CO)$_2$ by a Et$_3$SiH molecule. The initial solvation results in the final product η^5-CpRe(CO)$_2$(H)(SiEt$_3$) or the ethyl-solvate η^5-CpRe(CO)$_2$(Et$_3$SiH) on time scales of ~4.4 and ~2.2 ps, respectively. The former 4.4 ps product appearance time is indicative of a very small free energy barrier, if any, for activation of a Si–H bond. The η^5-CpM(CO)$_2$(Et$_3$SiH) ethyl solvate, for both the Mn and Re complexes, remains stable on the ultrafast time scale.

Having discussed the likeness in the reaction pattern of the Mn and Re complexes, we next turn to the differences in their photochemical properties. Figure 11 shows the fs-IR spectra of η^5-CpMn(CO)$_2$ in Et$_3$SiH following 325 nm pump. The 1892 and 1960 cm^{-1} bands are assigned to the ethyl-solvate η^5-CpMn(CO)$_2$(Et$_3$SiH) in the singlet electronic ground state, where the dicarbonyl intermediate is solvated through the ethyl moiety of the Et$_3$SiH (47). The other two bands at 1883 and 2000 cm^{-1} that appear in the early-time panels and decay away later are ascribed to the unsolvated triplet η^5-CpMn(CO)$_2$. These assignments are supported by quantum chemical computations. Energetically, density-functional theory using the B3LYP exchange correlation functional predicts a triplet η^5-CpMn(CO)$_2$ that is 8.1 kcal/mol more stable than a singlet one. This result is supported by a multiconfigurational SCF (MCSCF) calculation to the second-order perturbation (PT2), which also predicts a more stable triplet species by 9.9 kcal/mol. Furthermore, at the DFT/B3LYP level, the singlet η^5-CpMn(CO)$_2$ assumes a different geometry than a triplet one, where the singlet dicarbonyl exhibits a "bent" configuration (Fig. 12). In fact, the qualitative correlation in the molecular configuration and spin states for the 16-electron η^5-CpML$_2$ complexes has also been suggested by Hofmann and Padmanabhan from the results of extended Hückel calculations (48).

How does the nascent triplet η^5-CpMn(CO)$_2$ interact with a common hydrocarbon solvent? In general, high-spin unsaturated organometallic complexes do not interact very well with alkane solvents (49). In a dense liquid environment, however, the triplet species may undergo a rapid, concerted intersystem crossing and solvation to become a solvated complex in the singlet state. The time scale for such a process can be on the order of hundreds of picoseconds. Figure 13 shows the conversion of the unsolvated triplet η^5-CpMn(CO)$_2$ to the singlet alkyl solvate η^5-CpMn(CO)$_2$(alkane), the latter of which remains stable in the ultrafast regime. This process can be understood by a free-energy scheme depicted in Fig. 14, which mirrors Marcus's electron-transfer theory (50). The adiabaticity of the process depends upon the magnitude of the spin-orbit coupling. An excess internal energy may expedite the spin crossover/solvation process as demonstrated by the faster decay of the 295 nm trace (~90 ps) compared to the 325 nm

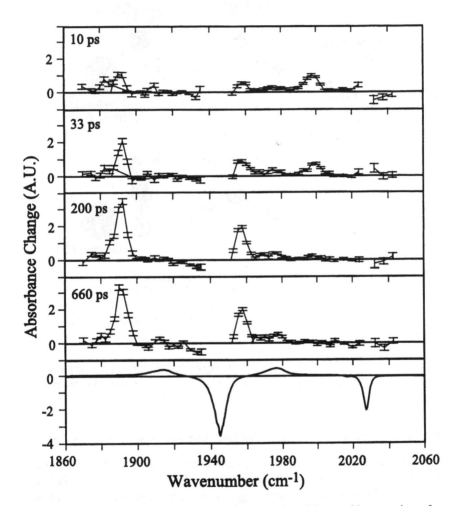

Figure 11 Transient difference spectra in the CO-stretching region for η^5-CpMn(CO)$_3$ in neat room-temperature triethylsilane at various time delays following 325 nm photolysis. Under the experimental conditions, the large cross section of the solvent Si–H band ($\sim$2100 cm^{-1}) and the parent CO bands (1947 and 2028 cm^{-1}) make it difficult to access some regions of the spectrum. The last panel is an FTIR difference spectrum before and after 308 nm photolysis. A broad, wavelength-independent background signal from CaF$_2$ windows has been subtracted. (Adapted from Ref. 45.)

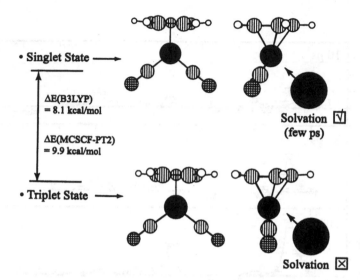

Figure 12   Computed molecular geometries of η^5-CpMn(CO)$_2$ in its singlet and triplet electronic manifolds at the DFT B3LYP/LANL2DZ level of theory. The left column shows their energy differences calculated using DFT (8.1 kcal/mol) or ab initio (9.9 kcal/mol) methods. The right column illustrates their interaction with an alkane solvent molecule.

one (~105 ps). A greater solvation energy also promotes the intersystem crossing rate as long as the system configuration is in the Marcus normal region.

B.   The Reaction Barrier — Solvent Molecule Rearrangement

The above discussion demonstrates that the activation barrier for the silane Si–H bond is relatively small compared to that for an alkane C–H bond. This is surprising in that one might have expected comparable energy barriers for activation of both the Si–H and C–H bonds based on the similar enthalpy of activation $\Delta H^{\ddagger}$ from macroscopic kinetic measurements (41–44). Clearly other mechanisms are at work to make up the energy barrier in the case of Si–H bond activation. To investigate if more intermediates are involved, which may provide an explanation for the reported apparent $\Delta H^{\ddagger}$ values, the reaction is followed extending into the nano- and microsecond regime. The experiments show that only the previously discussed ethyl-solvate appears on these time-resolved IR spectra. Its decay correlates very well with the product rise as displayed in Fig. 15.

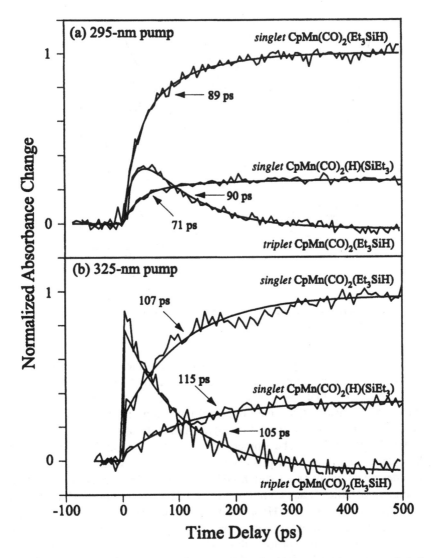

Figure 13 Ultrafast kinetics of η^5-CpMn(CO)$_3$ in neat room-temperature triethyl-silane after (a) 295 nm; and (b) 325 nm photolysis. In each panel, the transients are normalized against that of the singlet η^5-CpMn(CO)$_2$(Et$_3$SiH) to demonstrate a reduced singlet-to-triplet ratio when excited at a longer (325 nm) wavelength. The time constants for the exponential fits are also shown in the plots. (Adapted from Refs. 45 and 46.)

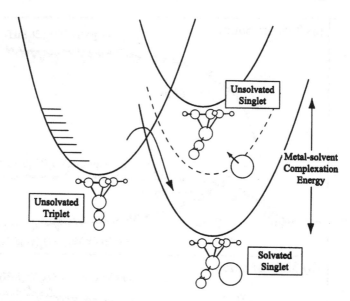

Figure 14 Schematic representation of solvation/spin-crossover process. The short horizontal bars in the unsolvated triplet surface denote vibrational levels.

The apparent free-energy barriers estimated from these time constants are 8.25 and 10.41 kcal/mol for the Mn and Re reactions, respectively.

The fact that no other intermediates are present in these reactions suggests that the rate-limiting step is the isomerization from the ethyl-solvate η^5-CpM(CO)$_2$(Et$_3$SiH) to the final product η^5-CpM(CO)$_2$(H)(SiEt$_3$). The isomerization mechanism can be ascribed to a dissociative intermolecular process. Evidence that substantiates this picture includes an enthalpy of activation $\Delta H^{\ddagger}_{act}$ comparable to that of dissociating an organometallic-alkane complex $\Delta H^{\ddagger}_{diss}$. The reaction of η^5-CpMn(CO)$_2$ with Et$_3$SiH shows a $\Delta H^{\ddagger}_{act} \approx 7.9$ kcal/mol (42), very close to that of metal-heptane complexation energy $\Delta H^{\ddagger}_{diss} \approx 8 - 9$ kcal/mol (51,52). Since for a dissociative process the $\Delta H^{\ddagger}_{diss}$ is expected to be a dominating factor in the isomerization rate, the measured $\Delta G^{\ddagger}$ difference for different metals should reflect the difference in the enthalpy of complexation. An MP2 theoretical calculation shows that the η^5-CpM(CO)$_2 \ldots$ (H$_3$CCH$_3$) binding energy for M = Mn is 3.5 kcal/mol less than that for M = Re (45). This is in good agreement with the measured 2.16 kcal/mol difference in $\Delta G^{\ddagger}$, thereby providing additional support for the picture that the apparent rate-limiting step is most likely composed of a dissociative process.

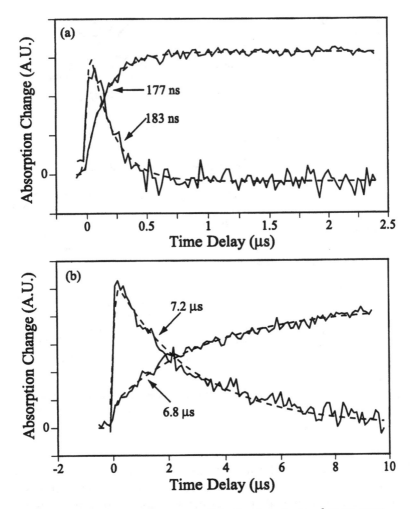

Figure 15 Nanosecond kinetics (solid lines) of (a) η^5-CpMn(CO)$_3$ and (b) η^5-CpRe(CO)$_3$ in neat room-temperature triethylsilane after 295 nm photolysis for the ethyl-solvate intermediates (the decaying traces) and the final products (the rising traces). (Adapted from Ref. 45.)

C. The Reaction Mechanism — Resolving a Convolved Chemical Reaction

A comprehensive reaction mechanism composed of the above-discussed elementary reaction steps is illustrated in Fig. 16. The proposed scheme

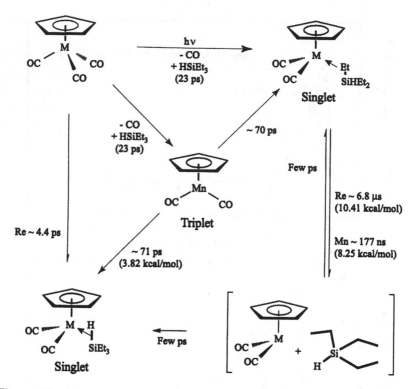

Figure 16 A proposed reaction mechanism for the silane Si–H bond activation by the η^5-CpM(CO)$_3$ (M = Mn, Re) covering the ultrafast dynamics to nanosecond kinetics.

allows for a direct experimental assessment of the Si–H bond-breaking step. The ~4.4 ps time scale of that step is indicative of a very small energy barrier, thus providing experimental support for a previous theoretical prediction put forth by Musaev and Morokuma (8). The marked difference in the bond-breaking barriers for the isoelectronic C–H and Si–H bond activation reflects their distinct bonding characteristics. The d orbitals on the Si atom make it easier for the Si–H σ bond to interact with the transition metal center both energetically and spatially.

Figure 16 also demonstrates the complexity of a chemical reaction in the liquid phase. A photochemical reaction such as those discussed in this section can easily span several orders of magnitude in time from its initiation to completion, as results of the intricate dynamical processes in solution. Following photoexcitation, the reaction may begin with a dynamical partitioning in the dissociative excited state that, on the $\sim$100 fs time scale, leads to reactive intermediates such as η^5-CpMn(CO)$_2$ in either the singlet or triplet spin states. The nascent singlet η^5-CpM(CO)$_2$ may interact either with the chemically inert ethyl site of Et$_3$SiH to form the ethyl solvated η^5-CpM(CO)$_2$(Et$_3$SiH) or with the reactive Si–H bond to form the final product. Under the dynamical influence of the solvent bath, the weakly coupled ethyl solvate together with the surrounding solvent shell may undergo various reorganization until the metal center encounters a reactive Si–H bond to complete the reaction. The time scale of such a procession, ranging from hundreds of picoseconds to a few microseconds, is expected to depend upon the specific metal-alkane interactions, the number of active sites in a solvent molecule, and steric interactions. Therefore, the macroscopic reaction rate is determined by the rearrangement process, which sets the time scales for the two solvation-partitioned product formation pathways three decades apart.

Another important aspect brought to light by this study is the realization of the dynamics of a high-spin, 16-electron transition metal center in a two-electron oxidative-addition reaction. As noted earlier, transition metal–mediated reactions normally occur at an unsaturated metal center that may potentially exist in more than one spin state. Conventional thinking advises that if such a reaction begins with an $S = 0$ metal center, for example, the system will follow a reaction coordinate in the same electron-spin manifold. This thinking, together with the prevailing postulate that most organometallic reactions can be understood by invoking 16- or 18-electron intermediates or transition states (14), has been influential in describing a reaction mechanism. Not until recently was the importance of spin-state changes in the reactivity of unsaturated transition metal complexes recognized (53,54). For instance, a recent study by Bengali, et al. shows that the photogenerated η^5-CpCo(CO) and η^5-Cp*Co(CO) do not form stable adducts with alkanes or rare gas atoms (Xe, Kr) but react readily with CO molecules at a diffusion-limited rate (55). Utilizing both DFT and ab initio computational methods, Siegbahn later attributed these observations to an η^5-CpCo(CO) species in the ground triplet manifold. The triplet η^5-CpCo(CO) then undergoes a rapid spin-flip to form the singlet η^5-CpCo(CO)$_2$ under the influence of an incoming CO, but not when the incoming ligand is a more weakly binding alkane molecule (49,56,57). Ultrafast infrared studies by Dougherty and Heilweil show that η^5-CpCo(CO) reacts very quickly ($<$vibrational cooling time) with the strongly-binding ligand 1-hexene to form presumably a singlet π complex, in which a 1-hexene molecule complexes to the Co metal through its C=C double bond (58). A similar reactivity has also been observed in the reaction of CO and N_2 with triplet η^3-$Tp^{i-Pr,Me}Co(CO)$ ($Tp^{i-Pr,Me}$ = HB-$Pz_3^{i-Pr,Me}$, $Pz^{i-Pr,Me}$ = 3-iso-propyl-5-methylpyrazolyl) (59) or η^5-$Cp^*MoCl(PMe_3)$ (60,61), in the oxidative addition of benzene or aldehydes C–H bonds to unsaturated η^5-$Cp^*Co(\eta^2$-H_2C=$CHSiMe_3)$ (62–64) and most recently in the silane Si–H bond activation by η^5-$CpV(CO)_4$ (65). In view of the above examples, it would seem that a stronger metal-ligand interaction tends to facilitate a high-spin to low-spin crossover in an organometallic compound. Although such an intermolecular process can be qualitatively described by Fig. 14, substantial efforts will be required, both in experiments and in theoretical development, to reach the same level of understanding as intramolecular intersystem crossing (66). The unique information provided by ultrafast infrared spectroscopy, which includes the dynamics of IVR and those of molecular morphology change, is expected to be crucial in future developments.

V.   C–Cl BOND ACTIVATION BY THE Re(CO)₅ RADICAL

In a one-electron oxidative-addition reaction, only one chemical moiety transfers to a transition metal center. Consequently, the formal charge of the metal is changed by +1 (oxidation). Prototypical examples include the abstraction of a halogen atom from a halogenated organic molecule by transition-metal radicals that have formally 17 valence electrons at the metal. Previous studies of such reactions have led to proposals that involve either an intermediate that has 19 valence electrons at the metal center or a charge-transfer intermediate (9–11). These two scenarios are illustrated in Fig. 17 using the reaction of Cl atom abstraction by the $(CO_5)Re$ radical as an example. Due to the strong metal-carbonyl coupling, the CO ligands

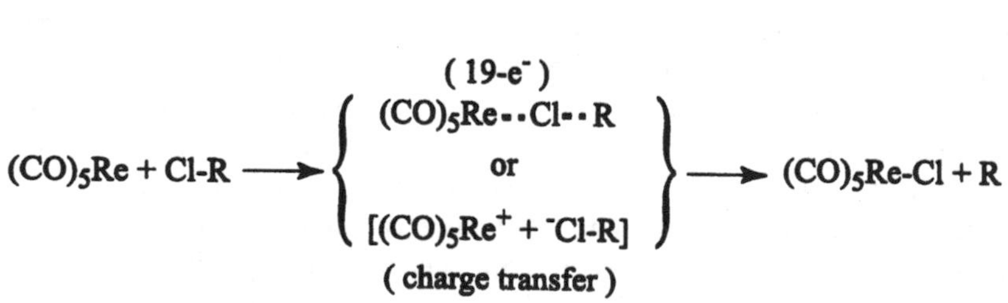

Figure 17   Previously proposed reaction schemes for Cl atom abstraction by $Re(CO)_5$.

may serve as a sensitive local probe for the charge density of the metal. This offers an opportunity to clarify the reaction mechanism by comparing the CO stretching frequencies in the reactive chlorinated methane solutions with those in the chemically inert hexane solution. If the reaction proceeds through a 19-electron intermediate, the increased metal electron density will result in a red shift of the CO stretching frequency. On the other hand, if the reaction proceeds through a charge-transfer intermediate of the form $[(CO)_5Re^+ + {}^-Cl-R]$, the positively charged rhenium pentacarbonyl will exhibit a substantial blue shift in the CO-stretching frequency.

A. Clarification of the Reaction Pathway

In the current study, the reaction is initiated by photochemically splitting the Re–Re σ bond of $(CO)_5Re - Re(CO)_5$ with UV pulses. The resulting $Re(CO)_5$ radical further reacts to abstract a Cl atom from a chlorinated methane molecule CH_nCl_{4-n} ($n = 0, 1, 2$) to form the final product $(CO)_5ReCl$. In Fig. 18d, $(CO)_5ReCl$ shows two CO-stretching bands at 1982 and 2045 cm^{-1} in CCl$_4$ solution (67). At shorter time delays (40 ns $< \Delta t <$ 2.5 μs) after photoexcitation, there appear five additional bands marked by asterisks in Fig. 18c. These bands are assigned to the equatorially solvated eq-Re$_2$(CO)$_9$(CCl$_4$), where a CCl$_4$ solvent molecule takes up the vacant site that is left behind by a leaving equatorial CO ligand, in accordance with low-temperature studies (68). On the ultrafast time scale, a broad feature comes into view at about 1990 cm^{-1}, as indicated by the down-pointing arrow in Fig. 18b. To assign this broad feature, one compares the spectrum in Fig. 18b with that taken in the chemically inert hexane solution shown in Fig. 18a. In Fig. 18a, the most intense peak at $\sim$1992 cm^{-1} is assigned to the weakly solvated $(CO)_5Re$ radical in room-temperature hexanes, in very good agreement with previously reported Re(CO)$_5$ band (1990 cm^{-1}) in cyclohexane solution (69). It follows that the $\sim$1990 cm^{-1} feature in Fig. 18b can be ascribed to CCl$_4$ solvated Re(CO)$_5$. The similarity in the CO-stretching frequencies of Re(CO)$_5$ in both hexanes and CCl$_4$ solutions suggests that the electron density of the Re center does not change appreciably in both solvents and that the interaction of Re(CO)$_5$(CCl$_4$) is comparable in magnitude to that of $(CO)_5Re$ (alkane). Both considerations are supported by DFT calculations. The computed Mulliken population of Re only increases 8% (or 4% using natural population) from Re(CO)$_5$/CH$_4$ to Re(CO)$_5$(CCl$_4$) as shown in Table 1 (70). The Re(CO)$_5$/CCl$_4$ interaction energy is calculated to $\sim$−0.6 kcal/mol at the DFT B3LYP level, comparable to $\sim$−0.6 kcal/mol

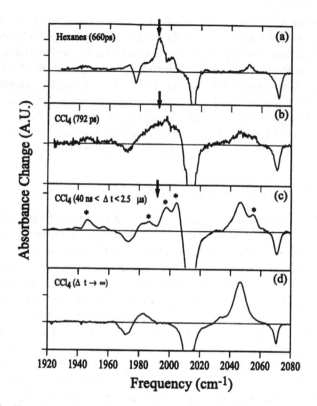

Figure 18 Transient difference spectra in the CO-stretching region for $Re_2(CO)_{10}$ (a) in hexane solution; and (b)–(d) in CCl_4 solution at various time delays following 295 nm photolysis. In panel (c), bands due to the solvated $Re_2(CO)_9(CCl_4)$ are marked by asterisks. Panel (d) is an FTIR difference spectrum before and after 308 nm photolysis. The down-pointing arrows in panels (a)–(c) indicate the CO stretch of the $Re(CO)_5$ radical. A broad, wavelength-independent background signal from CaF_2 windows has been subtracted. (Adapted from Ref. 72.)

for $Re(CO)_5/CH_4$. The calculated weak interaction indicates that the mean thermal energy $\sim$0.6 kcal/mol at room temperature is sufficient to disrupt the formation of a stable complex of the form $Re(CO)_5$(solvent). In other words, a dynamical equilibrium is established for $Re(CO)_5 \ldots$ (solvent) $\leftrightarrow$ $Re(CO)_5$ + solvent (10), the time scale of which is on the order of collision in liquids. This allows the chemically active Re center to undergo recombination reaction with another $Re(CO)_5$ radical to reform the parent $Re_2(CO)_{10}$ molecule.

Table 1 Summary of Gas Phase Electron Population Analysis for Re and Cl Centers Involved in Reaction of Cl Atom Abstraction by $Re(CO)_5$ Radical

	Mulliken charge[a]		Natural charge[b]	
Atomic center	Re	Cl	Re	Cl
$(CO)_5 Re$	−0.147	—	−0.412	—
Solvated Complex				
$(CO)_5 Re(CH_4)$	−0.175	—	−0.412	—
$(CO)_5 Re(CCl_4)$	−0.189	+0.086	−0.430	+0.022
Transition State				
$ts\text{-}(CO)_5 Re \ldots Cl \ldots CCl_3$	−0.222	−0.025	−0.481	−0.168
$ts\text{-}(CO)_5 Re \ldots Cl \ldots CHCl_2$	−0.244	−0.0925	−0.496	−0.260
$ts\text{-}(CO)_5 Re \ldots Cl \ldots CH_2Cl$	−0.264	−0.144	−0.508	−0.331
Product				
$(CO)_5ReCl$	−0.321	−0.209	−0.567	−0.477

[a] Mulliken charge population for a Cl atom in CCl_4 is $+0.090$ e^-.
[b] Natural charge population for a Cl atom in CCl_4 is $+0.068$ e^-.

Geminate-recombination dynamics of the parent molecule provide further experimental support for the weak $Re(CO)_5$/solvent interaction. Shown in Fig. 19 are kinetic traces of the 2071 cm^{-1} parent bleach in various solvents. They all exhibit a recovery of about 50% on two time scales. For example, the parent bleach of $Re_2(CO)_{10}$ in CCl_4 recovers on ~50 ps and ~500 ps if fitted to a bi-exponential function. The biphasic recovery can be described by a diffusion model (solid lines in Fig. 19) that takes into account the dynamics of geminate-pair recombination (71,72). The physical picture imbedded in the model is illustrated in Fig. 20. In the figure, the abscissa is the separation of the two $Re(CO)_5$ monomers, where the equilibrium Re – Re distance R_{eq} (≈ 3.0 Å), the contact distance R (≈ 6.3 Å), and the initial separation r_0 (≈ 8.6 Å) are marked by short vertical bars. The portion to the left of R is a schematic potential energy surface plot depicting dynamics that are not susceptible to the current model. To the right of R are plots of time-dependent spatial distribution of a monomer around a recombination center. At short time delays such as 1 and 5 ps, the distribution maxima still linger around the initial separation r_0, while the concentration of the recombining monomer builds up very quickly at R. This accounts for the experimentally observed fast recovery since the recombination rate k_r is defined to be proportional to the concentration

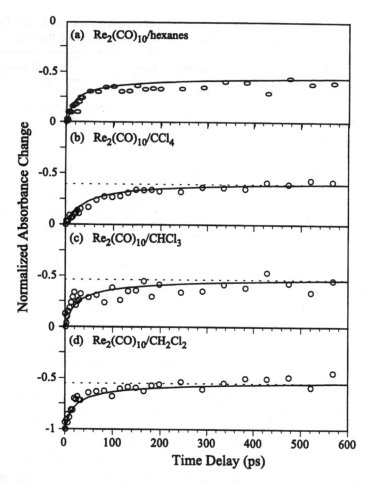

Figure 19 Parent bleach kinetics (open circles) of $Re_2(CO)_{10}$ in (a) hexanes, (b) CCl_4, (c) $CHCl_3$, and (d) CH_2Cl_2. Fits to a diffusion model to account for geminate recombination are shown as solid lines. Except for the macroscopic viscosity, identical molecular parameters (see Fig. 20) are used for all the solvents studied.

gradient at the contact distance R. In other words, the fast 50 ps component can be attributed to a convolution of vibrational relaxation and the probability of reactive collision of the $Re(CO)_5$ pair within the solvent cage to form the parent $Re_2(CO)_{10}$. Driven by the accumulating population around R, the distribution starts to diffuse away from the recombination center at time delays longer than ∼20 ps until the system reaches equilibrium. This

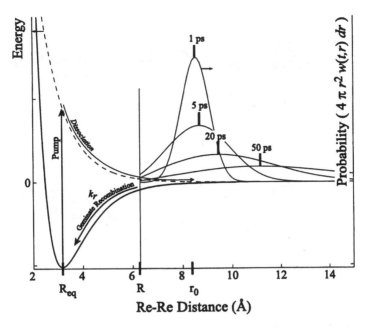

Figure 20 Illustration for the diffusion model for geminate recombination. Details are described in the text.

diffusive process is identified with the slower recovery (hundreds of ps) in the parent bleach data.

In consideration of the above experimental and theoretical evidence, it is concluded that the reaction, at least the ones that have been investigated, be viewed as proceeding through a weakly solvated 17-electron $Re(CO)_5$ radical instead of a 19-electron or a charge-transfer intermediate. The fact that no other intermediates are detected prior to the product formation suggests that the reaction involve only the rate-limiting Cl atom transfer step.

B. The Nature of the Reaction Barrier — Atom Transfer

This section discusses the correlation of the reaction rate to the chemical reactivity based on the nature of the transition state. The reaction is carried out in a series of chlorinated methane solutions, CH_nCl_{4-n} ($n = 0, 1, 2$). The free-energy barriers $\Delta G^{\ddagger}$, determined by the rise time of the product at 2045 cm^{-1}, are found to be 5.35 ± 0.03 ($n = 0$), 8.03 ± 0.20 ($n = 1$), and 8.48 ± 0.13 ($n = 2$) kcal/mol.

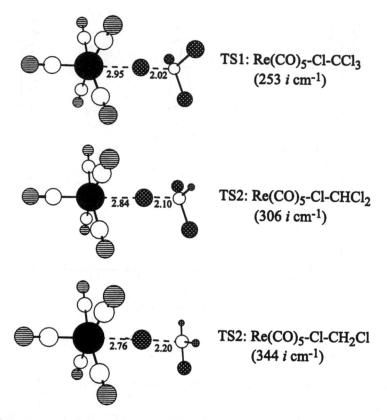

TS1: Re(CO)$_5$-Cl-CCl$_3$
(253 i cm^{-1})

TS2: Re(CO)$_5$-Cl-CHCl$_2$
(306 i cm^{-1})

TS2: Re(CO)$_5$-Cl-CH$_2$Cl
(344 i cm^{-1})

Figure 21 Transition-state structures computed at the DFT B3LYP level of theory. The Re–Cl and C–Cl bond lengths are in Å. Also shown are the imaginary frequencies associated with each transition state. (Adapted from Ref. 72.)

To better understand the reactivity, the transition states for the series of reactions were studied using DFT methods. As illustrated in Fig. 21, the results show that each transition state can be characterized by a single imaginary frequency that involves simultaneous dissociation of the C–Cl bond and formation of the Re–Cl bond. The structural variation along the series of chlorinated methanes suggests that the transition state becomes more product-like as the number of hydrogen n in CH$_n$Cl$_{4-n}$ increases. For example, the Re . . . Cl distance in (CO)$_5$ Re . . . Cl . . . CH$_n$Cl$_{3-n}$ decreases monotonically from 2.95 Å ($n = 0$) to 2.84 Å ($n = 1$) to 2.76 Å ($n = 2$), and finally to 2.52 Å for the product (CO)$_5$Re-Cl. The calculated electron density distribution also displays a similar trend as displayed in Table 1.

The DFT energy barriers $\Delta E_0^{\ddagger}$ in CCl_4, $CHCl_3$, and CH_2Cl_2 are, respectively, 0.86, 3.0, and 8.5 kcal/mol, where the solvent effects are treated self-consistently in the reaction-field approximation. The current level of theory, however, does not permit an immediate comparison of the experimental energy barriers to the calculated values. Nonetheless, comparison of the experimental trend and the qualitative trend from the theoretical calculations along the series of chloromethane provides greater insight in understanding the reactivity. The observed trends in the barrier and the transition-state structure are a clear demonstration of Hammond's postulate, which associates a later transition state with a higher energy barrier (73). This suggests that the valence-bonding picture of structure-reactivity correlation, which has enjoyed much success in the field of physical organic chemistry, may be applicable to this case (74–76). Figure 22 shows a linear correlation model that utilizes this concept (77,78). In this model, the reaction proceeds along the C–Cl bond potential energy curve, passes through the transition state, and forms the final product along the Re–Cl bond potential energy curve. The position of the transition state is approximated

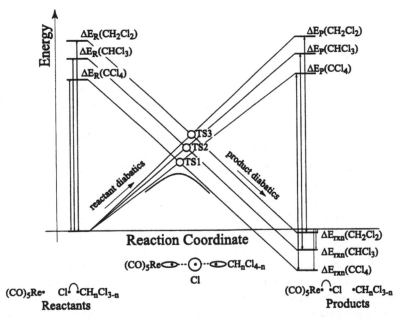

Figure 22 A correlation diagram for Cl atom abstraction from CH_nCl_{4-n} by the $Re(CO)_5$ radical.

by the crossing point of the two diabetic curves. Accordingly, the reaction profile can be understood by three factors: the C–Cl and Re–Cl bond strengths and the overall reaction enthalpy. The required parameters are available in the literature, obtained either through experiments or by electronic structure calculations. While this simple model correctly predicts the energetic and structural trends of the transition states, more case studies are necessary in order to arrive at a better, quantitative model for the atom-transfer step (79,80).

VI. CLOSING REMARKS

Our understanding of the reactivity of complex chemical systems such as organometallic compounds has greatly advanced in the past decade owing to the vast development of experimental and computational techniques. Basic principles born of these collective efforts have guided the thinking of reactions in fields including chemistry, biological sciences, and material sciences. Yet the critical examination of the postulates in the realistic solution phase has been a daunting challenge due to the intricate liquid dynamics, which render the relevant time scales to spreading several orders of magnitude. Femtosecond infrared spectroscopy, which is capable of "real-time" observation and characterization of a chemical reaction, has been utilized to follow oxidative-addition reactions of prototypical two-electron C–H and Si–H bonds and one-electron Cl atom to organometallic complexes. This technique affords the description of not only the static reaction mechanism but also the dynamics of each elementary step, including vibrational cooling, geminate-pair recombination, morphological reorganization, solvation-assisted intersystem crossing, and solvent rearrangement. The explicit reaction scheme composed of elementary steps allows an experimental assessment for the current understanding of the reactivity. For C–H and Si–H bond activation, the present results corroborate the prevailing picture of the reactivity. For Cl atom abstraction by the Re(CO)$_5$ radical, however, the present results caution the notion of a 19-electron species or a charge-transfer complex as reactive intermediates. In addition, in each of the reactions discussed, the rate-determining process has been determined to be the bond-breaking step for C–H activation, solvent rearrangement for Si–H activation, and atom-transfer step for C–Cl activation. They represent three stereotypical rate-limiting steps that are expected to be common to solution-phase reactions. More examples are anticipated to emerge as more efforts are devoted to the study of complicated chemical processes, which, in time, will improve our picture of reactions in liquids.

ACKNOWLEDGMENT

The following individuals have made indispensable contributions in bringing this project to the current state: T. Lian, K. T. Kotz, M. C. Asplund, H. Frei, S. E. Bromberg, P. T. Snee, W. J. Wilkens, C. K. Payne, J. S. Yeston, B. K. McNamara, and R. G. Bergman. The use of the static FTIR spectrometer of the C. B. Moore group and that of a UV-Vis spectrometer of the A. P. Alivisatos group are gratefully acknowledged. The use of specialized equipment under the Office of Basic Energy Science, Chemical Science Division, U.S. Department of Energy contract DE-AC03-76SF00098 is also acknowledged. This project is supported by a grant from the National Science Foundation.

REFERENCES

1. Collman JP, Hegedus LS, Norton JR, Finke RG. Principles and Applications of Organotransition Metal Chemistry. 1987.
2. Masters C. Homogeneous Transition-Metal Catalysis — A Gentle Art. London: Chapman and Hall, 1981.
3. Lippard SJ, Berg JM. Principles of Bioinorganic Chemistry. Mill Valley, CA: University Science Books, 1994.
4. Kubas GJ, Ryan RR, Swanson BI, Vergamini PJ, Wasserman HJ. Characterization of the first examples of isolable molecular hydrogen complexes, $M(CO)_3(PR_3)_2(H_2)$ (M = Mo, W; R = Cy, i-Pr). Evidence for a side-on bonded H_2 ligand. J Am Chem Soc 1984; 106:451–452.
5. Crabtree RH. Dihydrogen complexes: some structural and chemical studies. Acc Chem Res 1990; 23(4):95–101.
6. Kubas GJ. Molecular hydrogen complexes: coordination of a σ bond to transition metals. Acc Chem Res 1988; 21:120–128.
7. Hall C, Perutz RN. Transition metal alkane complexes. Chem Rev 1996; 96(8):3125–3146.
8. Musaev DG, Morokuma K. Ab initio molecular orbital study of the mechanism of H–H, C–H, N–H, O–H, and Si–H bond activation on transient cyclopentadienylcarbonylrhodium. J Am Chem Soc 1995; 117(2):799–805.
9. Stiegman AE, Tyler DR. Reactivity of seventeen- and nineteen-valence electron complexes in organometallic chemistry. Comments Inorg Chem 1986; 5(5):215–245.
10. Tyler DR. 19-Electron organometallic adducts. Acc Chem Res 1991; 24:325–331.
11. Baird MC. Seventeen-electron metal-centered radicals. Chem Rev 1988; 88:1217–1227.

12. Astruc D. Nineteen-electron complexes and their role in organometallic mechanisms. Chem Rev 1988; 88(7):1189–1216.

13. Ohkubo K, Kanaeda H, Tsuchihashi K. An MO-theoretical interpretation of the reductive cleavage of organic halides by pentacyanocobaltate(II). Bull Chem Soc Jpn 1973; 46(10):3095–3098.

14. Tolman CA. The 16 and 18 electron rule in organometallic chemistry and homogeneous catalysis. Chem Soc Rev 1972; 1:337–353.

15. van Leeuwen PWNM, Morokuma K, van Lenthe JH. In: Ugo R, James BR, eds. Theoretical Aspects of Homogeneous Catalysis. Vol. 18. Catalysis by Metal Complexes. Boston: Kluwer Academic Publishers, 1995.

16. Arndtsen BA, Bergman RG. Selective intermolecular carbon-hydrogen bond activation by synthetic metal complexes in homogeneous solution. Acc Chem Res 1995; 28:154–162.

17. Bengali AA, Arndtsen BA, Burger PM, Schultz RH, Weiller BH, Kyle KR, Moore CB, Bergman RG. Activation of carbon-hydrogen bonds in alkanes and other organic molecules by Ir(I), Rh(I), and Ir(II) complexes. Pure Appl Chem 1995; 67:281–288.

18. Bergman RG. Activation of alkanes with organotransition metal complexes. Science 1984; 223:902–908.

19. Bergman RG. A physical organic road to organometallic C–H bond oxidative addition reactions. J Organomet Chem 1990; 400:273–282.

20. Crabtree RH. The organometallic chemistry of alkanes. Chem Rev 1985; 85:245–269.

21. Janowicz AH, Bergman RG. C–H activation in completely saturated hydrocarbons: direct observation of M + R–H → M(R)(H). J Am Chem Soc 1982; 104(1):352–354.

22. Hoyano JK, Graham WAG. Oxidative addition of the carbon-hydrogen bonds of neopentane and cyclohexane to a photochemically generated iridium(I) complex. J Am Chem Soc 1982; 104(13):3723–3725.

23. Bromberg SE, Lian TQ, Bergman RG, Harris CB. Ultrafast dynamics of $Cp^*M(CO)_2$ (M = Ir, Rh) in solution: the origin of the low quantum yields for C–H bond activation. J Am Chem Soc 1996; 118(8):2069–2072.

24. Ghosh CK, Graham AG. Efficient and selective carbon-hydrogen activation by a tris(pyrazolyl)borate rhodium complex. J Am Chem Soc 1987; 109:4726–4727.

25. Ghosh CK, Graham AG. A rhodium complex that combines benzene activation with ethylene insertion — subsequent carbonylation and ketone formation. J Am Chem Soc 1989; 111:375–376.

26. Lees AJ, Purwoko AA. Photochemical mechanisms in intermolecular C–H bond activation reactions of organometallic complexes. Coord Chem Rev 1994; 132:155–160.

27. Purwoko AA, Lees AJ. Photochemistry and C–H bond activation reactivity of $(HBPz_3^*)Rh(CO)_2$ ($Pz^* = 3, 5$-dimethylpyrazolyl) in hydrocarbon solution. Inorg Chem 1995; 34:424–425.

28. Lee M, Harris CB. Ultrafast studies of transition-metal carbonyl reactions in the condensed phase: solvation of coordinatively unsaturated pentacarbonyls. J Am Chem Soc 1989; 111(24):8963–8965.

29. Lian T, Bromberg SE, Yang H, Proulx G, Bergman RG, Harris CB. Femtosecond IR studies of alkane C–H bond activation by organometallic compounds — direct observation of reactive intermediates in room temperature solutions. J Am Chem Soc 1996; 118(15):3769–3770.

30. Bromberg SE, Yang H, Asplund MC, Lian T, McNamara BK, Kotz KT, Yeston JS, Wilkens M, Frei H, Bergman RG, Harris CB. The mechanism of a C–H bond activation reaction in room-temperature alkane solution. Science 1997; 278:260–263.

31. Zaric S, Hall MB. Prediction of the reactive intermediates in alkane activation by tris(pyrazolyl borate)rhodium carbonyl. J Phys Chem A 1998; 102(11):1963–1964.

32. Schultz RH, Bengalli AA, Tauber MJ, Weiller BH, Wasserman EP, Kyle KR, Moore CB, Bergman RG. IR flash kinetic spectroscopy of C–H bond activation of cyclohexane-D_0 and cyclohexane-D_{12} by $Cp^*Rh(CO)_2$ in liquid rare-gases — kinetics, thermodynamics, and an unusual isotope effect. J Am Chem Soc 1994; 116(16):7369–7377.

33. Asbury JB, Ghosh HN, Yeston JS, Bergman RG, Lian TQ. Sub-picosecond IR study of the reactive intermediate in an alkane C–H bond activation reaction by $CpRh(CO)_2$. Organometallics 1998; 17(16):3417–3419.

34. Wick DD, Goldberg KI. C–H activation at Pt(II) to form stable Pt(IV) alkyl hydrides. J Am Chem Soc 1997; 119(42):10235–10236.

35. Gutiérrez-Puebla E, Monge Á, Nicasio MC, Pérez PJ, Poveda ML, Rey L, Ruíz C, Carmona E. Vinylic C–H bond activation and hydrogenation reactions of $Tp'Ir(C_2H_4)(L)$ complexes. Inorg Chem 1998; 37(18):4538–4546.

36. Jiménez-Cataño R, Niu S, Hall MB. Theoretical studies of inorganic and organometallic reaction mechanisms. 10. Reversal in stability of rhodium and iridium η^2-ethene and hydridovinyl complexes. Organometallics 1997; 16(9):1962–1968.

37. Moser WR, Slocum DW. Homogeneous Transition Metal Catalyzed Reactions. Washington, DC: American Chemical Society, 1992.

38. Schubert U. η^2 Coordination of Si–H σ bonds to transition metals. Adv Organomet Chem 1990; 30:151–187.

39. Crabtree RH. Aspects of methane chemistry. Chem Rev 1995; 95(4):987–1007.

40. Koga N, Morokuma K. SiH, SiSi, and CH bond activation by coordinatively unsaturated $RhCl(PH_3)_2$. Ab initio molecular orbital study. J Am Chem Soc 1993; 115(15):6883–6892.

41. Young KM, Wrighton MS. Temperature dependence of the oxidative addition of triethylsilane to photochemically generated $(\eta^5\text{-}C_5Cl_5)Mn(CO)_2$. Organometallics 1988; 8(4):1063–1066.

42. Palmer BJ, Hill RH. The energetics of the oxidative addition of trisubstituted silanes to photochemically generated $(\eta^5\text{-}C_5R_5)Mn(CO)_2$. Can J Chem 1996; 74:1959–1967.

43. Hart-Davis AJ, Graham WAG. Silicon-transition metal chemistry. VI. Kinetics and mechanism of the replacement of triphenylsilane by triphenylphosphine in hydridotriphenylsilyl(π-cyclopentadienyl)dicarbonylmanganese. J Am Chem Soc 1971; 94(18):4388–4393.

44. Hu S, Farrell GJ, Cook C, Johnston R, Butkey TJ. Rearrangement of η^5-CpMn(CO)$_2$(HSiEt$_3$): a missing step in the energy surface for the oxidative addition of silane to CpMn(CO)$_2$ (heptane). Organometallics 1994; 13(11):4127–4128.

45. Yang H, Asplund MC, Kotz KT, Wilkens MJ, Frei H, Harris CB. Reaction mechanism of silicon-hydrogen bond activation studied using femtosecond to nanosecond IR spectroscopy and ab initio methods. J Am Chem Soc 1998; 120(39):10154–10165.

46. Yang H, Kotz KT, Asplund MC, Harris CB. Femtosecond infrared studies of silane silicon-hydrogen bond activation. J Am Chem Soc 1997; 119(40):9564–9565.

47. Hill RH, Wrighton MS. Oxidative addition of trisubstituted silanes to photochemically generated coordinatively uncaturated species $(\eta^4\text{-}C_4H_4)Fe(CO)_2$, $(\eta^5\text{-}C_5H_5)Mn(CO)_2$, and $(\eta^6\text{-}C_6H_6)Cr(CO)_2$ and related molecules. Organometallics 1987; 6(3):632–638.

48. Hofmann P, Padmanabhan M. Electronic and geometric features of $(\eta^5\text{-}C_5H_5)ML$ 16-electron fragments. A molecular orbital study of ligand effects. Organometallics 1983; 2(10):1273–1284.

49. Siegbahn PEM. Comparison of the C–H activation of methane by $M(C_5H_5)(CO)$ for M = cobalt, rhodium, and iridium. J Am Chem Soc 1996; 118(6):1487–1496.

50. Sutin N. Theory of electron transfer reactions: insights and hindsights. In: Progress in Inorganic Chemistry, Vol. 30. Lippard SJ, ed. New York: John Wiley & Sons, Inc., 1983:441–498.

51. Klassen JK, Selke M, Sorensen AA, Yang GK. Metal-ligand bond dissociation energies in CpMn(CO)$_2$L complexes. J Am Chem Soc 1990; 112:1267–1268.

52. Hester DM, Sun J, Harper AW, Yang GK. Characterization of the energy surface for the oxidative addition of silanes to CpMn(CO)$_2$ (heptane). J Am Chem Soc 1992; 114(13):5234–5240.

53. Poli R. Open-Shell organometallics as a bridge between Werner-type and low-valent organometallic complexes. The effect of the spin state on the stability, reactivity, and structure. Chem Rev 1996; 96(6):2135–2204.

54. Shaik S, Filatov M, Schröder D, Schwarz H. Electronic structure makes a difference: cytochrome P-450 mediated hydroxylations of hydrocarbons as a two-state reactivity paradigm. Chem Eur J 1998; 4(2):193–199.

55. Bengali AA, Bergman RG, Moore CB. Evidence for the formation of free 16-electron species rather than solvate complexes in the ultraviolet irradiation

of CpCo(CO)$_2$ in liquefied noble gas solvents. J Am Chem Soc 1995; 117(13):3879–3880.

56. Poli R, Smith KM. Spin state and ligand dissociation in [CpCoL$_2$] complexes (L = PH$_3$, H$_2$C=CH$_2$): a computational study. Eur J Inorg Chem 1999; 877–880.

57. Smith KM, Poli R, Legzdins P. A computational study of two-state conformational change in 16-electron [CpW(NO)(L)] complexes (L = PH$_3$, CO, CH$_2$, HCCH, H$_2$CCH$_2$). Chem Eur J 1999; 5(5):1598–1608.

58. Dougherty TP, Heilweil EJ. Transient infrared spectroscopy of (η^5-C$_5$H$_5$)Co(CO)$_2$ photoproduct reactions in hydrocarbon solutions. J Chem Phys 1994; 100(5):4006–4009.

59. Detrich JL, Reinaud OM, Rheingold AL, Theopold KH. Can spin state change slow organometallic reactions? J Am Chem Soc 1995; 117(47):11745–11748.

60. Keogh DW, Poli R. Spin state change in organometallic reactions. Experimental and MP2 theoretical studies of the thermodynamics and kinetics of the CO and N$_2$ addition to spin triplet Cp*MoCl(PMe$_3$)$_2$. J Am Chem Soc 1997; 119(10):2516–2523.

61. Poli R. Molybdenum open-shell organometallics. Spin state changes and pairing energy effects. Acc Chem Res 1997; 30(12):494–501.

62. Lenges CP, White PS, Brookhart M. Mechanistic and synthetic studies of the addition of alkyl aldehydes to vinylsilanes catalyzed by Co(I) complexes. J Am Chem Soc 1998; 120(28):6965–6979.

63. Lenges CP, Brookhart M. Co(I)-catalyzed inter- and intramolecular hydroacylation of olefins with aromatic aldehydes. J Am Chem Soc 1997; 119(13):3165–3166.

64. Lenges CP, Brookhart M, Grant BE. H/D exchange reactions between C$_6$D$_6$ and C$_5$Me$_5$Co(CH$_2$=CHR)$_2$ (R = H, SiMe$_3$): evidence for oxidative addition of C$_{sp2}$-H bonds to the [C$_5$Me$_5$(L)Co] moiety. J Organomet Chem 1997; 528:199–203.

65. Snee PT, Yang H, Kotz KT, Payne CK, Harris CB. Femtosecond infrared studies of the mechanism of silicon-hydrogen bond activation by η^5-CpV(CO)$_4$. J Phys Chem 1999; J Phys Chem A 1999; 103(49):10426–10432.

66. Hauser A. Intersystem crossing in iron(II) coordination compounds: a model process between classical and quantum mechanical behavior. Comments Inorg Chem 1995; 17(1):17–40.

67. Wrighton MS, Ginley DS. Photochemistry of metal-metal bonded complexes. II. The photochemistry of rhenium and manganese carbonyl complexes containing a metal-metal bond. J Am Chem Soc 1975; 97(8):2065–2072.

68. Firth S, Klotzbuecher WE, Poliakoff M, Turner JJ. Generation of Re$_2$(CO)$_9$ (N$_2$) from Re$_2$(CO)$_{10}$: identification of photochemical intermediates by matrix isolation and liquid-noble-gas techniques. Inorg Chem 1987; 26(20):3370–3375.

69. Firth S, Hodges PM, Poliakoff M, Turner JJ. Comparative matrix isolation and time-resolved infrared studies on the photochemistry of $MnRe(CO)_{10}$ and $Re_2(CO)_{10}$: evidence for CO-bridged $MnRe(CO)_9$. Inorg Chem 1986; 25(25):4608–4610.

70. Glendening ED, Badenhoop JK, Reed AE, Carpenter JE, Weinhold F. NBO 4.0. Madison: University of Wisconsin, 1996.

71. Naqvi KR, Mork KJ, Waldenstrom S. Diffusion-controlled reaction kinetics. Equivalence of the particle pair approach of Noyes and the concentration gradient approach of Collins and Kimball. J Phys Chem 1980; 84(11):1315–1319.

72. Yang H, Snee PT, Kotz KT, Payne CK, Frei H, Harris CB. Femtosecond infrared studies of a prototypical one-electron oxidative-addition reaction: chlorine atom abstraction by the $Re(CO)_5$ radical. J Am Chem Soc 1999; 121(39):9227–9228.

73. Hammond GS. A correlation of reaction rates. J Am Chem Soc 1954; 77:334–338.

74. Agmon N, Levine RD. Energy, entropy, and the reaction coordinate: thermodynamic-like relations in chemical kinetics. Chem Phys Lett 1977; 52(2):197–201.

75. Levine RD. Free energy of activation. Definition, properties, and dependent variables with special reference to "linear" free energy relations. J Phys Chem 1979; 83(1):159–170.

76. Agmon N. From energy profiles to structure-reactivity correlations. Int J Chem Kinet 1981; 13:333–365.

77. Bernardi F, Bottoni A. Polar effect in hydrogen abstraction reactions from halo-substituted methanes by methyl radical: a comparison between Hartree-Fock, perturbation, and density functional theories. J Phys Chem A 1997; 101(10):1912–1919.

78. Bottoni A. Theoretical study of the hydrogen and chlorine abstraction from chloromethanes by silyl and trichlorosilyl radicals: a comparison between the Hartree-Fock method, perturbation theory, and density functional theory. J Phys Chem A 1998; 102(49):10142–10150.

79. Herrick RS, Herrinton TR, Walker HW, Brown TL. Rates of halogen atom transfer to manganese carbonyl radicals. Organometallics 1985; 4(1):42–45.

80. Lee K-W, Brown TL. On the nature of halogen atom transfer reactions of $Re(CO_4)L$ radicals. J Am Chem Soc 1987; 109(11):3269–3275.

3

Applications of Broadband Transient Infrared Spectroscopy

Edwin J. Heilweil
National Institute of Standards and Technology, Gaithersburg, Maryland

I. INTRODUCTION

The application of ultrafast transient infrared (IR) spectroscopy to the study of chemical, biochemical, and related physical phenomena has dramatically increased over the last decade. This surge of interest has largely been inspired by the increased availability of commercial, solid-state, table-top pulsed laser equipment and nonlinear frequency conversion methods for generating high-power, broadly tunable infrared pulses in the near- to mid-infrared wavelength regime (ca. 2–25 µm). Indeed, the accessibility and flexibility of these modern optical systems has far exceeded the expectations of many researchers working in this field. Early methodologies for generating ultrafast IR pulses used simple multipass optical parametric amplifiers (1) or visible difference frequency mixing techniques (2) to access the mid-IR spectral region. Dynamical measurement of vibrational energy flow within solute and adsorbate species begun in the late 1970s (3) has given way to highly sophisticated ultrafast methods (4), studies of ultrafast molecular reaction mechanisms (5), energy transfer within short-chain amino acids (6), at surfaces (7,8), and in liquids (9,10) and solids (11), to name a few examples.

Early ultrafast transient infrared measurements were typically performed using identical frequency picosecond infrared pulses for both pumping and probing a vibrational mode of condensed-phase molecules

(12). In this scenario, an intense, narrowband tunable infrared pump pulse excites population from the $v = 0$ to the $v = 1$ level of a molecular vibration (producing increased sample transmission), and a weaker probe pulse interrogates the recovery of ground state absorption to yield the $v = 1$ population relaxation time (T_1 lifetime). These simple "single-color" measurements were quickly superceded by "two-color" techniques since it was recognized that monitoring the transient infrared spectrum as a function of time yields much more detailed information about the system dynamics. For example, detection of mode-specific vibrational solute-to-solvent energy transfer (13) or the generation of new transient reaction intermediates (14) necessarily dictates that a tunable infrared probe pulse be used to observe new spectral features removed in frequency from the parent molecular absorptions. However, using tunable, narrowband probe pulses to obtain broadband spectra involves the arduous task of taking frequency-scanned spectra with single element detectors at multiple pump-probe time delays. This generic approach suffered from extended data acquisition times for low-repetition-rate laser systems and potential long-term drift problems that potentially distorted the spectral intensity information.

To circumvent these difficulties, broadband infrared detection using multichannel arrays was employed by our group (15–17) and others (18,19). In many respects this approach is complementary to the accepted technique for performing ultraviolet-visible transient absorption spectroscopy. By generating broadband mid-infrared probe pulses a few picoseconds or less in duration and detecting a portion of an infrared spectral region for each laser pulse, data acquisition rates are vastly enhanced and system stability issues are diminished. It should also be pointed out that picosecond or femtosecond broadband probe pulses can be used to detect narrowband transient absorption features as long as the transient absorber has lived on the order of the coherence (T_2) lifetime of the interrogated state (20). These typically narrowband features ($3-15$ cm^{-1} FWHM) are spectrally resolved by an up-converter crystal (with CCD detection) or through the spectrograph resolution independent of the inherent time-bandwidth characteristics of the probe pulse. Demonstrations of the general broadband IR detection approach were first made using nonlinear IR frequency down- and up-conversion with visible spectrograph-CCD detection (15,16). With Defense Department declassification and increased commercial availability of infrared focal-plane arrays for mid-infrared imaging use, direct broadband infrared spectroscopic detection covering the $1-12$ μm range is now practical (21).

This chapter will first describe the state of the art in using ultrafast picosecond and femtosecond lasers to generate tunable, broadband mid-infrared pulses for time-resolved spectroscopic applications. Methods developed for acquiring transient IR spectra using multielement array technologies is then presented. After discussing these techniques in detail, selected studies performed at NIST using these broadband techniques are reviewed. In these exploratory cases, transient infrared spectroscopy yielded new insight into the mechanisms and rates of vibrational energy transfer within complex condensed-phase hydrogen-bonded systems. It also provided "glimpses" of previously unknown transient species that exist during molecular chemical reactions and revealed injection rates of electrons transferred from adsorbed ruthenium-based dyes into conducting TiO_2 substrates. It is hoped that an overview of these different examples will inspire and promote various extensions of the broadband ultrafast IR technique for future research endeavors and applied spectroscopic applications.

II. EXPERIMENTAL TECHNIQUES

A. Ultrafast Broadband Infrared Pulse Generation

The production of ultrafast mid-IR pulses typically begins with a fixed-frequency laser oscillator that is pulse amplified from nanojoule energies to microjoule (μJ) and millijoule (mJ) levels. Examples of such laser systems include Nd^{+3}:glass or Nd^{+3}:YAG low rep-rate picosecond oscillators (running near 1.06 μm, <30 Hz) followed by multiple-pass or regenerative amplifiers. These amplified pulses are subsequently frequency-shifted or beat against another pulse (by frequency down-conversion steps) in an appropriate nonlinear crystal to reach the mid-IR range. Various frequency-shifting techniques include difference frequency generation between two synchronously pumped tunable dye lasers or optical parametric amplification (OPA) of a weak infrared continuum, which provides narrow-bandwidth tunability in the mid-IR region at moderate power levels (<20 μJ/pulse). Multistage $LiNbO_3$ OPA crystals pumped by an amplified Nd^{+3}:YAG laser were first used to generate <0.5 mJ picosecond pulses tunable between 1.6 and 4 μm (1). Difference frequency mixing of amplified tunable visible dye oscillators beat against a fixed-frequency pulse (e.g., the 532 nm second harmonic of the Nd^{+3}:YAG) in $LiIO_3$ crystals enabled generation of independently tunable IR pump and probe pulses (2). The above approaches produced tunable 1–20 ps pulses with relatively narrow

($3-10$ cm^{-1} FWHM) spectral bandwidths appropriate for single-frequency "point-by-point" spectral interrogation of condensed-phase systems. For reference, a schematic for a generic laser system with mixing optics and crystals capable of providing a tunable (UV to IR) pump and broadband IR probe is shown in Fig. 1.

Modifications to the above laser systems and frequency down-converting schemes are readily performed to generate broadband IR output. For example, synchronously pumped picosecond visible or near-IR dye oscillators deliberately run without intracavity tuning elements permit the lasers to run broadband with output center frequency dictated by the dye gain spectrum (15). Subpicosecond dye laser pulses can also be fiber frequency chirped, amplified (to <1 mJ) in three-stage dye amplifiers, grating compressed to <150 fs pulse duration, and then mixed against $5-10$ ps, 300 µJ dye pulses in LiIO$_3$ to yield 150–200 fs FWHM broadband (200–300 cm^{-1} FWHM) IR output (22). When beat against the second harmonic of the Nd:YAG amplifier or another amplified picosecond (i.e., narrowband) dye laser in LiIO$_3$ or AgGaS$_2$, this approach yields pulses in the $3-10$ µm spectral region of up to several hundred wavenumbers

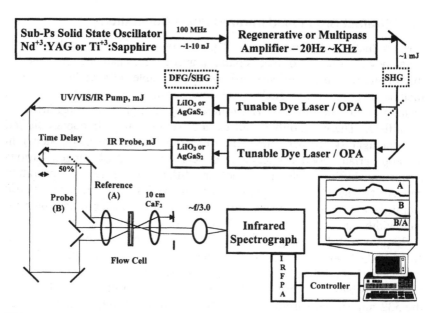

Figure 1 Transient broadband infrared system schematic using IR Focal Plane Array (IRFPA) dual track detection.

bandwidth (FWHM). Difference frequency generation between a broadband seed pulse and narrowband pump pulse ensures that the generated broadband IR pulse experiences minimal temporal dispersion or jitter while traversing the nonlinear crystal. Dye-based broadband sources (deliberately run without intracavity seeding) are inherently spectrally noisy and require shot-to-shot normalization and averaging during data collection (see below). However, there is no inherent reason why this general method cannot be extended to longer wavelengths for probing in the molecular "fingerprint" (6–12 μm) region. To date, investigations of organic C=O stretching modes in the 6-7 μm range have been conducted (6,19,23), but little if any known transient IR work has been performed in the 8 μm and longer wavelength range.

Most recently, laser systems based on femtosecond Ti:sapphire oscillators (80–100 MHz, pulse duration <150 fs FWHM, ~800 nm, 0.3–1.0 W average power) and multi-kilohertz regenerative or multipass amplifiers capable of delivering approximately 1 W of output power are being used as pulsed sources. These systems possess the desirable characteristics of delivering ultrafast (but only slightly tunable) pulses, which maintain their inherent near transform-limited broadband characteristics (>100 cm^{-1} FWHM) and high repetition rates for rapid data collection and signal averaging. Currently applied frequency down-conversion approaches employ BBO-based (β-Barium Borate) OPAs (with continuum seeding or multiple BBO passes) that "split" a nominally 800 nm, amplified Ti:sapphire input pulse into two longer-wavelength pulses (signal and idler tunable between 1.1 to 2.9 μm). These pulses are subsequently co-propagated and temporally recombined in a down-converting mixing crystal (e.g., Type I $LiIO_3$ or $AgGaS_2$) to produce tunable, relatively broad-bandwidth (>150 cm^{-1} FWHM, >1.8 μm) mid-IR probing pulses. The salient features of producing IR pulses in this fashion are straightforward tuning of the solid-state OPA to cover the desired mid-IR spectral range (accessible between 2 and 12 μm), natural short-pulse duration for time-resolved measurements, the inherent energy stability (a few percent rms) and smooth nearly transform-limited broadband spectral output. Generated IR pulse energies of a few μJ is more than sufficient for subsequent up-conversion with CCD detection or direct IR focal plane array detection schemes, discussed in the next section.

B. Broadband Up-conversion with CCD Detection

One of the first known demonstrations of broadband IR probe pulse generation and detection used a 30 ps Nd^{+3}:YAG laser system with two synchronously pumped, amplified dye lasers and dual solid-state $LiIO_3$ down- and up-converting crystals. Broadband detection was accomplished with a double-grating visible spectrograph and visible Si-CCD detector (15). In this configuration, IR pulses of about 20 ps FWHM duration with bandwidths on the order of $100-200$ cm^{-1} FWHM were generated and detected in the $4.8-5.0$ μm range. These pulses were produced by down-conversion mixing of the amplified, broadband Rhodamine-B or Rhodamine 610 dye lasers against the second harmonic (SHG, 532 nm) output of the amplified Nd^{+3}:YAG laser. Slightly broader bandwidth pulses were actually generated because phase-matched angle tuning of the up-converting crystal produced broader spectral output of the detected up-converted spectrum. It should be noted that this approach is capable of producing two independently tunable pump (UV-VIS or IR) and broadband probe pulses with approximately 20 ps instrumental time resolution.

Transient IR spectra were typically obtained by splitting the broadband IR pulse into two energy-balanced beams ("reference" and "probe"), which were up-converted after the sample by two independently timed and spatially overlapped Nd:YAG SHG pulses in a single $LiIO_3$ crystal (14). These up-converted pulses were dispersed by a f/4.0 double spectrograph (1200 g/mm holographic gratings) and imaged onto the focal plane of a 378×512 element Si:CCD array to yield two parallel spectral tracks (ca. 10 pixels high separated by 1-2 mm). The excitation pulse (e.g., a UV pulse at 289 nm generated via SHG of one of the dye laser pulses) was overlapped with the IR probe beam in a flowing liquid sample cell (ca. 1 mm pathlength, 0.1 mm beam diameter) placed between two matched 10 cm focal length CaF_2 lenses. Alternate on-off chopping of the pump beam was employed to monitor probe pulse spectral changes during the data collection time. This approach greatly reduced long-term drift of an acquired spectral baseline. Software was developed to collect the vertically integrated spectra and average the data to obtain a difference spectrum. By obtaining the ratio of the probe versus reference spectrum on each laser shot, baseline noise for 4000 averaged shots ($\sim$10 minute collection time at 20 Hz) was typically measured to be about 0.25% (1σ standard deviation case A, $k = 1$). The system exhibited about 4 cm^{-1} FWHM spectral resolution (2.8 cm^{-1}/pixel dispersion) with a spectral window of about 200 cm^{-1} in the 5 μm wavelength region. We note that under these collection conditions, one attains an effective "single wavelength" data acquisition repetition rate of 1.4 kHz with this system.

Spectral resolution and transient IR tests of this instrument were conducted using metal-carbonyl species dissolved in n-hexane. These systems were studied in part because they typically exhibit strongly absorbing (transition moment, ~ 1 Debye) and narrow CO-stretching features (ca. 3 cm^{-1} FWHM in n-hexane) in the 1800–2100 cm^{-1} region (24). A variety of experiments using the above picosecond apparatus and a shorter time resolution, subpicosecond (fiber chirped-amplified-compressed) synchronously pumped dual-dye laser system successfully monitored metal-carbonyl vibrational energy transfer (25) and reaction dynamics (26) for several systems in room-temperature solution. This set of experiments revealed vibrationally "hot" photoproducts (22), new transient metal-carbonyl solvated species (26), and helped extract their ultrafast reaction kinetics by this inherently straightforward broadband probing approach.

C. Direct Broadband Detection Using Infrared Focal Plane Arrays

After performing multiple experiments with the CCD up-conversion technique described above, it became apparent that several improvements could be made to the apparatus if the nonlinear up-converting step was eliminated. Most importantly, it was determined that the detectable up-conversion bandwidth was generally strongly dependent on the crystal acceptance angle and input focusing parameters. Because this angle is frequency and crystal length dependent, severe reduction of the detected generator output bandwidth was observed at wavelengths greater than 3–5 μm. For example, when performing electron injection studies near 6 μm using a 5-mm-long AgGaS$_2$ crystal, usable bandwidth windows of only 20–30 cm^{-1} were obtained (27). Additionally, it was felt that improved signal-to-noise difference spectra (over CCD detection) could be attained if direct multichannel infrared detection were possible. These alterations were potentially achievable if an infrared focal plane array detector (IRFPA or dual linear arrays on the same substrate) were used. Otherwise, the approach for generating two probe and reference beams (for single shot normalization) would be the same as for CCD detection except an IR spectrograph and beam delivery optics are employed.

Exploration of IRFPA sources in early 1994 revealed that several Defense Department contract manufacturers produced InSb (1–5.5 μm) and HgCdTe (or MCT for 2–12 μm) 256 × 256 and smaller arrays (<65, 536 total pixels on 30 μm centers) with built-in multiplexers for imaging projects (21). One requirement for pulsed spectroscopic applications is that an array be run in "snapshot" rather than normal imaging strobing readout format. This characteristic is necessary because a constantly strobing read-out device is incapable of capturing a single ultrafast laser pulse spectrum over the width of the array. For a snapshot array, all diode electron capacitive wells accept charge during the same integration period. After pulse integration (typically ~10 μs), the array is subsequently read out (using one or more analog outputs) by a clocking sequence and an image reconstructed from the analog-amplified and digitized chip output. Customized software was developed to vertically integrate the digitized spectral image to extract the two track spectra and to perform normalization, averaging, and automated data collection with user display functions. Many of the snapshot IRFPAs available at that time could support 20–50 Hz frame rates by using dual outputs for the formats described above. We elected to use two 256 × 256, 30 μm pixel pitch InSb (sensitive from 0.5 to 5.5 μm) and HgCdTe (sensitive from 3 to 12 μm) arrays produced by the same manufacturer (Santa Barbara Research Center)* and bonded to the same CD-585 dual output (maximum 1 MHz clock rate) multiplexer readout. These IRFPAs were installed in two interchangeable camera dewars (cooled by LN_2 at 77 K) so that the entire near- to mid-IR (1–12 μm) wavelength range could be covered using the same controller, readout electronics, image capture board, and custom software.

Currently produced two-dimensional IRFPA devices, similar to those described above, use multiple image windowing or have up to four video outputs to produce higher frame rates (reaching several khz) and are now available for related spectroscopic applications. New IRFPA detector arrays are constantly being designed (now with up to 2048 × 2048 HgCdTe pixel formats) for astronomical imaging, which could also be employed for ultrafast IR spectroscopy. However, it should be noted that while these devices would potentially advance ultrafast IR spectroscopic investigations, they are still difficult and extremely expensive to acquire. The challenge is to find readily available sources for these components from manufacturers interested in spectroscopic versus imaging capabilities. It should be mentioned here that to date, there is no known single chip composed of dual linear array, multiplexed IRFPA devices that would be best suited for dual-track, shot-to-shot normalization for spectroscopic users. Single linear arrays have been produced with interleaved InSb

---

* Certain commercial materials are identified in this paper in order to adequately specify the experimental procedure. In no case does such identification imply recommendation or endorsement by the National Institute of Standards and Technology, nor does it imply that the materials or equipment are necessarily the best available for the purpose.

and HgCdTe elements to enable complete mid-IR coverage. Linear 32-element HgCdTe arrays (50 μm wide by 1 mm high pixels with individual external preamplifier, sample-and-hold to parallel ADC circuits capable of very rapid read-out rates) without multiplexers are being used for broadband IR detection in conjunction with stable Ti^{+3}:sapphire kHz laser systems (6,21). However, this approach reduces broadband wavelength coverage and spectral resolution because of the limited array and pixel size. A proposed advanced and potentially more versatile IRFPA design for mid-IR spectroscopy would incorporate two parallel 1024-element linear interleaved InSb/HgCdTe arrays manufactured with multiplexed read-outs on the same substrate. This format would enable dual track normalization for unstable pulse output and provide extremely wide-band coverage for general-purpose use.

Examples of time-resolved studies using the InSb and HgCdTe arrays described above will be presented in the next section. Because this method extracts small difference signals riding on top of large-valued spectral counts, it was typically found that under nearly identical optical configurations for the CCD and IRFPA detector schemes, similar signal-to-noise spectra resulted. Apparently TE or LN_2-cooled CCDs with their inherently low but stable background count rates and high quantum efficiencies are optimal for extracting small visible signals on large spectral backgrounds. The IRFPAs, on the other hand, are still reasonably high quantum efficiency detectors (the QE is ∼0.8 for InSb near 5 μm) but are constantly flooded with black-body background radiation making constant background subtractions difficult and adding error to extracted difference spectra. In practice, the advantage of detecting the entire broadband probe pulse bandwidth by using an appropriately chosen spectrograph and IRFPA with optimized spectral dispersion is preferred to the more complicated optical arrangement and limited detected spectral bandwidth of the visible spectrograph/CCD approach.

III. APPLICATIONS OF BROADBAND INFRARED SPECTROSCOPY

We now discuss and review examples of studies conducted using the broadband probe pulse generation and detection methods described above. The examples presented here were chosen to provide the reader with a synopsis of the techniques and state-of-the-art performance of the tested array technologies. It is anticipated that these examples will generate an appreciation for the methods employed to extract detailed mechanistic and kinetic information about a variety of chemical, biochemical, and physical processes that occur in many research and applications areas.

A. Hydrogen Bond Dynamics in Model Systems—Motivation

Weak intramolecular or intermolecular interactions between proton donors (acids) or proton acceptors (bases) play distinct roles in many biological systems (28,29). These hydrogen bonds, with bond energies in the range of 6–25 kJ/mol (500–2100 cm^{-1}), determine protein folding conformations (30), DNA base pairing (31), and water-biopolymer interaction properties, to name only a few. The studies reviewed here were undertaken to show that ultrafast transient IR spectroscopy is a useful tool to monitor association or dissociation reaction mechanisms and rates of weakly hydrogen-bonded species in room-temperature solutions. Since hydrogen bonding involves pure vibrational motions between the constituent species, time-resolved infrared (TRIR) spectroscopy is naturally sensitive to the time-dependent changes of the vibrations closely involved in hydrogen bonding and to the concentrations of monomeric and complexed species as they are affected by nonequilibrium perturbations. Such studies may be compared to investigations where time-resolved visible fluorescence spectroscopy was used to monitor fluorescent chromophores of hydrogen-bonded complex species (32).

In our work, a room temperature (at equilibrium) hydrogen-bonded system is perturbed by direct infrared excitation of its high-frequency OH- or NH-stretching vibration. Pump pulse frequency tuning can either excite a monomeric OH- or NH-containing acid constituent or the 1:1 acid-base complex itself. Dilute 1:1 acid-base complexes were selected (generated by starting with dilute acid and excess base in a transparent solvent such as CCl$_4$ as a tertiary mixture) to remove the possibility of cooperative effects or unknown conformations between large agglomerates of monomers. Dilute systems were also used to reduce the possibility that high IR pulse energies could sufficiently raise the excitation volume temperature to "temperature jump" the system away from equilibrium rather than directly exciting and examining the 1:1 complex dynamics. Hydrogen-bonded species typically exhibit strong red-shifted absorptions relative to the unassociated monomer OH- or NH-stretching frequency and the magnitude of the red shift is typically proportional to the hydrogen bond strength (33). Infrared $v = 0 \rightarrow 1$ excitation essentially "tags" these species so that they are no longer in their ground vibrational state. Because modes involving proton-stretching motions generally have very large anharmonicities ($2\omega_e X_e =$

$100-200$ cm^{-1}), tagged monomers or complexes exhibit strong, isolated $v = 1 \rightarrow 2$ red-shifted overtone absorptions. Hydrogen-bond dynamics is subsequently revealed by comparing the recovery of $v = 0 \rightarrow 1$ ground state absorption or $v = 1 \rightarrow 2$ excited stated absorption decay versus the vibrational relaxation times (T_1) for the species measured independently (34).

Another motivation for these studies was to investigate whether single IR photon vibrational excitation of complexes in solution is sufficient to break hydrogen bonds directly. This process might be expected because hydrogen bond formation and dissociation constantly occurs in thermal, near room temperature environments. In addition, earlier gas phase studies demonstrated that if sufficient energy is deposited into an isolated, noncolliding complex (with $E >$ dissociation energy of $12-25$ kJ/mol or $1000-2100$ cm^{-1}), the H-bonded species can dissociate after internal energy redistribution occurs (typically on the nanosecond timescale) (35). With typical OH-stretch or NH-stretch excitation energies of about 3300 cm^{-1} (>30 kJ/mol) or higher, the 1:1 complex species in solution were expected to dissociate. Direct evidence for pump-induced dissociation would be the appearance of increased monomer (acid or base) ground state absorption after complex excitation. Unfortunately, *none* of the systems investigated to date and discussed below (e.g., 1:1 species of pyrrole, methanol, ethanol, or phenol complexed with various bases such as pyridine, acetonitrile, etc.) exhibited IR-induced dissociation for observation delay times of up to one nanosecond. This result is in stark contrast to multiple studies of highly associated or dimeric alcohol (36) and water (37,38) systems that do show dissociation on the picosecond timescale.

Perhaps these negative findings can be attributed to rapid and competing internal vibrational relaxation (IVR) or solute-to-solvent vibrational energy transfer (VET) processes that allow the energy to drain away from the hydrogen-bond dissociation coordinate before dissociation can occur. Another possibility is that dissociation does take place on much longer time scales (e.g., >1 ns) for these model species and with low quantum yield. This scenario would make it difficult to detect newly formed monomeric species by our approach. One must also consider that excess vibrational energy (above thermal excitation) in the hydrogen bond–stretching coordinate is required for the hydrogen bond to break. For example, eight or more quanta of OH-stretch energy must directly transfer to the AOH-B 100 cm^{-1} hydrogen bond–stretching mode (the H bond will have an average thermal excitation of 208 cm^{-1} at room temperature) if the dissociation energy is at least 1000 cm^{-1}. Such a multiple quantum transfer event is of relatively low probability and is expected to take orders of magnitude longer than the picosecond time scale to occur. In any case, this paradox requires further study to reveal the source or inconsistencies of our results compared to those of others.

B.   Dynamics of Hydrogen-Bonded $(Et)_3SiOH$ and Pyrrole Complexes

Initial picosecond transient IR studies of hydrogen-bonded 1:1 acid-base complexes were conducted on triethyl silanol and pyrrole because of their inherently long OH- and NH-stretch T_1 lifetimes (183 and 42 ps, respectively) in room temperature CCl_4 solution (34). In these cases, 1:1 complexes with a variety of proton acceptors (e.g., acetonitrile, acetone pyridine, tetrahydrofuran, diethyl-ether) were studied by exciting at the peak of the monomeric acid or complex NH- or OH-stretch bands and monitoring the anharmonically red-shifted $v = 1 \rightarrow 2$ excited state absorption as a function of time (see Fig. 2). For these investigations, the probe pulses used were narrow-band and the vibrational energy decay was monitored with single element reference and probe InSb detectors.

The important findings are that when the "free" acid is complexed to various hydrogen-bonding proton acceptors, the complex vibrational T_1 lifetime decreases monotonically and is inversely proportional to the basicity (Kamlet-Taft) of the corresponding base (34,39). Measured T_1 lifetimes are also found to be proportional to the inherent red shift (hydrogen bond energy increase) of the "free" monomer absorption upon complex formation. Because this trend was observed for bases with inherently different internal vibrational mode structure, it is suggested that localized intramolecular vibrational energy transfer from the high-frequency stretching mode to adjacent modes is the dominant relaxation pathway. Increased hydrogen-bonding strength between these constituents apparently enhances the coupling between excited high-frequency complex modes and the hydrogen-bond–stretching acceptor modes.

Measurements of the "free" acid T_1 vibrational lifetimes were also monitored as a function of base concentration (e.g., pyrrole with acetonitrile) to determine the effect of collisions and hydrogen-bond formation rates. Stern-Volmer plots of $1/T_1$ rates versus base concentration enabled extraction of a bimolecular rate constant (k_{bm}) for pyrrole:acetonitrile of $2.5 \pm 0.2 \times 10^{10} \text{ dm}^3/\text{mol-s}$, which is slightly larger than the estimated Stokes-Einstein diffusion coefficient ($0.73 \times$

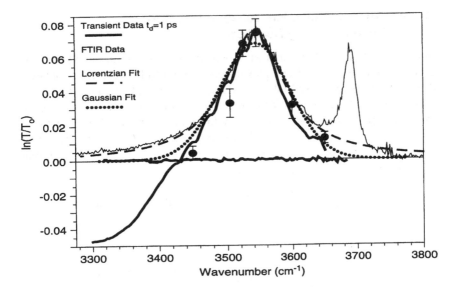

Figure 2 Transient IR spectrum (solid line) after IR excitation of the OH stretch (3550 cm^{-1}) of dilute 1:1 hydrogen-bonded triethylsilanol-acetonitrile in CCl$_4$ at 295 K. The positive feature is the complex ground state bleach while the negative feature is the OH stretch (v = 1 → 2) excited state absorption. These spectra are averages of three 4000 shot data acquisitions. The filled circles result from using discretely tuned narrowband probe pulses (10 cm^{-1} FWHM). The thin line is the ground state FTIR absorption spectrum for the sample while the dashed and dotted lines are Lorentzian and Gaussian fits to this absorption. A baseline scan yielding no transients (time delay = −10 ps) is also shown.

10^{10} dm^3/mol-s) for this system (39). An increased rate constant above the diffusion limited rate may be attributed to an enhanced interaction or affinity between the acid-base constituents within the solvent cage. Similar studies were conducted for triethylsilanol, (CH$_3$CH$_2$)$_3$SiOH, complexed with acetonitrile, pyridine, and tetrahydrofuran (34). In these cases, k$_{bm}$ = 1.2 ± 0.2 × 10^{10} dm^3/mol-s was found to be within experimental error for all three bases.* While this value is approximately half of that found for pyrrole, steric factors such as ethyl group hydroxyl shielding versus the planar geometry of pyrrole are believed to be responsible for this effect.

---

* All reported uncertainties are type B with k = 1 (i.e., ±1σ) or 10%, whichever is greater.

C. Vibrational Population Conservation During Hydrogen-Bonding Reactions

In the course of the above studies, it was argued that the measured transient absorption decays for 1:1 complex species do not truly represent population deactivation of the "free" OH- or NH-stretch. Instead, the high-frequency absorption could merely shift to lower frequencies. The prospect of experimentally detecting ground state or vibrationally excited $(Et)_3SiOH$:base complexes by probing at the red-shifted $v = 1 - 2$; OH stretch was then considered (40). The following kinetic scheme was assumed for this type of dynamical hydrogen-bonded system:

$$[(Et)_3SiOH]^* + [base] \xrightarrow[k_{BM}]{k_{diss}} [(Et)_3SiOH^*\text{---}base]$$

$$\Big\uparrow k_1 \qquad\qquad\qquad\qquad\qquad\qquad \Big\downarrow k_2$$

$$[(Et)_3SiOH] + [base] \xrightarrow[k_{BM}]{k_{diss}} [(Et)_3SiOH\text{---}base]$$

where the asterisk designates OH-stretch $v = 1$ excited state "free" or complexed species and $k_1 = 5.5 \times 10^{-3}$ ps^{-1} is the experimentally measured "free" OH-stretch vibrational relaxation rate with no base present in solution. The microscopic acid-base dissociation rate constant $(k_{diss} = k_{bm}/K_{eq})$ can be deduced from concentration-dependent FTIR measurements (with changing acetonitrile concentration) to extract the equilibrium constant $K_{eq} = [(Et)_3SiOH - base]/([(Et)_3)SiOH] + [base]) = 0.96$, which is assumed to be the same for ground and vibrationally excited species. The k_2 complex relaxation rate constant was deduced through a separate experiment by measuring the $v = 1 \rightarrow 2$ transient absorption decay time of the vibrationally excited complex species.

Modeling of the time-dependent populations for the $(Et)_3SiOH$:acetonitrile system for the above kinetic scheme was undertaken using first-order rate equations (40). Predictions for the transient concentrations (and concomitant absorption or bleaching signals) of vibrationally excited species were made for direct excitation of either the "free" or complexed species. Excitation of "free" OH stretch in $(Et)_3SiOH$ at 3692 cm^{-1} produced a slowly rising transient absorption at 3383 cm^{-1} arising from excited complex species formed as $(Et)_3SiOH$ hydrogen bonds to acetonitrile. This result, coupled to other measured signals for slowly recovering complex ground state bleaching and $v = 1$ excited state relaxation, exactly matched the modeled signal amplitudes and time dependence.

It was deduced that for this particular chosen system, excited state vibrational population is "conserved" throughout all of the possible picosecond time scale association or dissociation events described by the above simple kinetic scheme. The data closely fits the predicted kinetics and amplitudes, and no direct IR absorption dissociation pathways were included. Again, one concludes that the rate constant for direct photo-dissociation of an excited complex must be several orders of magnitude smaller than the largest rate constant participating in this scheme (i.e., k_1). This finding is in agreement with our inability to observe any newly produced "free" ground state silanol species after direct $OH(v = 1)$ excitation of the complex (implying the direct dissociation is slower than the first-order equilibrium association and dissociation rates at the studied concentrations).

D. IR Spectral Hole-Burning of 1:1 Hydrogen-Bonded Complexes

Broadband picosecond IR probe investigations of the systems described in the previous sections were conducted to monitor the spectral region between 3000 and 3800 cm^{-1} using an InSb focal plane camera (41). A primary motivation for this study was interest in comparing the dynamical changes in the spectrum taken with a broadband detection system and earlier use of discretely tunable probe pulses. The results of this comparison (see Fig. 2) gave strong evidence that the previously measured kinetics of hydrogen-bonded systems was not affected by overlapping or interfering spectral features. In addition, the ability to obtain an entire broadband IR transient spectrum enables one to monitor unusual properties of the transiently excited species and to look for potential sample inhomogeneities via unusually shaped or narrow transient absorption features.

Towards this end, several 1:1 hydrogen-bonded species were investigated including mixtures of weak acids (methanol, pyrrole, and triethylsilanol at <0.1 mol/dm^3) with various bases (in increasing strength: acetonitrile, tetrahydrofuran, and pyridine at about 2 mol/dm^3) in room-temperature CCl_4 solution. Measurements for triethylsilanol:acetonitrile and pyrrole:tetrahydrofuran mixtures yielded transient complex bleach bandwidths (200–300 cm^{-1} FWHM) that very closely matched their broad ground state absorption spectrum. Interestingly, the FTIR spectra of these two particular examples are very well fit to a Lorentzian bandshape function, indicating the system is homogeneous on the time scale of the measurement (e.g., 1 ps time delay between pump and probe).

Investigation of methanol-pyridine complexes, on the other hand, produced a relatively narrow (75 cm^{-1} FWHM) bleached "hole" that is burned into the 260 cm^{-1} FWHM OH-stretch ($v = 0 \to 1$) absorption band. The methanol:pyridine complex OH-stretch absorption band was better fit by a Gaussian function than with a Lorentzian bandshape, indicating this system is inhomogeneously broadened on the >1 ps timescale.

It is interesting to surmise from the above results that the most strongly hydrogen-bonded methanol:pyridine complex could have multiple conformations at room temperature while more "loosely bound" silanol and pyrrole species rotate freely around their hydrogen bond. While this picture is conjecture at this time, it would be extremely informative to perform structural calculations and molecular dynamics simulations of these and related systems to confirm or refute the results of these hole-burning experiments. Similar pictures may evolve for oligomeric alcohols or extended hydrogen-bonded systems with inhomogeneous spectral transients such as water and ice (38).

E. Vibrational Coherent Control with Chirped Picosecond Infrared Excitation

Investigations of "vibrational coherent control" of molecular vibrational populations are now attainable with specifically tailored high power, chirped infrared ultrafast pulses (42). Controlling the phase and amplitude of an infrared excitation pulse that is properly tuned to adiabatically sweep population up a molecular vibrational manifold has been theoretically shown to produce population in specific overtone states (43). By choosing pulse properties for coherently exciting vibrational population above a bond dissociation energy, for example, it may be possible to "control" the dissociation process and hence improve chemical reaction yields over conventional thermally activated processes. Ultrafast (picosecond or shorter pulsewidths) chirped excitation naturally spans multiple anharmonic shifts (allowing access to higher-energy vibrational states) as well as circumventing intrinsic population relaxation mechanisms. Broadband infrared probing is also ideal for monitoring population distributions as a function of time after coherent absorption of the excitation pump pulse.

We elected to study coherent up-pumping dynamics in solution-phase metal-hexacarbonyl systems because of their strong vibrational infrared absorption cross sections, relatively simple ground-state spectra, and small (ca. 15 cm^{-1}) anharmonic overtone shifts. It was felt that these systems are ideal candidates to demonstrate that population control could be achieved for polyatomic species in solution because the excited state population lifetimes (T_1) were already known to be long-lived (14,22,24,25,44) and the coherence lifetimes for lower-lying states are also relatively long ($T_2 >$ 4 ps). The first systematic investigation was conducted on the triply degenerate CO-stretch T_{1u} mode (v $= 0 \rightarrow 1$ at 1986 cm^{-1}) of W(CO)$_6$ in n-hexane solution at room temperature (45). It was known from previous work that the CO-stretch (v $= 1$) state has a T_1 lifetime of approximately 140 ps and that the system in n-hexane exhibits narrow (ca. 3 cm^{-1} FWHM) absorption features (14). Initial studies that used nonchirped, difference-frequency–generated (near transform-limited) 2 ps, <10 μJ pump pulses tuned across the v $= 0 \rightarrow 1$ absorption feature revealed that population distributions spanning up to the v $= 3$ overtone state could be altered by choice of center frequency. Generation of only CO(v $= 1$) population was achieved when the pump pulse was tuned and overlapped with the high-frequency side of the 1986 cm^{-1} absorption feature. However, up to v $= 3$ excitation with nearly equal v $= 1$ and v $= 0$ populations (saturation) was achieved by tuning the pump pulse center frequency to the low-frequency side of the same feature. Concomitant excited overtone state T_1 relaxation times were also deduced from the transient spectra, producing a monotonic lifetime reduction as one ascends the manifold (45).

More advanced tests of vibrational "coherent control" theory were conducted by using deliberately chirped picosecond excitation pulses (42). By deliberately mixing positively or negatively chirped 2 ps visible pulses (near 589 nm) with much longer, narrowband dye laser pulses (8 ps FWHM at 650 nm) in LiIO$_3$ crystals, one can generate chirped IR pulses that experience negligible group velocity delay distortion. The sign of the chirp was produced by either passing a subpicosecond dye pulse through a short length of optical fiber (red-to-blue or positive chirp) or by deliberately stretching a two-pass grating compressor to invert the phase (blue-to-red or negative chirp). Actual pulse chirp rates (~10 cm^{-1}/ps) and signs were deduced from time versus directly obtained IR spectral datasets (via a spectrograph and InSb IR focal plane detector) (46) that were analyzed by the FROG iterative fitting algorithm.

Results for up-pumping W(CO)$_6$ with chirped pulse excitation were compared to excitation of the T_{1u} manifold using transform-limited pulses with center frequencies all tuned to the peak vibrational mode absorption frequency. Figure 3 shows transient broadband absorption spectra taken at 40 ps time delay for the three different pulse types. As depicted, one readily observes that the relative population amplitudes in the CO-stretch v $= 1$ (at 1970 cm^{-1}) and v $= 2$ (at 1955 cm^{-1}) levels are strongly affected by the chirp of the excitation pulse. Excitation with negatively chirped

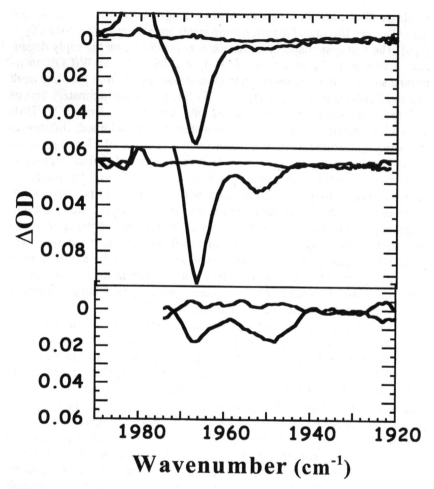

Figure 3 Examples of transient infrared spectra obtained at 40 ps time delay for $W(CO)_6$ in n-hexane using (top) positively chirped, (middle) no chirp, and (bottom) negatively chirped IR excitation pulses centered at 1983 cm^{-1}. Note the suppression and increase in $v = 1 \rightarrow 2$ excited state absorption near 1950 cm^{-1}.

pulses and 10 times lower pump pulse energy was found to produce higher-level populations than using non-chirped excitation. This general trend was confirmed by multilevel coherent up-pumping modeling using excitation pulse properties measured independently by the frequency resolved optical gating (FROG) technique (46).

The theoretically predicted chirped-pulse excitation effect on the up-pumping dynamics of $W(CO)_6$ and related theories for diatomic rotational population distributions in the gas phase (47) suggest that overtone population distributions can be "controlled" with carefully generated chirped IR pulses. It remains to be seen if deliberately formed broadband IR femtosecond pulses with inherently larger bandwidths and varying chirp rates can produce pure populations in specific vibrational overtones that lead to chemically interesting bond-breaking phenomena and ground state reactions (48).

F. Ultraviolet Photochemistry: Self-Association Reactions of $Mn(CO)_3CpR$ Species and $[CpFe(CO)_2]_2$ in Solution

High quantum yield photochemical reactions of condensed-phase species may become useful for future optical applications such as molecular switches, optical limiters, and read-write data storage media. Toward these ends, much research has been conducted on novel nonlinear chemical-based materials such as conducting polymers and metal-organic species. Monitoring the early time-dependent processes of these photochemical reactions is key to understanding the fundamental mechanisms and rates that control the outcome of these reactions, and this could lead to improved speed and efficiencies of devices.

To investigate prototypical reaction processes, studies of $Mn(CO)_3CpR$ systems [Cp = cyclopentadienyl or C_5H_5; R = $-COCH_3$ (**I**), $-COCH_2SCH_3$ (**II**), and $-CO(CH_2)_3SCH_3$ (**III**)] in room temperature n-hexane solution were conducted in collaboration with Prof. Theodore J. Burkey and his group at the University of Memphis (49). Self-closing ring reactions of these species are initiated by near-UV excitation (260–300 nm) of the metal-to-ligand charge transfer transition, which leads to ejection of a single CO ligand and unpaired radical metal center in solution. They found that the quantum yields for **I** reacting with the sulfur atom of tetrahydrothiophene (THT) and self-ring closure between the Mn radical center and the Cp ligand sulfur atoms of **II** and **III** were $\phi = 0.82$, 1.0, and 0.82, respectively. Microsecond photoacoustic calorimetry measurements on these reactions were unable to deduce mechanistic reasons for the differences in the observed quantum yields or whether transient species or structural effects at early times were controlling these reactions. Thus, it was decided that ultrafast transient infrared methods could be employed to directly monitor the early-time dynamics and mechanics of these intriguing molecular systems to try to extract reasons for the observed quantum yield differences.

Measurements of **I** with THT were conducted to test the capability of broadband IR for monitoring the disappearance of the parent species, identify any short-lived intermediates, and determine the appearance time of the final product (49). Since all parent and product species are stable, long-lived compounds, static FTIR spectra could first be used to identify and compare their time evolution from the transient IR results. In this case, bleaching bands of **I** (three IR-allowed fundamental CO stretches) were observed at 2060, 1961, and 1952 cm^{-1}, and their amplitude did not change significantly during the entire observation time window. Two bands appear at 1907 and 1969 cm^{-1} within the first 200 ps after excitation. These bands are attributed to the asymmetrical and symmetrical CO-stretch modes, respectively, of an n-hexane solvated and vibrationally relaxed transient intermediate. These transient absorption bands decay at the same rate ($2.3 \pm 0.6 \times 10^{-6}$ s^{-1}) as the appearance of the stable I-THT product bands at 1946 and 1885 cm^{-1}. Thus, it is possible to distinguish transiently solvated dicarbonyls from the stable (Acyl-Cp)(CO)$_2$Mn-THT product and that the fundamental reaction rate is much longer than the estimated average diffusion-limited bimolecular encounter rate of a few hundred picoseconds. In this case the barrier to reaction may be high enough and thus makes the transient species lifetime extremely long (ca. 435 ns). Under these circumstances, many thousands of collisions are required on average before reaction occurs (implying $\phi = 1.0$) or recombination with liberated CO molecules also competes with the THT reaction (producing the observed less than unity quantum yields).

Related measurements of internal ring-closure reactions of **II** and **III** were performed, but different rates and transients were identified (see Fig. 4). The reaction of **II** showed that within 200 ps of UV excitation, nearly equal populations of n-hexane solvent species and internal six-membered ring-closed species were created. The solvated species were subsequently found to self-react with a time constant of approximately 35 ns, indicating that this species has a much lower reaction barrier than **I** and that ring-closure reaction dominates over potential geminate recombination with liberated CO species. This could explain the reason for the near unity quantum yield for this species. However, reaction of **III** produced similar spectral features and transients as for the acyl compound (**I**) reacting with THT described above. Within 200 ps of excitation, an estimated ratio of solvated to eight-membered ring-closed species was found to be 3:1 (i.e., the propensity at early times is to form solvated species over internal ring-closure), and conversion of the solvated species to final product takes approximately 263 ns.

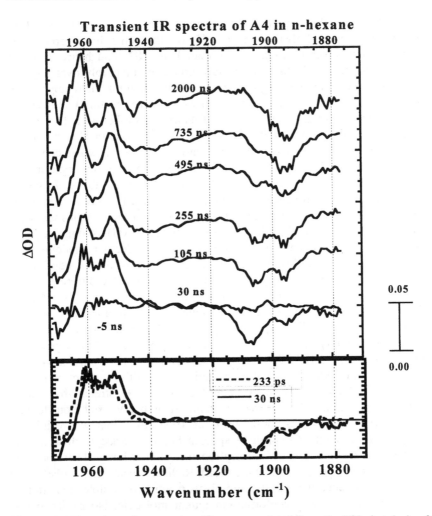

Figure 4 Time-dependent transient IR spectra arising from the UV photolysis of the self-ring-closing system $Mn(CO)_2Cp\text{-}CO(CH_2)SH_3$ (III) in *n*-hexane.

Considering all of the above observations and measured rate constants, a consistent picture of the reaction mechanisms emerges. Species **I** and **III** must have high enough reaction barriers (and hence several hundred nanosecond transiently solvated lifetimes) that reaction with an electron-rich sulfur center competes with CO recombination. Perhaps this scenario permits about 20% of the radical species enough time to recombine with CO while the remaining activated species either react with THT or self-ring close. Solvated species **II**, on the other hand, appears to have a much lower reaction barrier than **I** and **III** (and hence 10 times smaller lifetime) such that ring closure competes favorably over any CO recombination, and the ring-closure reaction quantum yield approaches unity for **II**. Calculations of lowest energy conformations for **II** and **III** also suggest that **II** spends about one-half of its time with the Mn radical center close to the sulfur atom, while **III** is approximately one-third in a reactive (close proximity) configuration for these atoms. This result is also consistent with the observed transiently solvated to ring-closed ratios measured at the earliest observation time delay (200 ps) discussed above. It should be added that the difference in reaction barrier heights may be controlled by steric factors (i.e., the propensity to form six- versus eight-membered ring compounds) and that the reaction enthalpy (ΔG) is dominated by entropic rather than enthalpic factors. Further molecular modeling and studies of these self-closing reaction rates as a function of equilibrium system temperature may help uncover the source of these differences.

Very similar studies were conducted in collaboration with Dr. Michael George of The University of Nottingham on the ultrafast UV reaction dynamics of the iron-dimer species $[CpFe(CO)_2]_2$ in n-hexane solution (50). Ultraviolet photolysis of this compound produces the triply-bridged intermediate $CpFe(\mu\text{-}CO)_3FeCp$ (with new absorption at 1824 cm^{-1}), which is formed within 10 ps. This feature exhibits bandwidth reduction attributed to vibrational cooling of modes coupled to the CO stretch that occurs with an approximate 60 ps time constant. At high UV excitation fluence, we observed a new absorption band at 1908 cm^{-1} that was originally assigned to a singly CO-bridged species but may arise from a species formed via multiphoton UV absorption to a higher-lying electronic state (M. W. George, private communication). Radical species [i.e., $CpFe(CO)_2$] were also produced through the homolysis reaction by visible excitation of both the *cis* and *trans* conformation parent molecule. No evidence was found for subsequent CO recombination or impurity reactions for delay times up to 560 ps. The most interesting result of this study is that the triply-bridged species is formed in about 10 ps, suggesting that the triplet state is produced and rearrangement occurs on this rapid time scale.

G. Primary Electron Transfer Dynamics of Dye-Sensitized Semiconductor Solar Cell Devices

As a final example, we discuss the use of broadband ultrafast infrared spectroscopy as a tool for monitoring electron transfer rates in dye-sensitized solar cell applications. Systems composed of Ru-bipyridine derivatives chemisorbed onto nanoparticle TiO_2 thin films exhibit highly efficient electron transfer to the substrate after photoexcitation of the adsorbed dye (27,51). When dye-impregnated films are sandwiched between transparent electrodes (e.g., tin oxide) and a redox couple electrolyte (typically I_2/I^- in propylene carbonate), these devices have been shown to produce currents with solar efficiencies up to 10% and up to 80% absorbed photon to current ratio under monochromatic irradiation (52). Because of the simplicity, reduced cost, and potentially high efficiency of these solar cells compared to conventional silicon-based cells, much recent effort has been expended to optimize and understand the fundamental electron transfer mechanisms responsible for improving these devices.

In early studies, transient ultraviolet and visible spectroscopies were employed to monitor the electron transfer rate from the adsorbed dye to the underlying substrate. Time-dependent emission or absorption measurements of the sensitizer (for dyes in solution or on ZnO_2 and TiO_2) and near-infrared absorption signals from injected electrons were measured (52). Electron injection times ranging from picoseconds to several nanoseconds were obtained, so it was felt that some of these kinetic rates could be affected by dye excited state interference or other intervening mechanistic processes. To eliminate these possibilities, investigations were initiated to determine whether transient broadband infrared spectroscopy would be sensitive to electrons directly injected into the nanoparticle semiconductor substrates and if vibrational modes of coordinated dye ligands could be used to monitor electrons transferring to the substrate.

We first employed picosecond time-resolved IR spectroscopy in the 6 μm spectral region to study the vibrational and electron dynamics of $[Ru(4,4'-(COOCH_2CH_3)_2-2,2'-bipyridine)(2,2'-bipyridine)_2]^{+2}$ and $[Ru(4,4'-(COOCH_2CH_3)_2-2,2'-bipyridine)(4,4'-(CH_3)_2-2,2'-bipyridine)_2]^{+2}$ in room-temperature dichloromethane (DCM) solution and anchored to nanostructured thin films of TiO_2 and ZrO_2 (51). Visible excitation of the dyes reveals a red shift of the CO-stretching mode ($1731\ cm^{-1}$) of the ester groups for the free molecules in solution (see Fig. 5) and similar spectral changes when attached to insulating ZrO_2 substrates. However, for these molecules attached to TiO_2 semiconductor films, an extremely broad transient absorption throughout the mid-infrared range and without any identifiable spectral features is observed. Our group and others (53,54) now attribute this broad signature to excited state absorption of electrons directly injected into the TiO_2 semiconductor substrate. Initial attempts to time resolve the appearance of injected

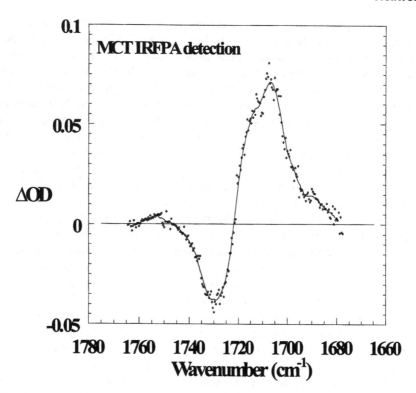

Figure 5 Transient IR difference spectrum of [Ru(dceb)(bpy)$_2$]$^{+2}$ in room temperature dichloromethane 35 ps after 532 nm excitation showing the parent CO stretch bleach and red-shifted CO stretch excited state absorption. The data points were obtained by direct broadband detection with a HgCdTe (MCT) 256 × 256 focal plane array system.

electrons were unsuccessful because the risetime of this absorption signal followed the instrumental cross-correlation between the visible and infrared pulses, but an upper limit for the injection rate of approximately 5×10^{10} s^{-1} was deduced from this study. Subsequent investigations using the related nonionic dyes Ru(4,4'-(COOH)$_2$-2,2'-bipyridine)$_2$(NCS)$_2$ (**IV**) and Ru(5,5'-(COOH)$_2$-2,2'-bipyridine)$_2$(NCS)$_2$ (**V**) covalently bound to TiO$_2$ were undertaken using higher time resolution (350 fs FWHM VIS-IR cross-correlation). Again, an instrumentally determined response was measured for both species (see Fig. 6), indicating the injection time is <350 fs (or 3×10^{12} s^{-1}) and that perhaps injection rates for other dyes could exceed this value (27).

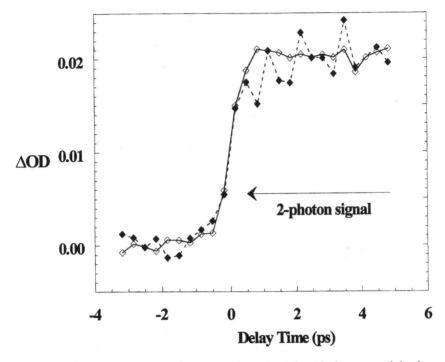

Figure 6 Transient infrared kinetics risetime for injected electrons originating from visible excitation (590 nm) $Ru(4,4'\text{-}(COOH)_2\text{-}2,2'\text{-bipyridine})_2(NCS)_2$ absorbed on nanostructured TiO_2 thin films. The filled points arise from the infrared transient absorption at 5.4 μm, while the open points represent the single-sided VIS-IR cross correlation. An upper limit for the injection risetime is 350 fs. The horizontal arrow indicates the level of infrared signal induced by two-photon excitation of the substrate.

Ultrafast injection of carriers into the TiO_2 substrate suggests that overall cell efficiency is not limited by this process but by intervening transfer mechanisms (e.g., trapped state populations reducing quantum yield) or longer time scale electron-dye recombination rates (typically taking microseconds). For example, it was found that the absorbed photon to current efficiency (APCE) is considerably reduced for **V** compared to **IV** under identical cell conditions (27,55). While **V** has a 350 mV lower electrochemical reduction potential than **IV** (and hence a red-shifted absorption spectrum), it is unclear why the redder absorbing dye (**V**) does not inject as efficiently (56). Recent visible excitation with broadband

UV-VIS nanosecond probe experiments for the above dyes on TiO_2 and ZrO_2 suggest that injection occurs only on TiO_2 but the yield for **V** is lower than that for **IV** (exhibiting larger parent bleach and transient ligand or injected electron features) (55). This corroborates the previously mentioned transient IR measurement that also gave approximately one-half IR signal amplitude for **V** versus **IV** (27). Recent transient UV-VIS and APCE measurements for these same dyes anchored to nanostructured SnO_2 films (having 400 mV lower redox potential than TiO_2) indicate that **IV** and **V** inject with nearly equal and enhanced total efficiencies (about 45%) across the entire dye absorption band (T. A. Heimer, unpublished). Perhaps both dyes are better matched to the SnO_2 bandgap and inject with higher quantum yield under these conditions. Continued broadband transient IR and UV-VIS studies for these and related dye systems are clearly warranted to better understand the electron transfer mechanisms and rates for these important solar energy converters.

IV. CONCLUSIONS AND FUTURE DIRECTIONS

A state-of-the-art description of broadband ultrafast infrared pulse generation and multichannel CCD and IR focal plane detection methods has been given in this chapter. A few poignant examples of how these techniques can be used to extract molecular vibrational energy transfer rates, photochemical reaction and electron transfer mechanisms, and to control vibrational excitation in complex systems were also described. The author hopes that more advanced measurements of chemical, material, and biochemical systems will be made with higher time and spectral resolution using multichannel infrared detectors as they become available to the scientific research community.

Extensions of these concepts to the study of low-frequency hydrogen-bond stretches, cooperative librational motions of biopolymers in the condensed-phase, and broadband surface sum-frequency spectroscopy (57) may now be explored. Novel broadband far-infrared generation and detection techniques (58) are being used to obtain high-resolution spectra in the 0.5–10 THz frequency range of DNA and protein films and pellets (59). It remains to be seen if intermolecular hydrogen-bond dynamics and functionally important biomolecular structural rearrangements can be identified using time-resolved THz spectroscopy. However, detailed spectroscopic investigations in this region of the spectrum may become an important venue for future broadband transient spectroscopy of complex biological systems.

ACKNOWLEDGMENTS

The author is indebted to the hard work and extreme experimental efforts of my many colleagues without whom this research would never have been possible. These include NIST/NRC postdoctoral research associates Drs. Steven Arrivo, Tom Dougherty (who passed away in 1997), Tandy Grubbs, Todd Heimer, and Andrea Markelz, guest researcher Dr. Valeria Kleiman, and collaborators Prof. Ted Burkey, Dr. Joe Melinger, and Dr. Michael George. I am also indebted to my esteemed colleague Dr. John Stephenson, Group Leader of the NIST Laser Applications Group, for his continual support of this research and invaluable scientific discussions. Research funding for most of this work was provided by internal NIST STRS support and the NIST Advanced Technology Program.

REFERENCES

1. Laubereau A, Kaiser W. Vibrational dynamics of liquids and solids investigated by picosecond light pulses. Rev Mod Phys 1978; 50(3):607–666.
2. Moore CA, Goldberg LS. Tunable UV and IR picosecond pulse generation by nonlinear mixing using a synchronously mode-locked dye laser. Optics Comm 1976; 16:21–25.
3. Laubereau A, Fischer SF, Spanner K, Kaiser W. Vibrational population lifetimes of polyatomic molecules in liquids. Chem Phys 1978; 31:335–344.
4. Diels J-C, Rudolph W. Ultrashort Laser Pulse Phenomena. San Diego: Academic Press, 1996.
5. Elsaesser T, Fujimoto JG, Wiersma DA, Zinth W, eds. Ultrafast Phenomena XI. Berlin: Springer-Verlag, 1998.
6. Hamm P, Lim M, Hochstrasser RM. Structure of the amide I band of peptides measured by femtosecond nonlinear-infrared spectroscopy. J Phys Chem B 1998; 102:6123–6138.
7. Heilweil EJ, Casassa MP, Cavanagh RR, Stephenson JC. Vibrational energy relaxation of surface hydroxyl groups on colloidal silica. J Chem Phys 1984; 81:2856–2859.
8. Heilweil EJ, Casassa MP, Cavanagh RR, Stephenson JC. Picosecond studies of vibrational energy transfer in surface adsorbates. Ann Rev Phys Chem 1989; 40:143–171.
9. Laenen R, Rauscher C, Laubereau A. Dynamics of local substructures in water observed by ultrafast infrared hole burning. Phys Rev Lett 1998; 80:2622–2625.
10. Nienhuys H-K, Woutersen S, van Santen RA, Bakker HJ. Mechanism for vibrational relaxation in water investigated by femtosecond infrared spectroscopy. J Chem Phys 1999; 111:1494–1500.

11. Heilweil E. Vibrational population lifetimes of OH(v = 1) in natural crystalline micas. Chem Phys Lett 1986; 129:48–54.

12. Heilweil EJ, Casassa MP, Cavanagh RR, Stephenson JC. Vibrational deactivation of surface OH chemisorbed on SiO_2: solvent effects. J Chem Phys 1985; 82:5216–5231.

13. Grubbs WT, Dougherty TP, Heilweil EJ. Vibrational energy redistribution in $Cp^*Ir(CO)_2$ ($Cp^* = \eta^5$-pentamethylcyclopentadienyl) studied by broadband transient infrared spectroscopy. Chem Phys Lett 1994; 227:480–484.

14. Dougherty TP, Heilweil EJ. Ultrafast transient infrared absorption studies of $M(CO)_6$ (M = Cr, Mo, W) photoproducts in n-hexane solution. Chem Phys Lett 1994; 227:19–25.

15. Heilweil EJ. Ultrashort-pulse multichannel infrared spectroscopy using broadband frequency conversion in $LiIO_3$. Optics Lett 1989; 14:551–553.

16. Dougherty TP, Heilweil EJ. Dual beam subpicosecond broadband infrared spectrometer. Optics Lett 1994; 19:129–131.

17. Arrivo SM, Kleiman VD, Dougherty TP, Heilweil EJ. Broadband femtosecond transient infrared spectroscopy using a 256×256 element indium antimonide focal-plane detector. Optics Lett 1997; 22(19):1488–1490.

18. Yang H, Snee PT, Kotz KT, Payne CK, Frei H, Harris CB. Femtosecond infrared studies of a prototypical one-electron oxidative-addition reaction: chlorine atom abstraction by the $Re(CO)_5$ radical. J Am Chem Soc 1999; 121(39):9227–9228.

19. Hamm P, Wiemann S, Zurek M, Zinth W. A dual-beam sub-picosecond spectrometer for the 5–11 μm spectral region and its application to vibrational relaxation in the S_1 state. In: Barbara PF, Knox WH, Mourou GA, Zewail AH, eds. Ultrafast Phenomena IX. Berlin: Springer-Verlag, 1994:152–153.

20. Beckerle JD, Cavanagh RR, Casassa MP, Heilweil EJ, Stephenson JC. Subpicosecond transient infrared spectroscopy of adsorbates: vibrational dynamics of CO/Pt(111). J Chem Phys 1991; 95:5403–5418.

21. Kidder LH, Levin IW, Lewis EN, Kleiman VD, Heilweil EJ. Mercury cadmium telluride focal-plane array detection for mid-infrared Fourier-transform spectroscopic imaging. Optics Lett 1997; 22:742–744.

22. Dougherty TP, Grubbs WT, Heilweil EJ. Photochemistry of $Rh(CO)_2$ (acetylacetonate) and related metal dicarbonyls studied by ultrafast infrared spectroscopy. J Phys Chem 1994; 98:9396–9399.

23. Chudoba C, Nibbering ETJ, Elsaesser T. Dynamics of site-specific excited-state solute-solvent interactions as probed by femtosecond vibrational spectroscopy. In: Elsaesser T, Fujimoto JG, Wiersma DA, Zinth W, eds. Ultrafast Phenomena XI. Berlin: Springer-Verlag, 1998:535–537.

24. Dougherty TP, Heilweil EJ. Dual track picosecond infrared spectroscopy of metal-carbonyl photochemistry. In: Lau A, Siebert F, Werncke W, eds. Proceedings in Physics. Berlin: Springer-Verlag, 1994:136–140.

25. Grubbs WT, Dougherty TP, Heilweil EJ. Vibrational energy redistribution in $Cp^*Ir(CO)_2$ studied by broadband transient infrared spectroscopy. Chem Phys Lett 1994; 227:480–484.

26. Dougherty TP, Heilweil EJ. Transient infrared spectroscopy of (η^5-C_5H_5) $Co(CO)_2$ photoproduct reactions in hydrocarbon solutions. J Chem Phys 1994; 100:4006–4009.

27. Heimer TA, Heilweil EJ. Sub-picosecond interfacial electron injection in dye sensitized titanium dioxide films. Proceedings for Ultrafast Phenomena XI. Berlin: Springer-Verlag, 1998:505–507.

28. Jeffrey GA, Saenger W. Hydrogen Bonding in Biological Structures. Berlin: Springer-Verlag, 1991.

29. Nossal R, Lecar H. Molecular and Cell Biophysics. Redwood City, CA: Addison-Wesley, 1991.

30. Hayward S, Go N. Collective variable description of native protein dynamics. Ann Rev Phys Chem 1995; 46:223–250.

31. Zhuang W, Feng Y, Prohofsky EW. Self-consistent calculation of localized DNA vibrational properties at a double-helix–single-strand junction with anharmonic potential. Phys Rev A 1990; 41:7033–7042.

32. Benigno AJ, Ahmed E, Berg M. The influence of solvent dynamics on the lifetime of solute-solvent hydrogen bonds. J Chem Phys 1996; 104:7392–7394.

33. Lippincott ER, Schroeder R. One-dimensional model of the hydrogen bond. J Chem Phys 1955; 23:1099–1106.

34. Grubbs WT, Dougherty TP, Heilweil EJ. Bimolecular interactions in $(Et)_3SiOH$:base:CCl_4 hydrogen-bonded solutions studied by deactivation of the "free" OH-stretch vibration. J Am Chem Soc 1995; 117:11989–11992.

35. Audibert M-M, Palange E. Vibrational predissociation of the hydrogen-bonded $(CH_3)_2O$-HF complex. Chem Phys Lett 1983; 101:407–411.

36. Laenen R, Rauscher C. Transient hole-burning spectroscopy of associated ethanol molecules in the infrared: structural dynamics and evidence for energy migration. J Chem Phys 1997; 106:1–7.

37. Graener H, Seifert G, Laubereau A. New spectroscopy of water using tunable picosecond pulses in the infrared. Phys Rev Lett 1991; 66:2092–2095.

38. Heilweil EJ. Ultrafast glimpses of water and ice. Science 1999; 283:1467–1468.

39. Grubbs WT, Dougherty TP, Heilweil EJ. Vibrational energy dynamics of H-bonded pyrrole complexes. J Phys Chem 1995; 99:10716–10722.

40. Arrivo SM, Heilweil EJ. Conservation of vibrational excitation during hydrogen bonding reactions. J Phys Chem 1996; 100:11975–11983.

41. Arrivo SM, Kleiman VD, Grubbs WT, Dougherty TP, Heilweil, EJ. Infrared spectral hole burning of 1:1 hydrogen-bonded complexes in solution. In: Proceedings for TRVS VIII, Oxford, UK: Laser Chemistry 1999; 19:1–10.

42. Kleiman VD, Arrivo SM, Melinger JS, Heilweil EJ. Controlling condensed-phase vibrational excitation with tailored infrared pulses. Chem Phys 1998; 233:207–216.

43. Melinger JS, McMorrow D, Hillegas C, Warren WS. Selective excitation of vibrational overtones in an anharmonic ladder with frequency- and amplitude-modulated laser pulses. Phys Rev A 1995; 51:3366–3369.

44. Beckerle JD, Cavanagh RR, Casassa MP, Heilweil EJ, Stephenson JC. Subpicosecond study of intramolecular vibrational energy transfer in solution-phase $Rh(CO)_2$acac. Chem Phys 1992; 160:487–496.

45. Arrivo SM, Dougherty TP, Grubbs WT, Heilweil EJ. Ultrafast infrared spectroscopy of vibrational CO-stretch up-pumping and relaxation dynamics in metal hexacarbonyls. Chem Phys Lett 1995; 235:247–254.

46. Richman BA, Krumbugel MA, Trebino R. Temporal characterization of mid-IR free-electron laser pulses by frequency resolved optical gating. Optics Lett 1997; 22:721–723.

47. Liu W-K, Wu B, Yuan J-M. Nonlinear dynamics of chirped pulse excitation and dissociation of diatomic molecules. Phys Rev Lett 1995; 75:1292–1295.

48. Trushin SA, Sugawara K, Takeo H. Formation of Cr atoms in the 5 μm multi-photon decomposition of $Cr(CO)_6$. Chem Phys 1996; 203:267–278.

49. Jiao T, Pang Z, Burkey TJ, Johnston RF, Heimer TA, Kleiman VD, Heilweil EJ. Ultrafast ring closure energetics and dynamics of cyclopentadienyl manganese tricarbonyl derivatives. J Am Chem Soc 1999; 121:4618–4624.

50. George MW, Dougherty TP, Heilweil EJ. UV photochemistry of $[CpFe(CO)_2]_2$ studied by picosecond time-resolved infrared spectroscopy. J Phys Chem 1996; 100:201–206.

51. Heimer TA, Heilweil EJ. Direct time-resolved infrared measurement of electron injection in dye-sensitized titanium dioxide films. J Phys Chem B 1998; 101(51):10990–10993.

52. Nazeerudin MK, Liska P, Moser J, Vlachopoulos N, Gratzel M. Conversion of light into electricity with trinuclear ruthenium complexes adsorbed on textured TiO_2 films. Helv Chim Acta 1990; 73:1788–1803.

53. Hannappel T, Burfeindt B, Storck W, Willig FJ. Measurement of ultrafast photoinduced electron transfer from chemically anchored Ru-dye molecules into empty electronic states in a colloidal anatase TiO_2 film. J Phys Chem B 1997; 101:6799–6802.

54. Ellington RJ, Asbury JB, Ferrene S, Ghosh HN, Sprague JR, Lian T, Nozik AJ. Dynamics of electron injection in nanocrystalline titanium dioxide films sensitized with $[Ru(4,4'-dicarboxy-2,2'-bipyridine)_2(NCS)_2]$ by infrared transient absorption. J Phys Chem B 1998; 102:6455–6458.

55. Argazzi R, Bignozzi CA, Heimer TA, Castellano FN, Meyer GJ. Enhanced spectral sensitivity from ruthenium (II) polypyridyl based photovoltaic devices. Inorg Chem 1994; 33:5741–5479.

56. Heimer TA, Heilweil EJ, Bignozzi CA, Meyer GJ. Electron injection, recombination, and halide oxidation dynamics at dye-sensitized metal oxide interfaces. J Phys Chem A 2000; 104(18):4256–4262.

57. Richter LJ, Petralli-Mallow T, Stephenson JC. Vibrationally resolved sum-frequency generation with broadband infrared pulses. Optics Lett 1998; 23:1594–1596.

58. Wu Q, Zhang X-C. Free-space electro-optics sampling of mid-infrared pulses. Appl Phys Lett 1997; 71:1285–1286.

59. Markelz AG, Roitberg A, Heilweil EJ. Pulsed terahertz spectroscopy of DNA, bovine serum albumin and collagen between 0.1 and 2.0 THz. Chem Phys Lett. 2000; 320(1–2):42–48.

4

The Molecular Mechanisms Behind the Vibrational Population Relaxation of Small Molecules in Liquids

Richard M. Stratt
Brown University, Providence, Rhode Island

I. INTRODUCTION

The question of how fast a vibrationally hot molecule loses its excess energy — or, better yet, how fast any one piece of a molecule cools down — is central to chemical reaction dynamics (1–8). Not only is the inverse, but closely related problem of how fast vibrational energy can be added to a molecule at the very core of what chemical reactions are all about, it is the ability to disperse the excess thermal energy produced in a reaction that prevents the product version of our molecules from promptly reverting to their nascent reactant forms (9–11).

The issues raised in pursuing these problems are particularly interesting because they highlight the striking differences (and the curious similarities) between isolated-molecule dynamics and that seen in liquids. In a gas-phase bimolecular reaction, the two colliding partners can dispose of their excess internal energy by converting it into translational or rotational kinetic energy. But are these same options going to be open to systems in the tightly packed confines of a dense liquid, where there is neither free translation nor free rotation? Worse still, will the inability of the products to escape physical contact until they diffuse apart prevent such simple kinematic mechanisms from operating (12)? Besides, even if our molecules found a way to rid themselves of their unwanted vibrational energy, how easily could the surrounding liquid absorb it? The energy of a single quantum of a CO stretch is 10 times what $k_B T$ is at room temperature. In effect, shedding a quantum of vibration is equivalent to plunging a red-hot iron into our cold liquid. Should we think of the process as instantaneously vaporizing the surrounding solvent?

The prevailing theoretical models used to confront these issues reflect this profound dichotomy between gas-phase and condensed-phase perspectives. Historically, the isolated binary collision (IBC) picture of energy transfer has been the most frequently invoked scheme for calculating rates in liquids (1,13–17), yet it has all the earmarks of a quintessentially gas-phase treatment: molecules are assumed to lose vibrational energy through discrete (binary) collisions with individual solvent molecules. Not every such collision is likely to be equally effective, but one might surmise, as is the case in the gas phase, that the fraction of collisions that do succeed in transferring energy might be largely independent of the features of the surrounding solvent. If one proceeds with this assumption and blithely goes on to regard the collisions as completely uncorrelated (18–20), the net result is that the rate of energy loss by a dissolved vibrating molecule can be written as the simple product of the collision rate — a condensed-phase quantity, but one independent of any specifics of vibrational dynamics — and the fraction of effective collisions — something reflecting the details of what the two-body dynamics would be in isolation.

Probably the most appropriate first response to this model is to regard it as contrary to any legitimate, microscopically detailed view of what liquids are about (21–23). The principal problem is that it is not at all clear that there is such a thing as a well-defined collision in a dense liquid. Each molecule in a liquid is constantly being jostled by on the order of a dozen neighboring molecules. Indeed, were we to try to define "collisions" strictly as changes in our solute's potential energy caused by motion of the solvent, we would be hard pressed to find a time when collisions were not happening (24).

We should hasten to note that these fundamental difficulties do not mean that this theory does not often "work." The most common application of IBC theory points to its particularly simple prediction for the dependence of relaxation rates on the thermodynamic state of the solvent: with the Enskog estimate of collision rates, the ratio of vibrational relaxation rates at two different liquid densities ρ_1 and ρ_2 is just the ratio of the local solvent densities $[\rho_1 g_1(R)/\rho_2 g_2(R)]$, where $g(r)$ is the solute-solvent radial distribution function and R defines the solute-solvent distance at which an energy-relaxing collision is presumed to take place (13,22,25). These kinds of ratios of rates are, in fact, often well predicted by such expressions. There is some evidence as well that ratios of rates stemming from different excited vibrational states can also be rationalized quite nicely (16). The fact remains though, that because of its gas-phase roots, the IBC theory per se could never provide us with the tools necessary to discover the genuine molecular mechanisms by which vibrational energy relaxes in liquids. The need to postulate an arbitrary solute-solvent collision distance R (and the extraordinary sensitivity to the precise value chosen) serves as a warning sign that the theory does not bear too close an examination (15).

By the same token, of course, just because a model has its home in the condensed-phase world does not mean that it is any more suited to our purposes. It is not impossible to think about vibrational energy relaxation from a diametrically opposite limit, to regard the relaxation as a dissipation of heat into the surrounding, more or less continuous, medium (26,27). Vibrational dephasing rates have actually been predicted based on the values of a variety of different macroscopic transport coefficients of the solvent (25). As with the IBC approach, such models will neatly circumvent the need to understand the microscopic details of dynamics in a liquid, but for the same reasons, these continuum models are going to be fundamentally incapable of telling us which solvent molecules are doing what or when they are doing it. To get at mechanistic questions this specific — and even to find out how to whether it is useful to try to be this molecularly detailed in the ever-changing environment of a liquid — we need to pursue a more broadly based statistical mechanical approach to liquid dynamics. This chapter is an attempt to summarize some of the recent progress that has been made in understanding the actual mechanisms of vibrational energy relaxation using one such approach.

We should emphasize that the work discussed in this chapter is rather limited in scope. With few exceptions it will be concerned with obtaining purely classical mechanical perspectives on the very simplest example of vibrational energy relaxation — that of diatomic solutes dissolved in simple atomic and molecular liquids. Such problems do not come close to spanning the range of interesting topics suggested by modern infrared and Raman spectroscopies, but they are among the first examples of solute relaxation processes for which it has been possible to elucidate molecular mechanisms. Besides, beginning at the beginning is not necessarily a bad approach. We will return in the final portion of the chapter to the prospects for taking a somewhat wider view.

II.   VIBRATIONAL FRICTION

A.   Vibrational Energy Relaxation and Vibrational Friction

A powerful way of simultaneously including molecular and macroscopic perspectives is to write the exact equation of motion for the interesting degree of freedom — for us, the normal coordinate of the relevant solute vibrational mode — allowing the influence of the solvent to make an appearance only through one of a number of different kinds of effective forces. Generalized Langevin equations, (e.g., see Reference 28) express the (classical) equation of motion for the special coordinate x in terms of the potential of mean force $W(x)$, the potential energy x would feel were the solvent equilibrated around the solute, and the residual forces resulting explicitly from the solvent dynamics:

$$m \, d^2x/dt^2 = -\partial W/\partial x - \int_0^t dt' \eta(t - t') v(t') + \mathcal{F}(t) \qquad (1)$$

Here m is the mass and v is the velocity (dx/dt) associated with the x coordinate. The remaining, dynamically induced solvent effects show up in $\mathcal{F}(t)$, the so-called fluctuating force, and $\eta(t)$, the dynamical friction. Indeed, the last two terms in Equation (1) are the key — without them there would be no mechanism for a solvent to accept energy from a solute (29,30).

　　One of the features that makes Equation (1) such a good starting point for our work is that it can be, in principal, exact. It is possible to show, without ever explicitly evaluating $\mathcal{F}$ and η, that these crucial functions really do exist and are well defined (31). These formal definitions are rarely, if ever, useful in practical numerical calculations, but one can also work backwards from the exact dynamics $x(t)$ (e.g., from a molecular dynamics simulation) to derive what the friction in particular must look like (32). The analysis tells us, moreover, that the exact $\mathcal{F}$ and η are actually related to one another (28,31). The requirement that the relaxed system must be in equilibrium at some temperature T can be shown to set the magnitude and correlations of the fluctuating force:

$$\langle \mathcal{F}(0)\mathcal{F}(t) \rangle = k_B T \, \eta(t) \qquad (2)$$

with k_B being Boltzmann's constant and the brackets representing an equilibrium average. The upshot is that if we can understand the specifics of the vibrational friction, we should be able to predict the desired vibrational relaxation rates.

　　Interestingly, this microscopic friction behaves much the way our macroscopic intuitions predict that it should. The manner in which the friction appears on the right-hand side of Equation (1) says that it leads to an effective force proportional to the mode velocity v, but opposing it — much as one would expect from some sort of frictional drag. The fact that this drag has a time delay, that the drag at time t results from a velocity at an earlier time t' [which makes Equation (1) a generalized rather than an ordinary Langevin equation], might seem a bit of a complication, but it too is eminently reasonable. One can think of any motion of the solute mode, v, as perturbing the solvent away from its preferred parts of phase space. The solvent, in its best LeChatelier fashion, reacts to restore the status quo by evolving in such a way as to penalize any subsequent motion of the mode — that is, it generates a frictional drag. However, in any genuinely molecular picture, the effects of this solvent back-reaction cannot be instantaneous; it has to have a time lag commensurate with the time scales on which the solvent moves. Much of our study of vibrational relaxation can therefore be interpreted as an investigation into just what these time scales are.

This conceptual link between the solvent vibrational friction and vibrational energy relaxation is actually mirrored by an important practical connection. Within the rather accurate Landau-Teller approximation, (29,33,34), the rate of vibrational energy relaxation for a diatomic with frequency ω_0 and reduced mass μ is given by

$$\frac{1}{T_1} = \mu^{-1} \, \eta_R(\omega_0) \tag{3}$$

where $\eta_R(\omega)$ is the cosine transform of the vibrational friction

$$\eta_R(\omega) = \int_0^\infty dt \, \cos \omega t \, \eta(t) \tag{4}$$

In other words, the ability of the solvent to absorb a quantum of energy $\hbar\omega_0$ (or its classical equivalent) is determined quite literally by the ability of the solvent to respond to the solute dynamics at a frequency $\omega = \omega_0$. One can derive this relation quantum mechanically by assuming that the solvent's effect on the solute can be handled perturbatively within Fermi's golden rule (1), but it is actually more general than that. Perhaps it is worth pausing to see how the same basic result appears in a purely classical context.

Quite generally we can imagine the Hamiltonian for our system as a sum of $\mathcal{H}_u$, a Hamiltonian for the solute vibration, $\mathcal{H}_v$, a Hamiltonian for the solvent, and V_c, the piece of the potential energy coupling the two:

$$\mathcal{H} = \mathcal{H}_u(p, x) + \mathcal{H}_v(\mathbf{p}, \mathbf{q}) + V_c(x, \mathbf{q}) \tag{5}$$

a mechanistic perspective. Its frequency range will provide our basic explanation as to why solvents have the natural time scales they do. Comparisons between the influence spectra for different kinds of solute relaxation processes will also be informative in letting us see the differences (and surprising similarities) between solvent responses to different kinds of solute motions. Nor, for that matter, are we limited to looking at average solvent responses. Instead of taking an average over liquid configurations, we can examine the response from any instantaneous configuration R_0:

$$[\rho_{\text{vib}}(\omega)]_{R_0} = \sum_\alpha c_\alpha^2(R_0)\,\delta(\omega - \omega_\alpha(R_0)) \tag{22}$$

and ask mechanistic questions stripped of the configuration-to-configuration inhomogeneous broadening that makes the dynamics seem so chaotic (50).

Even at the level of the averaged solvent response there is some fairly detailed analysis we can pursue. Because we know the molecular identity of each normal mode associated with a given instantaneous configuration by virtue of knowing its eigenvector e_α, we can project out of the influence spectrum the contributions of any desired subset of solvent molecules or geometries of solvent motion (49). The configurational average will then tell us how whether our candidate set of degrees of freedom — our candidate *mechanism* — is actually an important ingredient in the relaxation (51).

To pick an example, suppose we want to the look at the portion of our influence spectrum arising from the first solvent shell around the solute (the shell being defined however we care to). One can demonstrate quite formally that the projected influence spectrum will resemble the full influence spectrum, but with different weightings for the individual modes (51):

$$\rho_{\text{vib}}^{\text{proj}}(\omega) = \left\langle \sum_\alpha [c_\alpha^{\text{proj}}]^2\,\delta(\omega - \omega_\alpha) \right\rangle \tag{23}$$

While the unprojected weighting defined by Equation (16) can be written (for a solution of rigid molecules) in terms of explicit components of the mode eigenvectors as

$$c_\alpha = \sum_j \sum_{\mu=x,y,z,\hat{\Omega}} (\partial F_{\text{ext}}^{(0)}/\partial r_{j\mu})_{R(0)}\, (e_\alpha)_{j\mu}$$

with $j = 1, \ldots, N$ labeling the molecules of the solution and μ denoting their center of mass translational (x,y,z) and reorientational ($\hat{\Omega}$) degrees of freedom, the first-shell projected weightings will include out of each mode only the degrees of freedom of the molecules in the first solvent shell:

$$c_\alpha^{\text{1st shell}} = \sum_{j=\text{1st shell}} \sum_{\mu=x,y,z,\hat{\Omega}} (\partial F_{\text{ext}}^{(0)}/\partial r_{j\mu})_{R(0)}\, (e_\alpha)_{j\mu}$$

total energy transfer in terms of the time evolution of the external force.*

$$\Delta E_u(t) = \int_0^t dt_1\, v^{(0)}(t_1)\, F_{ext}(t_1)$$

$$+ \mu^{-1} \int_0^t dt_1 \int_0^{t_1} dt_2 \cos \omega_0(t_1 - t_2)\, F_{ext}(t_1)\, F_{ext}(t_2) \quad (10)$$

Unfortunately, evaluating this formula exactly would still require that we know the fully coupled solute-solvent dynamics because it calls for $F_{ext}(t) = F_{ext}(q((t))$, but since the solvent perturbs the solute vibration only weakly, a perturbative treatment suffices (just as it does quantum mechanically). To leading order, $F_{ext}(t) = F_{ext}^{(0)}(t)$, what the solvent force would be if the solute's vibrational mode were held fixed. Thus, the average rate of solute-solvent energy transfer in the steady state is

$$\lim_{t \to \infty} d\langle \Delta E_u(t) \rangle / dt = \mu^{-1} \int_0^\infty dt \cos \omega_0 t\, C^{FF}(t) \equiv \mu^{-1} C_R^{FF}(\omega_0) \quad (11)$$

$$C^{FF}(t) = \langle F_{ext}^{(0)}(0)\, F_{ext}^{(0)}(t) \rangle \quad (12)$$

proportional to the cosine transform of autocorrelation function for this frozen-mode force (36). Note that in writing Equation (11) we made use of the statistical independence of the solute and solvent degrees of freedom in the absence of coupling and we took advantage of the invariance of equilibrium behavior to anything but time intervals:

$$\langle v^{(0)}(t)\, F_{ext}^{(0)}(t) \rangle = \langle v^{(0)}(t) \rangle \langle F_{ext}^{(0)}(t) \rangle = 0$$

$$\langle F_{ext}^{(0)}(t_1)\, F_{ext}^{(0)}(t_2) \rangle = \langle F_{ext}^{(0)}(0)\, F_{ext}^{(0)}(t_2 - t_1) \rangle$$

The frozen-mode force correlation function $C^{FF}(t)$ not only closely resembles the vibrational friction [Equation (2)], it is often a rather accurate way of calculating it in practice (29,32). One reason for this fortunate circumstance is that in typical molecular vibrations the vibrational frequency is so large that the solvent hardly sees the effects of the dynamics on the forces (32). If we take this identification for granted, however,

$$C^{FF}(t) = k_B T\, \eta(t) \quad (13)$$

and remember that the equilibrated solute vibrational energy will be $k_B T$, we reach our final destination: the rate constant for vibrational energy transfer is

---

* A related expression for energy transfer appears when one studies the underdamped (energy diffusion) limit of chemical reaction dynamics in liquids. See, for example, Reference 11.

seen to be proportional to the value of the friction at the natural vibrational frequency:

$$\lim_{t\to\infty} \langle E_u(t)\rangle^{-1}(d\langle\Delta E_u(t)\rangle/dt) = \lim_{t\to\infty} d\ln\langle\Delta E_u(t)\rangle/dt = \mu^{-1}\eta_R(\omega_0)$$

in perfect agreement with Equation (3), the Landau-Teller formula.*

B. The Instantaneous Vibrational Friction and the Instantaneous Normal Modes of the Solvent

The problem before us now is not simply to evaluate the vibrational friction numerically for realistic examples of vibrational relaxation in liquids; Equation (13) has already been applied in the literature to a wide variety of interesting cases (33,34,37,38). What we are concerned with here is finding out what the *specific* molecular events are that contribute to this vibrational friction.

Having posed the problem in this fashion, we must deal with the knotty question of what it is that would constitute a satisfactory answer. The orientations, locations, and even identities of the participating solvent molecules are constantly changing. So how can there be any kind of definitive mechanism to find? The short answer is that for a general liquid process there is no such detailed mechanism, at least not one with any claim to generality. Ultrafast processes, however, are a different story — and vibrational friction is, in fact, an ultrafast process.

Actually, this last statement might seem quite counterintuitive, especially for small solutes. The vibrational energy relaxation of diatomics and triatomics, for example, can be quite slow when judged by the ps and sub-ps time scales of intermolecular motion. Though ps relaxation times have been seen (39,40), there are also some well-known examples of T_1s in the μs and even ms ranges (5). Yet, the vibrational friction apparently begins its work very quickly, even in these slowest of cases. As one can see from the behavior of a typical vibrational friction (Fig. 1) (Y. Deng and R. M. Stratt, unpublished), the solvent retains its memory of the solute dynamics for only a short time, so the time lag between solute motion and

---

*   This derivation is largely meant to be a schematic way of helping us see how the vibrational friction influences the rate of solute-solvent energy transfer. Notice, however, that we never specified the actual initial conditions of an experiment (in particular whether the solute was to be initially hot or cold with respect to the surrounding solvent). Without such a specification we cannot predict the net sign of the energy flow.

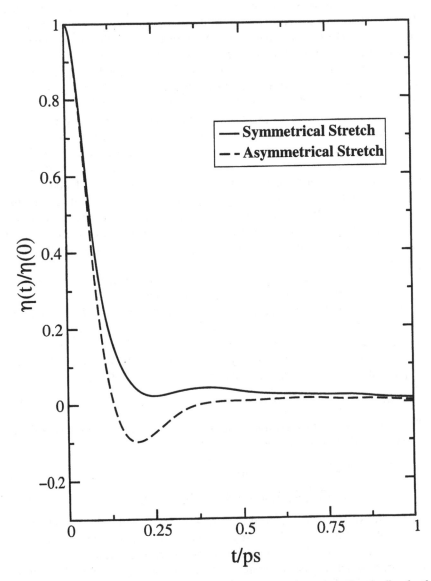

Figure 1 Vibrational friction on a symmetrical linear triatomic molecule dissolved in high-density supercritical Ar. The figure compares the differing frictions felt by the symmetrical and asymmetrical stretching modes of the triatomic.

the solvent response is remarkably brief. This brevity is in itself a clue that the critical solvent motions we are looking for cannot be all that complex, but the key point for us is that a short-time treatment of the liquid dynamics may be all that we need to perform the analyses we want to do.

At short enough times, the dynamics of liquids really is quite simple. There is little opportunity for the basic liquid structure to change over the course of a few hundred fs, so we can regard the molecules of our liquid as essentially vibrating in place (or more correctly, beginning to vibrate in place). The time evolutions of each of the $3N$ degrees of freedom of the liquid (with N the number of solvent atoms) are thus described by the $3N$ normal modes $q_\alpha(t)$, $\alpha = 1, \ldots 3N$, the *instantaneous normal modes* (INMs) of the instantaneous liquid configuration (41,42). Of course, any given experiment is likely to sample an ensemble of such configurations, meaning that our formulas will have to be averaged over the equilibrium distribution of configurations, but it will turn out to be a particular advantage for us to being able to think about the specifics of molecular mechanisms one configuration at a time.

To see how these harmonic solvent modes translate into vibrational friction, (43-46) consider how the correlation function for the solvent force on the frozen mode [Equation (12)], behaves at short times (47). The solvent modes themselves, $q_\alpha(t)$, are the displacements of the liquid along the $3N$-dimensional eigenvectors $\mathbf{e}_\alpha$ of each mode. Literally, if the $3N$-dimensional vector giving the position of every atom in the liquid at time t is $\mathbf{R}(t)$, the displacement from the time zero configuration $\mathbf{R}(0)$ is the sum

$$\mathbf{R}(t) - \mathbf{R}(0) = \sum_\alpha q_\alpha(t)\,\mathbf{e}_\alpha \tag{14}$$

Thus, at the shortest times, the evolution of the solvent force is just a linear function of these displacements.

$$F_{\text{ext}}^{(0)}(t) \approx F_{\text{ext}}^{(0)}(0) + \sum_\alpha c_\alpha q_\alpha(t) \tag{15}$$

$$c_\alpha \equiv (\partial F_{\text{ext}}^{(0)}/\partial q_\alpha)_{q=0} = (\partial F_{\text{ext}}^{(0)}/\partial \mathbf{R})_{\mathbf{R}(0)} \cdot \mathbf{e}_\alpha \tag{16}$$

where the coefficients c_α describe the efficiency with which each mode α modulates the force. On taking advantage of the independence of the modes, we see that the autocorrelation function of the time derivatives of the force, which is just the negative of the second derivative of the force autocorrelation function we want, becomes simply

$$-d^2 C^{FF}(t)/dt^2 = \langle [dF_{\text{ext}}^{(0)}/dt]_0 [dF_{\text{ext}}^{(0)}/dt]_t \rangle = \left\langle \sum_\alpha c_\alpha^2 v_\alpha(0) v_\alpha(t) \right\rangle \tag{17}$$

with $v_\alpha(t) = dq_\alpha/dt$ being the mode velocities.

The reason we work with the derivatives is related to the fact that the mode displacements are measured from their instantaneous, $t = 0$, position rather than some hypothetical harmonic minimum (48). We know that the global potential energy surface of a liquid is far from harmonic, so such a minimum would be a rather unphysical construct. What happens instead is that our modes do obey simple harmonic dynamics, but subject to somewhat unusual initial conditions (49):

$$q_\alpha(t) = [f_\alpha/\omega_\alpha^2](1 - \cos\omega_\alpha t) + v_\alpha(0)\sin\omega_\alpha t \qquad (18)$$

$$v_\alpha(t) = v_\alpha(0)\cos\omega_\alpha t + [f_\alpha/\omega_\alpha]\sin\omega_\alpha t \qquad (19)$$

Here the ω_α are the mode frequencies (technically, the ω_α^2 are eigenvalues of the mass-weighted dynamical matrix evaluated at the instantaneous liquid configuration), and $v_\alpha(0)$ and f_α are the initial velocities, dq_α/dt, and the initial forces, $-\partial V/\partial q_\alpha$, along the modes. We can avoid being tripped up by our ignorance of the distribution of f_αs (and take advantage of knowing that the $v_\alpha(0)$s are governed by a mass-weighted Maxwell-Boltzmann distribution) by making use of Equation (17) (47). On substituting Equation (19) and using the facts that

$$\langle v_\alpha(0)f_\alpha\rangle = 0, \quad \langle v_\alpha^2(0)\rangle = k_B T$$

we obtain

$$-d^2C^{FF}(t)/dt^2 = k_B T\left\langle\sum_\alpha c_\alpha^2\cos\omega_\alpha t\right\rangle = k_B T\int d\omega\,\rho_{vib}(\omega)\cos\omega t$$

implying, from Equations (4) and (13), that our desired frequency-domain vibrational friction itself is just

$$\eta_R(\omega) = (\pi/2)\rho_{vib}(\omega)/\omega^2 \qquad (20)$$

with what we shall call the *influence spectrum* for vibrational relaxation the equilibrium distribution of INM frequencies of the solvent weighted by the ability of each solvent mode to promote vibrational relaxation (50):

$$\rho_{vib}(\omega) = \left\langle\sum_\alpha c_\alpha^2\delta(\omega - \omega_\alpha)\right\rangle \qquad (21)$$

C.   Deducing Molecular Mechanisms from Instantaneous-Normal-Mode Theory

Equations (20) and (21) help us in a number of ways. For one thing, the average vibrational influence spectrum is itself going to be interesting from a mechanistic perspective. Its frequency range will provide our basic explanation as to why solvents have the natural time scales they do. Comparisons between the influence spectra for different kinds of solute relaxation processes will also be informative in letting us see the differences (and surprising similarities) between solvent responses to different kinds of solute motions. Nor, for that matter, are we limited to looking at average solvent responses. Instead of taking an average over liquid configurations, we can examine the response from any instantaneous configuration $\mathbf{R}_0$:

$$[\rho_{\mathrm{vib}}(\omega)]_{\mathbf{R}_0} = \sum_\alpha c_\alpha^2(\mathbf{R}_0)\,\delta(\omega - \omega_\alpha(\mathbf{R}_0)) \tag{22}$$

and ask mechanistic questions stripped of the configuration-to-configuration inhomogeneous broadening that makes the dynamics seem so chaotic (50).

Even at the level of the averaged solvent response there is some fairly detailed analysis we can pursue. Because we know the molecular identity of each normal mode associated with a given instantaneous configuration by virtue of knowing its eigenvector $\mathbf{e}_\alpha$, we can project out of the influence spectrum the contributions of any desired subset of solvent molecules or geometries of solvent motion (49). The configurational average will then tell us how whether our candidate set of degrees of freedom — our candidate *mechanism* — is actually an important ingredient in the relaxation (51).

To pick an example, suppose we want to the look at the portion of our influence spectrum arising from the first solvent shell around the solute (the shell being defined however we care to). One can demonstrate quite formally that the projected influence spectrum will resemble the full influence spectrum, but with different weightings for the individual modes (51):

$$\rho_{\mathrm{vib}}^{\mathrm{proj}}(\omega) = \left\langle \sum_\alpha [c_\alpha^{\mathrm{proj}}]^2\,\delta(\omega - \omega_\alpha) \right\rangle \tag{23}$$

While the unprojected weighting defined by Equation (16) can be written (for a solution of rigid molecules) in terms of explicit components of the mode eigenvectors as

$$c_\alpha = \sum_j \sum_{\mu=x,y,z,\hat{\Omega}} (\partial F_{\mathrm{ext}}^{(0)}/\partial r_{j\mu})_{\mathbf{R}(0)}\,(\mathbf{e}_\alpha)_{j\mu}$$

with $j = 1, \ldots, N$ labeling the molecules of the solution and μ denoting their center of mass translational (x,y,z) and reorientational $(\hat{\Omega})$ degrees of freedom, the first-shell projected weightings will include out of each mode only the degrees of freedom of the molecules in the first solvent shell:

$$c_\alpha^{\mathrm{1st\ shell}} = \sum_{j=\mathrm{1st\ shell}} \sum_{\mu=x,y,z,\hat{\Omega}} (\partial F_{\mathrm{ext}}^{(0)}/\partial r_{j\mu})_{\mathbf{R}(0)}\,(\mathbf{e}_\alpha)_{j\mu}$$

Should we wish to examine the contributions from solvent motions along some particular directions, we could do that as well. To monitor the solvent motions in an atomic solvent that are parallel to the unit vector $\hat{\Omega}$ describing a vibrating bond, for example, we would write

$$c_\alpha^\parallel = \sum_{j=\text{solvent}} \sum_{\mu=x,y,z} (\partial F_{\text{ext}}^{(0)}/\partial r_{j\mu})_{\mathbf{R}(0)} (\hat{\Omega})_\mu (\mathbf{e}_\alpha)_{j\mu}$$

Note that with these definitions the area under a projected influence spectrum gives the fraction of the entire liquid response associated with that mechanism (51). With the first shell projection, for example, the fraction is:

$$f_{\text{1st shell}} = \int d\omega\, \rho_{\text{vib}}^{\text{1st shell}}(\omega) \bigg/ \int d\omega\, \rho_{\text{vib}}(\omega)$$

$$= \left\langle \sum_{j=\text{1st shell}} \sum_\mu (\partial F_{\text{ext}}^{(0)}/\partial r_{j\mu})_{\mathbf{R}(0)}^2 \right\rangle \bigg/ \left\langle \sum_{j\mu} (\partial F_{\text{ext}}^{(0)}/\partial r_{j\mu})_{\mathbf{R}(0)}^2 \right\rangle$$

where we have used the fact that the normal mode transformation is orthogonal to write the last line. However, it is also worth bearing in mind that these kinds of fractional areas do not tell the whole story. It is possible, in particular, to define cross projections, such as the one between the first-shell and the outer-shell (non–first-shell) solvents (51):

$$\rho_{\text{vib}}^{\text{1st/outer}}(\omega) = 2 \left\langle \sum_\alpha [c_\alpha^{\text{1st shell}}][c_\alpha^{\text{outer shell}}]\, \delta(\omega - \omega_\alpha) \right\rangle$$

The area under any such cross projection is identically zero (because of the orthogonality of the normal mode transformation), yet there is a real physical meaning to the cross spectrum between any two candidate mechanisms. If the INMs themselves neatly separated into modes moving the first-shell solvents and modes moving the second shell, then the cross projections would vanish. The fact that it does not is therefore a real indication that coupled motion between the two different kinds of degrees of freedom contributes to vibrational relaxation. It is, of course, precisely this kind of detailed information that we need to have in order to pursue our search for molecular mechanisms.

III. HOW COLLECTIVE IS VIBRATIONAL ENERGY RELAXATION?

A. Basic Features

Our instantaneous-normal-mode theory of vibrational energy relaxation suggests a rather interesting physical picture. The solvent surrounding our solute apparently has a wide band of collective vibrations suitable for taking up the solute's excess energy — its INMs — something we can see by looking at the typical solvent density of states (the distribution of the vibrational frequencies of those modes) illustrated in Fig. 2.* Not all such modes are going to be equally proficient at accepting energy; the influence spectrum shown in Fig. 2, $\rho_{\text{vib}}(\omega)$, clearly indicates how

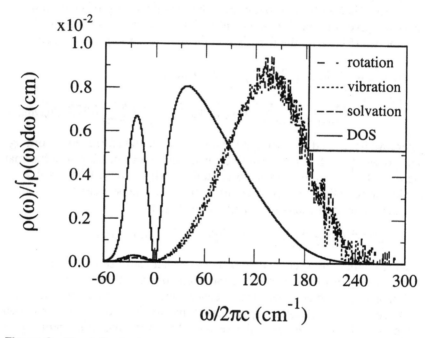

Figure 2 Normalized instantaneous-normal-mode spectra for high-density supercritical Ar. The overall density of states (DOS) is contrasted with three different INM influence spectra for a diatomic solute: for rotational friction, vibrational friction, and (nonpolar) solvation dynamics. Only the spectrum of modes for vibrational friction is of direct relevance to this chapter, but the other influence spectra show the strong similarities in the instantaneous solvent dynamics associated with different kinds of solute relaxation.

---

* An interesting feature of INM calculations for liquids is that some of the modes are imaginary, corresponding to unstable motion. These imaginary modes seem to contribute little to influence spectra, but they show up universally in densities of states. As is conventional, imaginary modes are plotted in spectra as if they were negative frequencies.

much better modes in the middle of the band are at this task. However, Equations (3) and (20) seem to be saying that all our solute has to do to rid itself of that excess energy is to find one such collective vibration whose frequency precisely matches the vibrational frequency of the solute. When this resonance condition is satisfied, the solvent mode will take up the energy, dissipating it eventually as heat.

The immediate question, then, is whether this scenario reflects what actually happens. Do the INM theories really work? There is, in fact, some evidence on this score (45,52). If we compare the vibrational friction predicted by INM theory, Equation (20), with that revealed by an exact molecular-dynamics evaluation of the force autocorrelation function, Equations (4) and (13), we see some reasonably impressive agreement (Fig. 3) (52).* Not only is the few hundred cm^{-1} spectral range of the friction predicted quite nicely, but the basic form of the response is as well. Each example shows that the friction diminishes as the frequency rises, beginning with a sharp drop from its maximum value at $\omega = 0$ and gradually going over to a much slower decay, behavior captured nicely by the INM formulas.

There are some difficulties we should be aware of just the same. The maximum that is supposed to appear at $\omega = 0$ shows up in the INM calculations as a full-blown divergence (43,44). Indeed this infinity is just one instance of the fundamental problems with INMs at zero frequency. It probably should not be a surprise that a theory that pretends that basic liquid structure does not change with time is going to be ill-suited to studying behavior at the lowest frequencies. The same level of theory predicts liquid diffusion constants to be identically zero, for example.[†] Fortunately, realistic molecular vibrational frequencies tend to be well outside this low-frequency regime, so the effects on predicted T_1s are likely to be minimal. Still, as we shall note in Section VI, not every aspect of vibrational spectroscopy will be quite so insulated from this basic issue.

---

[*]   For the purposes of this article, our equations, figures, and discussion will neglect the small centrifugal contributions to the friction arising from the bond-length dependence of the solute moment of inertia.

[†]   Some authors take a somewhat less conservative approach to INM theory, relaxing the strict short-time interpretation. They argue that INM results such as the density of imaginary modes supplies information on the potential hypersurface of the liquid that can be used as an ingredient in theories for the diffusion constant and other low-frequency aspects of dynamics. See References 42 and 53.

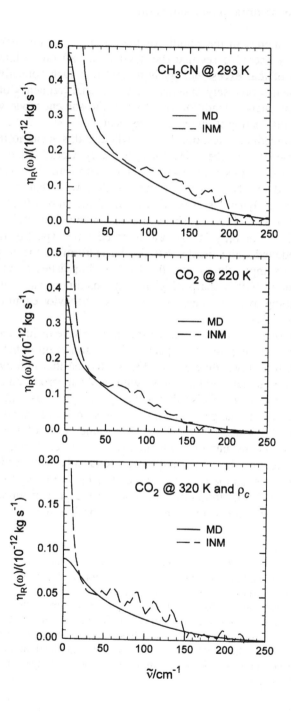

There is another difficulty with the INM results that is a little more difficult to see in Fig. 3, but it, too, is physically significant. A closer look at the high-frequency side of Fig. 3 would show that the INM band of vibrational frequencies comes to an end significantly more quickly than does the real friction. Beyond a few hundred cm^{-1} or so, even the correct values of the friction are quite small on the scale of the figure, but the INM predictions are orders of magnitude smaller still. We will return to this problem of how to understand what seems to be *nonresonant* vibrational energy relaxation in Section V.

B. Mechanistic Investigation

Aside from the difficulties at the upper and lower ends of the liquid's vibrational band, the INM ideas do seem to work and to work quantitatively. The ability of the liquid-mode concept to account for the absolute magnitude of the vibrational friction, including the factor of 2 difference between liquid and supercritical CO_2 solvents, is worth noting (52). But does this success mean that vibrational energy relaxation is really a collective process? To answer this question, we need to carry out precisely the kind of mechanistic investigation we discussed in Section II.C.

We can begin by asking the inverse question: How much of vibrational energy relaxation is fundamentally few-body in character? Using the methods of Section II.C we can project out of the vibrational influence spectrum the portion arising from the motion of the solute and the single most prominently contributing solvent (Fig. 4) (52). The answer is clear, but perhaps surprising in view of our comments about collective vibrational modes. Some three quarters of the contribution to the relaxation in these two examples stems from two-body dynamics. Vibrational energy relaxation is evidently not a collective phenomenon in this sense.

One way to understand this observation is to note that the full and the two-body projected spectra overlap the most closely at the highest frequencies inside the band. At these (relatively) high frequencies, a liquid's modes simply do not encompass many atoms. Indeed, careful studies of the upper band edge of an atomic liquid reveal that the modes there can

---

Figure 3 Frequency-domain vibrational friction felt by diatomic solutes dissolved in molecular fluids (52). The three panels show the friction for a model dipolar solute dissolved in acetonitrile (top), and for I_2 dissolved in liquid (middle) and supercritical (bottom) carbon dioxide. Each panel compares the exact molecular dynamics (MD) results with the linear INM predictions.

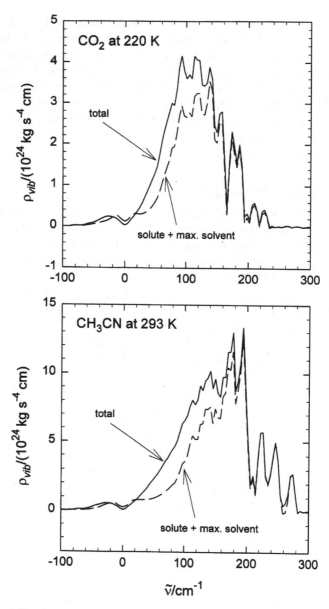

Figure 4 Vibrational influence spectra for two of the systems illustrated in Figure 3 (52). In each panel the total influence spectrum is compared with the portion of the spectrum arising from the combined motion of the solute and the maximally contributing solvent.

be understood quantitatively as vibrations of sufficiently closely spaced pairs of atoms, what we might call *instantaneous binary modes* (50). The frequencies of these modes ω_{bin} are given accurately by the instantaneous frequency the pair would have if it were vibrating against a backdrop of the remainder of the liquid held fixed in place:

$$\mu\omega_{bin}^2 = u''(r(0)) + \text{background correction} \tag{24}$$

where μ is the reduced mass of the pair, $u(r)$ is the pair potential, and $r(0)$ is the instantaneous pair separation. Similarly, the eigenvectors can be taken to be just those of the isolated pair. With these expressions it is therefore a simple matter to recompute Equation (21) pretending that only these kinds of modes are present and to compare the results with full vibrational friction (Fig. 5) (50).

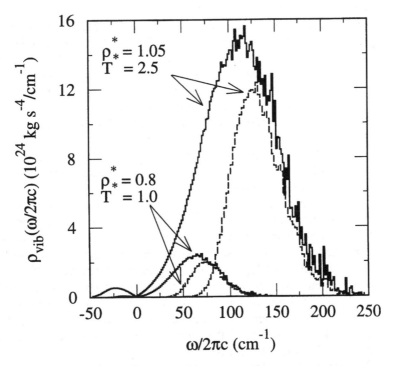

Figure 5 Vibrational influence spectra for a diatomic solute in Ar under two different thermodynamic conditions: a high-density supercritical state (upper curves) and a near-triple-point liquid (lower curves) (50). For each thermodynamic state, the total influence spectrum (solid) is contrasted with the portion stemming from binary modes (dashed).

The comparison seems to supports our hypothesis rather well. In the limit of an infinitely massive solute, some 57% of the area under the full friction spectrum can apparently be accounted for in this fashion, a significant proportion considering that only 9% of the density of the solvent's states comes from binary modes (50) — and the agreement with the binary-mode prediction is even better at the very highest frequencies. However, for an atomic liquid under these same conditions, an even larger percentage, 87%, of the influence spectrum is reproduced by incorporating no more than the single maximally contributing solvent (50). What is more, it seems unlikely on physical grounds that pairwise motion would not give way to more and more collective vibrational motion as one descends from the band edge towards the band center. There must be something more general behind the preeminence of few-body character, something beyond the mere presence of literal instantaneous binary modes.

To get at these more general, lower frequency, features it is helpful to be able to separate out the superposition of many different liquid configurations always present in our influence spectra — the inhomogeneous broadening — from the instantaneous (homogeneous) dynamics launched from each separate configuration. Following the dictates of Equation (22) then, consider what the vibrational friction influence spectrum looks like for several distinct liquid configurations (Fig. 6) (50). What we find is curious. Instead of the continuous spectra that we have become accustomed to, the single-configuration spectra are a series of discrete lines, corresponding to the contributions of individual INMs that are somehow being strongly selected in the relaxation process.

It is not implausible that certain INMs would more be much more efficient at fostering vibrational relaxation than others, but our question has now become what it is that makes the special modes special. The most prominent modes seem to be scattered throughout the band without any particular preference for high or low frequencies. Moreover, although a few modes often completely dominate the response (witness the modes with amplitudes 245 and 362 in the top and bottom panels, respectively) the median number of contributing modes is not a small number, typically scaling with the square root of the number of degrees of freedom in our simulation studies, (45,50,52). So what is going on?

Because we know the molecular composition of each of the eigen-vectors, the simplest resolution comes from just looking at the interesting modes (Fig. 7) (50). Taking, for example, the relatively low-frequency 63.8 cm^{-1} mode shown in Fig. 6, we can see in Fig. 7a that this mode moves quite a number of solvent atoms around the solute. Out of all of

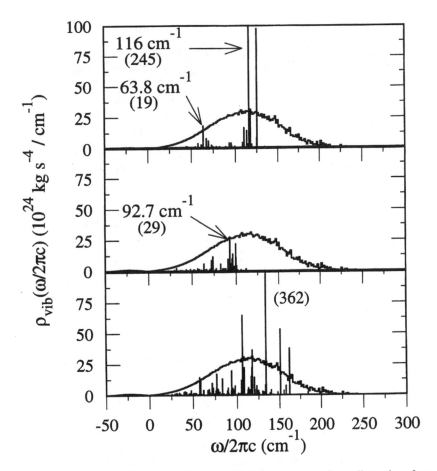

Figure 6 Single-configuration vibrational influence spectra for a diatomic solute dissolved in high-density supercritical Ar (50). Each panel presents the influence spectrum for a different liquid configuration; the solid histograms give the configurationally averaged spectrum (shown here for comparison). Each contributing mode is portrayed as a sharp line whose height (reported in parentheses for several of the modes) is the net contribution of the mode to the total solute-solvent coupling.

the solvent atoms, though, Fig. 7b tells us that only a half dozen atoms are coupled to the vibrational coordinate strongly enough to have much of an effect on vibrational relaxation in this instantaneous liquid configuration — and most of these are not among those displaced significantly by the liquid mode. Combining these two insights, then, yields our answers.

(a) Mode   **(b) Coupling**

(c) Effective Mode

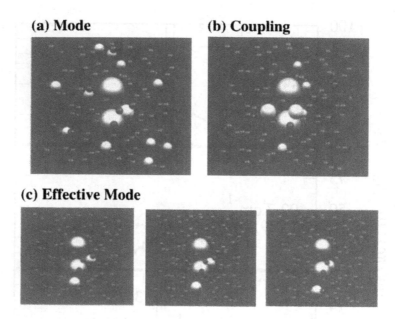

Figure 7 Solvent geometry, coupling, and dynamics for a diatomic solute in an atomic liquid (50). Shown here is the liquid configuration from the top panel in Fig. 6, with the two largest spheres representing the solute and the remaining spheres the solvent. All the atoms in the actual configuration have the same size; the size differences in the figure are used for visual clarity. Panel (a) illustrates the 63.8 cm^{-1} mode with the principal solvent atoms highlighted in size. Panel (b) depicts (as highlighted in size) the solvent atoms most strongly coupled to the solute vibration in this particular liquid configuration. The three frames in (c) show three consecutive snapshots of the dynamics of some of the most important solvent atoms in this configuration (again highlighted in size): namely, those atoms both moving in the 63.8 cm^{-1} mode and strongly coupled.

If we look only at the solvent atoms that are both moving and strongly coupled to the solute vibration, we never end up looking at more than one or two atoms, as shown in Fig. 7c. It is not that the important liquid modes themselves cannot be collective, but out of all the solvent atoms displaced by the modes, the only ones that will count are those few positioned so as to exert a strong force along the direction of the solute mode. Hence we can think of vibrational population relaxation as being carried out by *effective modes* — modes that are simply the visible few-body part of what may be a significantly more extended normal mode, but that have the characteristic frequency of the underlying normal mode (50).

This notion of effective modes explains, or at least has the potential to explain, a great deal. The apparent success of isolated binary collision (IBC) models in predicting the thermodynamic state dependence of relaxation rates could arise not from the independent collision idea doing a particularly good job of reflecting the real dynamics, but from the crucial importance of the equilibrium probability of there being a single critical solvent molecule in close proximity to the solute. Indeed, INM calculations for vibrational friction themselves manage to mimic the IBC density dependence without ever invoking discrete collisions, presumably because the equilibrium liquid structure is correctly built into the c_α coupling coefficients (45,52).

Returning to the single configuration level, we see that we can also understand the reasons for the special role of selected individual modes: the critical modes will invariably be those that have significant amplitude at the critical solvent(s). Since there is no reason for the number of such modes to be microscopic, this model might remove some of the mystery from the observation that we tend to see a mesoscopic number of contributing modes (though the occurrence of a specifically $\sqrt{N}$ dependence has yet to be explained) (45,50,52).

In a sense all of these ideas depend on the central point that the important solvent forces on the vibrational mode are sufficiently short-ranged that no more than a small number of solvents can contribute at a time. Certainly the fact that repulsive intermolecular forces are so sharply varying that the nearest solvent molecules will always dominate is consistent with this requirement. But what of longer ranged forces? What happens to this analysis with the longer-ranged electrostatic forces seen in polar solvents, for example (54,55)? As we shall see in the next section, the mechanism of vibrational relaxation is quite blasé about such pedestrian changes.

IV. HOW DOES DIELECTRIC FRICTION EFFECT VIBRATIONAL ENERGY RELAXATION?

A key experimental observation that reawakened much of the interest in vibrational population lifetimes was the discovery that aqueous ions such as CN^- and N_3^- could discharge their vibrational quanta, along with energies amounting to 2000 cm^{-1}, on ps time scales (56,57). Since these new examples of ultrafast energy relaxation were largely ions in protic solvents and therefore instances in which electrostatic interactions were prominent, it was natural to suspect that the vibrational relaxation was being accelerated dramatically by the dielectric medium. The friction dissipating the vibrational energy was, presumably, a kind of dielectric friction paralleling the dielectric friction thought to govern the translation of ions and the reorientation of dipoles in polar solvents (58,59).

That electrostatic forces could be crucial to vibrational energy relaxation was amply demonstrated by the liquid water simulations of Whitnell et al. (34). They noted that since the electrostatic portion of the force between their solvent and a dipolar solute was linear in the solute dipole moment, Equations (12) and (13) implied that the electrostatic part of the friction ought to scale as the dipole moment squared. When they then found that their entire relaxation rate scaled with the square of the solute dipole moment, it certainly seemed to be convincing evidence that electrostatics forces were indeed the primary ingredients generating ultrafast relaxation. Subsequent theoretical work on relaxation rates in such manifestly protic solvents as water and alcohols has largely served to reinforce this message (37,38,60,61).

But consistent with the overall theme of this chapter, we need to ask ourselves whether we really have to conclude that the *mechanism* of vibrational energy relaxation is fundamentally electrostatic just because we find the overall relaxation rate to be sensitive to Coulombic forces. Let us attempt to get at this question through another mechanistic analysis of the INM vibrational influence spectrum, this time looking at the respective contributions of the electrostatic part of the solvent force on our vibrating bond, the nonelectrostatic part (in most simulations, the Lennard-Jones forces), and whatever cross terms there may be.

The results of such a calculation, shown in Fig. 8 (52), seem to tell a very different story from the earliest studies. With nondipolar I_2 as a solute and CO_2 as a solvent, the complete domination of the solvent response by the Lennard-Jones forces is impressive, but perhaps not all that startling. One might surmise that the quadupole-quadrupole forces at work in this example are a bit too weak to accomplish much. Yet, when we have a dipolar solute dissolved in the strongly polar solvent CH_3CN, we get almost the same kind of complete control by Lennard-Jones forces. Electrostatics now seems totally unimportant.

We can begin to get at this discrepancy if we carry out an equivalent to the Whitnell-Wilson-Hynes calculation (34,38), looking at the effects of

Figure 8 Vibrational influence spectra for the three systems illustrated in Fig. 3 (52). In each panel the total influence spectrum is compared with the portion of the spectrum arising from purely Lennard-Jones coupling to the solute (LJ), from purely electrostatic coupling (elec.), and from cross coupling (LJ-elec.)

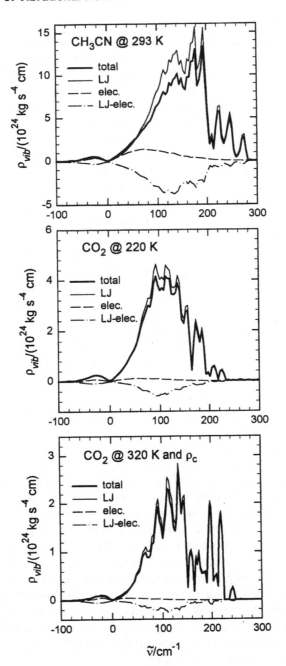

systematically varying the solute dipole on the vibrational friction. Rather than monitoring the friction itself, however, it is better for mechanistic purposes to look just at its form, the normalized friction $\eta(t)/\eta(0)$ (Fig. 9) (62). What we see is quite telling, especially in CH_3CN solutions. The magnitude of the friction [as measured by the initial friction $\eta(0)$] is, in fact, quite sensitive to electrostatics; it changes by more than a factor of 2 in going from the nondipolar "Br_2" solute to the almost 9 D dipole moment of the "d8" solute. The detailed time dependence of the vibrational friction, however, is virtually unchanged in the polar solvent. Hence we can see that it is the magnitude, not the mechanism, of the vibrational friction that is controlled by electrostatic forces. As Fig. 8 makes clear, the mechanism of the relaxation, at least in our example of a polar solvent, does rely almost completely on the solvent modulating the Lennard-Jones–like forces on the solute, regardless of the solvent's polarity.

That there might be such a crucial distinction between magnitude and mechanism was actually anticipated by several authors (63–65). But perhaps we can see it a little more clearly by adopting our familiar instantaneous perspective, distinguishing carefully between equilibrium phenomena that go into selecting a set of instantaneous liquid configurations and the subsequent dynamics that are what we really mean by the term "mechanism" (62). When we view things in this fashion, the reason why electrostatics forces are important in vibrational relaxation has to do solely with their equilibrium role. Coulombic forces are sufficiently powerful that they can position solvents much higher on the steep repulsive wall of the solute-solvent potential than, say, van der Waals forces. By the same token though, these forces are not rapidly varying enough to contribute much to frequency domain friction at anything but the lowest frequencies. In polar solvents, as in others, it is going to be the most rapidly varying forces, the sharp repulsive forces, whose time evolution is ultimately responsible for the relaxation. Polar solvents simply seem to amplify this repulsive-force friction by encouraging small solute-solvent distances (63–65).

We should probably keep in mind that these conclusions were drawn from studies of neutral solutes in aprotic solvents, whereas the most striking experiments have featured ionic solutes in hydrogen-bonding solvents (5,56,57). Presumably, the interactions with ionic solutes should not change matters much: the forces are even stronger than with neutrals, but the forces are even more slowly varying. It is not out of the question, though, that the directional character of hydrogen bonding makes it somehow unique in the context of vibrational relaxation (66,67). Further mechanistic analysis of this point might prove interesting.

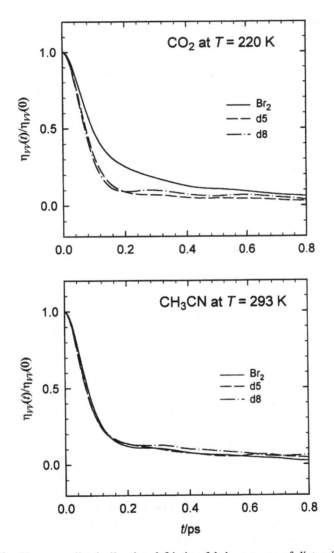

Figure 9 The normalized vibrational friction felt by a range of diatomic solutes dissolved in liquid carbon dioxide and liquid acetonitrile (62). The solutes are meant to represent the nondipolar molecule Br_2 itself and two bromine mimics differing only in the replacement of the bromine quadrupole by permanent dipoles of different strengths. The "d5" solute has a dipole moment of 5.476 D and the "d8" solute a dipole moment of 8.762 D. (The notation η_{vv} emphasizes the fact that only potential-energy contributions are included in the calculations; centrifugal force terms are neglected.)

V. VIBRATIONAL ENERGY RELAXATION AT HIGH FREQUENCIES

As we noted earlier, the fact that our instantaneous vibrational friction is built from a set of harmonic solvent modes suggests that it is natural to think of vibrational energy relaxation as taking place through resonant energy transfer to the solvent: relatively rapid energy transfer that occurs when the solute finds a solvent mode matching the solute's own vibrational frequency. We did discover in Section III that only a few key atoms in a mode are involved in providing the gateway necessary to access the mode, but this observation by itself does nothing to cast doubt on the basic resonant-transfer paradigm.

However, several items should make us continue to be wary. Most T_1s for neutral small-molecule vibrational relaxations are not ultrafast events, nor, for that matter, do they involve vibrational frequencies remotely close to the few hundred cm^{-1} bands of typical liquids (5).* More typical are the 940 ps it takes to relax the 3265 cm^{-1} C–H stretch in HCN (5) and the 4.4 ns required to relax the 2850 cm^{-1} HCl vibration (68) when either one of these hydrides is dissolved in CCl_4. Even slower relaxations are well known; the 3 ms relaxation times for O_2's 1556 cm^{-1} vibration in liquid oxygen, for example, (69) has been the subject of much recent discussion (70–72).

One way to put these problems into perspective is to look at what might be one of the simplest examples of a high-frequency vibrational energy relaxation, that of I_2 dissolved in Xe, (15,17,73–76) (Fig. 10). The 211 cm^{-1} vibrational frequency of iodine hardly qualifies it as high by most standards, but the weak interatomic forces and the high atomic weight of Xe cause its INM density of states to be increasingly small beyond 120 cm^{-1}. Indeed, experiments indicate that I_2 relaxes quite slowly in Xe, with vibrational lifetimes on the order of hundreds of ps (15–17,74). The difficulty with the INM theory we outlined in Section II is that the low density of solvent modes would lead to a calculated vibrational friction orders of magnitude lower still. Apparently our harmonic theory is correct in predicting significant slowdowns once the solute's vibration is out of

---

* To simplify our discussion, we will ignore the presence of the high-frequency intramolecular degrees of freedom present in most molecular solvents, though they do pose interesting conceptual and computational issues.

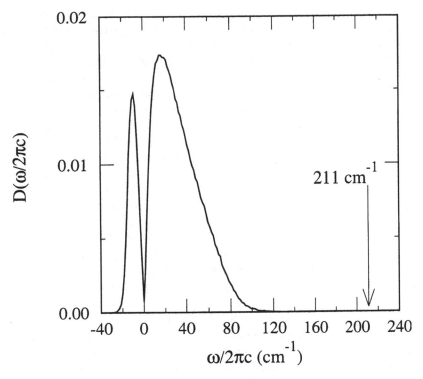

Figure 10 The situation confronted by a vibrationally excited I_2 molecule dissolved in liquid Xe (76). The solid line is the INM density of states for the liquid; the arrow indicates the vibrational frequency of I_2.

the solvent band, but it seriously underestimates the actual rates. The true relaxation mechanism, it would seem, must rely on some fundamentally anharmonic aspect of the dynamics.

Having to understand the full spectrum of anharmonic possibilities for liquids would pose a formidable challenge, but for our present purposes we need only come to grips with the highest-frequency dynamics (14,77). But high-frequency behavior is something we do understand, at least within INM theory. The very highest frequency modes inside the INM band are instantaneous binary modes, modes that, except for some perturbative corrections, behave as if they involve no more than a pair of atoms at a time (50). As we look for higher and higher frequencies, it becomes less and less likely to find pairs with a sufficiently sharp interatomic potential curvature, hence the appearance of a rapidly decaying edge to both the INM density of states and the influence spectrum. It is possible, however, that this instantaneous pair concept is much more general than this harmonic interpretation. For any kind of excitation, the effective mass has to be small in order get large frequencies, so at the very highest frequencies we would never expect more than a pair of atoms to be moving cooperatively. Conversely, if we had a pair of atoms that were so closely spaced that their forces on each other were significantly larger than any other forces acting on them — the prototypical high frequency situation — we could easily imagine the motion of the pair to be dynamically decoupled from the remainder of the liquid (76). From neither perspective do we have to invoke harmonic behavior in order to see that we should still be looking at solute-solvent pairs for our high-frequency analysis.

The anharmonicity we need to understand our relaxation could actually show up in either of two completely different guises within such a pair framework. The pair dynamics itself certainly could be deeply anharmonic. That is, the solute-solvent pair distance $q(t)$ could obey fundamentally anharmonic equations of motion. But it is also conceivable that the dynamics is not all that anharmonic (or more precisely, that its anharmonicity is not all that important). It might also be $F_{ext}^{(0)}(t)$, the force on the solute vibration being driven by the dynamics, that is fundamentally nonlinear. Either one of these kinds of anharmonicities (or both) could affect the instantaneous-pair (IP) vibrational friction (76):

$$\eta_{IP}(t) = \langle F_{ext}^{(0)}(0)\, F_{ext}^{(0)}(t) \rangle = \langle F_{ext}^{(0)}(q(0))\, F_{ext}^{(0)}(q(t)) \rangle \tag{25}$$

By contrast, our previous INM theory not only assumed that the underlying dynamics, the set of $q_\alpha(t)$, was harmonic [Equation (18)]; it took the force on the vibration to be linear [Equation (15)]. The theory did have the redeeming feature of incorporating a complete set of harmonic modes, $\{q_\alpha(t), \alpha = 1, \ldots\}$, instead of a single degree of freedom $q(t)$ (thereby representing the full many-body to few-body range of solvent dynamics), but it was completely incapable of allowing for either of the two anharmonic possibilities we now want to evaluate.

To test our new expression [Equation (25)], all we need to do is to identify the critical solvent for each instantaneous liquid configuration. Consistent with our previous discussion we look for solute-solvent pairs whose intrapair forces are larger than the forces imposed by the rest of the solvent, what we have called "mutual-nearest-neighbor pairs" (78,79). If the relevant solute-solvent reduced mass and pair potential are μ and $u(q)$, respectively, then the fully anharmonic equation of motion becomes

$$\mu \, d^2q/dt^2 = -u'(q) \tag{26}$$

subject to the requirement that the instantaneous liquid configuration determine the initial condition, $q(0)$.* The correctly nonlinear force on the solute vibration then follows as

$$F_{ext}^{(0)}(t) = u'(q(t)) \tag{27}$$

Equivalently, in the language of the correlation functions introduced earlier, the IP theory predicts that the vibrational friction obeys

$$-k_B T \, d^2\eta/dt^2 \equiv G(t) = \langle u''(q(t)) \, u''(q(0)) \, v(t) \, v(0) \rangle \tag{28}$$

$$\eta_R(\omega) = (k_B T)^{-1} G_R(\omega)/\omega^2 \tag{29}$$

with $v(t) = dq/dt$ and $G_R(\omega)$ the cosine transform of the $G(t)$ correlation function defined in Equation (28). By comparison, our earlier, linear, INM theory, Equation (17), took u'' to be a constant related to the assumed harmonic mode frequency via Equation (24) (76).

The numerical results of evaluating Equation (26)–(29) and its INM equivalent, Equation (17), are shown in Fig. 11 for a model diatomic solute dissolved in Xe (76). Consider first the behavior inside the INM band, the region below 120 cm^{-1} shown in the bottom panel. The original INM theory, which relies on the complete set of collective harmonic solvent modes, actually does rather well here, whereas the IP theory, for all its anharmonic enhancements, tremendously underestimates the vibrational friction. Liquid motion within the spectral range of the solvent band evidently has some profoundly collective features, despite the fact that the coupling to the solute is often funneled through a few key solvents.

Once we go beyond the band edge, the situation is quite different. As we see in the top panel, beyond 150 cm^{-1} the absolute scale of the friction is two orders of magnitude smaller than it is inside the band, but the friction there is well represented by the IP theory. Now it is the collective, linear INM theory that makes far too small a contribution. Even at this early stage we can conclude that the mechanism of vibrational energy relaxation must switch when we cross the band edge, and that the new mechanism must involve exciting highly localized solvent motion mediated in some essential

---

* For an atomic solvent of mass m_v and a solute of total mass M_u, Ref. 76 shows that the correct reduced mass is $m_v M_u/(m_v + M_u)$.

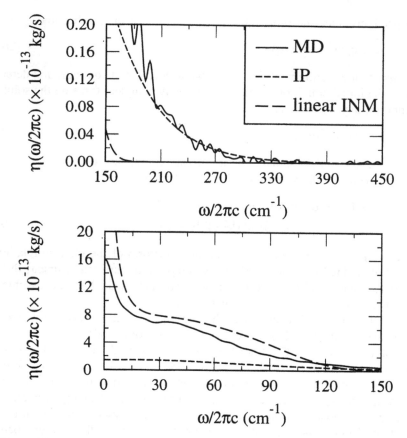

Figure 11 Frequency-domain vibrational friction felt by a diatomic solute dissolved in high-density supercritical Xe (76). The bottom panel shows the frequency region within the INM band, the top panel the region beyond the band edge. Note the differences in the ordinate scales. In both panels the solid line is the exact molecular dynamics (MD) result, the long-dashed line is the outcome of a traditional (linear) INM calculation, and the short-dashed line is calculated from instantaneous pair (IP) theory.

fashion by the system's anharmonicity. But precisely what role does the anharmonicity play?

Once we have gotten to this point, the easy way to answer this question is to try to undo the agreement between the IP theory and the exact molecular dynamics by putting back one of the harmonic approximations. Thus, while the IP formalism includes both anharmonic dynamics and nonlinear coupling in Equation (26) and (28), there is nothing preventing us from postulating that the instantaneous dynamics is still as harmonic as it is inside the band, and that it is only the nonlinearity of the coupling that we need to respect in order to obtain high-frequency relaxation. The end result of such a hypothesis would be a *nonlinear INM theory*, which could be evaluated numerically just by evaluating Equation (28) using the harmonic dynamics [Equation (18)] for the lone relevant mode q(t) [rather than solving Equation (26) for the true dynamics]. The outcome from this kind of calculation is shown in Fig. 12 (76).

The quantitative succcess of this revised model is unmistakable. There is virtually no difference between the exact molecular dynamics, the full instantaneous-pair theory, and the nonlinear INM version. The crucial anharmonicity then really must result from the importance of nonlinear

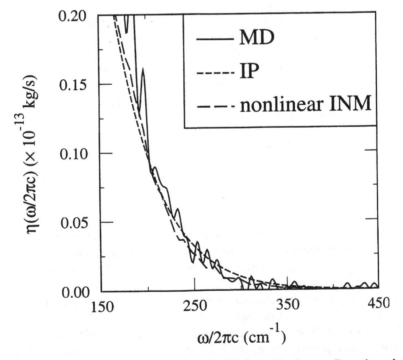

Figure 12 Frequency-domain vibrational friction felt by a diatomic solute dissolved in high-density supercritical Xe (76). The molecular dynamics (MD) and instantaneous pair (IP) calculations shown in Fig. 11 are compared here to the results from nonlinear INM theory for the region beyond the band edge.

coupling. Interestingly, this observation gives us a handle on what is happening physically. Our ability to retain the instantaneous harmonic dynamics suggests that it is still meaningful to talk about characteristic frequencies for liquid motion, but what nonlinear functions of this motion can generate that linear functions cannot are overtones (multiples) of fundamental harmonic frequencies. Thus this agreement tells us that we might be able to think about relaxation of solute frequencies outside the solvent band as a kind of resonant process after all; all it would take is for the nonlinear solute-solvent coupling to produce a sufficiently high-order harmonic of the solvent mode that it comes into resonance with the solute's own vibrational frequency. In the terminology of solid-state physics we are saying that high-frequency vibrational energy relaxation is basically a "multiphonon" processes (80–87).

Our findings lead us in a number of useful directions. One of these directions is a generalization of our basic instantaneous approach to dynamics. Our original linear INM formalism assumed the potential energy was instantaneously harmonic, but that the coupling was instantaneously linear [Equation (15)]. We still need to retain the harmonic character of the potential to justify the existence of independent normal modes (at least inside the band), but we are free to represent the coupling by any instantaneously nonlinear function we wish. A rather accurate choice for vibrational relaxation, for example, is the instantaneous exponential form:

$$u'(r(t)) = u'(r(0)) e^{-\alpha_0 [r(t) - r(0)]}$$

where α_0 is the instantaneous logarithmic derivative of the coupling:

$$\alpha_0 = -[d \ln u'(r)/dr]_{r=r(0)}$$

Note that this kind of assumption is not tantamount to assuming the coupling is literally exponential (any more than the overall INM approach is equivalent to taking the entire potential surface to be literally harmonic). All we are doing is assuming that these forms are correct instantaneously — for short enough times. In fact, this approach actually produces a useful analytical formula for the friction, because it allows us to sum all the contributions of all orders of the solvent overtones (76). The explicit result is a little too involved to reproduce here, but it is probably worth pointing out that the result is rather close to the familiar exponential gap law (84,85) for the high-frequency relaxation of a solute of frequency ω_0:

$$(T_1)^{-1} \sim \exp(-\omega_0 t_0)$$

where our theory identifies the characteristic nonlinear time scale t_0 as

$$t_0 = (2\mu/k_B T \langle \alpha_0 \rangle)^{1/2}$$

From a somewhat more physical perspective, our results also allow us to expand a bit on our previous comments about the venerable IBC theory of vibrational relaxation. Our numerical findings do point out, in a rather quantitative and forceful way, that the IBC idea of focusing on the interaction of a single solvent with the solute is quite sensible (88–91). But by postulating a well-defined collision rate, what the IBC theory does is to assume that the many- and few-body aspects are independent. Schematically, IBC relaxation rates are given by

(dynamical many-body problem) × (dynamical 2-body problem)

an assumption that invariably leads to arbitrariness in locating the dividing point. What our instantaneous pair results say is that these aspects are logically separable, but not independent. The IP prediction is that

(equilibrium $\xrightarrow{\text{initial conditions}}$ (dynamical

many-body problem) 2-body problem)

so that a separate, but perfectly well-defined dynamical 2-body calculation has to be undertaken for each instantaneous liquid configuration. It seems that there is no need — and the subject of vibrational relaxation provides no special justification — for trying to ascribe physical reality to discrete collisions in dense liquids.

VI. CONCLUDING REMARKS

The list of topics we have chosen to ignore in this chapter is long indeed. In choosing to focus on diatomic solutes dissolved in simple atomic and molecular solvents, leaving out any mention of pure dephasing and totally ignoring quantum mechanics, we have obviously resorted to the theorists' approach of seeing how much of the fundamentals we could understand before venturing into more complex arenas. The reader should be aware, though, that some of our omissions were dictated by basic difficulties in the formalism we have been using, whereas other topics slighted here should prove to be excellent candidates for much the same kind of instantaneous mechanistic analysis we have been doing. In either event, we should not conclude this chapter without noting how some of the missing topics could, or should, fit into the perspectives we have been presenting.

Our omission of the special features of polyatomic solutes was particularly unfortunate, because it meant that we had to leave out one of the most interesting and important avenues by which a molecule can rid itself of vibrational energy — what one might call solvent-induced intramolecular vibrational relaxation (IVR) (81,92–95). The long time it takes for a solute to donate anything more than a few hundred cm^{-1} of vibrational energy directly to the solvent (even with the participation of the overtones of the solvent band) means that a solute is ripe for an alternative way of dispersing its energy — and polyatomic solutes provide a number of such ways. Because both the solute and the solvent are anharmonic, it has been suggested that it is almost always going to be possible for an excited high-frequency vibrational mode in a polyatomic to funnel most of its energy into a slightly lower-lying vibrational mode (or into its overtones or combinations bands) within the same molecule. The solvent would then simply have to absorb the relatively small energy difference between the initial and final intramolecular states (81,96,97).

If polyatomic vibrational relaxation really takes place in this fashion, many of the same kinds of mechanistic questions we have posed for diatomics will obviously become much richer when posed for polyatomics: Which particular solvent motions will serve to foster IVR? Will some motions actually interfere with IVR? Does solvent dynamics play a role in deciding between alternative routes for intramolecular energy transfer? Although the specific issues are different from those in this chapter, there is no conceptual barrier to pursuing them with the same techniques. Indeed, our preliminary investigations indicate that many of these questions should be perfectly amenable to the kinds of approaches we have been discussing.

Our omission of pure dephasing and of quantum mechanical concerns, by way of contrast, reflect more fundamental issues. Modern experimental methods have certainly made it possible to divorce vibrational energy relaxation from the generic concept of vibrational dephasing. One can now directly monitor the time evolution of the population of individual vibrational quantum states (4,5) instead of having to infer the answer from an infrared or Raman line shape, a process that requires us to strip away the inhomogeneous broadening from the homogeneous broadening and to partition the residual homogeneous broadening into the separate kinds of dephasing caused by population (energy) relaxation and phase (pure-dephasing) relaxation (98). But the delicate issue of how a solvent, without either adding or subtracting energy, makes a molecular vibration lose its phase memory does poses some mechanistic questions we might want to understand in their own right, especially if we want to think about the long-lived vibrational coherences that have been seen in solution photodis-sociation experiments (99–101).

The difficulty we face is that our instantaneous approach seems inher-ently ill-suited to answering such questions. At the same level of treatment that produces the Landau-Teller formula for energy relaxation, one finds a pure dephasing rate proportional to $\eta_R(0)$, the vibrational friction at zero frequency (29,102). Even without the divergence in our INM form for $\eta_R(\omega)$ as $\omega \to 0$, we would know, as we remarked in Section III.A, that this kind of static limit is the last place we should be applying our instantaneous methodology. On the other hand, we also know that T_2^*s, the pure vibrational dephasing times, can fall into the ultrafast range (104,105). It certainly seems plausible, moreover, that much of this dephasing is driven by the same sorts of violent, highly localized, interactions with nearby solvents that accounts for high-frequency energy relaxation (24). So perhaps carrying out the right sort of instantaneous mechanistic study of pure dephasing would not be entirely foolhardy (106).

Leaving out quantum mechanics from our considerations is, of course, undoubtedly foolhardy. The size of any conceivable solute vibrational quantum we might care about is going to be at least of the order of $k_B T$ (≈ 200 cm^{-1} at 300 K) and, more than likely, significantly larger than $k_B T$. In fact, the situation for a classical study is not quite as dire as it might seem, not because the quantal corrections are small, but because they can sometimes cancel. Bader and Berne pointed out that for the not-so-hypothetical scenario of a harmonic solute linearly coupled to a harmonic bath, the exact quantum mechanical T_1 is identical to the classical T_1 (despite the fact that the same is not true for the individual state-to-state rate constants) (107). While subsequent work has shown that adding elements of anharmonicity does not necessarily remove the desirability of doing a fully classical calculation (in preference to, say, a mixed quantum-classical study) (82), the Bader-Berne theorem is hardly going to serve as a license to neglect quantum mechanics for all future work.

The real reasons that the discussion in this chapter has been entirely classical are in part that mechanisms seem to be most easily visualized in terms of classical mechanical concepts — especially the idea of the instantaneous positions of solvent molecules serving as the initial conditions for subsequent dynamics. Beyond that, though, we have limited ourselves to classical mechanics because the localized snapshots of the potential surface that INM ideas so depend on seem difficult to incorporate into standard quantum mechanical calculations. It is tempting to pretend that the INM harmonic modes derived from a classical calculation can be directly quantized (108), but there is nothing in the derivation of INM dynamics that legitimizes portraying the potential energy surface of a liquid as globally harmonic (which is what a naive introduction of quantum harmonic oscillators would seem to require). Whether any of the current attempts to introduce quantum mechanics into INM theories will turn out to be satisfactory (109–111) or whether the INM ideas themselves will have to be generalized remains to be seen.

Besides these thorny theoretical issues, perhaps the most obvious question we have left unanswered is what the prospects are for testing our mechanistic interpretations experimentally. The idea that if we could single out an individual liquid configuration, we would see individual liquid modes serving as repositories for the solute's vibrational energy is, after all, a fairly specific prediction. Even more specific are our notions of what these modes should be like: that there should be one or two crucial solvents that serve as gatekeepers for the rest of the liquid, and that for high enough frequencies the motion of these special solvents should comprise the entire energy-absorbing mode. There are therefore some real predictions we could imagine confronting in a simulation; the question is will any of these be accessible in an experiment?

The crux of the matter is the apparent need to select a single liquid configuration, obviously a rather daunting prospect. Nevertheless, we have already noted that examining a single liquid configuration is our way of removing the blurring created by an ensemble of liquid configurations superimposed upon one another — precisely what one does by removing inhomogeneous broadening via a photon echo (112,113) or through transient hole burning (114,115). The problem is that current generations of infrared and Raman echo experiments are largely about finding the amount of inhomogeneous broadening (itself no small feat). Perhaps determining what typical members of the ensemble look like is more of a job for single-molecule spectroscopy (116), which, at least as of this writing, has not been extended into the ultrafast realm. When and if this extension eventually does occur, we may well see some genuine tests of the usefulness of instantaneous perspectives on liquids and, more generally, on the whole idea of molecularly defined mechanisms in liquid dynamics.

ACKNOWLEDGMENTS

I am delighted to be able to thank my coworkers for sharing their insights and their talents. In particular, the molecular liquid studies presented here were carried out in conjunction with Professor Branka Ladanyi and most of the atomic liquid studies referred to here were performed in close collaboration with Dr. Grant Goodyear, Dr. Ross Larsen, and Dr. Edwin David. The friction illustrated in Fig. 1 was calculated by Yuqing Deng and the results in Fig. 2 were computed by Joonkyung Jang. The work in my research group was supported by NSF grants CHE-9417546, CHE-9625498, and CHE-9901095.

REFERENCES

1. Oxtoby DW. Adv Chem Phys 47 (part II):487, 1981.
2. Oxtoby DW. Annu Rev Phys Chem 32:77, 1981.
3. Chesnoy J, Gale GM. Adv Chem Phys 70 (part 2):297, 1988.
4. Heilweil EJ, Casassa MP, Cavanagh RR, Stephenson JC. Annu Rev Phys Chem 40:143, 1989.
5. Owrutsky JC, Raftery D, Hochstrasser RM. Annu Rev Phys Chem 45:519, 1994.
6. Miller DW, Adelman SA. Int Rev Phys Chem 13:359, 1994.
7. Morresi A, Mariani L, Distefano MR, Giorgini MG. J Raman Spectrosc 26:179, 1995.
8. Stratt RM, Maroncelli M. J Phys Chem 100:12981, 1996.
9. Hynes JT. In: Baer M, ed. Theory of Chemical Reaction Dynamics, Vol. 4. Boca Raton, FL: CRC Press, 1985. pp. 171–234.
10. Hanggi P, Talkner P, Borkovec M. Rev Mod Phys 62:251, 1990.
11. Pollak E. In: Wyatt RE, Zhang JZH, eds. Dynamics of Molecules and Chemical Reactions. New York: Marcel Dekker, 1996. pp. 617–669.
12. Harris AL, Brown JK, Harris CB. Annu Rev Phys Chem 39:341, 1988.
13. Davis PK, Oppenheim I. J Chem Phys 57:505, 1972.
14. Oxtoby D. Mol Phys 34:987, 1977.
15. Paige ME, Harris CB. Chem Phys 149:37, 1990.
16. Paige ME, Harris CB. J Chem Phys 93:3712, 1990.
17. Russell DJ, Harris CB. Chem Phys 183:325, 1994.
18. Zwanzig R. J Chem Phys 34:1931, 1961.
19. Zwanzig R. J Chem Phys 36:2227, 1962.
20. Fixman M. J Chem Phys 34:369, 1961.
21. Dardi PS, Cukier RI. J Chem Phys 89:4145, 1988.
22. Dardi PS, Cukier RI. J Chem Phys 95:98, 1991.
23. Simpson CJSM, Turnidge ML, Reid JP. Mol Phys 70:125, 1996.
24. Frankland SJV, Maroncelli M. J Chem Phys 110:1687, 1999.
25. Chesnoy J, Weis JJ. J Chem Phys 84:5378, 1986.
26. Scherer POJ, Seilmeier A, Kaiser W. J Chem Phys 83:3948, 1985.
27. Sukowski U, Seilmeier A, Elsaesser T, Fischer SF. J Chem Phys 93:4094, 1990.

28. Friedman HL. A Course in Statistical Mechanics. Englewood Cliffs, NJ: Prentice-Hall, 1985. Chapter 14, note problem 14.10 as well.
29. Tuckerman M, Berne B. J Chem Phys 98:7301, 1993. The classic paper that applies the Generalized Langerin equations to vibrational relaxation and critically tests the application.
30. Adelman SA, Stote RH. J Chem Phys 88:4397, 1988.
31. Zwanzig R. J Stat Phys 9:215, 1973.
32. Berne BJ, Tuckerman ME, Straub JE, Bug ALR. J Chem Phys 93:5084, 1990.
33. Whitnell RM, Wilson KR, Hynes JT. J Phys Chem 94:8625, 1990.
34. Whitnell RM, Wilson KR, Hynes JT. J Chem Phys 96:5354, 1992.
35. Rejto PA, Chandler D. J Phys Chem 98:12310, 1994.
36. Sibert III EL. Chem Phys Lett 307:437, 1999. A procedure for systematically taking this kind of development beyond leading order in perturbation theory is presented.
37. Benjamin I, Whitnell RM. Chem Phys Lett 204:45, 1993.
38. Ferrario M, Klein ML, McDonald IR. Chem Phys Lett 213:537, 1993.
39. Pugliano N, Palit DK, Szarka AZ, Hochstrasser RM. J Chem Phys 99:7273, 1993.
40. Pugliano N, Szarka AZ, Gnanakaran S, Triechel M, Hochstrasser RM. J Chem Phys 103:6498, 1995.
41. Stratt RM. Accts Chem Res 28:201, 1995.
42. Keyes T. J Phys Chem A 101:2921, 1997. A somewhat different approach toward instantaneous normal modes.
43. Goodyear G, Larsen RE, Stratt RM. Phys Rev Lett 76:243, 1996.
44. Goodyear G, Stratt RM. J Chem Phys 105:10050, 1996.
45. Goodyear G, Stratt RM. J Chem Phys 107:3098, 1997.
46. (a) Schvaneveldt SJ, Loring RF. J Chem Phys 102:2326, 1995; (b) Schvaneveldt SJ, Loring RF. J Chem Phys 104:4736, 1996. An alternative INM approach to vibrational dynamics.
47. Stratt RM, Cho M. J Chem Phys 100:6700, 1994.
48. Cho M, Fleming GR, Saito S, Ohmine I, Stratt RM. J Chem Phys 100:6672, 1994.
49. Buchner M, Ladanyi BM, Stratt RM. J Chem Phys 97:8522, 1992.
50. Larsen RE, David EF, Goodyear G, Stratt RM. J Chem Phys 107:524, 1997.
51. Ladanyi BM, Stratt RM. J Phys Chem 100:1266, 1996.
52. Ladanyi BM, Stratt RM. J Phys Chem A 102:1068, 1998.
53. (a) Li W-X, Keyes T, Sciortino F. J Chem Phys 108:252, 1998; (b) Li W-X, Keyes T. J Chem Phys 111:5503, 1999.
54. Cho M. J Chem Phys 105:10755, 1996.
55. Schvaneveldt SJ, Loring RF. J Phys Chem 100:10355, 1996.
56. Li M, Owrutsky J, Sarisky M, Culver JP, Yodh A, Hochstrasser RM. J Chem Phys 98:5499, 1993.
57. Hamm P, Lim M, Hochstrasser RM. J Chem Phys 107:10523, 1997.

58. Wolynes PG. Annu Rev Phys Chem 31:345, 1980.
59. Papazyan A, Maroncelli M. J Chem Phys 102:2888, 1995.
60. Rey R, Hynes JT. J Chem Phys 108:142, 1998.
61. Morita A, Kato S. J Chem Phys 109:5511, 1998.
62. Ladanyi BM, Stratt RM. J Chem Phys 111:2008, 1999.
63. Bruehl M, Hynes JT. Chem Phys 175:205, 1993.
64. Gnanakaran S, Hochstrasser RM. J Chem Phys 105:3486, 1996.
65. Gnanakaran S, Lim M, Pugliano N, Volk M, Hochstrasser RM. J Phys Condens Matter 8:9201, 1966.
66. Klippenstein SJ, Hynes JT. J Phys Chem 95:4651, 1991.
67. Staib A. J Chem Phys 108:4554, 1998.
68. Knudtson JT, Stephenson JC. Chem Phys Lett 107:385, 1984.
69. Faltermeier B, Protz R, Maier M. Chem Phys 62:377, 1981.
70. Everitt KF, Egorov SA, Skinner JL. Chem Phys 235:115, 1998.
71. Everitt KF, Skinner JL. J Chem Phys 110:4467, 1999.
72. Egorov SA, Everitt KF, Skinner JL. J Phys Chem A 103: 9494, 1999.
73. Egorov SA, Skinner JL. J Chem Phys 105:7047, 1996.
74. Harris CB, Smith DE, Russell DJ. Chem Rev 90:481, 1990.
75. Brown JK, Harris CB, Tully JC. J Chem Phys 89:6687, 1988.
76. Larsen RE, Stratt RM. J Chem Phys 110:1036, 1999.
77. (a) Rostkier-Edelstein D, Graf P, Nitzan A. J Chem Phys 107:10470, 1997; (b) Rostkier-Edelstein D, Graf P, Nitzan A. J Chem Phys 108:9598, 1998. (erratum).
78. Larsen RE, Stratt RM. Chem Phys Lett 297:211, 1998.
79. David EF, Stratt RM. J Chem Phys 109:1375, 1998.
80. Kenkre VM, Tokmakoff A, Fayer MD. J Chem Phys 101:10618, 1994.
81. Moore P, Tokmakoff A, Keyes T, Fayer MD. J Chem Phys 103:3325, 1995.
82. Egorov SA, Berne BJ. J Chem Phys 107: 6050, 1997.
83. Nitzan A, Mukamel S, Jortner J. J Chem Phys 60:3929, 1974.
84. Nitzan A, Mukamel S, Jortner J. J Chem Phys 63:200, 1975.
85. Egorov SA, Skinner JL. J Chem Phys 103:1533, 1995.
86. Egorov SA, Skinner JL. J Chem Phys 105:10153, 1995.
87. Egorov SA, Skinner JL. J Chem Phys 106:1034, 1997.
88. Liu HJ, Pullen SH, Walker II LA, Sension RJ. J Chem Phys 108:4992, 1998.
89. Shiang JJ, Liu H, Sension RJ. J Chem Phys 109:9494 1998.
90. Kalbfleisch TS, Ziegler LD, Keyes T. J Chem Phys 105:7034, 1996.
91. Biswas R, Bhattacharyya S, Bagchi B. J Chem Phys 108:4963, 1998.
92. Hofmann M, Graener H. Chem Phys 206:129, 1996.
93. Graener H, Zurl R, Hofmann M. J Phys Chem B 101:1745, 1997.
94. Deak JC, Iwaki LK, Dlott DD. Chem Phys Lett 293:405, 1998.
95. Deak JC, Iwaki LK, Dlott DD. J Phys Chem A 102: 8193 (1998).
96. Dlott DD. In: Yen W, ed. Laser Spectroscopy of Solids II. Berlin: Springer, 1989.
97. Rey R, Hynes JT. J Chem Phys 104:2356, 1996.

98. Tokmakoff A, Fayer MD. J Chem Phys 103:2810, 1995.

99. Banin U, Ruhman S. J Chem Phys 98:4391, 1993.

100. Scherer NF, Jonas DM, Fleming GR. J Chem Phys 99:153, 1993.

101. Pugliano N, Szarka AZ, Hochstrasser RM. J Chem Phys 104:5062, 1996.

102. Levine AM, Shapiro M, Pollak E. J Chem Phys 88:1959, 1988.

103. Kalbfleisch T, Keyes T. J Chem Phys 108:7375, 1998.

104. Vohringer P, Westervelt RA, Yang T-S, Arnett DC, Feldstein MJ, Scherer NF. J Raman Spectrosc 26:535, 1995.

105. Lindenberger F, Rauscher C, Purucker H-G, Lauberau A. J Raman Spectrosc 26:835, 1995.

106. Williams RB, Loring RF. J Chem Phys 110:10899, 1999. The subject of pure dephasing becomes more subtle and interesting once we leave the confines of low-order perturbation theory.

107. Bader JS, Berne BJ. J Chem Phys 100:8359, 1994.

108. Keyes T. J Chem Phys 106:46, 1997.

109. Cao J, Voth GA. J Chem Phys 101:6184, 1994.

110. Corcelli SA, Doll JD. Chem Phys Lett 263:671, 1996.

111. Chakravarty C, Ramaswamy R. J Chem Phys 106:5564, 1997.

112. Tokmakoff A, Fayer MD. Accts Chem Res 28:437, 1995.

113. Berg M, Vanden Bout DA. Accts Chem Res 30:65, 1997.

114. Laenen R, Rauscher C, Lauberau A. Phys Rev Lett 80:2622, 1998.

115. Chudoba C, Nibbering ETJ, Elsaesser T. Phys Rev Lett 81:3010, 1998.

116. Moerner WE, Orrit M. Science 283:1670, 1999.

5
Time-Resolved Infrared Studies of Ligand Dynamics in Heme Proteins

Manho Lim
Pusan National University, Pusan, South Korea

Timothy A. Jackson
Harvard University and Massachusetts Institute of Technology, Boston, Massachusetts

Philip A. Anfinrud
National Institute of Diabetes and Digestive and Kidney Diseases, National Institutes of Health, Bethesda, Maryland

I. INTRODUCTION

Many biochemical reactions are complex and involve consecutive equilibria along the reaction pathway (1). Further complicating matters, the rate of passage between neighboring free energy minima can be modulated by conformational changes of an enzyme. To understand the mechanism of reaction in such complex systems, it is crucial to probe, in detail, intermediates along the reaction pathway. Myoglobin (Mb) is an 18 kDa heme protein that has long served as a model system for probing protein control of ligand binding and discrimination (2). The active binding site in Mb consists of an iron(II)-containing porphyrin known as a heme that is embedded within the hydrophobic interior of the globular protein (see Fig. 1). This heme reversibly binds O_2 as well as biologically relevant [and infrared (IR) active] NO (3) and CO (4). According to sequence homologies among 60 mammalian species (5), most residues in the immediate vicinity of the distal

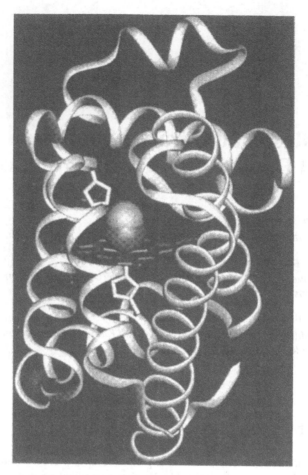

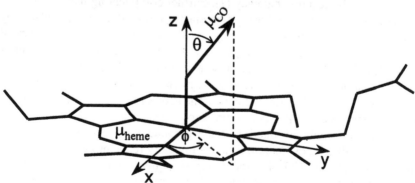

side of the heme are 100% conserved. These highly conserved residues evidently mediate the passage of ligands to and from the active binding site and modulate their binding affinities. For example, CO binds to free heme about $10^3 - 10^4$ times as strongly as O_2 (6) but binds to the heme in Mb only 30 times as strongly as O_2 (7). Because CO is produced endogenously by the metabolism of heme (8), discrimination against this toxic ligand is thought to be biologically important. The protein structure also influences the kinetics of geminate rebinding, with some rates being time-dependent and slaved corresponds to conformational changes of the protein.

To understand the functional role of highly conserved residues in heme proteins, it is crucial to know the time-dependent position and orientation of a ligand as it departs from the active binding site. Because ligands bound to the heme of Mb can be detached by illumination with light, ligand-binding intermediates can be investigated with time-resolved pump-probe spectroscopy. Furthermore, using ultrashort polarized mid-IR probe pulses, information about the ligand orientation can be determined as a function of time. Using a combination of near- and mid-IR time resolved spectroscopic methods, we have investigated heme and ligand dynamics in a variety of heme proteins under near ambient conditions. From these studies has emerged a clearer picture of the impact of photolysis on the heme and the ligand, the time-dependent orientation and population of ligand binding intermediates, and the functional role of a ligand docking site.

II. EXPERIMENTAL

A. Sample Preparation

Skeletal horse myoglobin (h-Mb), sperm whale myoglobin (sw-Mb), and human hemoglobin A (Hb) were prepared in deoxygenated D_2O buffered with 0.1 M potassium phosphate, pD 7.5. Skeletal h-Mb was also prepared in deoxygenated 75% glycerol/water (v/v) buffered with 0.1 M potassium phosphate, pH 7.1. h-Mb was obtained in its lyophylized form from Sigma, sw-Mb was obtained as a gift from Prof. John Olson from Rice University,

---

Figure 1 Myoglobin and the heme-CO coordinate system: θ is the angle between C–O and the heme plane normal and ϕ is the azimuthal angle about the normal. The heme absorbs visible light polarized in the plane of the heme. For certain wavelengths, including 527 nm, the absorbance in the x and y directions are equal and the heme is well described as a circular absorber. Around 590 nm, the absorbance in the x and y directions differs, and the heme becomes an elliptical absorber.

and Hb was prepared from the hemolysate of fresh red blood cells (7) and purified by chromatography on a DEAE-Sephacel column. The solutions were equilibrated with 1 atm of CO in the presence of a modest excess of sodium dithionite to produce carbon monoxy heme.

Because the mid-IR transmission of D_2O drops dramatically to the blue of 2100 cm^{-1}, all protein samples prepared in D_2O were liganded with ^{13}CO, thereby shifting the unbound CO spectrum into a region with greater IR transmission. The IR transmission through glycerol/water (G/W) is relatively flat around 2100 cm^{-1}, so samples prepared in G/W were liganded with ^{12}CO.

The protein concentration used in the near-IR measurements was about 7 mM in heme. The protein solution was filtered through either a 0.45 μm or a 0.22 μm membrane filter prior to being anaerobically transferred into a gas-tight 1-mm-path-length sample cell. The sample cell was rotated sufficiently quickly to ensure that each photolysis pulse illuminated a fresh volume within the protein sample. The temperature of the sample was about 28°C, which was elevated somewhat from ambient temperature due to heating by the motorized mount.

The protein concentration used in the mid-IR measurements was about 15 mM in heme. The protein solution was centrifuged at 5000 g for 15 minutes prior to being anaerobically transferred into a gas-tight 0.1-mm-path-length sample cell. The sample cell consisted of two 2-mm-thick calcium fluoride windows separated by a 0.1 mm spacer. The sample cell was mounted in a refrigerated enclosure, chilled to 10°C, and rotated fast enough to ensure that each photolysis pulse illuminated a fresh volume within the protein sample.

B. Time-Resolved Near- and Mid-IR Spectrometer

An ultrafast time-resolved near- and mid-IR absorption spectrometer was designed to achieve high sensitivity, ultrafast time resolution, and broad tunability in the near- and mid-IR regions (see Fig. 2). The details of this spectrometer are described elsewhere (9). Briefly, MbCO was photolyzed with a linearly polarized laser pulse, whose polarization direction was controlled electronically by a liquid crystal polarization rotator. The photolyzed sample was probed with an optically delayed, linearly polarized IR pulse whose transmitted intensity was spectrally resolved with a monochromator and detected with either a Si photodiode (near-IR; ≈10 cm^{-1} bandpass) or a liquid nitrogen–cooled InSb photodetector (mid-IR; ≈3 cm^{-1} bandpass). To measure the sample transmission, this signal was divided by a corresponding signal from a reference IR pulse

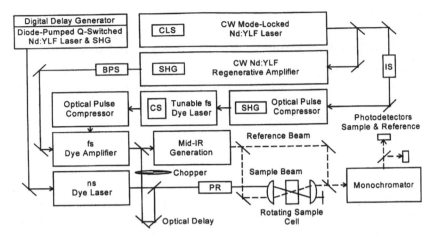

Figure 2 Schematic diagram of the time-resolved IR spectrometer. CLS, cavity length stabilizer; SHG, second harmonic generator; IS, intensity stabilizer; BPS, beam pointing stabilizer; CS, coherent seeder; PR, polarization rotator.

routed through the same monochromator and detected with a matched photodetector. A synchronous light chopper blocked every other pump pulse, thereby interleaving in time the pumped and unpumped sample transmittances. The photolysis-induced change of the sample absorbance, ΔA, was computed at each delay time and at each wavelength from the sample transmittances measured with and without the photolysis pulse. Polarization anisotropy studies were performed by alternately photolyzing the sample with parallel and perpendicular polarized pump pulses, thereby recording polarized absorbance spectra in a back-to-back fashion. The 1.2–1.5 kHz repetition frequency of the laser system permitted extensive signal averaging and led to the high-quality time-resolved spectra reported here. The shortest time resolution attained in this work was limited by the cross-correlation of the visible pump and broadband mid-IR probe pulses, which was <220 fs FWHM. The photolysis wavelength was 597 nm for the near-IR measurements and 592 for the mid-IR measurements. Experiments that required only picosecond time resolution were performed by photolyzing the sample with optically delayed picosecond pulses from the doubled output of a Quantronix 117 Nd:YLF regenerative amplifier (35 ps; 527 nm). Measurements at times beyond the ~4 ns limit imposed by the length of the optical delay line were accomplished by photolyzing the sample with pulses from an electronically delayed Spectra Physics TFR ns laser (4 ns; 523 nm).

III. THEORY

A. Vibrational Spectrum of Orientationally Constrained CO

When interpreting time-resolved mid-IR spectra, it is beneficial to consider the influence of rotational dynamics on the vibrational spectrum of a heteronuclear diatomic. It was shown more than 30 years ago that the vibrational absorption spectrum of a diatomic is related to its transition dipole correlation function $\langle \mu(0) \cdot \mu(t) \rangle$ through a Fourier transform (10):

$$\hat{I}(\omega) = \frac{1}{2\pi} \int_{-\infty}^{\infty} dt e^{-i\omega t} \langle \mu(0) \cdot \mu(t) \rangle \tag{1}$$

where $\hat{I}(\omega)$ is the frequency-dependent absorption normalized to unit area, $\mu(t)$ is a vector oriented along the direction of the transition dipole, and the brackets $\langle \cdots \rangle$ indicate an ensemble average. The magnitude of the vector $\mu(t)$ is unity at zero time but decays to zero due to vibrational dephasing. Therefore, the dipole correlation function reveals ensemble-averaged molecular dynamics consisting of both rotational motion and vibrational dephasing, and its Fourier transform reveals the spectral band shape that arises from those molecular dynamics.

To better understand the relationship between spectra and molecular motion, it is instructive to consider a quantum rotor, a hindered rotor in a disordered environment, and a hindered rotor in an ordered environment. To begin, our focus shifts to the orientational contribution to the dipole correlation function. The orientational correlation function lacks the effects of vibrational motion, but because the transition dipole is polarized along the bond axis, it is otherwise similar to the dipole correlation function. The orientational correlation function for a quantum rotor with angular velocity $\omega = (\hbar/I)\sqrt{J(J+1)}$ evolves according to $\cos(\omega t)\exp(-t/\tau)$, where J is the rotational quantum number, $\hbar$ is Planck's constant, I is the moment of inertia for the rotor, and τ is a phenomenological damping time that accounts for rotational dephasing. The Fourier transform of this orientational correlation function is a Lorentzian whose linewidth is inversely proportional to the damping time τ, i.e., faster damping times lead to broader spectral features. For ro-vibrational spectra with $J > 0$, the electric dipole selection rule $J \rightarrow J \pm 1$ leads to a pair of absorption lines that contribute to the well-known R and P branches in the IR spectrum of heteronuclear diatomic molecules. Therefore, the ro-vibrational spectrum of a Boltzmann-distributed ensemble of quantum rotors with $2J + 1$ degeneracy exhibits a series of equally spaced rotational lines (neglecting centrifugal distortion and rotation-vibration coupling) whose intensities are dictated by Boltzmann statistics and degeneracy factors (Fig. 3A). As the temperature is raised, the Boltzmann population distribution shifts to higher J values and the overall envelope of the P and R branches becomes broader. There are two relevant widths represented in the ro-vibrational spectrum: the width of the individual transitions and the overall width of the envelope containing the P and R branches. According to the Fourier transform relationship between spectra and dipole correlation functions, these two widths in the spectral domain transform as two different time scales in the time-dependent decay of the orientational correlation function. The corresponding orientational correlation function for a Boltzmann-distributed ensemble of quantum rotors is shown in Fig. 3B. The rapid initial decay of the orientational correlation function arises from rotational motion, whose Boltzmann-distributed angular velocities leads to rapid decorrelation of the ensemble of quantum rotors and causes the orientational correlation function to decay toward zero. Because the decorrelation is incomplete at a

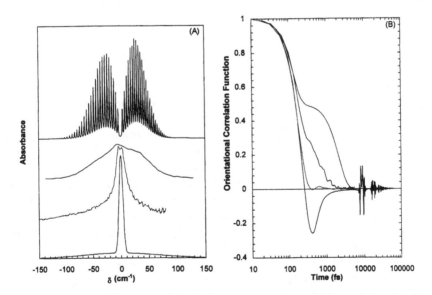

Figure 3 (A) Spectra of CO in a variety of environments and (B) corresponding orientational correlation functions. In (A), the four curves correspond to (top to bottom) CO in the gas phase, in cyclohexane, in water, and in an orientationally constrained environment (theoretical spectrum with $\alpha = 0.5$; see text). In (B), the corresponding ordering is bottom to top. Note the similarity of the fast, inertial contribution to the orientational correlation function at early times.

time corresponding to roughly one-half the average rotational period, the orientational correlation function becomes negative before decaying to zero. Clearly the decay dynamics are dictated by the average angular velocity of the rotor, which is related to temperature according to $\omega \approx \sqrt{2kT/I}$. For CO at room temperature, the rapid inertial contribution to the decay of the orientational correlation function transforms as a ~ 90 cm^{-1} spectral width, which is comparable to the overall width of the envelope containing the P and R branches of the ro-vibrational absorption spectrum of CO. Consequently, the overall breadth expected for an absorption band can be rationalized in terms of a rotor's temperature and its moment of inertia. How, then, do we rationalize the narrow features observed in the ro-vibrational spectrum of CO? Because the rotor is quantized, there are periodic recurrences of the orientational correlation function (see Fig. 3B), the amplitude of which decays according to a phenomenological damping time τ. Therefore, the orientational correlation function can be considered biphasic, with the faster inertial decay dictating how broad the overall absorption envelope must be and the slower damping time of the recurrences dictating how narrow the widths of individual features can be. This minimum width corresponds to the homogeneous width of the feature. Of course, environmental heterogeneity (inhomogeneous broadening) can broaden the feature beyond its homogeneous width.

A diatomic dissolved in a solvent can be thought of as a hindered rotor in a disordered environment with the diatomic rotating inertially between angular momentum–changing collisions with the surrounding solvent. Because the angular velocity of a diatomic is dependent on temperature and its rotational inertia, the presence of the solvent need not alter the velocity of its inertial rotation. Consequently, the inertial contribution to the decay of the orientational correlation function and the corresponding breadth of the absorption spectrum is expected to be nominally independent of environment. The influence of the solvent shows up at times corresponding roughly to the mean time between angular momentum–changing collisions, whereupon the inertial motion is interrupted and the orientation of the rotor can begin to evolve diffusively. The orientational correlation function should, therefore, exhibit a transition between inertial and diffusive motion. Where that transition occurs depends on the average angular rotation between angular momentum–changing collisions. Rotation by about 90 degrees between such collisions would cause the orientational correlation function to decay to zero in a single rapid phase. Rotation by about 180 degrees would cause the orientational correlation function to become negative before decaying to zero. Rotation by angles much less than 90 degrees would lead to a smaller amplitude inertial decay followed by a larger amplitude diffusive decay. How narrow a feature could appear in the spectral domain would be dictated by the time scale for the slower diffusive decay of the orientational correlation function. Moreover, the integrated absorbance contained within the narrow feature would be dictated by the relative amplitudes of the diffusive and inertial phases. Since angular momentum–changing collisions rapidly destroy any coherence in the rotational motion, the orientational correlation function would not exhibit periodic recurrences so the vibrational spectrum would not exhibit rotational structure. Therefore, the spectrum of CO in a disordered environment would exhibit a broad envelope with the possibility of a single narrower feature centered on top of that envelope. The amplitude of the narrower feature could be small or large, depending on the degree of angular rotation between angular momentum–changing collisions. For example, the IR spectrum of CO dissolved in cyclohexane, shown in Fig. 3B, is approximately 90 cm^{-1} broad and nearly featureless, whereas CO in water exhibits a ≈20 cm^{-1} FWHM feature centered on top of a broad pedestal with about one third of the integrated absorbance contained within the narrower feature. These spectra suggest that the average rotation of CO between collisions in cyclohexane is of the order of 90 degrees while that for CO in water is significantly less than 90 degrees.

A diatomic localized within a protein can be thought of as a rotor in an ordered environment. In this context, the distinction between ordered and disordered environments is that an ordered environment can constrain the rotor to point in a particular direction relative to the molecular frame. When orientationally constrained, the rapid inertial contribution to the decay of the orientational correlation function is reduced in amplitude and the decay to zero becomes biphasic. As in the disordered case, the absorption spectrum is expected to have a narrower feature centered on top of a broader pedestal with their relative integrated absorbances partitioned according to the relative amplitudes of the slow and fast decay. In addition, the environment anisotropy can cause the vibrational frequency to be dependent on orientation. If the rotational diffusion of the rotor is slow, the environment anisotropy will not be averaged out and the vibrational spectrum may exhibit multiple features on top of a broad pedestal. How narrow those features can become is limited by the time scale for interconversion among the preferred orientations. Generally speaking, the greater the orientational constraint, the greater the fraction of the oscillator strength appearing within the narrower feature(s) and the narrower those features can become. Consequently, CO localized within a highly ordered protein might be expected to exhibit an absorption spectrum with more than one narrow feature approximately centered on top of a broad pedestal. Unless the pedestal contains the majority of the integrated CO oscillator strength, it can easily "disappear" into the background noise of a measured spectrum. A phenomenological description for the decay of the orientational correlation function, based on a sum of Gaussians with differing amplitudes and variances, has been employed to illustrate how differing degrees of orientational constraints can influence the vibrational spectrum of a rotor (11). The result obtained when half of the decay is inertial ($\alpha = 0.5$) is included in Fig. 3. Note that when the amplitudes of the inertial and diffusive decays are the same, the broad pedestal can be quite small in amplitude relative to the narrower feature.

B. Orientation of CO via Photoselection

When measuring absorption spectra, one records a signal that is related to the wavelength-dependent probability of making a spectroscopic transition. From the molecular point of view, this probability is proportional to the dot product $\hat{\mu} \cdot \hat{p}$ where $\hat{\mu}$ is the molecular transition moment and $\hat{p}$ is the photon polarization direction. When the orientational distribution of the molecules is isotropic (not crystalline, liquid crystalline, or bound to a surface), its absorption spectrum represents the orientationally averaged probability of making a spectroscopic transition and the measured spectrum is independent of polarization direction. When the orientational distribution of the molecules is anisotropic, the probability of making a spectroscopic transition depends on the polarization direction, and that dependence can be exploited to deduce the direction of the transition moment relative to the laboratory frame. Because transition moments are often trivially related to the orientation of the molecule, structural information can be deduced from polarized absorption measurements on anisotropic samples.

The pump pulse in time-resolved pump-probe absorption spectroscopy is often linearly polarized, so photoexcitation generally creates an anisotropic distribution of excited molecules. In essence, the polarized light "photoselects" those molecules whose transition moments are nominally aligned with respect to the pump polarization vector (12,13). If the anisotropy generated by the pump pulse is probed on a time scale that is fast compared to the rotational motion of the probed transition, the measured anisotropy can be used to determine the angle between the pumped and probed transitions. Therefore, time-resolved polarized absorption spectroscopy can be used to acquire information related to molecular structure and structural dynamics.

The visible absorption spectrum of MbCO is dominated by the Q-band of the heme. Over a broad range of wavelengths, the transition moments along the x- and y-direction vectors of the heme (see lower panel of Fig. 1) are nearly identical and the heme behaves as a circular absorber (14). In contrast, near the red edge of the Q-band absorption, one of the two direction vectors has a stronger absorption probability and the heme becomes an elliptical absorber. In either case, photolysis of MbCO with linearly polarized visible light creates an anisotropic distribution of MbCO, Mb, and CO molecules. Measurements of the generated anisotropy can unveil the orientation of CO bound to and dissociated from Mb as well as the rotational dynamics of CO as it translocates from the binding site to the docking site.

First we focus on the CO orientation when bound to and after dissociation from a heme protein. When a solution containing a carbon monoxy-heme protein is illuminated with linearly polarized visible light, hemes whose planes are aligned with the polarization direction absorb light preferentially. The ligands bound to these "photoselected" hemes are dissociated with high quantum efficiency, leading to a loss of bound CO and the production of "free" CO. If the IR transition moment of CO is oriented at a particular angle θ relative to the heme plane normal, and if we assume the heme is a circular absorber, the ratio of its perpendicular and parallel polarized IR absorbance, $\Delta A^{\perp}/\Delta A^{\parallel}$, becomes a simple analytic function of θ (15):

$$\frac{\Delta A^{\perp}}{\Delta A^{\parallel}} = \frac{4 - \sin^2 \theta}{2 + 2\sin^2 \theta} \tag{2}$$

The measured polarization ratio can theoretically range from 2 ($\theta = 0$ degrees) down to 0.75 ($\theta = 90$ degrees). When polarized absorbance measurements are made in a solution, rotational tumbling of the protein randomizes the orientation of the photoselected hemes. Therefore, the measurement must be made on a time scale that is short compared to the rotational diffusion time, which is 8 ns for Mb in H_2O at 288 K (16). When measurements are made in low-temperature glasses, where the protein orientation is frozen and ligand rebinding is slow, the polarized IR spectra can be measured with conventional IR spectrometers (17). In either case, this equation is valid only in the small signal limit, i.e., the fraction of molecules photolyzed must be small (the measured polarization ratio asymptotically approaches unity as the fraction photolyzed becomes large). Moreover, the angle calculated using this equation assumes that the orientational distribution is a delta function in θ. Finally, what is determined is the orientation of the transition moment of CO, which is parallel to the CO bond axis when "free" but is not necessarily perfectly parallel to the C–O bond axis when CO is bound to the heme (18).

To investigate rotational dynamics, it is more natural to consider the polarization anisotropy, $r(t)$, a theoretical quantity that can be experimentally determined using:

$$r(t) = \frac{\Delta A^{\parallel}(t) - \Delta A^{\perp}(t)}{\Delta A^{\parallel}(t) + 2\Delta A^{\perp}(t)} \tag{3}$$

At photolysis wavelengths where MbCO is well described as a circular absorber, $r(t)$ can vary from -0.2 ($\theta = 0$ degrees) to 0.1 ($\theta = 90$ degrees). At photolysis wavelengths where MbCO is an elliptical absorber, $r(t)$ retains the lower limit -0.2, but the upper limit depends on several factors and the numerical value of $r(t)$ is no longer trivially related to the angle θ. Taken to the limit where the heme becomes a linear absorber, the upper limit of $r(t)$ becomes 0.4 (for $\theta = 90$ degrees *and* CO parallel to the heme transition moment). Clearly, quantitative determination of θ requires prior knowledge of the nature of the heme absorber at the photolysis wavelength.

IV. RESULTS

A. Laser Photolysis: A Sledgehammer or a Scalpel?

It has long been known that ligands axially bound to a heme can be detached by illumination with visible light. This phenomenon has been exploited in more than four decades of flash photolysis studies of ligand dynamics in respiratory proteins starting with the work of Gibson (19,20). Given the long history and widespread use of flash photolysis, it is pertinent to ask the following question: Does photon absorption by a heme surgically excise the ligand, or does it "blast" it away and open up reaction pathways that might not exist when the ligand is detached thermally? In other words, are results from flash photolysis studies physiologically relevant? This question warrants careful consideration and is tackled in this section. To begin this discussion, we consider the thermal consequences of photon absorption. If, for example, the energy of a visible photon is deposited in a heme and distributed among all 24 porphyrin skeleton atoms, its Boltzmann temperature jumps by more than 400°C (21). According to molecular dynamics simulations of Mb in vacuo, the "hot" heme cools biexponentially with each of the two time constants ($\tau_1 = 1$–4 ps; $\tau_2 = 40$ ps) contributing approximately equally to the decay (21). Therefore, the magnitude of the thermal perturbation is expected to be large and sustained for up to several tens of ps. Since that theoretical study, the electronic and thermal consequences of light absorption have been probed experimentally with both near- and mid-IR spectroscopy. The results of these experimental studies are now described.

1. Near-IR Study of Heme Relaxation

To experimentally probe the electronic and thermal consequences of flash photolysis, a femtosecond time-resolved near-IR study of photoexcited Mb was undertaken (22). This study probed the spectral evolution of band III, a weak ($\varepsilon_{max} \approx 100$ M$^{-1} \cdot$ cm^{-1}) near-IR charge transfer transition (14) centered near 13, 110 cm^{-1} that is characteristic of five-coordinate ferrous hemes in their ground electronic state (S = 2). Because band III is absent when the heme is electronically excited, the dynamics of its reappearance provides an incisive probe of relaxation back to the ground electronic state. Moreover, because the spectral characteristics of band III (integrated area; center frequency; line width) correlate strongly with temperature (23–26), the spectral evolution of band III also probes its thermal relaxation.

Time-resolved absorbance spectra are typically recorded as difference spectra with depletion of the ground state population appearing as negative-going ground state features and the photoexcited population appearing as positive-going features in absorbance and as negative-going features in stimulated emission. For photolyzed Mb, no stimulated emission is observed in the near-IR region, so the photoexcited population appears only as positive-going features. Because the ground state bleach and the photoproduct absorbance have features in the same spectral region, the spectral evolution of the photoexcited population cannot be measured without interference from the ground state bleach. Nevertheless, the photoproduct spectral evolution can be recovered from the transient absorption spectrum by adding an appropriately scaled ground state absorbance spectrum. The scale factor required is simply the fraction of Mb photoexcited by the pump pulse. Whereas this fraction is not easily determined by direct measurement, it can be determined by indirect measurement using a closely related system: MbCO. The integrated area of band III at equilibrium was compared with that of MbCO after photolysis under identical conditions (i.e., same heme concentration and pump energy). Assuming that the quantum yield for the photodissociation of MbCO is unity and correcting for small differences in the absorbance of Mb and MbCO at the pump wavelength, the fraction of photoexcited Mb within the probe-illuminated volume of the sample was determined (22). The spectra shown in Fig. 4 were recovered by adding an

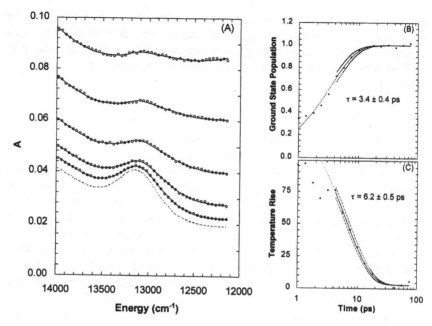

Figure 4 Time-resolved near-IR absorbance spectra of photoexcited Mb in 70/30 (w/w) glycerol/water at 1.00, 1.78, 3.16, 5.62, and 10 ps (open circles; top to bottom). The spectra were recovered by adding an appropriately scaled ground state Mb absorbance spectrum (dotted line) to the corresponding transient near-IR absorbance spectra. The complete series of spectra (8 per decade) were modeled (solid lines) with a temperature- and population-parameterized Gaussian function plus a cubic polynomial background. The least-squares parameters for the time-dependent population (B) and temperature rise (C) are plotted on a logarithmic time scale. The data beyond 4 ps are well described (solid lines) by the functions Population $= 1 - \exp(-t/3.4 \pm 0.4 \text{ ps})$ and $\Delta T = 140$ K $\exp(-t/6.2 \pm 0.5 \text{ ps})$. The dotted curves are extrapolations of these functions to early times, and the dash-dot curves reflect the estimated uncertainty. (Adapted from Ref. 22.)

appropriately scaled equilibrium band III spectrum (12.8%; illustrated by the dashed line) to each of the transient absorption spectra. Note that the spectra have not been offset from one another, i.e., the differences in the background are real.

Spectra at times earlier than 1 ps (not shown in Fig. 4) reveal a 1750 cm^{-1} broad (FWHM) feature centered at 12, 160 cm^{-1} that decays with a 1.0 ± 0.1 ps time constant (22). Because this band is red-shifted more than 5, 500 cm^{-1} from the Q-band, it cannot be a vibronic band associated with the ground electronic state of a "hot" heme. The integrated intensity of this feature is within a factor of two of the bleach of the Q-band, suggesting that this relaxation pathway is likely the dominant pathway. Finally, because this relaxation is much quicker than the reappearance of band III, the 1 ps time constant corresponds to internal conversion between two excited electronic states, not ground state recovery.

The spectral evolution in Fig. 4 reveals a broad, featureless background offset that decays in amplitude as band III both grows in integrated intensity and blue shifts in frequency. The growth of band III arises from excited state relaxation back to the ground electronic state, and its blue shift arises from cooling of the heme. To quantify this spectral evolution, each time-resolved near-IR spectrum was modeled with a Gaussian function added to a cubic polynomial background. The nonlinear least-squares parameters characterizing band III (integrated area; center frequency; FWHM) were found to evolve systematically beyond $\sim$3 ps but varied less systematically at earlier times where the amplitude of band III is small. To improve the analysis, an additional constraint was imposed on the parameters describing band III. This constraint is based on the fact that band III varies systematically with temperature. To implement this constraint, the equilibrium band III spectrum was measured about every 2 degrees from 0 to 70°C, and the temperature dependencies of its integrated area, center frequency, and FWHM were modeled as quadratic polynomials in temperature. Given this parameterization, a two-parameter characterization of band III (population and temperature) was employed to model the spectral evolution shown in Fig. 4 (22). This parameterization of the Gaussian function required one less parameter and therefore provided more robust estimates of those parameters. The experimental band III spectra are well described by this model at all delay times shown in Fig. 4, which also depicts the time dependence of the ground state population and the heme temperature rise. At 4.22 ps, the temperature of the heme is estimated to be approximately 100°C, a temperature modestly beyond the range over which band III was characterized. To minimize errors arising from extrapolating too far beyond reliable data, the time dependence of the ground state population and the heme temperature were modeled with the fitting range confined to times beyond 4 ps (solid lines). Beyond 4 ps, the dynamics were well described by exponential decays with the ground state recovering with a time constant of 3.4 ± 0.4 ps and the thermal relaxation proceeding with a time constant of 6.2 ± 0.5 ps.

The near-IR study of band III suggests that relaxation back to the ground electronic state proceeds with a 3.4 ps time constant. This value is consistent with a 4.8 ± 1.5 ps excited state lifetime determined using saturated resonance Raman spectroscopy (27) and a 3.2 ps decay seen in Soret band pump-probe studies (28). A recent Soret band pump-probe study reported a 3.6 ± 0.2 ps time constant for the ground state recovery and strengthened that assignment by measuring the time-dependent absorption anisotropy at 800 nm using diffractive optics-based heterodyned detection of the imaginary component of the material nonlinear susceptibility (29). The four different methods are all in excellent agreement with each other. Because the near-IR study was carried out with Q-band excitation while the other studies were carried out using Soret band excitation, it appears the dynamics of ground state recovery are nominally independent of excitation wavelength.

What are the consequences of the $\sim$3.4 ps electronic relaxation back to the ground state? Ligands such as O_2, CO, and NO bind to heme in its ground electronic state but are ejected when the heme becomes photoexcited. It is difficult to predict whether the intermediate excited electronic states have a propensity to bind ligands. However, the lack of ps geminate rebinding with CO suggests that these intermediate electronic states are not receptive to ligands. Therefore, the electronic excitation may hold the ligand at bay for a few ps, providing time for the surrounding protein to trap the ligand in a docking site. It is worth noting that the intermediate excited electronic states would have an influence on other ps time-resolved studies, i.e., spectra measured at times shorter than a few ps would be complicated by the presence of hemes in various electronic states.

The near-IR study of band III suggests that thermal relaxation of photoexcited Mb is exponential with a 6.2 ps time constant. In reality, this thermal relaxation is a nonequilibrium process, so any characterization of a time-dependent heme temperature is only approximate. Further complicating matters, electronic relaxation proceeds through more than one intermediate state so the photon energy is converted into heat in a sequence of steps that can be separated by as much as a few picoseconds. Consequently, the "temperature" of an individual photoexcited heme would be expected to jump suddenly with each step of electronic relaxation and would decay as the excess kinetic energy flows into the surroundings. The time-resolved absorbance spectra in Fig. 4 probe an ensemble of hemes, each having its own thermal history. In contrast, the parameterization of band III used to model the spectra in Fig. 4 was made with the ensemble of hemes in thermal equilibrium. Therefore, the calibration of temperature in Fig. 4 is only approximate. Nevertheless, one might expect the near-IR measurements to provide a reasonable ensemble-averaged estimate of the time-dependent heme temperature, especially for times beyond the electronic relaxation time of 3.4 ps. Therefore, the 6.2 ps thermal relaxation time constant should be reasonably accurate. This conclusion is supported by recent measurements of the real component of the material nonlinear susceptibility of photoexcited Mb (29). They assigned a decay component of 5–7 ps to thermal relaxation of the heme, which is in very good agreement with the 6.2 ± 0.5 ps estimate from the near-IR band III study.

Mizutani and Kitagawa measured the time-dependent Stokes and anti-Stokes Raman intensities of the heme v_4 band after photoexcitation and used the relative intensities to estimate its "temperature" and thermal relaxation dynamics (30). They found the population relaxation to occur biexponentially with 1.9 ps (93%) and 16 ps (7%) time constants. The dominant 1.9 ps population relaxation correlates with a 3.0 ps thermal relaxation, which is a factor of 2 faster than the ensemble averaged temperature relaxation deduced from the near-IR study of band III. The kinetic energy retained within a photoexcited heme need not be distributed uniformly among all the vibrational degrees of freedom, nor must the energy of all vibrational modes decay at the same rate. Consequently, a 6.2 ps ensemble-averaged estimate of the heme thermal relaxation is not necessarily inconsistent with a 3 ps relaxation of v_4.

The near-IR measurements were performed on photoexcited Mb in which all the photon energy is converted into heat. In contrast, when photolyzing MbCO, a portion of the photon energy is required to break the Fe–CO bond and the remainder is converted into heat. For example, if a 555 nm photon was used to photolyze the sample (corresponding to the center of the heme Q-band absorption) and if the bond dissociation energy of Fe–CO was 16.2 kcal/mol (31), then approximately two thirds of the photon energy would be available to heat the heme. Therefore, the magnitude of the heating effect in photolysis studies of MbCO would be only two thirds as great as that reported for photoexcited Mb. Because the heme cooling rate should be equally fast in photolyzed MbCO, the protein quickly loses memory of how the ligand was dissociated in the first place. Consequently, after the first 10–20 ps, the ligand dynamics should be independent of the detachment mechanism and photolysis studies of ligand rebinding and escape dynamics should be physiologically relevant.

2. Mid-IR Study of CO Relaxation

Whereas near-IR spectroscopy provides an incisive probe of the electronic and thermal state of the heme after flash photolysis, it tells us nothing about the state in which the ligand is created. Mid-IR spectroscopy, on the other hand, can probe the vibrational state of the dissociated ligand, provided it is not a homonuclear diatomic. The impact of flash photolysis on a dissociated ligand was recently investigated with time-resolved IR spectroscopy (32). A sample containing $Hb^{13}CO$ was photolyzed with picosecond laser pulses (35 ps; 527 nm) and probed with optically delayed femtosecond IR pulses (200 fs; $\sim$5 μm). The photolysis pulse dissociated approximately 35% of the $Hb^{13}CO$ within the probe-illuminated volume and time-resolved mid-IR absorbance spectra were measured at a series of times that were equally spaced on a logarithmic time scale (see Fig. 5). Four features are readily apparent in the early time spectra, but only two features, labeled B_1 and B_2 [following Alben et al. (33)], survive in the later time spectra. The surviving features correspond to ^{13}CO in its ground vibrational state ($\nu = 0$) and trapped within a protein docking site (34). The two satellite features labeled B_1^* and B_2^* decay exponentially with a 600 ± 150 ps time constant. B_1^* and B_2^* are well described as red-shifted replicas of B_1 and B_2, and the experimentally determined shift between $B_1(B_2)$ and $B_1^*(B_2^*)$ is 26 cm^{-1}, similar to the 25.3 cm^{-1} shift between the ground ($1 \leftarrow 0$) and hot band ($2 \leftarrow 1$) vibrational transitions of ^{13}CO in the gas phase (35). Therefore, the decaying features are assigned to ^{13}CO generated in its first excited vibrational state ($\nu = 1$).

The 600 ps vibrational cooling time constant for "docked" CO is far slower than the $\sim$18 ps cooling time constant measured for bound CO in MbCO (A_1-state) (36,37) and in HbCO (37) and is far slower than the 6.2 ps time constant for cooling of the heme (22). The sluggish cooling rate of photodetached CO demonstrates that coupling between the high-frequency CO vibrational motion and the lower-frequency acceptor modes of the heme and protein is weak. A theoretical study of the vibrational relaxation rate has been performed using the Landau-Teller model, and excellent agreement between experiment and theory was obtained (32). Moreover, that study included a normal mode analysis which identified those protein residues that act as the primary "doorway" modes in the vibrational relaxation of the oscillator. From that study, they concluded that the distal histidine plays an important role in the vibrational relaxation.

From the data in Fig. 5 one can also deduce the vibrational temperature of photodetached CO. Assuming the absorbance cross section for the first hot band transition ($2 \leftarrow 1$) is twice that for the ground state transition ($1 \leftarrow 0$) (38), the population of CO produced in its first excited vibrational state represents only 3.6% of the total photolyzed population. A 3.6% nascent yield in $\nu = 1$ corresponds to a Boltzmann temperature

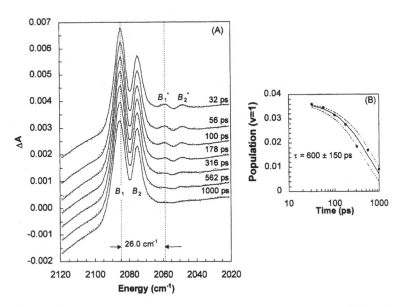

Figure 5 (A) Picosecond time-resolved mid-IR absorbance spectra of photolyzed $Hb^{13}CO$. Photolysis of $Hb^{13}CO$ detaches ^{13}CO from the heme, whereupon it becomes trapped in a docking site located near the heme-binding site. The docked ^{13}CO is produced predominately in its ground vibrational state ($\nu = 0$) and gives rise to two major features, B_1 and B_2. A small (3.6%) portion of the ^{13}CO is produced in its excited vibrational state ($\nu = 1$) and gives rise to two satellite features, B_1^* and B_2^*. Vibrational relaxation back to the ground state ($\nu = 1 \to 0$) causes the satellite features to disappear. The solid curves were obtained by least-squares fitting the experimental spectra with a constrained model involving Gaussian functions. (B) The time-dependent population of CO in its $\nu = 1$ vibrational state is also shown (reported as a percentage of the total population of "docked" CO). The nascent yield of vibrationally excited CO is about 3.6%, and this population decays exponentially with a time constant of 600 ± 150 ps. (Adapted from Ref. 32.)

of 825 K, which is hotter than the ≤ 478 K expected if the photon energy beyond that required to break the Fe–CO bond were distributed uniformly over all degrees of freedom of the heme-CO (21,22,39). Clearly, CO is not in thermal equilibrium with the heme when it is ejected. This result should come as no surprise because the time scale for CO photodetachment is within 100 fs of photoexcitation (28,34), a time too short for the CO to emerge in thermal equilibrium with the heme. Rather, the excited vibrational state population is dictated by the dissociation mechanism, which strongly favors production of ground state CO. This, too, should come as no surprise: one might expect the dominant pathway for generating vibrationally excited CO to be the impulsive half collision with Fe after the electronic potential becomes repulsive. Because the CO bond lengths in bound and unbound CO are similar, energy transfer through an impulsive half collision would be rendered inefficient (40,41). Because the nascent population of vibrationally excited CO is small, it is of little consequence to measurements of ligand rebinding and escape dynamics.

B. Evidence for a Ligand Docking Site in the Heme Pockets of Mb and Hb

Time-resolved mid-IR spectra of several photolyzed heme proteins are shown in Fig. 6. The negative-going features correspond to loss of bound CO (A states) and the positive-going features correspond to CO dissociated from the heme (B states). Note the transition frequencies and relative intensities of the A and B states. When CO is bound to heme, back-bonding between the CO π-orbitals and the iron d-orbitals weakens the CO bond and enhances its transition moment (42). Compared to "free" CO, the bound CO vibrational frequency is red shifted about 200 cm^{-1} and its integrated oscillator strength at 5.5 K is enhanced 21.7 ± 1.6 times (33).

The integrated areas, center frequencies, and line widths of the A and B states in Fig. 6 were characterized by modeling these features as a sum of Gaussians on a quadratic (B states) or cubic (A states) polynomial background (11). The center frequencies and line widths of the A states in sw-Mb^{13}CO:D$_2$O, which were recorded with ~200 fs duration mid-IR pulses, were found to be comparable to those reported previously (33,43). The agreement between the ultrafast time-resolved and static measurements of the A states should not be surprising: the time-resolved A state spectra measured at 100 ps simply recover the equilibrium A state spectra plotted in a negative-going fashion. On the other hand, the B states near ambient temperature are short-lived intermediates whose characterization requires time-resolved methods. The B state spectra shown in Fig. 6B reveal features that are significantly narrower than any feature ever assigned to non bonded CO in the condensed phase near room temperature. For example, spectra of CO dissolved in a host of organic solvents (44,45) are very similar to the experimental CO spectrum in cyclohexane shown in Fig. 3A, the FWHM of which is 98 cm^{-1}. In contrast, the B states in Fig. 6B are more than 10 times narrower. Clearly, the interior of the protein, which is composed primarily of hydrophobic residues, presents an environment quite different from that of liquid organic solvents. Moreover, this unique environment

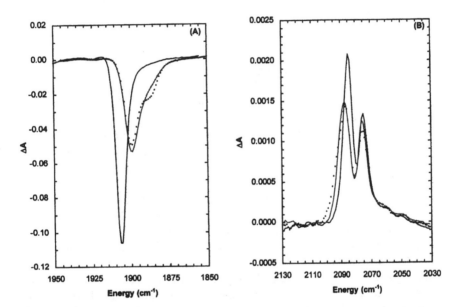

Figure 6 Time-resolved mid-IR spectra of photolyzed h-Mb^{13}CO:D$_2$O (gray line), sw-Mb^{13}CO:D$_2$O (filled circles), and Hb^{13}CO:D$_2$O (black line). The spectra were recorded at 100 ps and 283 K with the A-state region (A) and B-state region (B) collected independently. The A- and B-state designations follow the convention of Alben et al. (33). To facilitate comparison among the spectra, the constant and linear contributions to the polynomial fit of the background have been subtracted from the measured spectra.

appears to be conserved across the proteins h-Mb, sw-Mb, and the α and β chains of Hb, suggesting that this environment is functionally relevant. The relatively narrow B-state linewidths suggest that these proteins fashion a docking site that imposes orientational constraints on "docked" CO. Moreover, this docking site may mediate the binding and escape of ligands. To justify these conclusions, we discuss the temperature dependence of B-state spectra and the theoretical basis for interpreting the near-ambient temperature spectra.

1. Temperature Dependence of B-State Spectra

The B states have also been cryogenically trapped and characterized with static IR spectroscopy (33). The IR spectrum of Mb*CO, obtained by photolysis of sw-MbCO:G/W at 5.5 K, reveals three features denoted B_0, B_1, and B_2, whose integrated absorbance is 21.7 ± 1.6 times smaller than that for the A states. The B_0 state contains approximately 17% of the integrated B-state absorbance and has a vibrational frequency that is the same, within experimental error, as gas phase CO (2143.3 cm^{-1} for ^{12}CO; 2096 cm^{-1} for ^{13}CO). It was concluded that B_0 corresponds to CO that is "free" within the heme pocket. Because the center frequencies of B_1 and B_2 are modestly red-shifted from the gas phase CO frequency (12.8 and 24.3 cm^{-1} shifts for B_1 and B_2, respectively), the CO is not covalently bound but does interact weakly with residues in the heme pocket. The integrated B_1 absorbance was found to be about 1.96 ± 0.05 times that of B_2, but upon warming the photolyzed sample above $\sim$13 K, B_1 grew at the expense of B_2 while the total B-state integrated absorbance remained constant. The relative populations of B_1 and B_2 do not revert upon cooling the thermally annealed sample back down to 5.5 K. These results suggest that (1) B_2 is higher in energy than B_1, (2) B_1 and B_2 share the same oscillator strength, and (3) B_1 and B_2 do not correspond to different protein conformations, but rather to CO in a similar proximity but in a different orientation.

The B states generated by photolyzing various heme proteins at 283 K (11) are similar but not identical to those measured when photolyzing sw-MbCO:G/W at 5.5 K (33). The B_0 state is not seen in any of the three proteins shown in Fig. 6B. The integrated absorbance of B_1 and B_2 at 283 K is 32.8 ± 1 times smaller than that for the A states. In sw-MbCO:D$_2$O, the integrated absorbance of B_1 is approximately 1.94 times that for B_2, but in h-MbCO and HbCO that ratio is closer to 1.5-1.6. The linewidths of B_1 and B_2 at 283 K are approximately twice as broad as those at 5.5 K. The center frequencies of B_1 and B_2 in sw-MbCO:G/W at 283 K were estimated to be 1-2 cm^{-1} blue shifted compared to those at 5.5 K. [Although the B states for sw-MbCO:G/W at 283 K were not measured directly, we could predict their center frequencies by comparing the frequencies of sw-MbCO:D_2O vs. h-MbCO:D_2O (species-dependent shift) and those for h-MbCO:D_2O vs. h-MbCO:G/W (solvent-dependent shift) (11).]

The 1-2 cm^{-1} blue shift in the near-ambient spectra can be readily explained. Both B_1 and B_2 were found to blue shift nonexponentially with time after photolysis, with B_1 experiencing a larger shift. This time-dependent blue shift is a consequence of the conformational relaxation that can occur under ambient conditions (39,46–48) but is inhibited at temperatures below the ~185 K glass transition of the protein (24). Due to the similarity between the cryogenic and near-ambient temperature B-state center frequencies, the cryogenic B states appear to be the same as those measured near ambient temperature, but trapped within a conformationally unrelaxed protein.

The major differences between the cryogenic and near ambient temperature B_1 and B_2 states are their linewidths and integrated absorbances. In discussing integrated absorbance, a distinction must be made between the integrated absorbance and the integrated oscillator strength for the transition. The integrated absorbance is a measured quantity that relates to experimentally resolved features, while the integrated oscillator strength refers to a theoretical quantity that is intrinsic to the vibrational transition. For example, the measured integrated absorbance will be deficient when a portion of the oscillator strength resides in a feature broad enough to be subsumed by the background.

Because absolute absorbances are difficult to determine experimentally, we confine ourselves to relative absorbances and use the A states as an internal standard. The ratio of integrated A- to B-state absorbance at 5.5 K was reported to be 21.7 ± 1.6 (33). What should the measured ratio be at 283 K? In general, the integrated oscillator strength is intrinsic to a vibrational transition, and if it is only weakly coupled to other oscillators, it can be assumed to be temperature independent (provided the population resides primarily in its ground vibrational state). Therefore, we assume the integrated B-state absorbance at 283 K should be the same as that at 5.5 K. In contrast, the A-state integrated absorbance exhibits a weak temperature dependence with the integrated absorbance at $\approx$283 K being 0.8 of that measured at 30 K (49). It has been suggested that the loss of integrated oscillator strength at elevated temperature is due to a temperature-dependent change in electron-nuclear coupling (23). If one assumes the B-state integrated absorbance measured at low temperature to be equivalent to its integrated oscillator strength (a reasonable assumption given the spectral broadening expected due to thermal motion at 5.5 K is less than the measured linewidth), then the integrated A- to B-state ratio at 283 K should be approximately 17. Instead, the average of the integrated A- to B-state ratios measured for Mb and Hb at 283 K is 32.8 ± 1. Consequently, the B-state features of Fig. 6B contain only about 53% of the integrated absorbance expected. Where is the "missing" half of the integrated B-state absorbance at 283 K? The answer becomes clear in the next section, where the relationship between motional dynamics and spectra is considered.

2. Relationship Between B-State Spectra and CO Motional Dynamics

Because the rapid initial decay of the orientational dipole correlation function is independent of environment, the spectrum of CO trapped in a protein at 283 K should exhibit a feature approximately as broad as the envelope containing its gas phase P and R branches. One would also expect to see a narrower feature or features on top of the broad pedestal with the integrated oscillator strength of CO partitioned between the broad and narrow features in proportion to the relative amplitudes of the fast and slow decay of the orientational correlation function. Finding approximately half of the integrated CO oscillator strength under the narrow features suggests that the relative amplitudes of the fast and slow decay of the orientational dipole correlation are comparable. This conclusion is in near-quantitative agreement with the dipole correlation function of CO in Mb obtained from molecular dynamics simulations using CHARMM (50). The "missing" half of the integrated oscillator strength is not missing at all, but is contained within a broad pedestal under the narrow features. Because the width of the pedestal is more than an order of magnitude broader than the narrow features, its amplitude is approximately an order of magnitude smaller and is, therefore, difficult to extract from the solvent background in the experimental time-resolved IR absorbance spectra.

The relatively large amplitude of the slow-decaying contribution to the orientational correlation implies that the protein environment imposes relatively severe constraints on the orientational freedom of CO. Indeed, the mean angular displacement from equilibrium has been estimated to be ≈38 degrees (11). Consequently, we have concluded that the highly conserved protein residues on the distal side of the heme fashion a "docking" site where ligands can become trapped (11). This docking site mediates the transport of ligands to and from the active binding site and may function to discriminate between ligands such as O_2, CO, and NO.

C. Orientation of Bound and "Docked" CO

Time-resolved polarized IR absorbance spectra of photolyzed MbCO are shown in Fig. 7. The A-state spectra reveal two overlapping features, denoted A_1 and A_3 after Ormos et al. (17), with A_1 blue-shifted relative to A_3. The ratio of the polarized absorbance, $\Delta A^\perp / \Delta A^\parallel$, is nearly constant

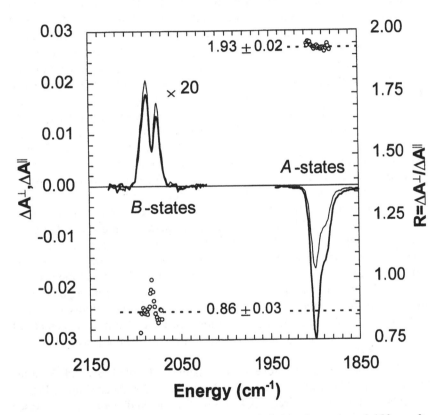

Figure 7 Polarized IR absorption spectra and their ratio measured 100 ps after photolysis of Mb^{13}CO in D$_2$O. The left axis corresponds to the photolysis-induced absorbance changes, $\Delta A^\perp$ (thick lines) and $\Delta A^\parallel$ (thin lines); the right axis corresponds to their ratio $\Delta A^\perp / \Delta A^\parallel$ (open circles). The ratio is plotted where the absorbance exceeds $\sim$25% of its maximum and has been corrected for fractional photolysis. The dashed lines correspond to the average A- and B-state ratios. The A- and B-state spectra were collected at 10.8% and 20% photolysis of Mb^{13}CO, respectively. The background and hot band contributions to the B-state spectra have been removed. (Adapted from Ref. 51.)

(1.931 ± 0.02) across these features, demonstrating that the transition moments for the two A states are oriented at a similar angle. From the measured polarization ratio, the equilibrium angle θ_{eq} for the transition moment was estimated to be $\leq 7°$ (51). The upper limit of $7°$ obtains in the limit where the heme is a perfectly flat circular absorber, and this value is in excellent agreement with static measurements of the polarization anisotropy in ambient temperature MbCO crystals (52–54). The two spectroscopic results are significantly different from a 1.5 Å crystal structure of $P2_1$ MbCO, where the highly conserved distal histidine purportedly caused CO to bind in a nonoptimal bent geometry of $39°$ and thereby inhibited the binding of CO (55). In contrast to that result, a 2 Å crystal structure of P6 MbCO reported an angle of $19°$ (56). Spiro and Kozlowski (18) suggested that the discordant results between IR spectroscopy and x-ray crystallography might be rationalized by density functional theory, which was used to explore the relationship between the direction of the CO transition moment and the C–O bond axis when bound to Mb. They found that a $7°$ angle between the heme plane normal and the transition moment of bound CO could correspond to a C–O angle as large as $15°$, but certainly not $39°$.

The discrepancy between IR spectroscopy and x-ray crystallography became negligible when Bartunik and coworkers reported a 1.15 Å resolution structure of MbCO at ambient temperature and found the C–O angle to be approximately $12°$ with respect to the heme plane normal (57). The near-quantitative agreement obtained between IR spectroscopy and atomic resolution x-ray crystallography leaves little doubt that the primary source of ligand discrimination between CO and O_2 is something other than steric hindrance. Rather than suppression of the binding affinity of CO, it has been suggested that ligand discrimination arises from enhancement of the binding affinity of O_2 by formation of a hydrogen bond with the distal histidine (58).

The B-state spectra reveal two features, denoted B_1 and B_2 after Ormos et al. (17), with B_1 blue-shifted relative to B_2. According to Fig. 7, the polarized absorbance ratio $\Delta A^{\perp}/\Delta A^{\parallel}$ for "docked" CO is much closer to 0.75 than it is to 2, demonstrating that CO rotates substantially upon dissociation from the heme iron. According to Fig. 7, the two B states reveal a similar ratio, $\Delta A^{\perp}/\Delta A^{\parallel} = 0.856 \pm 0.03$, which was found to be consistent with $\theta_{eq} \sim 90°$ (11,51). X-ray structures of Mb*CO at cryogenic temperatures reveal electron density assigned to unbound CO that is displaced ≤ 2 Å from the binding site (59,60). The CO orientation from those structures is not inconsistent with the polarized IR results.

Because the electrostatic field in the vicinity of the docking site is anisotropic, the vibrational frequency of "docked" CO should be Stark shifted and the direction of that shift should depend on the CO orientation. Consequently, the two B-state features are interpreted as Stark-shifted spectra with B_1 and B_2 corresponding to CO pointing in opposite directions. Unlike the polarized absorbance ratio measured for the A states, the ratio for the B states is vibration-frequency dependent with the polarization ratio a maximum at the midpoint between the two features. Because the center frequency of CO depends on its orientation, its vibrational frequency should shift smoothly from one B state frequency to the other as it undergoes end-to-end rotation. Consequently, one might expect the vibrational frequency of CO near the transition state for end-to-end rotation to be centered between the peaks of the two B states. At the midpoint, the ratio $\Delta A^{\perp}/\Delta A^{\parallel}$ exhibits a maximum of ~ 1, demonstrating that the trajectory for end-to-end rotation passes through a transition state that has a component out of the plane of the heme. If this trajectory were to lie in a plane, the transition state would be oriented at least $55°$ from the heme plane normal (51). Such a trajectory would maintain a ligand orientation far from that for bound CO, even at the transition state for end-to-end rotation, thereby inhibiting CO binding while permitting end-to-end rotation.

D. Ligand Translocation Trajectories

To experimentally probe the CO trajectory after dissociation, ultrafast time-resolved polarized mid-IR spectra of photolyzed h-MbCO in G/W were recorded (34), the results of which are plotted in Fig. 8A. This study was performed in G/W primarily because the flatness of the solvent absorbance spectrum near 2100 cm^{-1} minimizes temporal distortion of the transmitted femtosecond IR probe pulse, thereby maximizing the effective time resolution of the measurement. Two features are already apparent at 0.2 ps, the earliest time shown, and these features rapidly develop into the "docked" states denoted B_1 and B_2. The development of the "docked" CO spectrum is further quantified by the time dependence of the polarization anisotropy, as defined in Equation (2). The B_1 and B_2 polarization anisotropies, plotted in Fig. 8B, evolve exponentially with time constants of 0.20 ± 0.05 ps and 0.52 ± 0.10 ps, respectively, and converge to the same anisotropy of approximately 0.2. According to Fig. 8C, ligand translocation is accompanied by a 1.6 ± 0.3 ps growth of the integrated isotropic B-state absorbance.

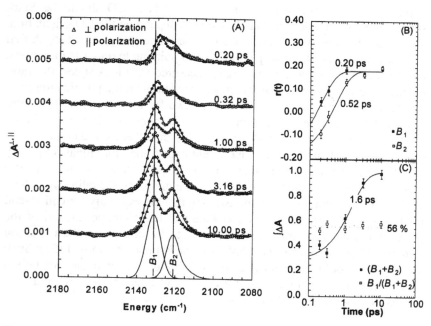

Figure 8 (A) Femtosecond time-resolved IR absorbance spectra of CO measured after photodissociation from the heme of h-Mb in G/W. Spectra were recorded with the photolysis and probe pulses polarized parallel ($\Delta A^{\parallel}$) and perpendicular ($\Delta A^{\perp}$) to one another. The polarized absorbance spectra reveal the time dependence of the ligand orientation as well as the protein surroundings (see text). The features labeled B_1 and B_2 (bottom) correspond to the least-squares fit of the $\Delta A^{\parallel}$ spectrum measured at 10 ps. For clarity, the background and hot band contributions to the time-resolved spectra have been removed and the spectra have been offset from one another. (B) Time dependence of the polarization anisotropy, $r(t) = [B_i^{\parallel}(t) - B_i^{\perp}(t)]/[B_i^{\parallel}(t) + 2B_i^{\perp}(t)]$, after photodissociation from the heme of Mb. $B_i(t)$ represents the integrated absorbance under state i at time t with the polarization denoted by a superscript. The polarization anisotropies of B_1 and B_2 appear to evolve exponentially (solid lines) with time constants of 0.2 ps and 0.52 ps, respectively. (C) Time dependence of the isotropic B-state absorbance (filled squares). The relative contribution of B_1 to the total absorbance is time independent out to 10 ps (open squares) and averages 56% (dashed line). To generate these data, the isotropic absorbance was synthesized from the polarized absorbance spectra according to the "magic" angle prescription: $\Delta A^{MA}(t) = [\Delta A^{\parallel}(t) + 2\Delta A^{\perp}(t)]/3$. (Adapted from Ref. 34.)

How do we rationalize these observations? Photoexcitation of MbCO renders the Fe–CO coordinate repulsive, causing CO to acquire translational kinetic energy as it moves up and away from the heme iron. Before the CO rotates, its polarization anisotropy should be approximately −0.19, and its mid-IR absorbance should be confined to a single, broad feature. Upon colliding with the surrounding protein, the ligand rebounds back toward the heme and proceeds along one of two different trajectories. Because ligand translation and rotation are not slow compared to 0.2 ps, the polarization anisotropy measured at 0.2 ps is not the theoretical minimum of −0.19, nor is the spectrum manifested as a single broad feature. As the CO translates and rotates, the two trajectories become spectroscopically distinguishable, owing to the vibrational Stark shift that arises from the electrostatic field surrounding the ligand. The fact that B_1 and B_2 evolve at different rates demonstrates that the trajectories leading to B_1 and B_2 are distinguishable kinetically as well as spectroscopically. The polarization anisotropy for both B_1 and B_2 converges within a few ps to ~0.2. Had the polarization anisotropy been measured using a wavelength where the heme is a perfectly flat circular absorber, the polarization anisotropy would rise to only about 0.05. By photolyzing at a wavelength where the heme is an elliptical absorber, the range of the polarization anisotropy was enlarged, thereby improving the signal-to-noise ratio of the measurement. Moreover, the fact that the polarization anisotropy exceeds 0.05 suggests that the major axis of the elliptical heme absorber must be at least partially aligned with the C–O axis. If the orientation of the major axis of the heme transition moment were known as a function of wavelength, polarization anisotropy measurements at more than one wavelength would permit a determination of both the azimuthal CO orientation and its angle with respect to the heme plane normal.

The prompt appearance and independent development of the two B states suggest that the two trajectories are deterministic in nature, with the outcome (B_1 or B_2) established promptly after photodetachment. What do the two limiting states correspond to structurally? From a kinematic argument, it was rationalized that the faster B_1 trajectory has CO sliding into the docking site with the O end of C–O pointing toward the heme iron (34). This structural assignment is supported by geminate rebinding studies of photolyzed MbCO at 20 K, where B_1 predominates: the geminate rebinding of $^{13}C^{16}O$ was found to be slower than $^{12}C^{18}O$, in spite of the latter being heavier (61). Because geminate rebinding at 20 K is dominated by tunneling (61) and the tunneling rate depends on distance as well as mass, this surprising isotope effect can be rationalized by orienting C–O such that the O end is pointing toward the heme iron. The orientation of the

CO in the docking site is too fine a detail to be extracted from the electron density maps of photolyzed MbCO (59,60).

Interestingly, the proportion of the integrated absorbance ascribed to B_1 (plotted in Fig. 8B) remains largely unchanged during the protein conformational reorganization. Assuming the integrated area under each B state is proportional to its population, an assumption for which there is experimental support (33), the partitioning between B_1 and B_2 is 1.3:1. This ratio is not far from the statistical (1:1) distribution expected with two possible CO orientations. The time independence of this ratio (out to 10 ps) provides added support for the suggestion that the ligand dissociation trajectories are deterministic. The ratio does not remain constant out to longer times, however, but increases to 1.7:1 by 100 ps (51), demonstrating that B_1 is lower in energy than B_2 and that end-to-end rotation between 10 and 100 ps renders the distribution thermodynamic rather than statistical. If the integrated areas under B_1 and B_2 are indeed proportional to population, then the ratio of 1.7:1 corresponds to a free energy difference of about 1.2 kJ/mol with B_1 lower in free energy. Evidently, there is a modest preference for the trajectory leading to B_1, which also turns out to be the more stable state thermodynamically.

Because the interior of Mb is densely packed, ligand translocation from the active binding site to the docking site requires some degree of protein rearrangement, a process that should affect the vibrational spectrum of "docked" CO. Moreover, the conformational response of the protein should be more sluggish than the motion of the ligand. Might the 1.6 ps growth of the integrated isotropic B-state absorbance be assigned to protein rearrangement, or might it arise from other causes? One often equates changes in integrated absorbance with changes in population, however, that is not the case here: all CO produced photolytically is generated in less than 0.2 ps, the time resolution of the measurement. Might the growth be due to thermal cooling of the CO and its environment? Because "docked" CO is in contact with the heme, its kinetic temperature would be expected to cool at a rate similar to the heme, which was found to thermally relax with a time constant of 6.2 ± 0.5 ps (22). Because the 1.6 ps growth of the integrated B-state absorbance is longer than the 0.2 and 0.5 ps rotation times and shorter than the 6.2 ps cooling rate, it cannot be ascribed to population or cooling dynamics. Rather, it most likely arises from reorganization of the neighboring protein residues about the nascent "docked" CO. Recall that the integrated absorbance of CO is partitioned between the narrow B states and a broad unresolved pedestal, the partitioning of which is determined by the orientational constraints imposed on CO by the surrounding protein. The more constrained the CO, the smaller the amplitude of its librational motion, and the greater the integrated absorbance measured under the narrow B-state features. A 1.6 ps time constant for protein rearrangement that serves to constrain "docked" CO appears to be quite reasonable. Evidently, this process helps to establish a steric barrier to the reverse rebinding process. That this steric barrier arises before the electronic state of the heme relaxes to its ligand-receptive ground state (3.4 ps) may explain the absence of ultrafast geminate ligand rebinding.

E. Origin of the Barrier to CO Rebinding

Photolysis of MbNO and MbO$_2$ is followed by substantial geminate recombination (28,62–64) with NO rebinding on the sub-ns time scale. On the other hand, geminate recombination of CO in Mb is minimal and occurs on the few hundred ns time scale (65). The lack of significant CO rebinding is remarkable considering the ligand remains "docked" ≤ 2 Å away from the binding site for several hundred ns. It had been suggested that the kinetic differences among these ligands arise from differences in the electronic barrier to binding, with CO having the highest electronic barrier and NO having the lowest (66). The discovery of a ligand docking site that can constrain the orientation of "docked" ligands as small as CO (11) lead us to consider another possibility. The docking site might slow the rebinding rate of CO by strongly hindering access to the transition state for CO rebinding. Because O$_2$ and NO both bind in a bent configuration, access to their transition state for rebinding is far less hindered.

To explore the possibility that slow CO rebinding is a consequence of a steric, not an electronic barrier, the geminate rebinding dynamics of CO to Mb and microperoxidase were compared (67). Microperoxidase is an enzymatically digested cytochrome c oxidase that consists of a heme with a "proximal" histidine that is part of an 11-peptide fragment. This peptide renders the heme soluble under the neutral conditions used in the Mb studies. When reduced to Fe(II), ^{13}CO binds to microperoxidase and a vibrational stretch near 1908 cm^{-1} appears. This frequency is virtually identical to that found in Hb^{13}CO and is similar to the 1900 cm^{-1} transition found in Mb^{13}CO. Consequently, extracting the heme out of the protein appears to have only a minor effect on the heme-CO interaction and, one might assume, the electronic barrier to ligand binding. Because the peptide is not long enough to wrap around and fashion a docking site on the distal side of the heme, photodissociated CO will be surrounded by disordered solvent, not a highly organized docking site. Any differences in the rates of geminate rebinding to Mb and microperoxidase might, therefore, be ascribed primarily to the steric constraints imposed by the docking site in Mb.

The geminate-rebinding dynamics measured after photolysis of MbCO and microperoxidase-CO are shown in Fig. 9. The survival fraction denotes the fraction of photolyzed hemes that remain in the deoxy form after CO dissociation. The population was determined by measuring the time dependence of the vibrational absorbance of bound CO. According to Fig. 9, CO rebinds to microperoxidase much more rapidly than to Mb.

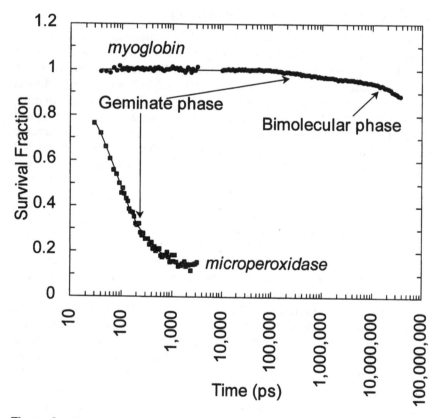

Figure 9 Geminate recombination after photolysis of MbCO (●) and microperoxidase-CO (■). The survival fraction refers to the population that remains unbound after photolysis. The population was determined by measuring the IR absorbance at frequencies corresponding to the peak of the bound CO stretch. (Adapted from Ref. 67.)

The rebinding dynamics to microperoxidase are nonexponential due to a solvent cage effect. To deduce the time constant for rebinding CO from the solvent cage, the recombination kinetics were modeled according to the scheme A $\underset{}{\overset{k_{BA}}{\longleftarrow}}$ B $\underset{k_{CB}}{\overset{k_{BC}}{\longleftrightarrow}}$ C $\underset{k_{SC}}{\overset{k_{CS}}{\longleftrightarrow}}$ S where A represents the population of bound CO, B represents the population of CO trapped within the solvent cage surrounding the heme, C represents the population of CO trapped just outside the first solvent shell, and S represents the population of CO that has escaped beyond C. With this model, $k_{BA} = (110 \text{ ps})^{-1}$. Modeling the Mb dynamics with a similar kinetic scheme leads to $k_{BA} \leq (3 \text{ μs})^{-1}$. Consequently, the rate of CO binding to microperoxidase is more than 27,000 times faster than the corresponding rate in Mb. In fact, the rate of geminate recombination in microperoxidase is not much slower than the $(27.6 \text{ ps})^{-1}$ rate observed for NO rebinding to Mb (28). It appears that the reason CO rebinds slowly to Mb is not because of a large electronic barrier, but because the docking site inhibits access to the transition state for CO binding.

V. CONCLUSIONS

Ultrafast time-resolved near- and mid-IR spectra of ligand-binding heme proteins have unveiled numerous details that have contributed to our understanding of the relations between protein structure, dynamics, and function. These studies showed that carbon monoxide binds to Mb to form nearly linear Fe–C–O. Upon dissociation from the heme iron, CO becomes trapped in a docking site located ≤ 2 Å from the heme-binding site. This docking site constrains CO to lie nominally parallel to the plane of the heme, an orientation approximately orthogonal to that of bound CO. Ligand translocation proceeds along one of two pathways, with the faster, 0.2 ps pathway leading to B_1 and the slower, 0.5 ps pathway leading to B_2. Of the two states, B_1 is lower in energy and is assigned to a structure with the O end of CO pointing toward the heme iron. The conformational response of the protein to ligand translocation proceeds with a 1.6 ps time constant and appears to tighten the orientational constraint imposed on the docked CO. The photoexcited heme was found to relax electronically with a 3.4 ps time constant and relax thermally with a 6.2 ps time constant. A modest amount of "docked" CO appears vibrationally hot ($\sim$4%) but relaxes back to its ground state with a 600 ± 150 ps time constant. The sluggish geminate rebinding rate in Mb is approximately 27,000 times slower than the geminate rebinding to a heme that lacks a docking site. Consequently, most

of the docked CO manages to escape from the docking site on the time
scale of a few hundred ns (>98% at 32°C).

The use of photolysis to explore ligand dynamics in ligand-
binding heme proteins appears to be well justified, as the heme
quickly loses memory of the dissociation pathway. The orientational and
spatial constraints imposed on "docked" CO have the effect of slowing
dramatically the rate of CO rebinding and facilitate efficient expulsion
of this toxic ligand from the protein. Evidently, the highly conserved
residues circumscribing the heme pocket of Mb fashion a docking site
that orientationally constrains the dissociated ligand and thereby influences
the rates and pathways for ligand binding and escape. A docking site near
an active site may be a general property among proteins that must shuttle
ligands to and from an active site in an oriented fashion. To probe more
deeply the role of the residues that fashion the docking site will require
additional time-resolved IR studies involving mutants of Mb.

REFERENCES

1. Loyd CR, Eyring EM, Ellis J. J Am Chem Soc 117:11993–11994, 1995.
2. Springer BA, Sligar SG, Olson JS, Phillips GN, Jr. Chem Rev 94:699–714, 1994.
3. Snyder SH, Science 257:494–496, 1992.
4. Verma A, Hirsch DJ, Glatt CE, Ronnett GV, Snyder SH, Science 259:381–384, 1993.
5. Bethesda, MD: National Center of Biotechnology Information, 1994.
6. Collman JP, Brauman JI, Halbert TR, Suslick KS. Proc Natl Acad Sci USA 73:3333–3337, 1976.
7. Antonini E, Brunori M. Hemoglobin and Myoglobin in Their Reactions with Ligands. London: North-Holland Publishing Company, 1971.
8. Landaw SA, Callahan EW, Jr., Schmid R. J Clin Invest 49:914–925, 1970.
9. Anfinrud PA, Lim M, Jackson TA. Proc SPIE-Int Soc Opt Eng (Longer Wavelength Lasers and Applications) 2138:107–115, 1994.
10. Gordon RG. J Chem Phys 43:1307–1312, 1965.
11. Lim M, Jackson TA, Anfinrud PA. J Chem Phys 102:4355–4366, 1995.
12. Ansari A, Szabo A. Biophys J 64:838–851, 1993.
13. Ansari A, Jones CM, Henry ER, Hofrichter J, Eaton WA. Biophys J 64:852–868, 1993.
14. Eaton WA, Hofrichter J. Methods Enzymol 76:175–261, 1981.
15. Moore JN, Hansen PA, Hochstrasser RM. Proc Natl Acad Sci USA 85:5062–5066, 1988.
16. Albani J, Alpert B. Chem Phys Lett 131:147–152, 1986.

17. Ormos P, Braunstein D, Frauenfelder H, Hong MK, Lin SL, Sauke TB, Young RD. Proc Natl Acad Sci USA 85:8492–8496, 1988.

18. Spiro TG, Kozlowski PM. J Am Chem Soc 120:4524–4525, 1998.

19. Gibson QH. J Physiol 134:123, 1956.

20. Gibson QH. J Physiol 134:112, 1956.

21. Henry ER, Eaton WA, Hochstrasser RM. Proc Natl Acad Sci USA 83:8982–8986, 1986.

22. Lim M, Jackson TA, Anfinrud PA. J Phys Chem 100:12043–12051, 1996.

23. Cupane A, Leone M, Vitano E, Cordone L. Biopolymers 27:1977–1997, 1988.

24. Steinbach PJ, Ansari A, Berendzen J, Braunstein D, Chu K, Cowen BR, Ehrenstein D, Frauenfelder H, Johnson JB, Lamb DC, Luck S, Mourant JR, Nienhaus GU, Ormos P, Philipp R, Xie A, Young RD. Biochemistry 30:3988–4001, 1991.

25. Srajer V, Champion PM. Biochemistry 30:7390–7402, 1991.

26. Nienhaus GU, Mourant JR, Frauenfelder H. Proc Natl Acad Sci USA 89:2902–2906, 1992.

27. Li P, Sage JT, Champion PM. J Chem Phys 97:3214–3227, 1992.

28. Petrich JW, Poyart C, Martin JL. Biochemistry 27:4049–4060, 1988.

29. Goodno GD, A Astinov, Miller RJD. J Phys Chem A 103:10630–10643, 1999.

30. Mizutani Y, Kitagawa T. Science 278:443–446, 1997.

31. Rudolph SA, Boyle SO, Dresden CF, Gill SJ. Biochemistry 11:1098–1101, 1972.

32. Sagnella DE, Straub JE, Jackson TA, Lim M, Anfinrud PA. Proc Natl Acad Sci 96:14324–14329, 1999.

33. Alben JO, Beece D, Bowne SF, Doster W, Eisenstein L, Frauenfelder H, Good D, McDonald JD, Marden MC, Mo PP, Reinisch L, Reynolds AH, Shyamsunder E, Yue KT. Proc Natl Acad Sci USA 79:3744–3748, 1982.

34. Lim M, Jackson TA, Anfinrud PA. Nature Struct Biol 4:209–214, 1997.

35. Guelachvili G, Rao KN. Handbook of Infrared Standards. Boston: Academic Press, Inc., 1986.

36. Hill JR, Tokmakoff A, Peterson KA, Sauter B, Zimdars D, Dlott DD, Fayer MD. J Phys Chem 98:11213–11219, 1994.

37. Owrutsky JC, Li M, Locke B, Hochstrasser RM. J Phys Chem 99:4842–4846, 1995.

38. Wilson EBJ, Decius JC, Cross PC. Molecular Vibrations: The Theory of Infrared and Raman Vibrational Spectra. New York: Dover Publications Inc., 1955.

39. Jackson TA, Lim M, Anfinrud PA. Chem Phys 180:131–140, 1994.

40. Simons JP, Tasker PW. Mol Phys 26:1267, 1973.

41. Anfinrud PA, Han C, Hochstrasser RM. Proc Natl Acad Sci USA 86:8387–8391, 1989.

42. Cotton FA, Wilkinson G. Advanced Inorganic Chemistry. New York: Wiley-Interscience Inc., 1988.

43. Makinen MW, Houtchens RA, Caughey WS. Proc Natl Acad Sci USA 76:6042–6046, 1979.

44. Richon D, Patterson D, Turrell G. Chem Phys 24:227–234, 1977.

45. Richon D, Patterson D. Chem Phys 24:235–243, 1977.

46. Lim M, Jackson TA, Anfinrud PA. Springer Series in Chemical Physics (Ultrafast Phenomena VIII) 55:522–524, 1992.

47. Lim M, Jackson TA, Anfinrud PA. Proc SPIE-Int Soc Opt Eng (Laser Spectroscopy of Biomolecules) 1921:221–230, 1993.

48. Lim M, Jackson TA, Anfinrud PA. Proc Natl Acad Sci USA 90:5801–5804, 1993.

49. Ansari A, Berendzen J, Braunstein DK, Cowen BR, Frauenfelder H, Hong MK, Iben IET, Johnson JB, Ormos P, Sauke TB, Scholl R, Schulte A, Steinbach PJ, Vittitow J, Young RD. Biophys Chem 26:337–355, 1987.

50. Straub JE, Karplus M. Chem Phys 158:221–248, 1991.

51. Lim M, Jackson TA, Anfinrud PA. Science 269:962–966, 1995.

52. Sage JT. Appl Spect 51:329, 1997.

53. Ivanov D, Sage JT, Keim M, Powell JR, Asher SA, Champion PM. J Am Chem Soc 116:4139–4140, 1994.

54. Sage JT, Jee W. J Mol Biol 274:21–26, 1997.

55. Kuriyan J, Wilz S, Karplus M, Petsko GA. J Mol Biol 192:133–154, 1986.

56. Quillin ML, Arduini RM, Olson JS, Phillips GN, Jr. J Mol Biol 234:140–155, 1993.

57. Kachalova GS, Popov AN, Bartunik HD. Science 284:473–476, 1999.

58. Olson JS, Phillips GN, Jr. J Biol Inorg Chem 2:544–552, 1997.

59. Schlichting I, Berendzen J, Phillips GN, Jr., Sweet RM. Nature 371:808–812, 1994.

60. Teng TY, Srajer V, Moffat K. Nature Struct Biol 1:701–705, 1994.

61. Alben JO, Beece D, Browne SF, Eisenstein L, Frauenfelder H, Good D, Marden MC, Moh PP, Reinisch L, Reynolds AH, Yue KT. Phys Rev Lett 44:1157–1163, 1980.

62. Jongeward KA, Magde D, Taube DJ, Marsters JC, Traylor TG, Sharma VS. J Am Chem Soc 110:380–387, 1988.

63. Petrich JW, Lambry JC, Kuczera K, Karplus M, Poyart C, Martin JL. Biochemistry 30:3975–3987, 1991.

64. Carver TE, Rohlfs RJ, Olson JS, Gibson QH, Blackmore RS, Springer BA, Sligar SG. J Biol Chem 265:20007–20020, 1990.

65. Henry ER, Sommer JH, J Hofrichter, Eaton WA. J Mol Biol 166:443–451, 1983.

66. Cornelius PA, Hochstrasser RM, Steele AW. J Mol Biol 163:119–128, 1983.

67. Lim M, Jackson TA, Anfinrud PA. J Biol Inorg Chem 2:531–536, 1997.

6

Infrared Vibrational Echo Experiments

Kirk D. Rector* and M. D. Fayer
Stanford University, Stanford, California

I. INTRODUCTION

The advent of the nuclear magnetic resonance spin echo experiment in 1950 began a new era in spectroscopy (1). The spin echo was the first spectroscopic experiment to take advantage of coherent interactions of a radiation field with the system to obtain information not available in a straight absorption measurement. The spin echo, which involves the application of two radio frequency pulses, is the simplest of all pulsed magnetic resonance experiments. Since 1950, a large number of complex pulse sequences have been developed and applied to the study of magnetic spin systems (2). All of these have direct lineage to the spin echo experiment.

   In 1964, the spin echo experiment was extended to the optical regime by the development of the photon echo experiment (3,4). The photon echo began the application of coherent pulse techniques in the visible and ultraviolet portions of the electromagnetic spectrum. Since its development, the photon echo and related pulse sequences have been applied to a wide variety of problems including dynamics and intermolecular interactions in crystals, glasses, proteins, and liquids (5–8). Like the spin echo, the photon echo and other optical coherent pulse sequences provide information that is not available from absorption or fluorescence spectroscopies.

---

* *Current affiliation*: Los Alamos National Laboratory, Los Alamos, New Mexico

Today, radio frequency coherent pulse sequences are used extensively in nuclear magnetic resonance and in electron spin resonance spectroscopies. Visible light coherent pulse sequence techniques, while not as ubiquitous as magnetic resonance, are widely used. In contrast, the use of coherent pulse sequences in the infrared (IR) portion of the spectrum to study vibrational states of molecules, rather than spin states or electronic states, is just beginning. The delay in applying coherent pulse sequences in the IR has been mainly caused by technological difficulties. The first applications of coherent IR pulse sequences to probe molecular vibrations occurred in the early 1970s (9,10). Because of limitations imposed by electronic switching of CW lasers to create pulses, experiments were restricted to small molecules, long times scales, and low-pressure gases. These novel experiments did not find general applicability and were not useful in studying condensed matter systems because of the lack of time resolution.

In 1993, the first ultrafast vibrational echo experiments on condensed matter systems were performed using a free electron laser as the source of temporally short, tunable infrared pulses (11). Recently, the development of Ti:sapphire laser-based optical parametric amplifier (OPA) systems has made it possible to produce the necessary pulses to perform vibrational echoes using a tabletop experimental system (12,13). The development and application of ultrafast, IR vibrational echoes and other IR coherent pulse sequences are providing a new approach to the study of the mechanical states of molecules in complex molecular systems such as liquids, glasses, and proteins (14–20). While the spin echo, the photon echo, and the vibrational echo are, in many respects, the same type of experiment, the term vibrational echo is used to distinguish IR experiments on vibrations from radio frequency experiments on spins or vis/UV experiments on electronic states. In this chapter, recent vibrational echo experiments on liquids, glasses, and proteins will be described.

The vibrational levels of a molecule in a condensed matter system are influenced by the surrounding medium through intermolecular interactions. The time-averaged forces exerted by the solvent on a molecular oscillator cause a static shift in the vibrational absorption frequency. The frequency shifts of the vibrational transitions of a molecule between the gas phase and a condensed matter environment is an indicator of the effect of the solvent on the internal mechanical degrees of freedom of a solute.

The fluctuating forces that a medium exerts on a solute molecule produces fluctuations in the molecular structure, time-dependent vibrational eigenstates, and, thus, time-dependent vibrational energy eigenvalues. Time evolution of the vibrational energy eigenvalues produces fluctuations in the vibrational transition energies. Fluctuating forces are involved in a wide variety of chemical and physical phenomena, including thermally induced chemical reactions, promotion of a molecule to a transition state, electron transfer, and energy flow into and out of molecular vibrations. The extent and time dependence of fluctuations of a solute's vibrational energy levels are sensitive to the nature of the dynamics of the condensed matter environment and the strength of intermolecular interactions.

In principle, information on dynamical intermolecular interactions of an oscillator with its environment can be obtained from vibrational absorption spectra. The forces experienced by the oscillator determine the vibrational line shape and width. The line shape and width depend on temperature and the nature of the solvent. However, a vibrational absorption spectrum reflects the full range of broadening of the vibrational transition energies, both homogeneous and inhomogeneous. In glasses, liquids and proteins, inhomogeneous broadening often exceeds the homogeneous linewidth. Under these circumstances, measurement of the absorption spectrum does not provide information on vibrational dynamics.

The medium containing solute molecules of interest is referred to as a bath. The bath includes bulk solvent degrees of freedom arising from the solvent's translational and orientational motions, the internal vibrational degrees of freedom of the solvent, and the solute's vibrational modes other than the oscillator of interest. In a glass, bath fluctuations range from very high frequency to essentially static. For a pair of energy levels, e.g., $v = 0$ and $v = 1$, coupling of the vibrational transition to the fast fluctuations produces homogeneous pure dephasing, which, in the frequency domain, is a source of homogeneous spectral broadening. Pure dephasing, which results from the time evolution of the vibrational transition energy, is an ensemble average property, which for an exponential decay of the off diagonal density matrix elements (Lorentzian homogeneous line shape) can be characterized by an ensemble average pure dephasing time, T_2^*. The total homogeneous dephasing time, T_2, (total homogeneous linewidth) also has contributions from the vibrational lifetime, T_1 and, possibly, orientational relaxation (21). While the fast fluctuations give rise to the dynamical homogeneous dephasing, the static structural disorder of a glass makes one molecule's environment, and, therefore, vibrational transition energy, different from another. These static differences are the source of inhomogeneous broadening of an absorption spectrum.

Unlike a glass, a liquid does not have essentially static structures that give rise to inhomogeneous broadening. Nonetheless, liquids can have fast time scale fluctuations that give rise to homogeneous broadening and much slower time scale structural evolution. Evolution of the system on time scales substantially slower than the homogeneous dephasing time, T_2, appears as inhomogeneous broadening. Since the absorption spectrum measures the transitions on all time scales, if the inhomogeneous broadening is significant compared to the homogeneous broadening, an absorption spectrum will reflect the inhomogeneous linewidth, which does not provide information on vibrational dynamics. Thus, vibrational echoes are useful in the studies of liquids as well as more static structures like glasses and proteins.

In this chapter, the first detailed studies performed using ultrafast vibrational echo experiments are described. The experiments examine dynamics in condensed matter systems as a function of temperature and other system parameters. First, the vibrational echo method, including some details of the experimental techniques, is described. Then vibrational echo experiments, used to probe vibrational dynamics in liquids and glasses, are presented. In addition, protein dynamics are studied using vibrational echo measurements on the CO ligand bound to the active sites of the proteins myoglobin and hemoglobin. In studies of liquids, glasses, and proteins, the vibrational echo experiment is used as a time domain probe of dynamical intermolecular interactions. A new two-dimensional spectroscopy, vibrational echo spectroscopy (VES), is also described. In VES experiments, vibrational echoes are used to suppress unwanted background in a vibrational spectrum and to enhance one peak over another in a manner akin to the methods used in NMR. The combination of the experiments demonstrates that a new era of IR ultrafast coherent vibrational spectroscopy has begun.

II. THE VIBRATIONAL ECHO METHOD AND EXPERIMENTAL PROCEDURES

A. The Vibrational Echo Method

The vibrational echo experiment is a time domain, degenerate, four-wave mixing experiment that extracts the homogeneous vibrational line shape even from a massively inhomogeneously broadened line. Vibrational line shapes contain the details of the dynamical interactions of a vibrational mode with the motions of the environment (22–24). However, the vibrational line shape can also include low-frequency, structural perturbations associated with the distribution of the vibrational oscillators' local environmental configurations, i.e., inhomogeneous broadening. The presence of inhomogeneous broadening in a wide variety of condensed matter systems makes the vibrational echo a useful experimental tool.

Vibrational echo experiments permit the use of optical coherence methods to study the dynamics of the mechanical degrees of freedom of condensed phase systems. Because vibrational transitions are relatively narrow, it is possible to perform vibrational echo experiments on well-defined transitions and from very low temperature to room temperature or higher. Further, vibrational echoes probe dynamics on the ground state potential surface. Therefore, the excitation of the mode causes a minimal perturbation of the solvent.

For experiments on vibrations, a source of ps IR pulses is tuned to the transition of interest. The vibrational echo experiment involves a two-pulse excitation sequence. The experiment is illustrated schematically in Fig. 1A. Initially, all of the vibrations are in the ground state, $|0\rangle$. This is represented by an arrow pointing down in the first circle. The first pulse excites each solute molecule's vibration into a coherent superposition state of the molecule's ground vibrational state and the first excited vibration, the $|0\rangle$ and $|1\rangle$ vibrational states. This is represented by an arrow in the plane shown in the second circle. Each molecule in a superposition state has associated with it a microscopic electric dipole, which oscillates at the vibrational transition frequency. Immediately after the first pulse, all of the microscopic dipoles in the sample oscillate in phase. Because there is a distribution of vibrational transition frequencies, the dipoles will precess with some distribution of frequencies. Thus, the initial phase relationship is very rapidly lost. This is represented in the third circle by the arrows fanning out. The molecules with lower transition frequencies fall behind the average, and the molecules with higher frequencies get ahead of the average. This effect is the free induction decay and occurs on a time scale related to the inhomogeneous line width. After a time, τ, a second pulse, traveling along a path making an angle, θ (see Fig. 1B), with that of the first pulse, passes through the sample. This second pulse changes the phase factors of each vibrational superposition state in a manner that initiates a rephasing process. This is illustrated in the fourth circle. The fan of arrows flips over so that the arrows that were moving apart are now moving toward each other. At time 2τ, the ensemble of superposition states is rephased. This is shown in the fifth circle as the reformed single arrow. The phased array of microscopic electric dipoles behaves as a macroscopic oscillating electric dipole, which acts as a source term in Maxwell's equations and gives rise to an additional IR pulse of light, the vibrational echo. A free induction decay again destroys the phase relationships, so only a short pulse of light is generated. As shown in Fig. 1B, the vibrational echo pulse propagates along a path that makes an angle, 2θ, with that of the first pulse.

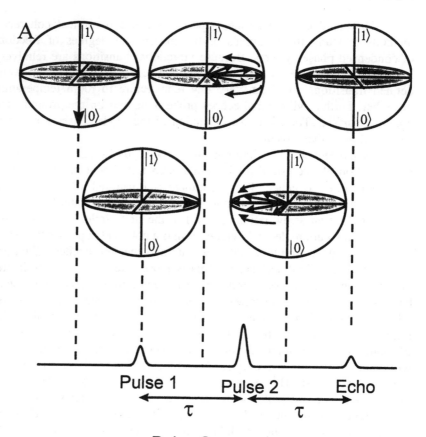

Pulse Sequence

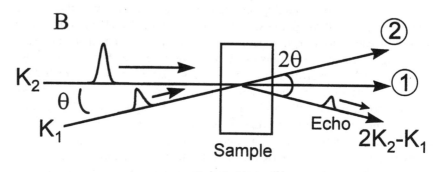

Spatial Profile

The signal intensity is proportional to the intensity of the first pulse and the intensity of the second pulse squared.

The rephasing at 2τ removes the effects of inhomogeneous broadening (25). The spread in frequencies responsible for the inhomogeneous linewidth and, in the time domain, the free induction decay is eliminated by the rephasing that gives rise to the echo pulse. However, fluctuating forces generated by interactions of the vibrational oscillator with the dynamical solvent environment produce fluctuations in each oscillation's frequency. At 2τ the rephasing is imperfect. As τ is increased, the fluctuations produce increasingly large accumulated phase errors among the microscopic dipoles at 2τ, and the vibrational echo signal amplitude is reduced. Thus, the vibrational echo decay is related to the homogeneous linewidth, i.e., the fast vibrational frequency fluctuations, not the inhomogeneous spread in frequencies.

A plot of the vibrational echo intensity change with delay between the pulses is a vibrational echo decay curve. Vibrational echo decays are frequently exponential, although intrinsically nonexponential dynamics (non-Lorentzian homogeneous line shapes) are also seen (26–28). In all of the data presented below, the echo decays are exponential or exponentials modified by laser pulse duration affects. At low temperature, when the dynamics are slow, the data can be fit well with a simple exponential. At high temperatures, the dynamics approach the time scale of the laser pulses. For these data, a more complex fitting routine is employed that takes into account the finite duration pulses. The signal is calculated from the three time-ordered interactions of the sample with the radiation fields. An example of a low-temperature vibrational echo decay curve measured on CO asymmetrical stretching mode (2010 cm^{-1}) of (acetylacetonato)dicarbonylrhodium(I) ($Rh(CO)_2acac$) in dibutylphthalate at 3.4 K and a fit to an exponential are shown in Fig. 2. This measurement was performed at the Stanford free electron laser, as discussed below. As can be seen, high-quality data can be obtained in vibrational echo experiments.

Figure 1 (A) Semiclassical Bloch picture of a vibrational echo in a frame rotating at the center frequency of the vibrational line. Vertical axis in circles represents the population axis of the $|0\rangle$ to $|1\rangle$ vibrational transition. The other two axes represent the coherence plane. The relationship of the diagram to the vibrational echo experiment is discussed in the text. (B) Schematic of the vibrational echo pulse sequence. The two excitation pulses are crossed and focused in a sample at a small angle, θ. The vibrational echo is emitted from the sample at an angle, 2θ, from that of the first pulse.

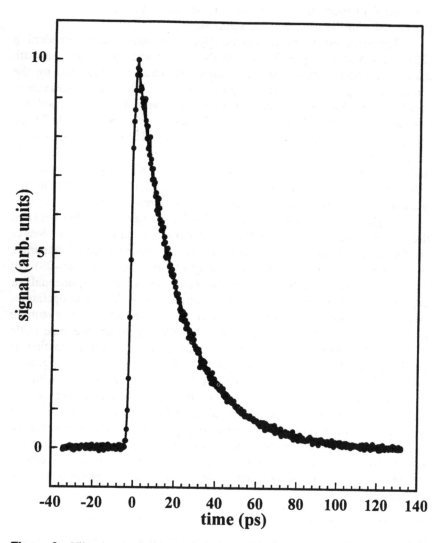

Figure 2 Vibrational echo decay data for the asymmetrical CO-stretching mode of Rh(CO)$_2$acac in DBP ($\sim$2000 cm^{-1}) at 3.4 K and a fit to a single exponential function. The data were taken using the Stanford Free Electron Laser. The decay constant is 23.8 ps, which yields a homogeneous linewidth of 0.11 cm^{-1}. The absorption spectrum has a linewidth of $\sim$15 cm^{-1} at this temperature, demonstrating that the line is massively inhomogeneously broadened.

The vibrational echo decay signal, $S(\tau)$, is given by

$$S(\tau) = S_0 e^{-4\tau/T_2} \tag{1}$$

where T_2 is the homogeneous dephasing time. The Fourier transform of the echo decay is directly related to the homogeneous lineshape (25). For systems in which orientational relaxation is not significant,

$$\frac{1}{T_2} = \frac{1}{T_2^*} + \frac{1}{2T_1} \tag{2}$$

where T_2^* is the homogeneous pure dephasing time and T_1 is the vibrational lifetime. T_2 is determined from the echo decay constant. T_1 is measured with pump-probe experiments. Measurements of T_2 and T_1 permit the determination of T_2^*, the pure dephasing contribution to the linewidth. An exponential vibrational echo decay corresponds to a Lorentzian lineshape with a linewidth, Γ, given by

$$\Gamma = \frac{1}{\pi T_2} = \frac{1}{\pi T_2^*} + \frac{1}{2\pi T_1}. \tag{3}$$

Pure dephasing describes the adiabatic modulation of the vibrational energy levels of a transition caused by fast fluctuations of its environment (29,30). Measurement of this quantity, and how this quantity changes with temperature, solvent, viscosity, or other experimental parameter, provides detailed insight into the dynamics of the system.

In addition to vibrational pure dephasing and vibrational population relaxation (lifetime), another contribution to the homogeneous dephasing time is orientational relaxation. The role of orientational relaxation in vibrational echo experiments of $W(CO)_6$ has been previously discussed in detail (21). In the experiments presented below on $Rh(CO)_2acac$ in dibutyl phthalate (DBP) and on myoglobin and hemoglobin proteins, orientational relaxation does not occur on the time scale of the vibrational echo experiments because of the samples' high viscosities (16). This fact was confirmed by magic angle pump probe experiments. Therefore, orientational relaxation is not discussed further.

B. Experimental Procedures

The vibrational echo experiments require tunable IR pulses with durations of ~1 ps, energies of ~1 μJ. Most of the experiments described below were performed using IR pulses of wavelength near ~5 μm generated by the Stanford superconducting-accelerator-pumped free electron laser (FEL). The FEL has been described in detail elsewhere (14,31,32). As stated above, an example vibrational echo scan performed using the FEL is shown in Fig. 2, which required approximately 15 minutes of averaging time. Signal-to-noise ratios of this quality are typical for these experiments and enable the resolution of dephasing mechanisms, as detailed below.

More recently, a commercial table top Ti:sapphire-based OPA system has been used to perform vibrational echo experiments. Several years ago, the FEL made it possible to perform the first ultrafast vibrational echo experiments. The advent of tabletop systems now makes it possible to perform vibrational echo experiments more routinely. Briefly, the Ti:sapphire-based system uses 5 W from a diode pumped doubled Nd:VO$_4$ laser to pump a Ti:sapphire oscillator, which produces fast, high-repetition-rate, 1 W, $\sim$800 nm pulses. These pulses are used as a seed for a Ti:sapphire regenerative amplifier (regen). The seed pulses are temporally stretched using conventional techniques. The bandwidth of the pulses is limited by slits in the stretching system. For the experiments described here, the bandwidth was limited to $\sim$18 cm^{-1}. The seed is then injected into the regen's cavity. The regen is pumped with 9.5 W from an intercavity doubled Nd:YLF laser. The regen cavity is triggered and amplifies the pulses to >1 W at 1 kHz. The regen output is compressed back to $\sim$1 ps, 18 cm^{-1}, 1 W at 1 kHz for pumping an OPA.

In the OPA, part of the incoming light from the regen is used to generate a white light continuum. This light is mixed with the rest of the regen beam in two passes through a BBO crystal. After the first pass through the BBO, a grating is used to wavelength select and narrow the broad bandwidth that is the output of the BBO. The narrowed idler from the first pass is amplified in the second pass. The signal and idler output of the BBO OPA become the pump and signal in a final AgGaS OPA, which generates midinfrared light. At 5 μm, the OPA typically produces 6 − 7 μJ/pulse at 1 kHz. Substantially more energy can be obtained when fs pulses rather than ps pulses are used. The IR output of the OPA is directed into the experimental set up for performing vibrational echo and vibrational pump-probe experiments.

One of the difficulties in performing the IR vibrational echo experiments is the fact that the IR beam is invisible. Unlike a UV beam, which can be viewed with a fluorescing card, there is no really good simple method for visualizing the IR beam. To overcome this problem, a coalignment system is used. The coalignment system efficiently coaligns the IR beam with a visible (HeNe) beam. The HeNe beam is brought into the system by reflection off of a Ge plate set at Brewster's angle for the IR. The IR beam passes through the plate. The HeNe and IR beams are made collinear. All subsequent optics are achromatic, e.g., off-axis parabolic reflectors are used instead of lenses, so the visible and IR beams remain aligned. It is then possible to align the experimental system using the visible HeNe beam.

The entire mid-IR part of the experiment is be enclosed in a purged (with dry air or N_2) compartment to eliminate the substantial atmospheric water absorptions. Fifteen percent of the IR beam is split off with a ZnSe beamsplitter and directed to the sample. The remaining 85% of the beam passes down a computer-controlled 0.1 μm step stepper motor delay line and is then sent into the sample. For the echo experiments, the probe beam is chopped at 500 Hz; for the pump-probe experiments the pump is chopped. Two matched 6" f.l. 90° off-axis parabolic reflectors are used to focus to ~100 μm and then recollimate the IR beams. The sample is contained in continuous flow cryostat. After the focused IR beams pass through the sample and are recollimated, either the probe or echo beam is directed into a HgCdTe detector. The signal from the detector is sampled by a gated integrator, the output of which is measured using a lock in amplifier. The 500 Hz signal from the lock-in is digitized for storage by a computer. To switch between a vibrational echo and pump probe experiments, only the delay line scanning direction, the beam that is chopped, and the detector that is sampled are changed.

A vibrational echo scan taken with the OPA system is shown in Fig. 3. These data, on hemoglobin-CO in EgOH/H_2O at 40 K, decay exponentially at 11.0 s. These echo data were taken on a sample in which the protein has a very strong background absorption compared to the CO peak under study. In addition, the sample is somewhat turbid to the eye. Nonetheless, it is possible to obtain high-quality echo data. The data took approximately 10 minutes to acquire.

III. VIBRATIONAL ECHO STUDIES OF DYNAMICS IN LIQUIDS AND CLASSES

In this section, a detailed vibrational echo study of $Rh(CO)_2$acac in DBP above and below the solvent's glass transition temperature ($T_g = 169$ K) is presented (17,18). The asymmetrical CO-stretching mode of $Rh(CO)_2$acac near 2000 cm^{-1} is examined over a wide range of temperatures. The temperature dependence of the pure dephasing time, T_2^*, which reflects the magnitude of the perturbations of the transition energy caused by fluctuations of

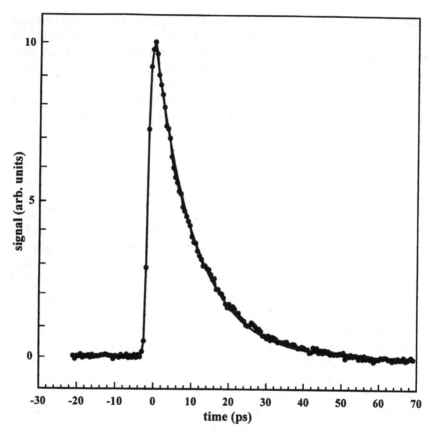

Figure 3 Vibrational echo decay data for the CO-stretching mode of the protein hemoglobin-CO ($\sim$1950 cm^{-1}) at 40 K and a fit to a single exponential function. The data were taken using a Ti:sapphire-based optical parametric amplifier system. The decay constant is 11.0 ps, corresponding to a homogeneous linewidth of 0.24 cm^{-1}. In contrast, the absorption linewidth is $\sim$9 cm^{-1}.

the bath, shows two clear temperature ranges in which different dynamics are responsible for the vibrational dephasing.

A. Liquid/Glass Results

Vibrational echo and vibrational pump-probe experiments were conducted on the CO asymmetric stretching mode of $Rh(CO)_2acac$ (2010 cm^{-1}) in DBP from 3.4 to 250 K. Figure 2 shows vibrational echo data taken at 3.4 K and an exponential fit [Equation (1)]. Within experimental uncertainty, the decay shown in Fig. 2 is a single exponential, indicating that a Lorentzian homogeneous line shape. The T_2 time is 95.2 ps, yielding a homogeneous linewidth of 0.11 cm^{-1}. For comparison, the absorption spectrum has a linewidth of $\sim$15 cm^{-1} at this temperature, demonstrating that the absorption line is massively inhomogeneously broadened. The absorption spectrum measures the inhomogeneous width and cannot provide information on the underlying homogeneous dephasing. The vibrational echo experiments show that the absorption line is inhomogeneously broadened at all temperatures studied, including 250 K ($1/(\pi T_2) = 1.5$ cm^{-1}). Above T_g, the sample is a liquid, but the vibrational spectrum is still inhomogeneously broadened.

Figure 4 displays the temperature-dependent vibrational echo (triangles) and pump-probe (squares) experimental results. The pump-probe experiments measure T_1. The data is plotted as $2T_1$, since this is the

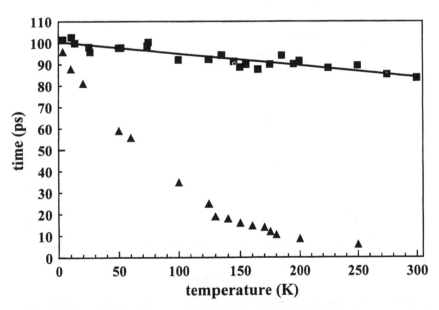

Figure 4 Vibrational echo (triangles) and pump-probe (squares) data for the asymmetrical CO-stretching mode of Rh(CO)$_2$acac in DBP. The pump-probe results are plotted as $2T_1$, for use with Equation (2). The solid line through the T_1 data is the best fit to the temperature dependence. Using these results, the temperature-dependent pure dephasing times, T_2^*, can be calculated from Equation (2).

relevant quantity [see Equation (2)]. As often is the case, the temperature dependence of $2T_1$ is very mild, and the temperature dependence of T_2 is much steeper. The pure dephasing, T_2^* is obtained using Equation (2) and the $2T_1$ and T_2 values obtained from the experiments.

Figure 5 displays the values of the pure dephasing width versus temperature on a log plot (17,33). The solid line through the data is a

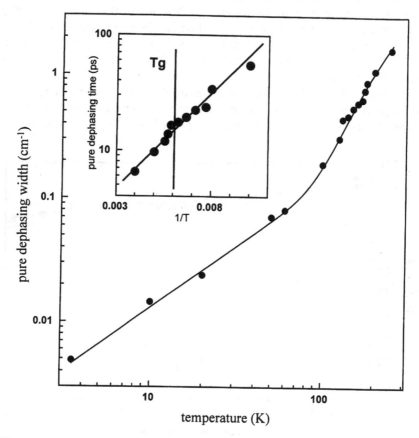

Figure 5 Pure dephasing widths, $1/(\pi T_2^*)$, of the asymmetrical CO-stretching mode of Rh(CO)$_2$acac in DBP versus temperature on a log plot. The solid line through the data is a fit to Equation (4), the sum of a power law and an exponentially activated process. The inset is an Arrhenius plot at higher temperatures showing that the process is activated. Note that there is no break at the experimental glass transition temperature, 169 K. The best fit has the power law exponent, $\alpha = 1.0$, and the activation energy, $\Delta E = 385$ cm^{-1}.

fit to the form

$$\frac{1}{T_2^*} = a_1 T^\alpha + a_2 e^{-\Delta E/kT} \tag{4}$$

with $\alpha = 1.0 \pm 0.1$ and $\Delta E = 385 \pm 50$ cm^{-1}. The inset is an Arrhenius (time vs. inverse temperature on semilog) plot of the high-temperature data showing that the data are exponentially activated at higher temperatures and that there is no break in the temperature dependence at $T_g = 169$ K.

B. Liquid/Glass Dephasing Mechanisms

1. Low-Temperature Pure Dephasing of $Rh(CO)_2 acac$

Pure dephasing of the form T^α where $\alpha \approx 1$ has been observed for homogeneous pure dephasing of electronic transitions of molecules in low-temperature glasses using photon echoes (5–7) and hole-burning spectroscopy (7,8,34–37). The electronic dephasing has been described using the two-level system (TLS) model of low-temperature glass dynamics (7,34,38,39).

The TLS theory was originally developed in the early 1970s to explain the anomalous heat capacity of low-temperature glasses, which is approximately linear in temperature (40,41). Glasses are continuously undergoing structural changes, even at low temperatures. The complex potential surface on which local structural dynamics occur is modeled as a collection of double wells. Only the lowest energy levels are involved, at low temperatures, so these are referred to as TLS. The mechanism of TLS-induced pure dephasing is illustrated in Fig. 6. Phonon-assisted tunneling can cause transitions between the two energy levels. At very low temperatures, the uptake of energy in going from a lower energy structure to a higher energy structure dominates the heat capacity. A glass is modeled as having many TLS with a broad distribution of tunnel splittings, E. If the probability, P(E), of having a splitting E is constant, P(E) = C, (all Es are equally probable), then the heat capacity is T^1.

The description of electronic dephasing in low temperature glasses is based on the TLS dynamics (7,34,38,39). We propose that identical considerations can apply to the vibrational dephasing of Rh(CO)$_2$acac in DBP at low temperature. For those TLS with E not too large (E < ~2 kT), the TLS are constantly making transition between the levels with a rate dependent upon E and the tunneling parameter (42). This is illustrated in the bottom part of Fig. 6. Transitions from one side of the double well to the other correspond to changes in the local glass structure. The molecular oscillator

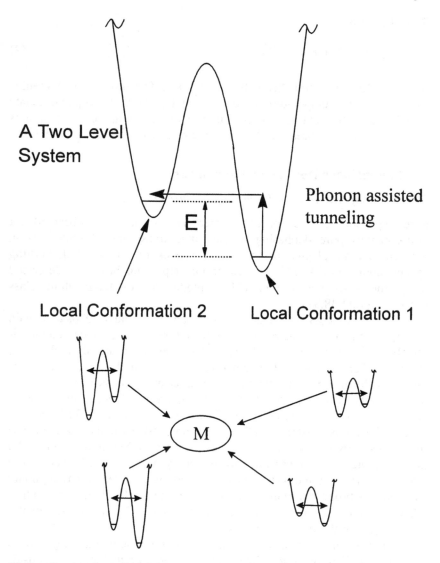

Figure 6 A schematic of the two-level system dynamics dephasing mechanism proposed to explain the observed T^1 temperature-dependent vibrational dephasing of the asymmetrical CO-stretching mode of $Rh(CO)_2$acac in DBP at low temperature. Phonon-assisted tunneling of the two-level systems (TLS) causes local structural fluctuations in the glass. The molecular oscillator, M, is coupled to many TLS. The structural fluctuations of the TLS produces fluctuating forces at M, resulting in pure dephasing.

is coupled to many TLS. The structural changes produce fluctuating strains, resulting in fluctuating forces on the CO oscillator. Thus, the vibrational pure dephasing can be caused by TLS dynamics. The uncorrelated sudden jump model predicts that for $P(E) = CE^\mu$, the temperature dependence of the pure dephasing is $T^{1+\mu}$ (7,8,43). Therefore, for the flat distribution, $\mu = 0$, the pure dephasing temperature dependence is T^1. T^1 and somewhat steeper temperature dependences, e.g., $T^{1.3}$, have been observed in electronic dephasing experiments in low-temperature glasses (5,7,8,37,43). Recent theoretical work, which has examined the problem in more detail, suggests that even the apparent superlinear temperature dependences may arise from an energy distribution $P(E) = C$ (28). Other theoretical work has investigated the influence of coupled TLS (44). Regardless of the theoretical approach, the qualitative results are the same. Coupling of a transition to a distribution of tunneling TLS can produce pure dephasing, which is essentially T^1.

The success of the TLS model in describing a large variety of distinct experiments adds weight to the proposition that the vibrational pure dephasing is produced by coupling to TLS. In electron excited-state photon echo experiments and hole-burning experiments, TLS dynamics have been observed a temperatures $\cong 10$ K. At higher temperatures, other processes with steeper temperature dependences dominated the pure dephasing, as well as other observables, such as heat capacities. In most systems, manifestations of TLS dynamics cannot be observed above a few K. In the vibrational dephasing experiments, the T^1 temperature dependence manifests itself to ~ 80 K. Additional experiments on this system and other low-temperature glassy systems are currently in progress. These will add additional information on the nature of vibrational dynamics at low temperatures.

2. High-Temperature Pure Dephasing of $Rh(CO)_2 acac$

Above ~ 80 K, the T^1 vibrational pure dephasing is dominated by the exponentially activated process. Electronic dephasing experiments have also shown power law temperature dependences that go over to activated processes at higher temperatures (7,8,45). However, in the electronic experiments, power law behavior is observed only to a few degrees K because in the electronic dephasing experiments it is found that $\Delta E = \sim 15$–30 cm^{-1}. Therefore, the activated process [arising from coupling of the electronic transition to low-frequency modes of the glass (46,47).] begins to dominate the power law pure dephasing at lower temperatures than is observed for the CO vibrational pure dephasing of Rh(CO)$_2$acac. In the vibrational pure dephasing experiments, the $\Delta E = {\sim}400$ cm^{-1}. Thus, the power law component of the temperature dependence is not masked until higher temperature.

In the Rh(CO)$_2$acac in DBP system, the temperature dependence of the pure dephasing changes rapidly above ${\sim}80$ K. By 100 K the temperature dependence is well described by the activated process alone (see inset in Fig. 5). There is no break in the pure dephasing data as the sample passes through T$_g$.

The activation energy, $\Delta E = {\sim}400$ cm^{-1}, is well above the phonon modes of organic solids (48,49). Furthermore, the far-IR absorption spectra of neat DBP show no significant transitions in the region around 400 cm^{-1}, indicating that there is no specific mode of the solvent that might couple strongly to the CO mode. These facts suggest that the high-temperature Arrhenius pure dephasing process is not caused by a motion associated with the glass/liquid solvent, but rather that the pure dephasing arises from coupling of the CO mode to another internal mode of Rh(CO)$_2$acac. If an internal low-frequency mode is excited, the combination band frequency can be different from the sum of the two vibrational frequencies. Therefore, excitation of a low-frequency mode can shift the frequency of the CO mode by an amount $\Delta\omega$. The proposed mechanism is shown schematically in Fig. 7.

For the proposed mechanism to account for the observed high-temperature pure dephasing, a mode of ${\sim}400$ cm^{-1} must couple nonnegligibly to the asymmetric CO stretch so that $\Delta\omega$ is significant. The Rh-C asymmetric stretching mode has an transition energy of 405 cm^{-1} (50). The closest other modes of Rh(CO)$_2$acac are outside of the error bars on the activation energy (50). Rh-C stretch couples more strongly to the CO mode than modes of lower frequency, which become populated at lower temperature. Rh(CO)$_2$acac has significant back donation of electron density from the Rh d$_\pi$ to the CO p$_{\pi*}$ antibonding orbital (back bonding) that weakens the CO bond and red-shifts the transition energy. Thus, the magnitude of back bonding plays a significant role in determining the transition frequency. When the Rh-C mode is thermally excited from the $v = 0$ state to the $v = 1$ state, the average bond length will increase. The increase in the sigma bond length will decrease the Rh d$_\pi$-CO p$_{\pi*}$ orbital overlap and, therefore, decrease the magnitude of the back bonding. Thus, excitation of the Rh-C mode causes a blue shift of the CO-stretching frequency by decreasing the back bonding (33).

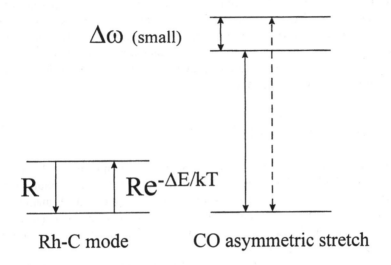

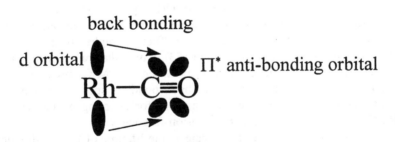

Figure 7 A schematic intramolecular low-frequency vibrational dynamics mechanism proposed to explain the observed exponentially activated temperature-dependent vibrational dephasing of the asymmetric CO-stretching mode of Rh(CO)$_2$acac in DBP at high temperature. Excitation and relaxation of the low-frequency Rh-C stretching mode causes the CO stretch frequency to shift $\Delta\omega$. The Rh-C mode is strongly coupled to the CO stretch through back bonding. Excitation of the Rh-C stretch increases the bond length, reducing and back bonding, and causing a blue shift of the CO stretch by $\Delta\omega$.

The proposed mechanism is thermal excitation of the Rh-C stretch mode causes the CO-stretching mode transition frequency to shift a small amount, $\Delta\omega$, as shown in Fig. 7A. During the time period in which the Rh-C mode is excited, the initially prepared CO superposition state precesses at a higher frequency, as indicated by the dashed arrow in Fig. 7A. Thus, a phase error develops. For a small $\Delta\omega$ and a short τ, the phase error is on the order of $\tau\Delta\omega < 1$. In the slow-exchange, weak coupling limit, the pure dephasing contribution to the linewidth from repeated excitation and relaxation of the low-frequency mode is (51,52):

$$\frac{1}{\pi T_2^*} = \frac{1}{\tau}\left[\frac{(\Delta\omega\tau)^2}{1 + (\Delta\omega\tau)^2}\right] e^{-\Delta E/kT} \tag{5}$$

Equation (5) shows that the contribution to the homogeneous linewidth from the excitation of the low-frequency mode will be exponentially activated. The right-hand side of Equation (5) is consistent with Equation (4) which was used to fit the data. The factor multiplying the exponential is the constant a_2 in Equation (4). This term dominates the temperature dependence at high temperature.

Although data is not available for $Rh(CO)_2acac$, IR absorption measurements on $M(CO)_6$ ($M = Mo, Cr$) support this mechanism (53). The frequency of the combination band of the M-C asymmetrical stretch and the CO asymmetrical stretch is ~ 20 cm^{-1} higher than the sum of the two fundamental energies alone (53). Thus, the transition of the CO stretch is 20 cm^{-1} higher in energy when the M-C mode is excited ($\Delta\omega \cong +20$ cm^{-1}). The change in back bonding upon excitation of the Rh-C mode provides a direct mechanism for coupling excitation of the Rh-C stretch to the CO stretch transition frequency.

It should be possible to estimate τ, the lifetime of the low-frequency Rh-C stretch using a_2 and $\Delta\omega$ in Equation (5). The values for $a_2 = 1.2$ THz and $\Delta\omega = 20$ cm^{-1} yield a value for τ of 0.75 ps. For a low-frequency mode that has a number of lower-frequency internal modes and the continuum of solvent modes to relax into, 0.75 ps is not an unreasonable value for the lifetime. A measurement of the Rh-C stretching mode lifetime would provide the necessary information to determine if the proposed dephasing mechanism is valid. In principle, the same mechanism will produce a temperature-dependent absorption line shift. However, other factors, particularly the change in the solvent density with temperature strongly influence the line position. Therefore, temperature-dependent line shifts cannot be used to test the proposed model.

IV. VIBRATIONAL ECHO SPECTRA

A. Vibrational Echo Spectroscopy Theory

In this section we present theoretical and experimental demonstrations of a vibrational spectroscopic technique, vibrational echo spectroscopy (VES) (54,55). The VES technique can generate a vibrational transition spectrum with background suppression using the nonlinear vibrational echo pulse sequence. In contrast to the previous results, VES is a utilization of vibrational echoes to measure spectra rather than dynamics. In a standard vibrational echo experiment, the wavelength of the IR light is fixed, and the delay, τ, between the excitation pulses is scanned. In VES, τ is fixed and the wavelength is scanned.

Background suppression in VES is in some respects analogous to NMR background suppression techniques (56,57). In both types of spectroscopy, coherent pulses sequences are used to remove unwanted spectral features.

Nuclear magnetic resonance and other magnetic resonance spectroscopies have had an enormous impact on the understanding of molecular structure and dynamics in the last 50 years (2,58–60). IR spectroscopy is inherently faster than NMR and can yield information about molecular motions and interactions in the fs-ns time ranges, while NMR yields information on far longer time scales. Infrared absorption spectroscopy has a much longer history, dating back to Newton's discovery of infrared radiation in the early 1700s (61). However, because of the relative difficulty in obtaining ultrafast IR pulses, coherent pulsed IR spectroscopy is a relatively new field.

Infrared absorption spectroscopy is a powerful technique for obtaining molecular structural information. In the midinfrared, all but the smallest molecules have a large number of transitions, which arise from fundamental, overtone, and combination modes. An absorption spectrum provides information about bonding, the shape of the molecular potential surface, solvent interactions, and dynamics. However, even moderate-sized molecules can generate spectra with a large number of peaks. For a large molecule, such as a protein, a solute in a complex solvent, tissue, or cells, the spectrum may become so crowded that clean observation of the spectral feature of interest can become difficult. The FTIR measurements, like the NMR spectroscopy before the spin echo, is a useful technique, but its utility falls off rapidly with molecular size. The VES technique may extend many of the useful observations of FTIR to far larger and more complex structures by improving line contrast and background suppression.

In VES, spectral selectivity can be achieved through two mechanisms: transition dipole selectivity and homogeneous dephasing (T_2) selectivity. If the background absorption, which can be a broad, essentially continuous absorption of undesired peaks, has homogeneous dephasing times, T_2^b (where the superscript b indicates background), short compared to the T_2 of the lines of interest, then VES can use the time evolution of the system to discriminate against the unwanted features. The time, τ, between the pulses in the vibrational echo sequence is set such that it is long compared to T_2^b but short compared to T_2. The VES signal from the background will have decayed to zero while the signal from the desired peaks will be nonzero. Scanning the IR wavelength of the vibrational echo excitation pulses and detecting the vibrational echo signal versus frequency will generate a spectrum in which the background is removed. If the background is composed of essentially a continuum of overtones and combination bands, while the peak of interest is a fundamental, it is likely that $T_2^b < T_2$.

It is also possible to discriminate against the unwanted signals based on the relative strengths of the transitions even when $T_2^b \cong T_2$. Absorbance is proportional to $m\mu^2$ while the vibrational echo signal is proportional to $m^2\mu^8$, where m is the concentration of the species and μ is the transition dipole matrix element. When background is composed of a high concentration of weak absorbers (m large, μ small) and the spectral features of interest are in low concentration but are strong absorbers (m small, μ large), the background absorption can overwhelm the desired features while the vibrational echo spectrum suppresses the background and reveals the relevant peaks.

Each spectral line can arise from a species with a particular concentration and transition dipole moment matrix element and a particular linewidth determined by the extent of homogeneous and inhomogeneous broadening. The magnitude of absorption as a function of frequency is given by Beer's law:

$$A(\omega) = \sum_{i,j} \varepsilon_{ij}(\omega) m_i l \tag{6}$$

where $A(\omega)$ is the absorption at frequency ω and $\varepsilon_{ij}(\omega)$ is the molar absorbtivity or the extinction coefficient of the jth transition of the ith species. ε has units of M^{-1} cm^{-1} and is related to the transition dipole matrix element squared (62). m_i is the concentration of the ith species in the sample, and l is the length of the sample. For the jth transition of the ith species, the absorption is

$$A = \varepsilon_{ij} m_i l \propto |\mu^{ij}|^2 m_i l \tag{7}$$

where μ^{ij} is the transition dipole matrix element of the jth transition of the ith species.

B. Model Calculation

To perform the VES calculations it is necessary to consider a finite duration pulse, which has a finite bandwidth. In addition, the actual shape of the vibrational echo spectrum depends on the bandwidth of the laser pulse and the spectroscopic line shape. Several species with different concentrations, transition dipole moments, line shapes, and homogeneous dephasing times can contribute to the signal. Therefore, VES calculations require determination of the nonlinear polarization using procedures that can accommodate these properties of real systems.

To calculate the vibrational echo intensity as a function of laser wavelength, using the details of the sample and realistic laser pulses, an efficient numerical algorithm for computing the vibrational echo signal is employed (63). The vibrational echo spectrum is calculated by numerically evaluating all of the rephasing and nonrephasing terms for the third-order nonlinear polarization, $P^{(3)}$, that contribute to the signal in the vibrational echo geometry (63).

To calculate the vibrational echo observable for a fixed laser frequency, ω_1, $P^{(3)}$ must be integrated over the spectroscopic line, $g(\omega)$, or the laser bandwidth, whichever is narrower, and then the modulus square of the result must be integrated over all time since the observable is the integrated intensity of the vibrational echo pulse,

$$I_s(\tau, \omega_\ell) \propto \int_{-\infty}^{\infty} dt_s \left| \int_0^{\infty} d\omega g(\omega) P_{tot}^{(3)}(\omega, t_s, \omega_\ell) \right|^2 \tag{8}$$

τ is the separation between the two laser pulses. This is the situation for a single transition of a single species. In general, there are two or more spectroscopic lines with independent $P^{(3)}$. The contribution from each transition of each species must be summed at the polarization level and squared

$$I_s(\tau, \omega_\ell) \propto \int_{-\infty}^{\infty} dt_s \left| \sum_{i,j} \left[\int_0^{\infty} d\omega^{i,j} g_{i,j}(\omega^{i,j}) P_{tot,i,j}^{(3)}(\omega^{i,j}, t_s, \omega_\ell) \right] \right|^2 \tag{9}$$

where i is the label for the species and j is the label for the jth transition of the ith species. It is necessary to distinguish between transitions on different species since the species may have different concentrations as well as the transitions having distinct line shapes, $g_{i,j}(\omega^{i,j})$, and transition dipole matrix elements, $\mu^{i,j}$.

Calculations were performed using Equation (9). Figure 8 displays model calculations for a system with a broad, high optical density (OD) solvent absorption and a narrow, low OD solute absorption. The abscissa is centered about the peak of the solute spectrum. Figure 8A is a model absorption spectrum. The parameters have been selected so that the broad solvent absorption has a 100 times larger optical density than the solute absorption. The inset shows a magnified view of the solute absorption. Figures 8B and 8C show background free vibrational echo spectra calculated using Equation (9), which demonstrate the two mechanisms for solvent background suppression. In Fig. 8B, the spectrum is calculated with $\tau = 0$, and the suppression occurs through transition dipole selectivity because the solute has a large μ but low concentration relative to the solvent. The suppression arises from the $m^2\mu^8$ dependence of the vibrational echo spectrum versus the $m\mu^2$ dependence of the absorption spectrum. This situation may be encountered frequently in real systems in which the peak of interest is a solute fundamental while the background consists of overtones and combination bands of the solvent. In Fig. 8C, an example of T_2 suppression is shown. The spectrum is calculated with $\tau = 5$ ps. The solvent has $T_2 = 1.0$ ps and the solute has $T_2 = 10$ ps.

Figure 8 Model calculations of vibrational echo spectroscopy (VES) for a system with a broad, strongly absorbing solvent background and a narrow, weakly absorbing solute. The abscissa is centered about the peak of the solute spectrum. (A) Absorption spectrum. The parameters have been selected so that the broad solute absorption has a 100 times larger optical density than the solute absorption. The inset shows a magnified view of the solute absorption. The solute is a strong absorber but low in concentration. The background is composed of weak absorbers at high concentration. (B) The VES spectrum calculated with $\tau = 0$. The background suppression occurs because the solute has a large transition dipole matrix element relative to the solvent even though it is low in concentration. (C) An example of T_2 suppression. The solute and solvent transition dipole matrix elements and concentrations were selected to give similar vibrational echo signals at $\tau = 0$. However, in this case, the solute $T_2 = 10$ ps and the solvent $T_2 = 1$ ps. The pulse delay is $\tau = 5$ ps. Because T_2 for the solvent is fast compared to T_2 for the solute, the solvent vibrational echo decay is essentially complete while the solute vibrational echo signal is still significant. Background suppression occurs because of differences in dynamics.

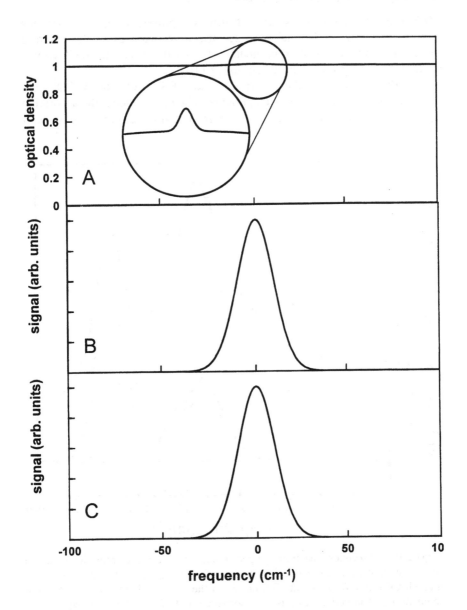

C.   Experimental Demonstrations of VES

VES requires a tunable source of infrared pulses. In the experiments presented below, vibrational echo spectra were taken using the Stanford FEL employing the same experimental setup used to perform the vibrational echo decay experiments discussed above (16).

Figure 9 displays the absorption and VES spectra of $W(CO)_6$ and $Rh(CO)_2 acac$ in DBP at room temperature. The upper trace is the absorption spectrum. The peak at 1976 cm^{-1} is the asymmetric CO-stretching mode of

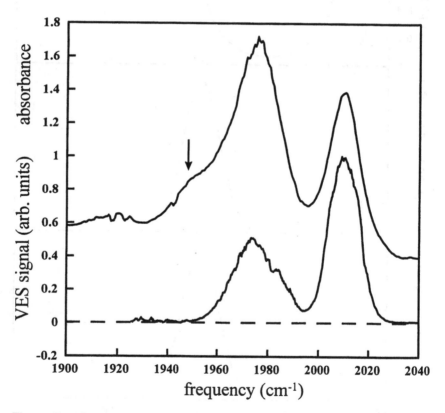

Figure 9   Absorption spectrum (upper trace) and VES spectrum (lower trace) of the asymmetrical CO-stretching modes of $W(CO)_6$ (centered at 1976 cm^{-1}) and $Rh(CO)_2 acac$ (centered at 2010 cm^{-1}) in the solvent dibutylphthalate at room temperature. The vertical axis is in absorbance units for the upper trace. For the VES spectrum, the vertical axis is in arbitrary units. In the VES spectrum, the solvent background and solvent peaks are eliminated.

$W(CO)_6$ and the peak at 2012 cm^{-1} is the asymmetric CO-stretching mode of $Rh(CO)_2$acac. The $Rh(CO)_2$acac mode is the same as the one mentioned in the previous section. The $W(CO)_6$ mode has been studied previously, although not specifically mentioned here (14,15,64). The vertical axis is in absorbance units for the upper trace. The DBP has a very broad absorption in the region giving a background absorbance of $\sim$0.5. In addition there is at least one solvent peak at $\sim$1948 cm^{-1}, which is indicated by an arrow. The lower trace is the VES spectrum. For the VES spectrum, the vertical axis is in arbitrary units. Two features are immediately clear. First, the background is zero, and second, the solvent peak at 1948 cm^{-1} is not visible.

As discussed above, there are two mechanisms by which VES can eliminate background and spectral peaks, T_2 selectivity, and transition dipole matrix element selectivity. The VES scan in Fig. 9 was taken with $\tau = 0$. Nonetheless, both selectivity mechanisms can be active. The pulses have finite duration of $\sim$1 ps. The vibrational echo signal arises from three interactions with the fields:the first interaction is with the first pulse, and the second and third interactions are with the second pulse. The interactions do not have to be time coincident, only time ordered, i.e., the second interaction must come after the first, and the third interaction must come after the second. A transition with a T_2 that is longer than the pulse will produce a polarization that involves the integral of the time ordered interactions throughout the pulses. Since the intensity of the signal is related to the absolute value squared of the polarization, the signal grows dramatically during the pulse duration. However, if T_2 is very short, the three interactions must occur almost simultaneously, and the polarization does not increase integrally throughout the pulse, greatly reducing the signal.

While T_2 selectivity may contribute to the elimination of the solvent background, it is clear that in this sample transition dipole matrix element selectivity will eliminate the background. Using round numbers, the DBP concentration is $\sim$10 M and its absorbance is $\sim$1. The metal carbonyl concentrations are $\sim$10^{-3} M and their absorbances are $\sim$1. Therefore, the metal carbonyl extinction coefficients are $\sim$10^4 larger than the solvents and their concentrations are $\sim$10^4 smaller than the solvents. In terms of the extinction coefficient, ε, and the concentration, m, the VES signal, $I_s \propto m^2\varepsilon^4$. Therefore, I_s should be on the order of $\sim$10^8 greater for the metal carbonyls than for the DBP solvent. The result is the observed zero solvent background spectrum.

Figure 10 illustrates T_2 selectivity between the metal carbonyl peaks. Two scans were taken, with zero delay time and with 1 ps delay time

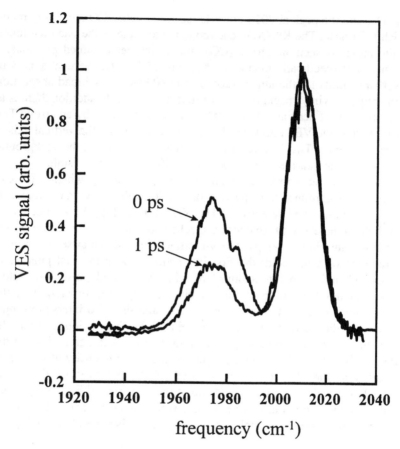

Figure 10 Two VES spectra of the asymmetric CO-stretching modes of $W(CO)_6$ and $Rh(CO)_2acac$ in the solvent dibutylphthalate at room temperature taken with delay times of 0 ps and 1 ps between the excitation pulses in the vibrational echo pulse sequence. The spectra are normalized at the peak of the $Rh(CO)_2acac$ spectra. When the delay is increased, the relative sizes of the $W(CO)_6$ and $Rh(CO)_2acac$ peaks change because the $W(CO)_6$ homogeneous dephasing time is shorter than that of $Rh(CO)_2acac$.

between the vibrational echo excitation pulses. The scan with zero delay contains the same data that is displayed in Fig. 9. The two spectra have been normalized to make the $Rh(CO)_2acac$ peaks the same size. The change in the relative peak heights is clear. $W(CO)_6$ has a shorter T_2 than $Rh(CO)_2acac$ in DBP at room temperature (17,64). With a longer delay, it would be possible to eliminate the $W(CO)_6$ completely from the spectrum. Subtracting the 1 ps trace from the 0 ps trace can eliminate the $Rh(CO)_2acac$ peak. Figure 10 shows that it is possible to manipulate peaks that appear in the VES spectrum in addition to eliminating a broad solvent background.

To demonstrate that the VES technique is not only useful with ideal samples, the VES experiment was conducted on CO bound to myoglobin in the solvent mixture (95:5) glycerol:water. Figure 11A displays the absorption spectrum of MbCO in the region of the CO stretch transition. The CO peaks at $\sim$1950 cm^{-1} are on top of a background with optical density $\sim$1. The A_1 peak is the largest peak, with the A_0 peak barely discernible. A_3 cannot be seen in this spectrum, but it has been observed in different types of samples (65,66). The A_1 peak has an optical density of $\sim$0.2 above the background. The background is composed of both protein and solvent absorptions.

Figure 11B displays VES data for MbCO along with a theoretical calculation of the vibrational echo spectrum. The solid points are the data. VES measurements were made at a number of fixed frequency points rather than continuously scanning the FEL. The amplitude of each point was determined from the magnitude of the vibrational echo signal at zero delay ($\tau = 0$). The square root of the vibrational echo spectrum is presented for direct comparison to the absorption spectrum. As discussed above, the vibrational echo spectrum at the polarization level is directly related to the absorption spectrum. The height of the spectrum has been scaled to 1. The vibrational echo spectrum has zero background as the protein and solvent do not contribute to the vibrational echo spectrum in the vicinity of 1950 cm^{-1}. The width of the vibrational echo spectrum is wider than the absorption spectrum because the bandwidth of the laser (13 cm^{-1}) is comparable to the spectral linewidth. Like any spectroscopic measurement, if the instrument resolution function is comparable to the linewidth, the spectrum will be broadened. The dashed line in Fig. 11B is the calculated vibrational echo spectrum using the procedures briefly outlined above and presented in detail elsewhere (55). While the A_3 line is not visible as a distinct peak, it was found that without including it in the calculation, the high-energy side of the calculated spectrum fell off much faster than the data. Also, the A_0 line is emphasized in the VES spectrum because it has a longer T_2 than the A_1 line.

VES is a type of coherent infrared two-dimensional spectroscopy. The two dimensions are frequency and time. The VES results demonstrate the potential of using VES to enhance vibrational spectra and potentially

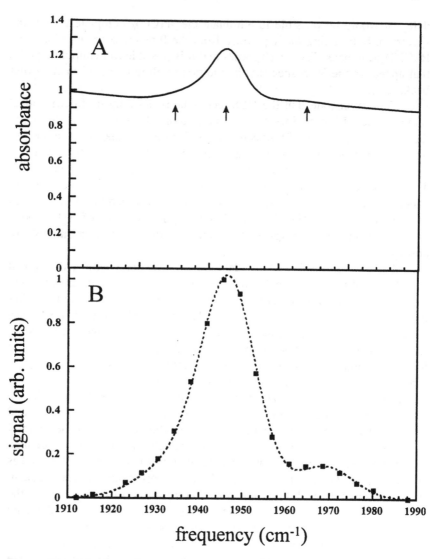

Figure 11 (A) Absorption spectrum of myoglobin-CO in the region of the CO-stretching mode. Only the A_1 conformer (center arrow) is clearly discernable. The A_0 peak is indicated by the arrow on the right, and the A_3 peak is indicated by the arrow on the left. The spectrum has a background (solvent + protein) optical density of ~ 1. (B) Example of myoglobin-CO VES data and fit. The dots are the square root of the experimental vibrational echo intensities at zero pulse delay with the laser wavelength varied. See text for details of the calculation.

observe peaks that are completely lost in a broad, highly absorbing background. The many powerful pulses sequences used in NMR to enhance spectra have been developed over a number of decades. The VES results may be the precursor to an equivalent approach, using coherent pulse sequences in vibrational spectroscopy.

V. VIBRATIONAL ECHO STUDIES OF PROTEIN DYNAMICS

In this section, the first applications of vibrational echoes to the detailed study of dynamics in proteins are discussed. Vibrational echo experiments are used to examine protein dynamics in myoglobin and hemoglobin as sensed by the CO ligand bound at the active sites of the proteins. The understanding of protein dynamics is fundamental in understanding the connection between protein function and protein structure, as determined by x-ray crystallography (67–69), NMR spectroscopy (70), or other experimental techniques (71–77), and theory (78). Degenerate four-wave mixing experiments, such as vibrational echoes (14–20), photon echoes (79), hole burning (37), and other ultrafast techniques (74,80–86) have shown great promise in obtaining crucial information about ultrafast protein motions unobtainable with other methods.

To date, the vibrational echo experiments have been used to study hemoglobin and myoglobin, small respiratory proteins that have the primary biological function of the reversible binding and transport of O_2 in the blood stream and in muscle tissues. The proteins' ability to bind O_2, and other biologically relevant ligands, such as CO or NO, is due to a nonpeptide prosthetic group, heme, which is covalently bound at the proximal histidine of the globin. Heme is a porphyrin-like structure with Fe at its center. The interior of both proteins consists almost entirely of nonpolar amino acids, while the exterior part of the protein contains both polar and nonpolar residues. The only internal polar amino acids are two histidines (87). The proximal histidine is covalently bonded to the Fe, forming the fifth coordination site of the heme. The sixth coordinate site of the heme is the active site of the protein where the ligand binds. The distal histidine is physically near the sixth coordinate site of the heme but not directly covalently bonded to it.

A. Vibrational Echo Results and Dephasing Mechanisms

Vibrational echo and pump-probe measurements were performed as a function of temperature on the CO stretch of MbCO in a variety of solvents

(54). In all cases, the pure dephasing rates were calculated from these results using Equation (2). Figure 12 shows the pure dephasing contribution to the linewidth, $1/\pi T_2^*$, versus temperature on a log plot for MbCO in trehalose. Trehalose is a sugar that is a glass over the entire temperature range of the study. As can be seen in Fig. 12, between 11 and $\sim$200 K the functional form of the data is a power law,

$$\frac{1}{\pi T_2^*} = aT^{1.3} \tag{10}$$

where the prefactor $a = 3.5 \times 10^7 \pm 0.1 \times 10^7$ Hz/K$^{1.3}$. The error bar on the power law exponent is ± 0.1. There is a change in the functional form of the data at $\sim$200 K. The points above $\sim$200 K can be fit with

$$\frac{1}{\pi T_2^*} = 3.3 \times 10^{12} e^{\frac{-650}{k_B T}} \text{ Hz} \tag{11}$$

where k_B is Boltzmann's constant, $k_B T$ has units of cm^{-1}, and the error bars on the prefactor and activation energy are $\pm 0.2 \times 10^{12}$ Hz and ± 25 cm^{-1}, respectively. However, it is important to emphasize that the form of Equation (1) is not unique given the small number of points. If this is done, the value of the exponent changes, but the power law is identical. A very good fit is obtained with a power law plus a Vogel-Tammann-Fulcher (VTF)–type equation (88–90):

$$\frac{1}{T_2^*} = a^* \exp\left(\frac{-E}{T - T_0}\right) \tag{12}$$

A VTF function often describes the temperature dependence of properties of glass-forming liquids. T_0 is referred to as the ideal glass transition temperature and is typically a few tens of degrees below the laboratory T_g. A fit to the data with Equation (10) plus Equation (12) yields a T_0 of $\sim$180 K and an E corresponding to a temperature of $\sim$230 K. These parameters can vary somewhat about the given values. However, the power law is always identical, independent of the form used to fit the points above $\sim$200 K. If the exponential fit and the VTF fit are extended to higher temperatures, they do not become distinguishable below 500 K. Therefore, experiments at temperatures below the Mb denaturation temperature cannot distinguish these two forms. Regardless of the form that is used to fit the data, it is clear that there is a sudden change in the nature of the temperature dependence of the pure dephasing.

The myoglobin dynamics near $\sim$200 K have been the subject of considerable investigation. There have been many experiments show a

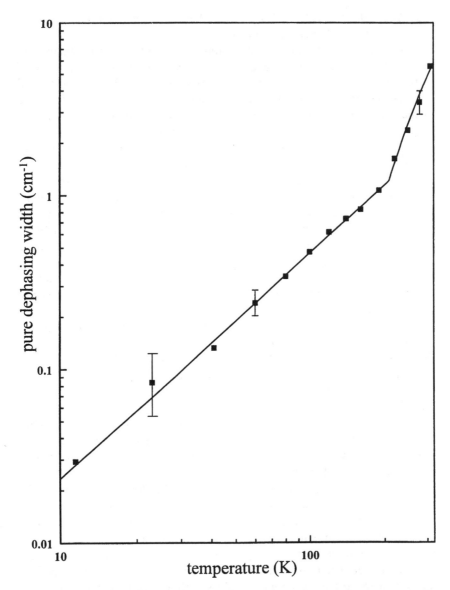

Figure 12 Log plot of the pure dephasing width, $1/\pi T_2^*$, versus temperature for native MbCO in trehalose. Below ~200 K the temperature dependence is dominated by a power law, $T^{1.3}$, which appears linear on the log plot. Above 200 K the data can be fit with an exponentially activated process or a VTF process.

break near 200 K (71,91–97). However, there have been other experiments that do not see a change near 200 K (93,98–101).

Given the many other types of studies that have observed changes in the Mb dynamics at ~200 K and that have been interpreted as evidence of a protein glass transition (16), it is reasonable to assume that the vibrational echo data do, indeed, display a manifestation of a change in the basic nature of the protein dynamics, i.e., they reflect the protein glass transition. The protein glass transition is not a true glass transition but, rather, reflects a change in the nature of protein dynamics that is akin to the liquid/glass transition. As discussed below, MbCO pure vibrational dephasing arises from global fluctuations of the protein structure (16,20,102). The CO dephasing is caused by electric field fluctuations produced by overall protein motions, rather than very local protein dynamics near the CO. A change in the nature of the protein dynamics, which influences the various observables that have been studied previously, can also produce a change in temperature dependence of the vibrational pure dephasing.

Figure 13 shows pure dephasing linewidths as a function of temperature for MbCO in three solvents. The circles are the trehalose data shown in Fig. 9; the diamonds are data for MbCO in the solvent 95:5 glycerol:water, (16,20) and the squares are data for MbCO in the solvent 50:50% ethylene glycol:water (54). The line through the trehalose data is the fit from Fig. 12. The lines through the other data are guides to facilitate discussion.

In all three solvents, at temperatures below their respective break points, the data fall on the same $T^{1.3}$ power law line. The fact that the vibrational dephasing comes solely from protein fluctuations, and not from the solvent, has been discussed in detail previously (16,20,102). The identical power law temperature dependences, which have the same dephasing rates in three solvents, is a demonstration that the vibrational pure dephasing is a measure of protein dynamics.

The $T^{1.3}$ temperature dependence observed at lower temperatures for the MbCO vibrational dephasing is reminiscent of the vibrational dephasing and other experimental observables measured in true glasses. In the MbCO vibrational dephasing, the $T^{1.3}$ temperature dependence is observed to much higher temperatures than in true glasses. Like the results in Rh(CO)$_2$acac, one possible explanation of the power law temperature dependence is thermally assisted tunneling among slightly different protein configurations. Small internal protein structural changes might be described in terms of protein two-level systems (PTLS) (16,20). The PTLS are the equivalent of the two-level systems discussed above, except the protein energy landscape would have to be such that tunneling is the dominant process at

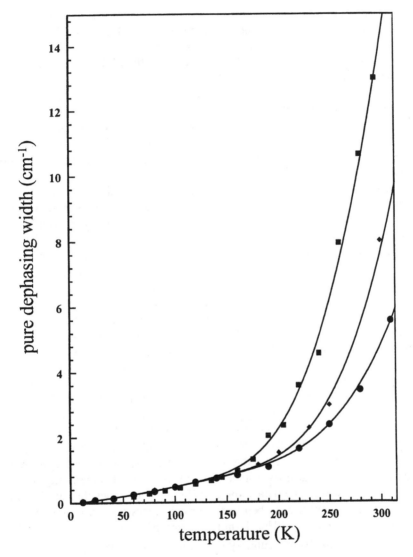

Figure 13 Pure dephasing of MbCO in three solvents. The circles are the trehalose data shown in Fig. 12. The diamonds are data taken in 95:5 glycerol:water (16). The squares are data taken in 50:50 ethylene glycol:water. Below ~150 K the dynamics for all three solvents are fit with the identical $T^{1.3}$ power law. Above ~200 K the trehalose data are fit with an exponentially activated process. The glycerol:water and ethylene glycol:water data have both a temperature dependence and a viscosity dependence at the higher temperatures.

temperatures up to 200 K. If this is the case, the same statistical mechanics machinery used to describe the low-temperature ($\sim$1 K) optical dephasing of electronic transitions of chromophores in low-temperature glasses (7,103) can be used to describe the PTLS induced vibrational dephasing of MbCO at much higher temperatures ($\sim$100 K).

In Fig. 13 it can be seen that the discontinuities in the dephasing in the three solvents do not occur at the same temperature and that the temperature dependences are not the same in the three solvents at higher temperatures. At all temperatures studied in trehalose, the viscosity is effectively infinite since it is a glass. The observed temperature-dependent pure dephasing comes from protein fluctuations in a solid medium in which the topology of the protein/surface interface is fixed. In the two liquid solvents, the situation is quite different. As the temperature is increased, the viscosity of the solvents decrease. Recent vibrational echo studies on MbCO conducted at room temperature as a function of solvent viscosity (data not shown here) show that there is a strong viscosity dependence to the MbCO pure dephasing (104). At constant temperature, as the viscosity of the solvent is decreased, the MbCO vibrational pure dephasing rate increases. Therefore, the temperature dependences observed in the glycerol:water and ethylene glycol:water solvents are actually combinations of a pure temperature dependence and a viscosity dependence. For this reason, the data in these solvents were not fit with the functions given in Equation (11) or (12). The trehalose data characterize the protein dynamics responsible for the pure dephasing with the protein/solvent boundary condition static.

The order from lowest temperature to highest temperature of the break in the functional form of the temperature dependences displayed in Fig. 13 is ethylene glycol:water ($\sim$150 K), glycerol:water ($\sim$180 K), and trehalose ($\sim$200 K). This is also the order of the solvents' glass transition temperatures. From Fig. 12 it is clear that the dynamical transition displayed in the vibrational echo data does not depend on the solvent undergoing a glass transition. However, the data show that if the solvent goes through its glass transition at a temperature below the protein glass transition temperature, T_g^P, then T_g^P is reduced. This is not a slaved protein glass transition, but, rather, a protein/solvent boundary condition influence on T_g^P (19).

B. Coupling of Protein Fluctuations to the CO Ligand at the Active Site

For vibrational dephasing of CO bound to the active site of Mb to occur, the fast motions of the protein must be coupled to the vibrational states of the CO in a manner that causes fluctuations in the CO vibrational transition energy. Two models have been proposed to explain the dephasing in Mb (20,102). One involves global electric field fluctuations and the other local mechanical coupling.

In the global electric field model, motions of polar groups throughout the protein produce a time-dependent electric field. The fluctuating electric field causes modulation of the electron density of the heme's delocalized π-electron cloud. Fluctuations of the heme π electron density modulate the magnitude of the back bonding to the CO π^* orbital, causing time-dependent shifts in ν_{CO}, or pure dephasing. In essence, the protein acts as a fluctuating electric field transmitter. The heme acts like an antenna, which receives the signal of protein fluctuations and communicates it to the CO ligand bound at the active site via the back bonding.

In the local mechanical fluctuation model, the local motions of the amino acids on the proximal side of the heme are coupled to the heme through the side group of the proximal histidine. The side chain of the proximal histidine is covalently bonded to the Fe. This bond is the only covalent bond of the heme to the rest of the protein. Thus, motions of the α-helix that contains the proximal histidine are directly coupled the Fe. These motions can push and pull the Fe out of the plane of the heme. Since the CO is bound to the Fe, these motions may induce changes in the CO vibrational transition frequency causing pure dephasing.

To test these models, we have performed a temperature-dependent vibrational echo and pump-probe study on two myoglobin mutants, H64V-CO and H93G(N-MeIm)-CO, and compared the results to those of the wild-type protein. To test the global electric field model, we studied H64V, a myoglobin mutant in which the polar distal histidine is replaced by a nonpolar valine (105). If the global electric field model of the dephasing is operative, then the decrease in the electric field in the mutant should reduce the magnitude of the frequency fluctuations, producing slower pure dephasing. To test the local mechanical model of pure dephasing, we studied H93G(N-MeIm), a myoglobin mutant in which the proximal histidine is replaced by a glycine (106). This mutation severs the only covalent bond between the heme and the globin and leaves a large open pocket on the proximal side of the heme. Inserted into this pocket and bound to the heme at the Fe is an exogenous N-methylimidizole, which has similar chemical and electrostatic properties as the side group of the histidine. Effectively, the proximal bond has been severed without changing significantly the electrostatic properties of the protein. If dynamics of the f α-helix are causing the pure dephasing by producing Fe motions via the proximal histidine, then the dephasing of this mutant should be less than that of the native protein.

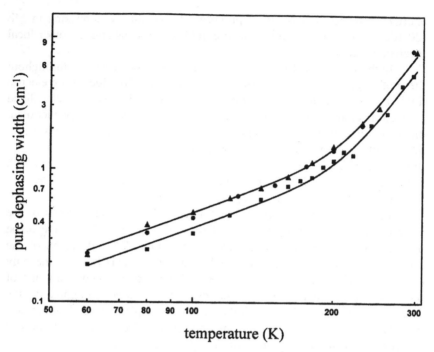

Figure 14 Pure dephasing rate versus temperature for native MbCO in ethylene glycol/H_2O (circles, same as in Fig. 13). Also plotted is pure dephasing for two mutants of myoglobin, H64V-CO (squares) and H93G(N-MeIm)-CO (triangles). The native MbCO and the mutant H93G(N-MeIm)-CO have identical pure dephasing temperature dependences. The H64V-CO has identical form of the pure dephasing but with a 21 ± 3% decrease in the pure dephasing rate at all temperatures studied.

Figure 14 shows the pure dephasing rates versus temperature on a log plot of the wild-type protein and the two mutants studied, all in 95:5% glycerol:water (102). The circles are the values for the native protein, which are the same as in Fig. 13. The triangles are the mutant H93G(N-MeIm)-CO pure dephasing rates. Clearly these values are identical to the native protein within experimental error, indicating that the proposed local mechanical dephasing model is not active in myoglobin. The squares are the pure dephasing rates for the mutant H64V-CO in 95:5% glycerol:water. The data have the same temperature dependence as the wild type. However, the dephasing is 21 ± 3% slower than that of the wild type at all temperatures. The functional form of the temperature dependence is unchanged because modification of one amino acid does not significantly change the global dynamics of the protein. However, replacing the polar distal histidine with a nonpolar valine removes one source of the fluctuating electric fields, reduces the coupling of the protein dynamics to the CO vibration, and slows dephasing. These results support the global electric field model of pure dephasing in myoglobin and suggest that the distal histidine contributes $21 \pm 3\%$ of the fluctuating electric fields felt at the heme. Recent molecular dynamics simulations (107) lend support to the $\sim 20\%$ electric field fluctuation produced by the distal histidine.

Figure 15 shows pure dephasing of Hb-CO in EgOH/H$_2$O, performed with the OPA and a comparison to MbCO in the same solvent mixture. The line through the MbCO data is the same line as in Fig. 13. The line through the Hb-CO data is the same as the Mb line multiplied by 0.73. On a log plot multiplication by a constant corresponds to a linear shift. It is clear from the data that the functional forms of the data on Mb and Hb are identical. This suggests that both the low-temperature TLS dynamics and the high-temperature combination Arrhenius and viscosity dependences active in Mb are also active in Hb. Considering the implications of Fig. 12, it is possible that the difference in the pure dephasing rates in the two proteins is caused by differences in the fluctuating electric field magnitudes felt at the heme in the two proteins. This proposal would suggest that the fluctuating electric field magnitude felt at the heme and coupled to the CO is 27% lower in Hb than in Mb. Qualitative agreement with this concept has been seen in mutant protein studies (107). The nature of the dephasing in MbCO and HbCO is under continuing experimental and theoretical study.

VI. CONCLUDING REMARKS

Vibrational echo experiments have made it possible to perform a detailed examination of the dynamics of inter- and intramolecular interactions that give rise to the homogeneous linewidths and pure dephasing of the asymmetric CO-stretching mode of Rh(CO)$_2$acac in liquid and glassy solvents and of the stretching mode of CO bound at the active site of the globin proteins, myoglobin and hemoglobin. At low temperature (3.5 to ~ 80 K), Rh(CO)$_2$acac temperature-dependent pure dephasing has the function form, T^1. This is interpreted as the result of coupling between the vibrational mode and the dynamical two-level systems of the glassy DBP solvent. Above ~ 80 K, the pure dephasing becomes exponentially activated with an activation energy of ~ 400 cm^{-1}. There is no change in the functional form of the temperature dependence in passing from the glass to the liquid.

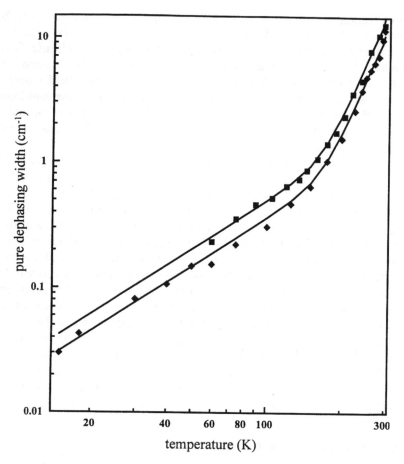

Figure 15 Pure dephasing rate of MbCO and HbCO both in ethylene glycol/H_2O versus temperature. The MbCO data (squares) are the same as those shown in Fig. 13. The HbCO data (diamonds) have the same functional form of the temperature dependence, but the dephasing is consistently slower at all temperatures.

These results suggest that the activated process arises from coupling of the high-frequency CO stretch to the internal 405 cm^{-1} Rh-C asymmetrical stretching mode.

The vibrational echo spectrum method was demonstrated theoretically and experimentally. In the VES technique, the delay between the two pulses in the vibrational echo pulse sequence is fixed and the laser frequency is scanned across the transition of interest. The VES technique can selectively remove unwanted spectral features, such as a broad background absorption or undesired peaks, using either differences in homogeneous dephasing or transition dipole moments. The method was demonstrated on a mixture $Rh(CO)_2acac$ and $W(CO)_6$ in DBP and on myoglobin-CO near 1950 cm^{-1}. Peak selectivity was clearly demonstrated in the inorganic system, and high optical density background suppression was demonstrated in the protein system.

Vibrational echo experiments have also been applied to the CO-stretching mode of myoglobin-CO, mutant myoglobins, and hemoglobin-CO. Temperature-dependent vibrational echo and lifetime measurements have been performed on CO bound to the active site of native Mb in a variety of solvents and two mutants of myoglobin and HbCO in the same solvent as a Mb study. In addition, an isothermal (300 K) viscosity dependence of MbCO has been recorded.

Temperature-dependent pure dephasing rates of MbCO in three solvents show identical power law behavior at low temperatures. At intermediate temperatures there is a break in the power law arising from the solvent-influenced protein glass transition. Above this point the data in glassy trehalose are exponentially activated. The other solvents, which at elevated temperatures are liquids, have additional solvent viscosity-dependent contributions to the pure dephasing rate.

The temperature-dependent vibrational echo results show that the pure dephasing of H64V is ~$21 \pm 3\%$ slower than native Mb, with no change in the functional form of the temperature dependence. The temperature dependence of the pure dephasing of H93G(N-MeIm) is identical to the native Mb. The general mechanism proposed (16) to explain the coupling of conformational fluctuations of the protein to the vibrational transition energy of CO bound at the active site is supported by the H64V results. The model states that protein motions produce fluctuating electric fields, which are responsible for the CO pure dephasing. Replacing the polar distal histidine with the nonpolar valine removes one source of the fluctuating electric fields, thus reducing the coupling between the protein fluctuations and the measured pure dephasing. The picture that emerges is that the heme acts as an antenna that receives and then communicates protein fluctuations to the CO ligand bound at the active site.

The vibrational echo results on HbCO in EgOH/H$_2$O show an identical functional form of the temperature dependence as MbCO in the same solvent mixture. We therefore concluded that the same dephasing mechanisms active in Mb are also active in Hb. In both proteins, a power law, $T^{1.3}$, is observed at low temperatures. This temperature dependence, observed in two proteins, strongly suggests that the protein behaves in a dynamical manner that is similar to a glass at low temperatures. The magnitude of the pure dephasing is 27% less in Hb than in Mb, suggesting that the magnitude of the electric field fluctuations is lower by this amount in Hb.

The ultrafast infrared vibrational echo experiment and vibrational echo spectroscopy are powerful new techniques for the study of molecules and vibrational dynamics in condensed matter systems. In 1950, the advent of the NMR spin echo (1) was the first step on a road that has led to the incredibly diverse applications of NMR in many fields of science and medicine. Although vibrational spectroscopy has existed far longer than NMR, the experiments described here are the first ultrafast IR vibrational analogs of pulsed NMR methods. In the future, it is anticipated that the vibrational echo will be extended to an increasingly diverse range of problems and that the technique will be expanded to new pulse sequences, including multidimensional coherent vibrational spectroscopies such as the vibrational echo spectroscopy technique describe above.

ACKNOWLEDGMENTS

A large number of individuals participated in the work discussed in this review. We would like to thank Dr. Alfred Kwok, Dr. Camilla Ferrente, Dr. David Thompson, Kusai Merchant, Dr. David Zimdars, Dr. Rick Francis, Stanford University, and Prof. Dana Dlott and Dr. Jeffrey Hill, University of Illinois at Urbana-Champaign for significant contributions. We would also like to thank Professors Alan Schwettman and Todd Smith and their research groups, especially Dr. Christopher Rella and Dr. James Engholm, at the Stanford Free Electron Laser Center, whose efforts made these experiments possible. We thank Prof. Stephen Boxer, Stanford University, and Prof. Steven Sligar, University of Illinois at Urbana-Champaign, for providing the protein mutants, H93G(N-Melm) and H64V, respectively. This research was supported by the Office of Naval Research, (N00014-92-J-1227-P00006, N00014-94-1-1024) and the National Science Foundation, Division of Materials Research (DMR93-22504, DMR-9610326)

REFERENCES

1. Hahn EL. Phys Rev 80:580, 1950.
2. Schmidt-Rohr K, Spiess HW. Multidimensional Solid-State NMR. London: Academic Press, 1994.

3. Kurnit NA, Abella ID, Hartmann SR. Phys Rev Lett 13:567, 1964.

4. Abella ID, Kurnit NA, Hartmann SR. Phys Rev Lett 14:391, 1966.

5. Molenkamp LW, Wiersma DA. J Chem Phys 83:1, 1985.

6. Vainer YG, Personov RI, Zilker S, Haarer D. In: Small GJ, ed. Fifth International Meeting on Hole Burning and Related Spectroscopies: Science and Applications, Vol. 291. Brainerd, Mn: Gordon and Breach, 1996.

7. Narasimhan LR, Littau KA, Pack DW, Bai YS, Elschner A, Fayer MD. Chem Rev 90:439, 1990.

8. Berg M, Walsh CA, Narasimhan LR, Littau KA, Fayer MD. J Chem Phys 88:1564, 1988.

9. Brewer RG, Shoemaker RL. Phys Rev Lett 27:631, 1971.

10. Shoemaker RL, Hopf FA. Phys Rev Lett 33:1527, 1974.

11. Zimdars D, Tokmakoff A, Chen S, Greenfield SR, Fayer MD. Phys Rev Lett 70:2718, 1993.

12. French PMW. Contemp Phys 37:283, 1996.

13. Reed MK, Shepard MKS. IEEE J Quantum Elect 32:1273, 1996.

14. Tokmakoff A, Fayer MD. J Chem Phys 103:2810, 1995.

15. Tokmakoff A, Zimdars D, Urdahl RS, Francis RS, Kwok AS, Fayer MD. J Phys Chem 99:13310, 1995.

16. Rella CW, Rector KD, Kwok AS, Hill JR, Schwettman HA, Dlott DD, Fayer MD. J Phys Chem 100:15620, 1996.

17. Rector KD, Fayer MD. J Chem Phys 108:1794, 1998.

18. Rector KD, Fayer MD. Intl Rev Phys Chem, 17(3):261, 1998.

19. Rector KD, Engholm JR, Rella CW, Hill JR, Dlott DD, Fayer MD. J Phys Chem A 103:2381, 1999.

20. Rector KD, Rella CW, Kwok AS, Hill JR, Sligar SG, Chien EYP, Dlott DD, Fayer MD. J Phys Chem B 101:1468, 1997.

21. Tokmakoff A, Urdahl RS, Zimdars D, Francis RS, Kwok AS, Fayer MD. J Chem Phys 102:3919, 1994.

22. Gordon RG. J Chem Phys 43:1307, 1965.

23. Gordon RG. Adv Magn Reson 3:1, 1968.

24. Berne BJ. Physical Chemistry: An Advanced Treatise. New York: Academic Press, 1971.

25. Farrar TC, Becker DE. Pulse and Fourier Transform NMR. New York: Academic Press, 1971.

26. Tokmakoff A, Kwok AS, Urdahl RS, Francis RS, Fayer MD. Chem Phys Lett 234: 289, 1995.

27. Rector KD, Kwok AS, Ferrante C, Tokmakoff A, Rella CW, Fayer MD. J Chem Phys 106:10027, 1997.

28. Geva E, Skinner JL. J Chem Phys, 109(12):4920, 1998.

29. Schweizer KS, Chandler D. J Chem Phys 76:2240, 1982.

30. Oxtoby DW. Annu Rev Phys Chem 32:77, 1981.

31. Bai YS, Greenfield SR, Fayer MD, Smith TI, Frisch JC, Swent RL, Schwettman HA. J. Opt Soc Am B 8:1652, 1991.

32. Schwettman HA. Nucl Instr Meth A 375:632, 1996.

33. Rector KD, Kwok AS, Ferrante C, Francis RS, Fayer MD. Chem Phys Lett 276:217, 1997.

34. Lee HWH, Huston AL, Gehrtz M, Moerner WE. Chem Phys Lett 114:491, 1985.

35. Hayes JM, Stout RP, Small GJ. J Chem Phys 74:4266, 1981.

36. Macfarlane RM, Shelby RM. Opt Commun 45:46, 1983.

37. Thijssen HPH, Dicker AIM, Völker S. Chem Phys Lett 92:7, 1982.

38. Macfarlane RM, Shelby RM. J Lumin 36:179, 1987.

39. Friedrich J, Wolfrum H, Haarer D. J Chem Phys 77:2309, 1982.

40. Phillips WA. J Low Temp Phys 7:351, 1972.

41. Anderson PW, Halperin BI, Varma CM. Philos Mag 25:1, 1972.

42. Harrer D. Photochemical Hole-Burning in Electronic Transitions. Berlin: Springer-Verlag, 1988.

43. Hayes JM, Jankowiak R, Small GJ. In: WE Moerner, ed. Persistent Spectral Hole Burning: Science and Applications, Vol. 44. Berlin: Springer-Verlag, 1988.

44. Kassner K, Silbey R. J Phys C 1:4599, 1989.

45. Selzer PM, Huber DL, Hamilton DS, Yen WM, Weber MJ. Phys Rev Lett 36:813, 1976.

46. Elschner A, Narasimhan LR, Fayer MD. Chem Phys Lett 171:19, 1990.

47. Greenfield SR, Bai YS, Fayer MD. Chem Phys Lett 170:133, 1990.

48. Kitaigorodsky AI. Molecular Crystals and Molecules. New York: Academic Press, 1973.

49. Keyes T. J Phys Chem A 101:2921, 1997.

50. Adams DM, Trumble WR. JCS Dalton 690, 1974.

51. Hsu D, Skinner JL. J Chem Phys 83:2097, 1985.

52. Shelby RM, Harris CB, Cornelius PA. J Chem Phys 70:34, 1978.

53. Jones LH, McDowell RS, Goldblatt M. Inorg Chem 8:2349, 1969.

54. Rector KD, Fayer MD, Engholm JR, Crosson E, Smith TI, Schwettmann HA. Chem Phys Lett, 305(1–2):51, 1999.

55. Rector KD, Zimdars DA, Fayer MD. J Chem Phys 109:5455, 1998.

56. Mani S, Pauly J, Conolly S, Meyer C, Nishimura D. Magn Res Med 37:898, 1997.

57. Yang X, Jelinski LW. J Magn Res B 107:1, 1995.

58. Clore GM, Gronenborn AM. Prog Nucl Magn Res Spectr 23:43, 1991.

59. Pelton JG, Wemmer DE. Ann Rev Phys Chem 46:139, 1995.

60. Tsuda S. Crystallogr Soc Jpn 38:84, 1996.

61. Newton I. Treatise of Reflections, Refactions, Inflections, and Colours of Light. New York: Dover, 1704.

62. Wilson EB, Jr., Decius JC, Cross PC. Molecular Vibrations: The Theory of Infrared and Raman Vibrational Spectra. New York: McGraw-Hill, 1955.

63. Zimdars DA. Stanford, CA: Stanford University, 1996. Thesis.

64. Rector KD, Thompson DE, Kusai M, Fayer MD. Chem Phys Lett 316(1–2): 122, 2000.

65. Austin RH, Beeson K, Eisenstein L, Frauenfelder H, Gunsalus IC, Marshal VP. Phys Rev Lett 32:403, 1974.

66. Ansari A, Beredzen J, Braunstein D, Cowen BR, Frauenfelder H, Hong MK, Iben IET, Johnson JB, Ormos P, Sauke T, Schroll R, Schulte A, Steinback PJ, Vittitow J, Young RD. Biophys Chem 26:337, 1987.

67. Kuriyan JW, Karplus M, Petsko GA. J Mol Biol 192:133, 1986.

68. Quillin ML, Arduini RM, Olson JS, Phillips GN, Jr. J Mol Biol 234:140, 1993.

69. Barrick D. Biochemistry 33:6546, 1994.

70. Havel HA. Spectroscopic Methods for Determining Protein Structure in Solution. New York: VCH Publishers, 1996.

71. Doster W, Cusack S, Petry W. Nature 337:754, 1989.

72. Chen SH, Bendedouch D. Meth Enzymol 130:79, 1986.

73. Braunstein DP, Chu K, Egeberg KD, Frauenfelder H, Mourant JR, Nienhaus GU, Ormos P, Sligar SG, Springer BA, Young RD. Biophys J 65:2447, 1993.

74. Jackson TA, Lim M, Anfinrud PA. Chem Phy 180:131, 1994.

75. Janes SM, Dalickas GA, Eaton WA, Hochstrasser RM. Biophys J 54:545, 1988.

76. Oldfield E, Guo K, Augspurger JD, Dykstra CE. J Am Chem Soc 113:7537, 1991.

77. Surewicz WK, Mantsch HH. In: Havel HA, ed. Spectroscopic Methods for Determining Protein Structure in Solution. New York: VCH Publishers, Inc., 1996.

78. Elber R, Karplus M. Science 235:318, 1987.

79. Leeson DT, Wiersma DA. Phys Rev Lett 74:2138, 1995.

80. Owrutsky JC, Li M, Locke B, Hochstrasser RM. J Phys Chem 99:4842, 1995.

81. Hill JR, Dlott DD, Rella CW, Peterson KA, Decatur SM, Boxer SG, Fayer MD. J Phys Chem 100:12100, 1996.

82. Hill JR, Dlott DD, Rella CW, Smith TI, Schwettman HA, Peterson KA, Kwok AS, Rector KD, Fayer MD. Biospec 2:227, 1996.

83. DeBrunner PG, Frauenfelder H. Ann Rev Phys Chem 33:283, 1982.

84. Frauenfelder H, Parak F, Young RD. Ann Rev Biophys Biophys Chem 17:471, 1988.

85. Petrich JW, Martin JL. In: RTH Clark, RE Hester, eds. Time-Resolved Spectroscopy. New York: Wiley, 1989.

86. Friedman JM, Rousseau DL, Ondrias MR. Ann Rev Phys Chem 33:471, 1982.

87. Stryer L. Biochemistry, 3rd ed. New York: Freeman WH and Co., 1988.

88. Angell CA. J Phys Chem Solids 49:863, 1988.

89. Angell CA, Smith DL. J Phys Chem 86:3845, 1982.
90. Fredrickson GH. Annu Rev Phys Chem 39:149, 1988.
91. Loncharich RJ, Brooks BR. J Mol Biol 215:439, 1990.
92. Steinbach PJ, Brooks BR. PNAS 90:9135, 1993.
93. Doster W, Bachleitner A, Dunau R, Hiebl M, Luscher E. Biophys J 50:213, 1986.
94. Hong MK, Draunstein D, Cowen BR, Frauenfelder H, Iben IET, Mourant JR, Ormos P, Scholl R, Schulte A, Steinbach PJ, Xie A, Young RD. Biophys J 58:429, 1990.
95. Mayer E. Biophys J 67:862, 1994.
96. Cordone L, Cupane A, Leone M, Vitrano E. J Mol Bio 199:213, 1988.
97. Parak F, Knapp EW, Kucheida D. J Mol Biol 161:177, 1982.
98. Frauenfelder H et al. Biochemistry 26:254, 1987.
99. Hartmann H, Parak F, Steigemann W, Petsko GA, Ponzi DR, Frauenfelder H. Proc Natl Acad Sci USA 79:4967, 1982.
100. Sartor G, Hallbrucker A, Mayer E. Biophys J 69:2679, 1995.
101. Sartor G, Mayer E, Johari GP. Biophys J 66:249, 1994.
102. Rector KD, Engholm JR, Hill JR, Myers DJ, Hu R, Boxer SG, Dlott DD, Fayer MD. J Phys Chem B 102:331, 1998.
103. Leeson DT, Wiersma DA, Fritsch K, Friedrich J. J Phys Chem B 101:6331, 1997.
104. Rector KD, Berg M, Fayer MD. J Phys Chem B (submitted).
105. Springer BA, Sligar SG. Proc Natl Acad Sci USA 84:8961, 1987.
106. Decatur SM, DePillis GD, Boxer SG. Biochemistry 35:3925, 1996.
107. Ma J, Huo S, Straub JE. J Am Chem Soc 119:2541, 1997.
108. Karavitis M, Fronticelli C, Brinigar WS, Vasquez GB, Militello V, Leone M, Cupane A. J Biol Chem 273:23740, 1998.

7

Structure and Dynamics of Proteins and Peptides: Femtosecond Two-Dimensional Infrared Spectroscopy

Peter Hamm
Max-Born Institut, Berlin, Germany

Robin M. Hochstrasser
University of Pennsylvania, Philadelphia, Pennsylvania

I. INTRODUCTION

A complete and predictive understanding of biological processes will require descriptions of the structures, and the dynamics. Knowing the three-dimensional structures of peptides and proteins is necessary in order to understand the selectivity and specificity of biological reactions. X-ray diffraction and nuclear magnetic resonance NMR spectroscopy (1–4) are extremely powerful spectroscopic tools with the ability to determine structures of proteins with hundreds (NMR) or even thousands (x-ray) of amino acids. The tremendous progress in understanding biological reactions of all types has been made possible by the detailed knowledge of the secondary, tertiary, and quaternary structures of the participating biomolecules. X-ray and NMR studies can also provide information on the distributions of structures and whether portions of the molecule are partially or fully disordered. However, the time scales of the fluctuations within ordered or disordered structures are not so readily obtained. Vibrational energy migration and energy relaxation within a protein are of crucial importance to the energetics of biological reactions. Energy dissipation by nuclear motions must strongly influence the rates, pathways, and efficiencies of chemical reactions involving proteins exactly as it does in solution phase chemistry. Some of the many types of processes that involve coupling to protein modes include (1) barrier crossings, which are important in enzyme reactions and conformational dynamics, (2) electron transfers, which are ubiquitous in biology, and (3) energy transfer, which is vital in photosynthesis. Furthermore, the possible roles of peptide nuclear motions in storing and transporting energy are not known. For example, vibrational energy transport through a peptide could occur entirely by thermal diffusion following very fast intramolecular vibrational relaxation or by directed coherent transport.

It is generally believed that in many cases the equilibrium fluctuations around the time-averaged structure are essential for the functionality of proteins. Presumably the evolution process of nature has optimized not only the structures of biomolecules but, as a consequence, also their dynamics. The classic example is myoglobin, whose binding site would never be reached by oxygen if the structure obtained from x-ray spectroscopy were static. Molecular dynamical (MD) simulations, however, reveal that ultrafast fluctuations of the peptide backbone and side chains open transient paths to the binding site (5,6). Presumably a similar situation arises in liquids in which, although they have close-packed structures, solutes can diffuse and chemical reactions can occur as a result of the motions of solvent molecules. Another interesting example, published only recently (7), demonstrated with the help of computer simulations how structural fluctuations of the binding pocket of acetylcholinesterase (AchE) determine the selectivity of this enzyme. Likewise, all-atom MD simulations of the protein-folding problem emphasize the substantial role of fast structural fluctuations in peptides (much slower) path towards the folded state (8–10). However, despite the importance of structural fluctuations, most of the knowledge on protein dynamics originates from computer simulations, along with rather indirect or no experimental verification.

The prevalence of structural compared with dynamical information arises from a scarcity of appropriate spectroscopic techniques. For example, the measurement process in NMR spectroscopy takes place on a millisecond timescale. The dominating part of the fluctuation correlation function responsible for dephasing processes, on the other hand, decays on the picosecond timescale, so that spin transitions are strongly in the motional narrowing limit and are, to a very good approximation, homogeneously broadened. It is exactly these rapid fluctuations in the solution phase that make the observation of sharp lines possible. This is in contrast to solid-state NMR spectroscopy, where special techniques such as magic angle spinning must be implemented. However, any detailed information on fast protein dynamics is completely lost in the homogeneous limit, and only very slow processes, such as hydrogen exchange, can be observed directly either in the time or the frequency domain.

Traditionally, x-ray spectroscopy measures an inhomogeneous distribution of structures, represented by the nuclear Debye Waller factors, and yields no information on the time scales of their rearrangements. Collective protein motions after fast optical triggers, on the other hand, have been studied with the help of pulsed synchrotron radiation with nanosecond time resolution (11). (See also Ref. 12 for a collection of review articles on time-resolved diffraction techniques.)

It is fortunate that the time scale of the measurement process in optical and vibrational spectroscopy incorporates the time ranges of major parts of protein fluctuations. This match makes these spectroscopic techniques ideal for investigating protein dynamics. Stimulated (three-pulse) photon echoes in the optical regime measure the fluctuations in the transition frequency due to movements of charged parts of the environment. Such experiments on optical chromophores in solution (13–21) and in a protein environment (22–26) have revealed that the correlation function that characterizes the fluctuations of the frequency of the optical transition decays quite differently in these two media. While the faster, subpicosecond, components of the decay are essentially equivalent in both surroundings, long-lived components are found only in proteins, proving that proteins fluctuate on a wide spread of time scales (26–28). The experimental properties of spectral diffusion and dynamical Stokes shifts are directly related to each other (17) in electronic transitions of polar molecules. They are both attributed to the coupling of the polar surroundings to changes in charge distribution that occur on photoexcitation. This coulombic coupling is sufficiently unspecific and long ranged that simulations are needed to evaluate its microscopic origins (20,29). We will see that in the case of vibrational transitions, it is the short-ranged anharmonic interaction that dominates the coupling of a vibrational mode to the fluctuating surrounding so that such fluctuations can be sensed much more locally (30,31) and experiments should guide theory in search of atomic level descriptions of the dynamics.

The tremendous success of NMR spectroscopy as a structural tool in biology originated mainly from its extension beyond a single dimension (1–4,32), allowing the resolution of congested spectra by spreading them out into a second dimension, to correlate connected states using cross-peaks, and to measure excitation transfer between states. A few examples of applying similar principles with femtosecond pulses have been proposed theoretically (33–35) and realized experimentally on vibrational transition with the help of impulsive fifth-order Raman experiments on liquids (36–40) or semi-impulsive third-order IR experiments on proteins and peptides (41,42). The same principle also has been demonstrated on electronic transitions (43). A review article on these topics has been published (44). Optical and vibrational transitions in solution dephase very rapidly compared with spin transitions, so these types of experiments have become feasible only very recently as a result of significant improvements in ultra-fast laser technology. IR spectra also have been spread into two dimensions by the variation of external conditions such as temperature or pressure (45). However, this approach differs in principle from nonlinear spectroscopy that only involves light fields as the perturbation so that a unified theory of the system-bath interaction can be used to describe the phenomena.

Multidimensional nonlinear methods enable a more complete determination of the multi–time point correlation function $\langle \vec{\mu}(t_3)\vec{\mu}(t_2)\vec{\mu}(t_1)\vec{\mu}(0)\rangle$, on which all third-order spectroscopic techniques are based. In the case of an isolated transition, such as an electronic two-level system (46) or a vibrational three-level system coupled to a bath (41), the multi–time points in the correlation function are often redundant so that both the third-order response and the linear response can be described in a unified theory with the help of *one* single line shape function g(t), which is related to the single time point correlation function $\langle \vec{\mu}(t)\vec{\mu}(0)\rangle$ (46). Nevertheless, as convincingly demonstrated in the photon echo peak shift experiments (13,14,16–18,21–25,31,41), nonlinear multidimensional methods are needed to isolate certain phenomena such as spectral diffusion, even though linear spectroscopy is adequately described by the same line shape function g(t). However, the multi–time point correlation function is necessary for the case of coupled transitions, such as spin systems or the amide I band of proteins, so that multidimensional methods are not just extremely advantageous, but are required when one wants to resolve the underlying coupling mechanisms. Spectroscopic techniques, such as three-pulse photon echoes, heterodyne or frequency resolved photon echoes, time-gated photon echoes, or dynamical hole-burning experiments, are all sensitive to the details of the coupling of multilevel systems.

We are beginning to develop a detailed understanding of these methods (18,21,30,33,34,37–40,42,44,47–49), many of which are described in this book. We have recently demonstrated a series of novel nonlinear all-IR spectroscopic techniques (IR-pump-IR-probe, IR-three-pulse photon echoes, IR-dynamic hole burning, IR-2D spectroscopy), all of them utilizing intense femtosecond IR pulses, with the intention to develop new multidimensional spectroscopic tools to study the structure *and* the dynamics of proteins (30,31,41,42,50–53). We shall summarize in this contribution our work, its underlying principles, and its applications.

II. IR LIGHT SOURCE

The experiments reported in this chapter have become possible only very recently with the progress in femtosecond laser technology, in particular the generation of high-power femtosecond IR-pulses. Even though IR light sources with pulse energies that are adequate to pump or probe vibrational transitions have been described in the literature over the last few years (54–57) and are moreover available commercially (Quantronix: Topas, Spectra Physics: OPA-800), there is a paucity of all-IR (i.e., IR-pump-IR-probe, IR-photon echoes, etc.) experiments below 2000 cm^{-1} on solution phase molecular systems. This limit arises from the fact that 4.5–5 µm is the highest possible wavelength reachable in one frequency conversion step, starting from dye, Ti:sapphire or Nd:Ylf laser systems and using nonlinear crystals such as KTP (58–64) or LiIO$_3$ (65–68).* Beyond this limit, other nonlinear crystals such as AgGaS$_2$, AgGaSe$_2$, and GaSe (69) have to be utilized. Since these crystals have band gaps in the visible range, they cannot be pumped directly with an intense pulse from a Ti:sapphire amplifier because of multiphoton absorption across the band gap (70), so that a two-step frequency conversion setup (57,71) has to be implemented. It was the high energy (1–3 µJ) and particularly the extremely good energy stability (<1% rms) of our instrument, that made possible the spectroscopy reported in this contribution, and it is therefore worthwhile to briefly describe it.

The instrument was based on a standard Ti:sapphire laser-amplifier system generating 80 fs pulses at 800 nm at a repetition rate of 1 kHz.

---

*   Intense picosecond pulses from free electron lasers, which are essentially freely tunable, have been used to study vibrational relaxation and dephasing of intense C=O modes of metal carbonyls in the 2000 cm^{-1} regime (73,75,98)

The generation of intense femtosecond mid-IR pulses was performed in two frequency conversion steps (see Fig. 1), similar to the setup described in Ref. 57:

1. A single-filament white light continuum was used to seed a double-stage BBO (Type II, 4 mm thick) optical parametric amplifier (OPA). The OPA setup was tunable between 1.2 and 2.3 μm. The output pulse duration was of the same order as the pump pulse (80 fs), and the pulses were close to the bandwidth limit. Total energy conversion efficiencies of up to 25% were achieved, corresponding to a total energy of up to 150 μJ in the signal and idler pulses.
2. In the second step, the signal and idler pulses were separated by a dichroic mirror, passed through a variable delay line, recombined, and

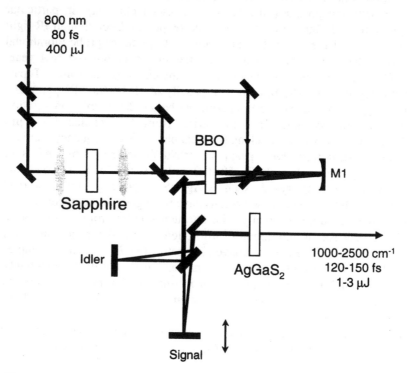

Figure 1 The setup used to generate intense ultrastable IR pulses. In a first frequency conversion step in a BBO optical parametrical amplifier, 800 nm pulses from a Ti:sapphire amplifier are split into signal and idler pulse, which subsequently are difference frequency mixed in a AgGaS$_2$ crystal.

focused into a type I AgGaS$_2$ crystal (2 mm) for difference frequency mixing. Energies of up to 3 µJ (at 2000 cm^{-1}) have been measured, corresponding to an overall quantum efficiency of 3%. The IR radiation was collected by a curved mirror and filtered by a Ge long pass filter. A typical pulse duration of 120–150 fs and a bandwidth of 150 cm^{-1}, corresponding to an almost transform limited time bandwidth product, were obtained in the whole spectral range of interest for the studies reported here (1600–2300 cm^{-1}).

The features that made this setup useful for fs IR spectroscopy were an almost transform limited focus ability, which enabled high intensities in the sample and the excellent energy stability of the IR light pulses. The former was achieved by mode matching between the first and the second pass in the BBO crystal. For the first pump beam approximately 10 µJ were focused tightly (f = 400 mm) into the crystal so that it acted as a spatial mode filter for the seeding white light continuum. The pump intensity was kept slightly below the threshold where white light generation and optical parametric fluorescence occurs in the nonlinear crystal. The generated divergent beam was collimated by a curved mirror (M1, see Fig. 1) such that the back-reflected beam had the same beam size (1 mm) as the second pump beam, for which an energy of ca. 400 µJ was used. Although the intensity of both pump beams in the BBO crystal was in a strong saturation regime and thus supported good energy stability, it was in fact the spatial and energy stability of a single filament white light continuum that was absolutely required to obtain IR pulses with a stability in the same range as that of the Ti:sapphire amplifier (<1% rms).

With IR light sources like this one, a technology is available which, in terms of day-to-day reliability and long-term and short-term stability, is entirely comparable with Ti:sapphire regenerative amplifiers. As shown in this article, it was possible to perform femtosecond experiments on all kinds of condensed phase phenomena involving vibrational transitions (such as energy relaxation, dephasing, spectral diffusion, coupled systems) with essentially the same facility and accuracy as can be achieved in visible and near-infrared experiments.

III. SPECTRAL DIFFUSION OF VIBRATIONAL TRANSITIONS

It is of great interest to obtain molecular-level descriptions of the structural fluctuations occurring in solutions phase systems. In this section we discuss how a solute vibrational transition is influenced by the fluctuating forces from the surrounding molecules. The solute could be a diatomic molecule or an isolated vibrational mode of a more complex system. At each instant of time there will exist a distribution of solute structures and, as a consequence, a distribution of vibrational frequencies of the test molecules. This distribution may be changing with time with the kinetics stretched over a wide time regime. Frequently the dynamics of this equilibrium distribution is thought to occur in only two time regimes — one that is short, and another that is long compared with the observation time. Then the rapid fluctuations give rise to a homogeneous or motionally narrowed line shape characterized by a time constant T_2, whereas the fixed distribution creates an inhomogeneous contribution to the linewidth. This so-called Bloch model relies on the time scales of homogeneous and inhomogeneous broadening being strictly separated.

Photon echo experiments can yield information about vibrational dynamics that is not evident from the linear absorption spectrum. To observe a conventional two-pulse delayed photon echo it is necessary to have an inhomogeneous distribution of frequencies to bring about the dephasing and rephasing of the macroscopic polarization. For example, in gases the Doppler broadening provides a fixed inhomogeneous distribution, whereas collisions and spontaneous emission contribute to homogeneous broadening of spectral lines. Conventional two-pulse photon echoes of the vibrational transitions of gaseous molecules (72) yield a direct measure of the homogeneous contribution. At low temperatures in the condensed phase, where there is certainly a fixed inhomogeneous distribution, the homogeneous dephasing time, T_2, can also be measured by two-pulse photon echo experiments (73–75). Even in solutions under certain conditions the Bloch assumptions give a reasonable description of the vibrational dynamics, as shown in the work by Fayer and coworkers, who have studied two-pulse vibrational echoes using picosecond pulses from a free electron laser (73–75). However, when the fluctuations are neither very fast nor very slow compared with the measurement process, the dynamics cannot be simply characterized by a single T_2 parameter. This is likely to be a common situation in fluids being examined with ultrafast laser pulses. In the solution phase and, in particular, in protein environments, fluctuations occur on many time scales, from tens of femtoseconds to seconds (22–28). It is easy to see that if the distribution of frequencies homogenizes during the measurement, a process often referred to as spectral diffusion, a more sophisticated approach would have to be applied to investigate the dynamics. The technique of choice to resolve this non-Markovian dynamics of the inhomogeneous distribution is the three-pulse (stimulated) photon echo experiment. Recently this coherent infrared technique was demonstrated for vibrational transitions of molecules and ions in ambient temperature solutions (41) and for test molecules bound to enzyme pockets (31). The three-pulse photon echo work involved intense femtosecond infrared laser pulses in a number of examples where a Bloch picture was completely inadequate to describe vibrational dephasing.

In a qualitative picture the two-pulse echo (the so-called Hahn-Purcell echo) arises because the contribution to the off-diagonal density matrix element from each part of the inhomogeneous frequency distribution propagates under a different Hamiltonian. Hence mean values that depend on these matrix elements will vanish on the time scale of the inverse width of the inhomogeneous distribution. Thus in the two-pulse echo the first pulse creates a vibrational coherence that begins to vanish by this inhomogeneous dephasing. After a delay time the second pulse transfers the surviving individual coherences into their conjugates, which then freely propagate toward the initial ensemble coherence to yield a macroscopic polarization and hence the echo signal. In a Bloch picture these two free propagation steps would occur with a fixed distribution of frequencies and the relative echo signal strength depends only on the homogeneous dephasing. However, if the distribution is not fixed, both propagation steps depend on the dynamics of the inhomogeneous distribution and the two-pulse echo results no longer have a simple interpretation in terms of a homogeneous dephasing time. In the three-pulse echo the second step of the two-pulse echo experiment is divided into two parts. The first part terminates the evolution of the initial coherence by creating a population. The second part converts this population into the conjugate coherences, which can then freely propagate towards a macroscopic polarization. The echo signal now depends on whether the inhomogeneous distribution survived during the time the system was in a population state. Since this period can be varied experimentally, the deviation from the Bloch model can be quantitatively examined.

The three-pulse experiments contain more information than two-pulse methods when the direction and timing of all three pulses is controlled. We have seen that this additional information cannot be interpreted within a Bloch picture. We will therefore outline in the following a more detailed theory, which includes spectral diffusion and which simultaneously explains the linear response (absorption spectrum) and the nonlinear response (four wave mixing, photon echo, transient grating, pump-probe) of vibrational transitions.

Through experimental and theoretical studies of the dynamics of electronic transitions (13,14,17,18,22,46), it was established that a three-pulse echo reveals the correlation function of the fluctuations of the electronic energy gap, which signifies the energy changes of the two electronic level system induced by nuclear motions. Although the experimental design based on the nonlinear field equations presented here for vibrational transitions is similar to those used for electronic transitions, we will see that the underlying dynamics and physical concepts are considerably different for vibrational transitions.

A. Theory of Vibrational Third-Order Nonlinear Spectroscopy

The system Hamiltonian is written in the form

$$H = H_S + H_B + V_{SB} - \mu E \tag{1}$$

where H_S is the Hamiltonian of the isolated vibrational mode, H_B the bath Hamiltonian, V_{SB} the interaction between system and bath, and $-\mu E$ the coupling of the system-bath complex to the light field. The first-order response is given by $\langle \mu(t)\mu(0)\rangle$ (76) with

$$\mu(t) = \exp_- \left(+\frac{i}{\hbar} \int_0^t V_{SB}(\tau)\,d\tau \right) \exp\left(+\frac{i}{\hbar}H_S t \right) \cdot \mu \cdot \exp\left(-\frac{i}{\hbar}H_S t \right)$$
$$\times \exp_+ \left(-\frac{i}{\hbar} \int_0^t V_{SB}(\tau)\,d\tau \right) \tag{2}$$

where $V_{SB}(\tau)$ is the system-bath coupling in the interaction representation. Rotational degrees of freedom, which also contribute to total dephasing, will be neglected at this point and will be introduced at a later stage through the time dependence of the dipole vectors in the laboratory frame. This approximation assumes the system-bath interaction to be independent of orientation.

The matrix elements of $\exp_-(+i/\hbar \int_0^t V_{SB}(\tau)\,d\tau)$ and $\exp_+(-i/\hbar \int_0^t V_{SB}(\tau)\,d\tau)$ are diagonal when vibrational relaxation between level pairs is neglected. With a classical (stochastic) approximation for the coupling term we then obtain:

$$\mu_{ij}(t) = \mu_{ij} e^{-i\omega_{ij}t} \exp\left(-i \int_0^t \delta\omega_{ij}(\tau)\,d\tau \right) \tag{3}$$

where $\delta\omega_{ij}(\tau)$ is the time-dependent fluctuation of the frequency separation of levels i and j. With the help of the cumulant expansion, the well-known result for the linear absorption spectrum of the vibrational ground state is obtained (46,76,77):

$$\langle \mu_{01}(\tau)\mu_{10}(0)\rho_{00}\rangle = |\mu_{10}|^2 \exp(-i\omega_{10}\tau - g(\tau)) \tag{4a}$$

with the line shape function, $g(\tau)$, given by:

$$g(\tau) = \int_0^\tau d\tau' \int_0^{\tau'} d\tau'' \langle \delta\omega_{10}(\tau'')\delta\omega_{10}(0)\rangle \tag{4b}$$

The corresponding double-sided Feynman diagram is shown in Fig. 2a. The double-sided diagrams, which have been used for many years, (78) have a very simple interpretation for spectroscopic or resonant processes because of the applicability of the rotating wave approximation in that case. The vertical lines are time lines with the earliest time at the bottom. The left and right time lines display properties of the light fields as they would appear on the ket and bra sides of the initial density operator ρ_{00} in a conventional expansion of the density matrix as a power series in the fields. The indices of the density matrix element created by the interaction are displayed horizontally. By convention the arrow pointing upward from left to right signifies a positive wavevector, whereas upward from right to left is a negative wave vector. Since the complex field is assumed to have the form $\exp[i\omega t - kr]$, a negative wave vector field implies a positive frequency, which causes a transition from a lower to a higher energy state on the bra time line or from a higher to lower energy state on the ket time line. The last arrow refers to the generated field, which we will always take as being emitted as the system undergoes a downward transition on the ket side. Thus in the correlation function containing $\mu_{01}(t)\mu_{10}(0)\rho_{00}$ the first interaction is on the left (ket) side of the initial density operator and the wave vectors of the outgoing and input fields cancel. After the first interaction the system is in a 1-0 coherence. The sign of the term is $(-1)^n$, where n is the number of interactions on the right (bra) side.

Within the same formalism, the response functions of the third-order Feynman diagrams in Fig. 2b can be written down immediately. For example, the response function of the first (rephasing) diagram in Fig. 2b is:

$$\begin{aligned}
R_1 &= \langle \mu_{01}(\tau_3)\mu_{10}(\tau_1)\rho_{00}\mu_{01}(0)\mu_{10}(\tau_2)\rangle \\
&= |\mu_{10}|^4 e^{-i\omega_{10}(\tau_3-\tau_1-\tau_2)} \cdot \left\langle \exp\left[-i\int_0^{\tau_3}\delta\omega_{10}(\tau')\,d\tau'\right.\right. \\
&\quad \left.\left. +i\int_0^{\tau_1}\delta\omega_{10}(\tau')\,d\tau' + i\int_0^{\tau_2}\delta\omega_{10}(\tau')\,d\tau'\right]\right\rangle
\end{aligned} \tag{5}$$

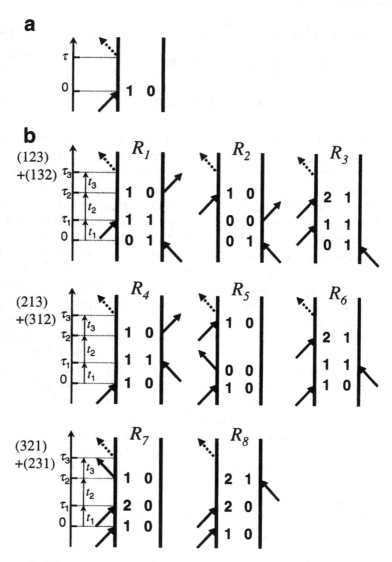

Figure 2 The double-sided Feynman diagrams, which have to be considered in (a) a linear absorption experiment and (b) a nonlinear third-order experiment such as photon echo, pump-probe, transient grating. The diagrams are arranged according to the possible time orderings, as discussed in the text and illustrated in Fig. 4.

Again with the help of the cumulant expansion (46) we obtain:

$$R_1 = |\mu_{10}|^4 e^{-i\omega_{10}(\tau_3 - \tau_1 - \tau_2)} \cdot \exp(+g(\tau_3) - g(\tau_1) - g(\tau_2)$$
$$- g(\tau_3 - \tau_1) - g(\tau_3 - \tau_2) + g(\tau_2 - \tau_1)) \tag{6}$$

with the same line shape function $g(\tau)$ as the one used for the linear absorption spectrum [Equation (4)]. By changing the absolute times τ_1, τ_2, and τ_3 to the time intervals $t_1 = \tau_1$, $t_2 = \tau_2 - \tau_1$, and $t_3 = \tau_3 - \tau_2$ corresponding to the delay times that are under experimental control, we obtain the final result for the response function:

$$R_1 = R_2 = |\mu_{10}|^4 e^{-i\omega_{10}(t_3 - t_1)} \exp(-g(t_1) + g(t_2) - g(t_3) - g(t_2 + t_1)$$
$$- g(t_3 + t_2) + g(t_1 + t_2 + t_3)) \tag{7}$$

In order to calculate the response function of the Feynman diagram R_3, it is further assumed that the transition frequency ω_{12} is anharmonically shifted with respect to the ground states transition frequency so that, $\omega_{12} = \omega_{01} - \Delta$. Another assumption that can be made (see later for a discussion of these assumptions) is that the fluctuations between both level pairs are strictly correlated $\delta\omega_{12} = \delta\omega_{01}$. This implies that only the harmonic part of the potential surface is perturbed by the bath fluctuations and the anharmonicity of the vibrator is unaffected. We then obtain for R_3:

$$R_3 = -|\mu_{10}|^2 |\mu_{21}|^2 e^{-i((\omega_{10} - \Delta)t_3 - \omega_{10}t_1))} \cdot \exp(-g(t_1) + g(t_2) - g(t_3)$$
$$- g(t_2 + t_1) - g(t_3 + t_2) + g(t_1 + t_2 + t_3)) \tag{8}$$

The response functions for the nonrephasing Feynman diagram (see Fig. 2b) are calculated accordingly:

$$R_4 = R_5 = |\mu_{10}|^4 e^{-i\omega_{10}(t_3 + t_1)} \cdot \exp(-g(t_1) - g(t_2) - g(t_3) + g(t_2 + t_1)$$
$$+ g(t_3 + t_2) - g(t_1 + t_2 + t_3)) \tag{9}$$
$$R_6 = -|\mu_{10}|^2 |\mu_{21}|^2 e^{-i((\omega_{10} - \Delta)t_3 + \omega_{10}t_1))} \cdot \exp(-g(t_1) - g(t_2)$$
$$- g(t_3) + g(t_2 + t_1) + g(t_3 + t_2) - g(t_1 + t_2 + t_3)) \tag{10}$$

The remaining time orderings, for which there is no corresponding process in a photon echo experiment on an electronic two-level system, are:

$$R_7 = |\mu_{10}|^2 |\mu_{21}|^2 e^{-i\omega_{10}(t_3 + 2t_2 + t_1)} e^{i\Delta t_2} \cdot \exp(+g(t_1) - g(t_2) + g(t_3)$$
$$- g(t_2 + t_1) - g(t_3 + t_2) - g(t_1 + t_2 + t_3)) \tag{11}$$
$$R_8 = -|\mu_{10}|^2 |\mu_{21}|^2 e^{-i\omega_{10}(t_3 + 2t_2 + t_1)} e^{i\Delta(t_3 + t_2)} \cdot \exp(+g(t_1) - g(t_2)$$
$$+ g(t_3) - g(t_2 + t_1) - g(t_3 + t_2) - g(t_1 + t_2 + t_3)) \tag{12}$$

As expected, we find that the total response function $\sum_{l=1}^{3} R_l = \sum_{l=4}^{6} R_l = \sum_{l=7}^{8} R_l = 0$ (i.e., for each possible time ordering) vanishes exactly in the harmonic case, defined by $\Delta = 0$ and $|\mu_{21}|^2 = 2|\mu_{10}|^2$. Furthermore, it can be easily seen that in the case of a strict separation of time scales of homogeneous and inhomogeneous broadening, the line shape function becomes $g(t) = t/T_2 + \sigma^2 t^2/2$ and the total response function reduces exactly to the result obtained within a Bloch picture (see, for example Refs. 52 and 75), e.g.,

$$\sum_{l=1}^{3} R_l = 2|\mu_{10}|^4 e^{-i\omega_{10}(t_3 - t_1)} \cdot (1 - e^{i\Delta t_3}) \cdot \exp(-(t_1 + t_3)/T_2)$$

$$\times \exp(-\sigma^2(t_1 - t_3)^2/2) \qquad (13)$$

In particular, the time coordinate t_2 disappears in Equation (13), which emphasizes the fact that three-pulse photon echo experiments are necessary to characterize effects that are beyond the Bloch picture.

Vibrational energy relaxation T_1 is taken into account phenomenologically through the multiplicative factors:

$$\exp\left(-\frac{t_1/2 + t_2 + t_3/2}{T_1}\right) \quad \text{for } R_1, R_2, R_4, \text{ and } R_5 \qquad (14a)$$

and

$$\exp\left(-\frac{t_1/2 + t_2 + 3t_3/2}{T_1}\right) \quad \text{for } R_3 \text{ and } R_6 \qquad (14b)$$

which assumes that vibrational relaxation from the $\nu = 2$ to the $\nu = 1$ level is twice as fast as from the $\nu = 1$ to the $\nu = 0$ level and that it is constant throughout the inhomogeneous distribution. Orientational relaxation for R_1 to R_6 is taken into account through the factor:

$$\langle \mu_z(0)\mu_z(t_1)\mu_z(t_1 + t_2)\mu_z(t_1 + t_2 + t_3) \rangle = \frac{2}{3}\left(\frac{1}{6} + \frac{2}{15}e^{-6Dt_2}\right)e^{-2D(t_1 + t_3)}$$

$$(15)$$

where $\langle \ldots \rangle$ is an orientational average, $\mu_z(t)$ is the projection of the transition dipole at t onto the laboratory z-axis, and D is the rotational diffusion coefficient. Thus, during the period t_1 and t_3, where the system is in a coherence, the four-wave mixing signal decays like the first-order Legendre polynomial $P_1(\cos\theta)$ just as is found in a two-pulse echo (74). However, during the period t_2, where the system is in a population state, it decays fractionally like $P_2(\cos\theta)$. The formulas for other polarization conditions and for nonspherical rotors are easily obtained by standard methods (79,80). Generally such simple diffusional models are not correct for small molecules in solutions. Their orientational correlation functions are not usually simple exponentials with parameters that depend on the molecular size, but they also involve varying amounts of inertial motion, and there is no general form for the intermediate region. Furthermore, there is in general no relation between the decays of P_1 and P_2 such as occurs when there is diffusive motion. However, if the orientational correlation functions could be measured independently it would be straightforward to introduce them in place of Equation (15) and then to solve the problem numerically.

The third-order polarization is obtained by convolution of the response functions with the electric fields of the three laser pulses:

$$P^{(3)}(t) = \int_0^\infty dt_3 \int_0^\infty dt_2 \int_0^\infty dt_1 \sum_1 R_1(t_1, t_2, t_3) \cdot E_3(t - t_3)e^{-i\omega(t-t_3)}$$

$$\times E_2(t - t_3 - t_2)e^{-i\omega(t-t_3-t_2)}E_1{}^*(t - t_3 - t_2 - t_1)e^{i\omega(t-t_3-t_2-t_1)} \quad (16)$$

where the $E_{1,2,3}$ are the field envelopes of the laser pulses with wave vectors vector k_1, k_2, and k_3, respectively, and ω is the carrier frequency of the light pulses. When the laser pulses are not overlapping in time, the time ordering is the same in the whole integration area and the sum in Equation (16) runs over those Feynman diagrams relevant for that particular time ordering ($1 = 1, 2, 3, 1 = 4, 5, 6,$ or $1 = 8, 9$, respectively). In the case of overlapping laser pulses, the set of response functions is interchanged each time the integration in Equation (16) switches the time ordering.

In a homodyne detection scheme, such as in the stimulated photon echo experiments described in the next paragraph, the detector measures the t-integrated intensity of the square of the third-order polarization

$$S(T, \tau) = \int_0^\infty |P^{(3)}(T, \tau, t)|^2 \, dt \quad (17)$$

where τ and T are the time delays between the peaks of pulse k_1 and k_2 and pulse k_2 and k_3, respectively (see Section III.D).

B.   Limitations of the Stochastic Model

The response function R_2, which develops on the vibrational ground state after the second light interaction, is the same as R_1 [see Equation (7)]. This is a consequence of the stochastic ansatz in Equation (3), which implies that the bath influences the vibrational frequency of the solute but excitation of the solute does not perturb the bath. In optical spectroscopy there is a Stokes shift which is due to the response of the bath molecules to the changed charge distribution of the solute after electronic excitation (81). This effect causes a nonzero imaginary part in the line shape function g(t) and the response functions R_1 and R_2 to become different (46). Vibrational transitions might undergo dynamical spectral shifts analogous to the Stokes shifts of electronic transitions. The nuclear motion dependence of the dipole or higher electric multipoles could induce a dynamical vibrational shift that will influence the interpretations of vibrational dynamics and cooling in general. Such dynamics is not included in, for example, Redfield theory (82–86) of multiple vibrational level systems coupled to bath. The usual electronic Stokes shift is caused by the response of the bath molecules to the changed charge distribution of the solute after excitation (87,88). For high-frequency vibrations the change in the static, averaged electric moment would be sensed by the solvent. However, when the solvent responds comparably or faster than the solute motion, both the curvature and the absolute energy of the potential surface may be changed, and any shift of a vibrational frequency will not simply record the change in free energy computed for a change in electric moment. Clearly the dipole moment as a function of the internuclear distance is already partly incorporated into the potential of mean force, which determines the mean frequencies in the Born-Oppenheimer approximation. Nevertheless, the vibrational state dependence of the dipole moment is well known (89–92) and changes of 0.03–0.2 D between $v = 0$ and $v = 1$ have been reported. Although the dipole (or electric quadrupole for nearest neighbor effects) changes are much smaller than for electronic transitions, the spectral resolution is improved in the infrared by a similar factor; therefore, when such dynamical shifts are present they might be observable. In summary, excitation of $v = 1$ by a short pulse should create a nonequilibrium distribution of solvent/solute configurations because of the mixed mode anharmonicity. Both the negative stimulated emission peak and the new ($1 \rightarrow 2$ transition) absorption in a IR-pump–IR probe experiment could shift and change width in time, while the bleaching peak should remain constant. It should be possible to measure shifts of less than one twentieth of the bandwidth. In some cases, specific modes (93) may dominate the anharmonicity while in others more complex solvent motions may be involved. In any event, further experimental work is needed to evaluate the assumption that $R_1 = R_2$.

   Another simplifying assumption included at this point in our model so far is the strict correlation of the fluctuations between the 0-1 and the 1-2 level pairs $\delta\omega_{12} = \delta\omega_{01}$. This assumption is based on the notion that the forces of the surroundings acting on the molecule change the potential energy surface that determines the frequencies. If only the harmonic force constants were changed under the influence of these forces, all the vibrational level pairs would exhibit correlated frequency fluctuations. As a result, a unified expression, e.g., for the response function of R_1, R_2 and R_3, based on one fluctuation correlation function $\langle \delta\omega_{10}(\tau'')\delta\omega_{10}(0)\rangle$ is obtained:

$$\sum_{l=1}^{3} R_l = 2|\mu_{10}|^4 (1 - e^{i\Delta t_3}) \cdot R_1 \tag{18}$$

For the dynamical distribution it will in general be necessary to consider both the auto and cross time correlation functions of the 0-1 and the 1-2 frequencies (117). For example, if the fluctuations, $\delta\Delta(t)$, in the anharmonicity are statistically independent of the fluctuations in the fundamental frequency, the oscillating term $(1 - e^{i\Delta t_3})$ in Equation (18) would be damped. In a Bloch model the fluctuations in anharmonicity translate into different dephasing rates for the 0-1 and 1-2 transitions that were discussed previously for two pulse echoes of harmonic oscillators. Thus we see that even if Δ vanishes, the third-order response can be finite (94).

Finally there are effects of electrical anharmonicity incorporating higher derivatives of the dipole moment to consider. The intensities of transitions between successive levels of the vibrator are not necessarily given accurately by harmonic or anharmonic matrix elements. For example, a ratio of $(|\mu_{12}|^2/2|\mu_{01}|^2) = (1 + (\Delta/\omega))$ is obtained for a Morse oscillator.

Fortunately in a number of cases there is experimental information on these points from broad band pump/probe experiments when the anharmonicity Δ is larger than the linewidth but much smaller than the bandwidth $\delta\omega$ of the laser. Then the 0-1 transition is seen as a bleaching signal and the 1-2 (66,67,71) as well as the 2-3 and often higher quantum number transitions (68,95) appear as new absorptions to an extent that depends on the pump intensity. A direct comparison of the total linewidths $(1/T_2)$ of these transitions, and the population relaxation times for the $v = 1$, $v = 2$ and perhaps higher levels can be obtained from such data. For N_3^- we found that ratio of the state to state relaxation from $v = 2$ to $v = 1$ was 1.8 times that for $v = 1$ to $v = 0$, not far from the harmonic value of 2 (50,95). However, the bandwidth of both transitions was roughly the same.

C. Comparison of Stimulated Photon Echoes of Vibrational and Electronic Transitions

From an operational standpoint photon echo spectroscopy has changed very little as the time scales have become shorter (96). However, there are some significant differences between vibrational and electronic properties that are important in understanding the echo signals. For electronic transitions the two-level model has been widely used and largely explains the observations (46,97). The system consists of two electronic states, which are coupled by the light fields. Nuclear motions of the target molecules and of the bath are coupled to both electronic states, and the experiment yields their spectral densities. However, the infrared echo response is intrinsically a multilevel problem and in general requires the involvement of vibrational levels up to $\nu = (n + 1)/2$ where n is the order of the nonlinearity. Hence the third-order stimulated echo requires the three vibrational states discussed earlier, a fifth-order echo response could involve levels up to $\nu = 3$, and so on. Of course, it is possible to manipulate the spectral bandwidth, $\delta\omega$, of the excitation source to exclude levels, if the vibrator is sufficiently anharmonic. For example, in the two-pulse echo experiments by Fayer using picosecond light pulses, the vibrator was often treated as a two-level system (73,74,98). This is equivalent to omitting the response R_3 of Fig. 2 from the echo signal by the condition $\delta\omega < \Delta$. However, some dynamics must be overlooked by this procedure.

Since the vibrational echoes arise from a multilevel system interacting with infrared radiation, it is necessary to understand the dynamical relationships between transitions involving different vibrational quantum numbers. For example, in order to average R_3 [see Equation (8)] over a fixed inhomogeneous distribution of frequencies, one will have to know the distributions of both ω_{21} and ω_{10} as well as how they are correlated (see Section III.B).

Another essential difference between vibrational and electronic stimulated echoes is that in the vibrational case, the nuclear modes of the bath that are coupled to the transition must be able to alter the potential function of the target molecule. This suggests that relevant spectral density is composed of bath motions that are highly localized to the target molecule. By contrast, in the case of electronic transitions where excitation alters the charge distribution significantly, the energy of the test molecule probably can be changed considerably by motions of solvent charges at larger separations. The foregoing properties suggest a number of possible applications of stimulated echo experiments on the vibrations.

Finally, all examples discussed below have in common the property that all the phenomena are ultrafast: the time scale of observation is limited by the T_1 time of the vibrational manifold. In the case of electronic transition echoes this is not such a limitation since the lifetimes of electronically excited states are generally long compared with the various dephasing processes that are considered.

D. The Three-Pulse Photon Echo Experiment

In a typical three-pulse (stimulated) photon echo (peak shift) experiment, the signal obtained under rephasing conditions ($\sum_{l=1}^{3} R_l$, the "real" photon echo) is compared with that obtained under nonrephasing conditions ($\sum_{l=4}^{6} R_l$, the "reverse" photon echo). If the system were in the homogeneous limit, the two signals corresponding to the two relevant time orderings $\sum_{l=1}^{3} R_l$ and $\sum_{l=4}^{6} R_l$ would be identical and decay with the dephasing time T_2 as a function of t_1. However, in the presence of inhomogeneity a photon echo delayed by a time equal to τ (i.e., the delay time between first and second pulse k_1 and k_2, see Fig. 3) arises after the

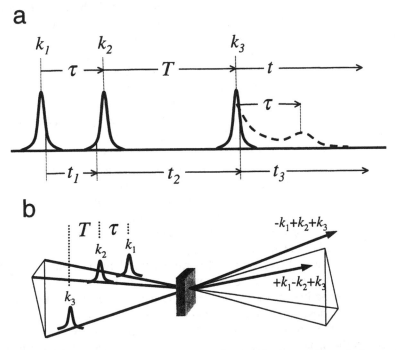

Figure 3 (a) Time sequence of the (stimulated) three-pulse photon echo experiment. The times t_1, t_2, and t_3 represent the time coordinates used in the response functions [Equations (7)–(12)] while τ, T, and t measure the delay times with respect to the peak positions of the light pulses. For δ-shaped light pulses, both sets of times would be equivalent. (b) The so-called box configuration, (101) which allows the spatial separation of the third-order polarization generated in the $-k_1 + k_2 + k_3$ and the $+k_1 - k_2 + k_3$ phase matched directions.

third light interaction. This photon echo is formed only under rephasing conditions and thus gives rise to a t-integrated signal which exceeds that in the nonrephasing direction. As delay time T increases, spectral diffusion destroys the phase memory, and eventually the response functions $\sum_{l=1}^{3} R_l$ and $\sum_{l=4}^{6} R_l$ become identical. In other words, the difference between the rephasing and the nonrephasing signals is a measure of the time variation of the inhomogeneous distribution.

The extraction of the required information from the data over the full ranges of τ and T shall be explained in some detail here. In the experimental setup used by us (31,41), the pulses k_1 and k_3 traversed computer controlled optical delay stages, while pulse k_2 was held fixed in time. Thus, one of the delay stages controlled the delay time τ' between the peaks of pulses k_1 and k_2, while the second one controlled delay time T' between the peaks of the pulses k_2 and k_3 (see Fig. 3). The three pulses can be time ordered in six ways (see Fig. 4). The Feynman diagram for the six permutations are given in Fig. 2. The permutations (123), corresponding to T', $\tau' > 0$, and (132), corresponding to $T' + \tau' > 0 > T'$, give a rephased part of the echo signal since interchanging pulses 2 and 3 does not change the signal under the $-k_1 + k_2 + k_3$ direction. Likewise, the response functions for the nonrephased parts of the echo signal [(213) and (312)] and those for (321) and (231), which are neither rephasing nor nonrephasing and which have no correspondence in a photon echo experiment on a electronic two-level system, have the same responses. When the delay times are controlled by variations in T' and τ', the six time orderings can be arranged in the quadrants spanned by these time coordinates as shown in Fig. 4. It is readily seen from Fig. 4 that independent continuous scanning of T' and τ' in this space over the complete time ranges encounters the unwanted permutations (231) and (321), and there is no simple continuous scan mode that just picks out the desired rephasing and nonrephasing Feynman diagrams. This is why we use instead the coordinate set τ and T, where τ denotes the positive or negative separation between the peaks of pulses k_1 and k_2, and T denotes the positive separation between the second and third pulses. Thus T measures the separation between k_3 and k_2 if $\tau > 0$, while if $\tau < 0$ it measures the separation between k_3 and k_1. The coordinate transformation between T', τ' and T, τ can be made either during the experiment by moving only one stepping motor for $\tau > 0$ and both for $\tau < 0$ or by rearranging the data afterwards in the computer. The former method was used by the Fleming group (19) and the latter method by us.

If the laser pulses were infinitely short (δ-shaped), τ and T would be identical to t_1 and t_2 [see Equation (16)] and the discontinuous coordinate

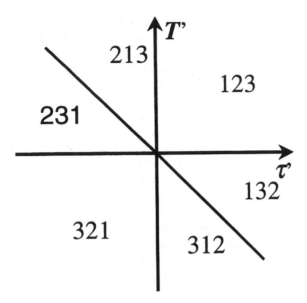

Figure 4 Distribution of all six possible time orderings, labeled (ijk), onto the four quadrants spanned by time coordinates τ' and T' (shown for the $-k_1 + k_2 + k_3$ phase matching direction).

transformation would not cause any difficulties. However, in reality, when the pulse width is finite, this transformation is somewhat arbitrary, since it introduces a discontinuity at $\tau = 0$ into a process, which, of course, is perfectly continuous. This representation of the data is nevertheless chosen throughout this paper [and also by other authors (19)], since the transformed coordinate system emphasizes the equivalence between the set of time coordinates τ, T and t_1, t_2. The latter set of time variables is the one used when representing nonlinear processes in the form of Feynman diagrams (see Fig. 2), so that using the transformed coordinate system τ and T is physically intuitive. However, it should be kept in mind that these sets of time coordinates are not identical, but they are connected by the convolution of the response functions with the electric fields [i.e., Equation (16)].

The photon echo (rephasing) signal can be compared with the reverse photon echo (nonrephasing) signal by scanning the delay time τ from negative to positive delay times or by collecting the signals in the $-k_1 + k_2 + k_3$ and the $+k_1 - k_2 + k_3$ phase-matching directions (13,14,16–18,31,41). Both procedures switch from the sets of graphs $R_4 - R_6$ to $R_1 - R_3$, and both procedures are often used simultaneously. In the work on electronic photons, (13,14,16–19,99,100) the asymmetry between the rephasing and the nonrephasing signal usually has been measured with the help of the so-called peak shift, which is defined as half the time between the peak of the $-k_1 + k_2 + k_3$ signal for $\tau > 0$ and the $+k_1 - k_2 + k_3$ signal for $\tau < 0$. This procedure is perfectly adequate in the case of electronic transitions, since there is always a peak shift >0 when there is inhomogeneity. The peak shift occurs because the $-k_1 + k_2 + k_3$ signal is larger for a given $\tau > 0$ (rephasing condition) than it is for $-\tau$ (nonrephasing condition), so that the $-k_1 + k_2 + k_3$ signal peaks at $\tau > 0$, given the signal is continuous at $\tau = 0$. The same argument holds for the $+k_1 - k_2 + k_3$ signal for $\tau < 0$.

However, the discontinuous coordinate transformation described above can introduce difficulties in the peak shift measurement, particularly when vibrational energy relaxation T_1 occurs on a similar timescale as dephasing, which is often the case for vibrational transitions, but generally not for electronic transitions. In that case, the peak shift can be zero at the discontinuous point $\tau = 0$, although the overall $-k_1 + k_2 + k_3$ and the $+k_1 - k_2 + k_3$ signals are not at all identical. Therefore, we have chosen to use the normalized first moment of the $-k_1 + k_2 + k_3$ signal

$$M_1(T) = \int_{-\infty}^{\infty} \tau \cdot S_{-k_1+k_2+k_3}(T, \tau)\, d\tau \Big/ \int_{-\infty}^{\infty} S_{-k_1+k_2+k_3}(T, \tau)\, d\tau \quad (19)$$

as a measure of the asymmetry between the rephasing and the nonrephasing signal. If the signal $S(T, \tau)$ were symmetrical with respect to its peak (which is commonly the case in optical photon echo experiments because of the fact that the time resolution of even the fastest laser pulses instrument response time is of the same order as electronic dephasing processes), the peak shift and the first moment would be identical. However, neither the first moment nor the peak shift have much fundamental physical meaning for vibrational echoes other than they are convenient measures of the asymmetry of the signal versus τ at a given value of T. They assess qualitatively whether an inhomogeneous distribution still exists after time T, so the approach of the photon echo signals to a symmetrical form centered at $\tau = 0$, at which point $M_1 = 0$ mimics in some respects the evolution of the spectral diffusion and gives an idea of the time scales of the spectral diffusion processes. For electronic two-level systems that the peak shift follows the energy gap fluctuation correlation function in certain limits (19,99,100).

E.   Spectral Diffusion of Small Molecules in Water

The asymmetrical stretching mode of the linear azide ion (N_3^-) exhibits a very strong infrared absorption ($\sigma = 5 \times 10^{-18}$ cm^2) and was therefore chosen as our first example of IR-stimulated echo spectroscopy. Our expectation was that N_3^- could be used as a probe of the dynamics of the solvent, which in this case was water. Its energy relaxation rate ($T_1 = 2.3$ ps), orientational diffusion constant ($D = 0.023$ ps^{-1}), and anharmonicity ($\Delta/2\pi c = 25$ cm^{-1}) are known from pump-probe and anisotropy experiments (66).

Figure 5a shows the stimulated photon echo signal from azide ion in the $-k_1 + k_2 + k_3$ and the $+k_1 - k_2 + k_3$ directions, which were collected by two independent IR detectors. The beam directions were arranged in the so-called box configuration, (101) (see Fig. 3). The experiments show clearly that neither the $-k_1 + k_2 + k_3$ nor the $+k_1 - k_2 + k_3$ signals are symmetrical with respect to $\tau = 0$. Moreover, the asymmetry diminishes with T and has mostly disappeared by T $\approx$ 5 ps. These results show that the vibrational coherence introduced by pulse 1 and the corresponding phase information stored in the population created by pulse 2 continues to be detectable after rephasing is induced by pulse 3, albeit in an ever-decreasing manner, as T increases. The rephasing process that generates the echo only occurs when there is some memory of the original inhomogeneous distribution of frequencies. In other words, the difference between both signals is due to a delayed photon echo, which is formed only under rephasing conditions ($\tau > 0$ for the $-k_1 + k_2 + k_3$ signal and $\tau < 0$ for the $+k_1 - k_2 + k_3$ signal), so that these results clearly prove that a certain amount of inhomogeneity is present for small T. The inhomogeneity is diminished on the picosecond time scale by spectral diffusion processes. Clearly, the asymmetrical stretching mode of the azide ion in water is not in the motional narrowing limit, which is what is generally assumed for vibrational dephasing. The normalized first moment $M_1(T)$ of the $-k_1 + k_2 + k_3$ signal (see Fig. 6) indicates that the inhomogeneity decays on at least two time scales. Furthermore, the fact that $M_1(T)$ is still finite at the longest measured times implies the existence of a small inhomogeneous contribution, which is essentially static on the time scale of these experiments. Consequently, the following model for the transition frequency correlation function was used to fit the photon echo data (Fig. 5):

$$\langle \delta\omega_{10}(\tau)\delta\omega_{10}(0) \rangle = \sum_{i=1}^{2} \Delta_i^2 e^{-t/\tau_i} + \Delta_0^2 \tag{20}$$

A global fit was performed by applying the formalism outlined in Section III. A [Equation (7)–(12), (14)–(17)], which connects the transition frequency fluctuation correlation function $\langle \delta\omega_{10}(\tau)\delta\omega_{10}(0) \rangle$ to the three-pulse photon echo signal $S(T, \tau)$ and by varying the five parameters in the right-hand side of Equation (20). The fit and the resulting first

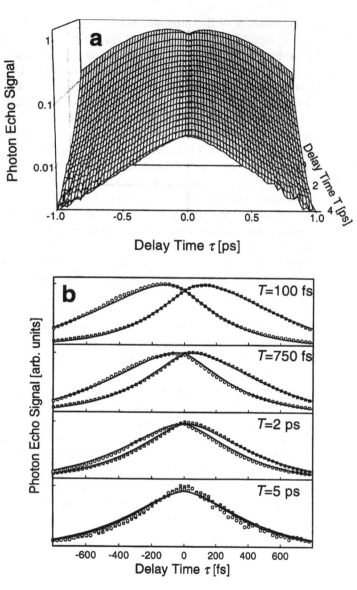

Figure 5 (a) Stimulated photon echo signal of the azide ion (N_3^-) in D_2O at 2043 cm^{-1} as a function of τ and T for the $-k_1 + k_2 + k_3$ (gray surface) and the $+k_1 - k_2 + k_3$ (white surface) phase matching directions. (b) Representative traces for four selected values of T. The solid lines represent a global fit of all the scans to the model correlation function Equation (20). (From Ref. 41.)

moment are compared with the corresponding experimental data in Fig. 5b and Fig. 6, respectively. An excellent agreement is obtained. The derived transition frequency fluctuation correlation function $\langle\delta\omega_{10}(t)\delta\omega_{10}(0)\rangle$ is also shown in Fig. 6 (insert). The initial fast decay gives rise to a dephasing contribution, which is essentially in the motional narrowing limit ($\tau_1 \cdot \Delta_1 = 0.2 \ll 1$), corresponding to a pure dephasing time of $(\Delta_1^2 \cdot \tau_1)^{-1} = 1.8$ ps and a homogeneous linewidth of 6 cm^{-1}, respectively. It is followed by a slow tail, which decays on a 1 ps time scale. In order to verify the self-consistency of the entire formalism, the linear absorption spectrum was calculated from this transition frequency correlation function according to Equation (4b) and

$$I(\omega) = 2 \ \text{Re} \int_0^\infty e^{i(\omega-\langle\omega_{10}\rangle)t} e^{-g(t)-t/2T_{10}-2Dt} \, dt \tag{21}$$

As seen in Fig. 7, the calculated absorption spectrum agrees quite well over three orders of magnitudes with a measured spectrum.

A classical molecular dynamics simulation of N_3^- dissolved in water revealed that six to seven water molecules are hydrogen bonded to the negatively charged terminal nitrogen atoms, (102) with an estimated hydrogen

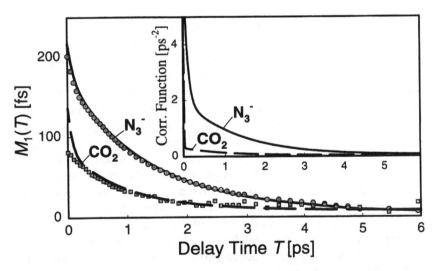

Figure 6 The first moment $M_1(T)$ of the data measured in the $-k_1 + k_2 + k_3$ direction (dots) of N_3^- and CO_2. The solid line is the first moment $M_1(T)$ as obtained from global fits. The insert shows the correlation functions obtained from global fits.

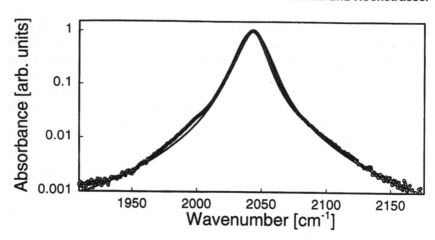

Figure 7 Linear IR absorption spectrum of NaN_3 in D_2O (dots). The solid line shows the line shape as calculated with the parameters obtained from the global fit of the photon echo signal. The small shoulder at the low frequency side of the spectrum appears to be the absorption of $^{14}N^{15}N^{14}N^-$. (From Ref. 41.)

bond lifetime of 1.3 ps. Similar conclusions were drawn from MD simulations of CN^- in water (103,104). Since the long-term tail in the fluctuation correlation function decays on exactly that time scale, the spectral diffusion process is interpreted as the dynamics of the formation and breaking of these hydrogen bonds. This interpretation is supported by the significant influence of the solvent on the vibrational frequency of the asymmetrical stretching mode of azide, as seen from the large blue shift from 1987 cm^{-1} in gas phase, (105) to 2043 cm^{-1} in solution. Consequently, intermittent H-bond interactions cause a fluctuating transition frequency, giving rise to the observed spectral diffusion process. The vibrational frequency of carbon dioxide, which is isoelectronic to the azide ion, is much less shifted by the solvation process (2349 cm^{-1} in gas phase, 2345 cm^{-1} in water), even though it also forms strong hydrogen bonds to water. As a consequence, the amplitude of the slow tail of the fluctuation correlation function is considerably smaller than in the case of the azide ion (see Fig. 6) (95), although, interestingly, the total line width of CO_2 in water (7 cm^{-1}) is essentially the same as the homogeneous contribution to the total linewidth of $N_3^-(\Delta_1{}^2\tau_1/c = 6\ cm^{-1})$.

The slow tail of the spectral diffusion process (1.3 ps) of azide in water is not the same as the relaxation of bulk water (800 fs), which has been measured with the help of optical dynamical Stokes shift experiments

(106). We have argued that the correlation function obtained from the vibrational photon echo experiment senses especially the hydrogen bonds between azide ions and water molecules, in contrast to long-ranged Coulomb interaction between the dipole moment of the optical chromophor and the fluctuating charges of the solvent. In that sense, the vibrational photon echo experiment is considered to be more sensitive to local structure fluctuations.

F. Spectral Diffusion of Vibrational Probes in Enzyme-Binding Pockets

It is this local character of vibrational dephasing that is utilized in the experiments described in this section. In these experiments, spectral diffusion of test molecules bound to enzymes has been investigated in order to study the fluctuations of the reactive sites. The local character of these interactions can be pictured in great detail since in many cases high-resolution x-ray structure of the complexes are available. One example we have studied, shown in Fig. 8, is azide bound to carbonic anhydrase ($CA-N_3^-$) (107). Carbonic anhydrase is a zinc enzyme that catalyzes the interconversion of

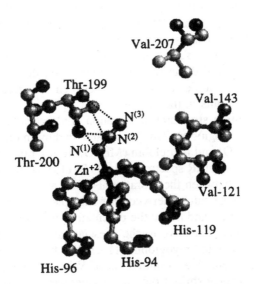

Figure 8 The structure of the binding pocket of azide bound to carbonic anhydrase ($CA-N_3^-$) (107). The atoms of the azide ion, which is bound to the active site (Zn^{+2}), is in close contact with Thr-199 (indicated by the dotted lines). (From Ref. 31.)

CO_2 and bicarbonate. The azide ion, which is a competitive inhibitor of this reaction and isoelectronic with CO_2, binds at Zn^{2+} without compromising the three-dimensional structure of the active site (107). The azide nitrogen ($N^{(1)}$) closest to the Zn^{2+} ion and the central nitrogen ($N^{(2)}$) have short contacts (3.3 Å) to the hydroxyl oxygen of Thr-199. In addition, $N^{(2)}$ (at 3.7 Å) and $N^{(3)}$ (at 3.5 Å) must sense the amide nitrogen of Thr-199. Therefore, it is natural to invoke the nearby Thr-199 as the group that modifies the potential energy function and controls the charges of Zn-bound azide. Recent calculations have shown that the charges on the azide nitrogen atoms are changed considerably when the effect of the enzyme environment, mainly the Thr-199, is taken into account (108). These simulations and quantum chemical calculations suggest that the coupling to the protein environment increases the admixtures of the triple-bonded valence bond structure $N \equiv N^{+1} - N^{-2}$ compared with the symmetrical form $N^{-1} = N^{+1} = N^{-1}$ of the azide ion. Since each admixture corresponds to a different potential energy function of the ground state, the vibrational frequency should be structure sensitive. The presence of a small admixture of the triple-bonded structure causes the averaged frequency to increase significantly, in agreement with the azide vibrational frequency in the enzyme being increased by ca. 50 cm^{-1} compared with azide in water and by 110 cm^{-1} compared with azide in the gas phase (105). In other words, azide is a very sensitive sensor to local charges. As a consequence, fluctuation of the contact between Thr-199 and the azide ion appears to be a likely mechanism for dephasing and spectral diffusion. Put another way, the three-pulse photon echo exposes the dynamics of couplings between the azide ion and nearby partially charged atoms in the protein.

High-resolution structural information is also available for the azide ion bound to hemoglobin ($Hb-N_3^-$) and carbon monoxide bound to hemoglobin (HbCO) and myoglobin (MbCO) from x-ray diffraction studies (109,110) and transient IR spectroscopy (111–113) so that detailed pictures of the interaction between the vibrational probe molecule and the binding pocket of the enzyme can be derived. For example, it is known that the CO vibrational frequency, and hence the potential energy for the CO-stretching motion, is extremely sensitive to the presence and positioning of His-E7 in hemoglobin (114) and myoglobin (115,116). On the other hand, the CO frequency is less sensitive to mutations of Val-E11, another heme pocket residue. The frequency shifts from the variations in the relative positioning of the polar His-E7 and the CO can be rationalized as resulting from the mixing of the valence bond structures $Fe = C = O$ and $Fe - C \equiv O$. The latter structure is stabilized by the hydrogen bond to E7, resulting in an increase in the vibrational frequency.

These complexes (CA–N_3^-, Hb–N_3^-, and Hb–CO) therefore seem to be ideal candidates for investigating local structural fluctuation of enzyme reactive sites. The corresponding photon echo signals are shown in Fig. 9 together with global fits, which were obtained in the same way as described in the previous paragraph for azide dissolved in water. The corresponding first moment data $M_1(T)$ and the initial decay of the transition frequency fluctuation correlation function $\langle \delta\omega_{10}(\tau)\delta\omega_{10}(0)\rangle$ obtained from global fits are shown in Fig. 10 and Fig. 11, respectively. The vibrational energy relaxation rate of carbon monoxide in hemoglobin (Hb–CO) (67) is considerably longer than that of azide, so that a time window in excess of 40 ps is opened for studies of spectral diffusion processes (note the different scale of the time axis in Fig. 10c).

Nevertheless, the responses of the test molecules embedded to the proteins differ significantly from that in solution (see Fig. 11). The fluctuation correlation function amplitude is considerably larger in solution for small times, but it decays much more quickly, so that only a very small quasi-static inhomogeneity (0.1 ps^{-2}) remains within the observation window available in this experiment (6 ps). A much larger inhomogeneity remains for even longer times in the case of a protein environment. This result is consistent with the interpretation of optical photon echoes in protein environment (22–25,27), where a quasi-static contribution of the energy gap correlation function has been observed up to 100 ps.

The time dependence of the inhomogeneous distribution has a special significance in the case of proteins, since it measures the changes in the structure of those parts of the protein that influence the probe vibrational spectrum. As outlined before, the potential energy surface of the probe molecule is changing as a result of the interaction with the protein. The potential will be sensitive to forces that can influence the charge distribution in the probe. Therefore, this method is a probe of the local structure. Long-range interactions can also cause frequency changes by shifting of vibrational transition in response to the fields from the fluctuating charges of the medium. However, the dipole needed to couple to those fields vary only slightly with the quantum number of vibrational states because of the small anharmonicity of the oscillators, so these perturbations will be small. However, an important effect of longer-range interactions on the spectral diffusion would be to cause energy and nuclear position fluctuations of those parts of the protein that are involved in direct coupling to the probe

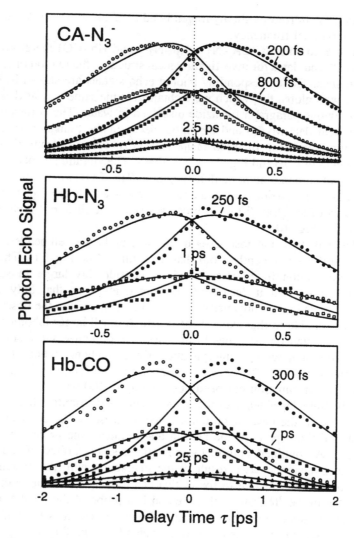

Figure 9 Stimulated photon echoes from various test molecules in different enzymes: CaN_3^- (azide bound to carbonic anhydrase), $Hb–N_3^-$ (azide bound to hemoglobin), and Hb–CO (carbon monoxide bound to hemoglobin). The signal is plotted against delay time τ for selected delay times T together with global fits (solid lines). The oscillatory part in the experimental data, which is not reproduced by these fits, reflects the anharmonicity of the transition and is due to interference between fifth and third order nonlinear polarization term (52). (From Ref. 31.)

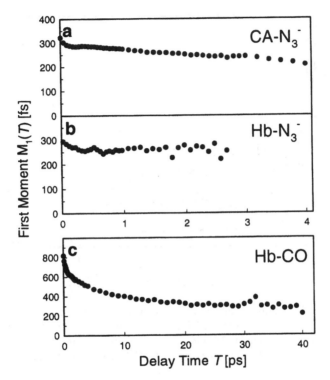

Figure 10 The normalized first moments of the photon echo signal of (a) $CA-N_3^-$, (b) $Hb-N_3^-$, and (c) $Hb-CO$ as a function of delay time T. Note the extended time axis of the. $Hb-CO$ data. The inhomogeneity decays with time T due to conformational fluctuations of the proteins.

atoms. By this mechanism the local interactions can sense the bulk fluctuations of the protein. The local forces can be changed either by fluctuations in the local structure or by the local structure responding to changes in other parts of the protein. This picture suggests a plausible interpretation of the time sequence of events in the evolution of the inhomogeneous distribution around a local region.

Because of the small size of the utilized test molecules and the fact that the molecules remain in their electronic ground states, accurate calculations of the transition frequency fluctuation correlation function of these systems fluctuations would seem to be achievable with state-of-the-art quantum dynamics calculations.

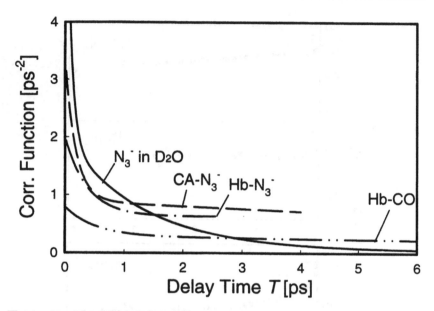

Figure 11 The initial decay of the transition frequency fluctuation correlation function $\langle \delta\omega_{10}(\tau)\delta\omega_{10}(0)\rangle$ obtained from global fits of the stimulated photon echo data of azide dissolved in D_2O (solid line, same data as in Fig. 6), $CA-N_3^-$ (dashed line) $Hb-N_3^-$ (dashed-dotted line), and $Hb-CO$ (dashed-double-dotted line).

It is evident from these data that the transition frequency fluctuation correlation function of samples that have the same probe molecule (azide) embedded into two different proteins (hemoglobin and carbonic anhydrase), or of the sample with different probe molecules (azide, carbon monoxide) embedded to one protein (hemoglobin), all differ considerably. This, we believe, is a consequence of sensitivity of this spectroscopic technique to the local structure, which is different in each case. This result must be contrasted with electronic dephasing, where it was found that the energy gap fluctuation correlation function reflects the response of the bulk solvent and is essentially independent of the chromophore used as a probe (81).

G. Spectral Resolution of the Echo

The properties of the various Feynman diagrams that contribute to the vibrational echo can often be measured separately by means of time resolution of the spectrally resolved echo. The time t_3 can be experimentally controlled by a variety of methods. One way is to time gate the echo field, which has been done for electronic transitions and poses no major experimental difficulties in the infrared. Another approach would be to examine the echo by spectral interferometry. Finally, in systems where there is inhomogeneous broadening the inhomogeneous averaging naturally creates a time gate that restricts t_3 to a range around t_1 [see Equation (13)]. The larger the inhomogeneous width, the smaller is this range. In all cases the field from R_3 dominates the dispersed signal at ω_{10}-Δ while that from R_1 and R_2 dominates the signal at ω_{10}. Equation (13) assumes that the dephasing rates for the 0-1 and 1-2 transitions are both $1/T_2$. The dephasing time for the 1-2 transition can be replaced by $1/T_2' = (\gamma + 1/T_2)$, where γ can be regarded as the contribution to the damping from the fluctuations in the anharmonicity. The dispersed gated signal is then the absolute square of the Fourier transform of the response in Equation (13). It is easy to see that the decay of the signal along the t_1 axis (or t_3 axis in an experiment with a gating pulse) at the 0-1 frequency depends only on T_2, while that at the 1-2 frequency depends on both T_2 and T_2'. In the limit that the inhomogeneity is larger than $1/T_2$ but smaller than Δ, both decays are exponential, as is also the case when the signals are time gated by an additional field. Of course in a system undergoing spectral diffusion the situation is more complex, but still the dephasing patterns for the 1-0 and 1-2 coherences contribute to different extents to a given spectral component of the generated field. An example of a frequency resolved echo (117) is shown in Fig. 12. The 0-1 and 1-2 coherences are clearly separated in frequency space. In this example the two relaxation times are similar.

IV. STRUCTURE AND DYNAMICS OF THE AMIDE I BAND OF SMALL PEPTIDES

Vibrational spectroscopy has been used in the past as an indicator of protein structural motifs. Most of the work utilized IR spectroscopy (see, for example, Refs. 118–128), but Raman spectroscopy has also been demonstrated to be extremely useful (129,130). Amide modes are vibrational eigenmodes localized on the peptide backbone, whose frequencies and intensities are related to the structure of the protein. The protein secondary structures must be the main factors determining the force fields and hence the spectra of the amide bands. In particular the amide I band (1600–1700 cm^{-1}), which mainly involves the C=O-stretching motion of the peptide backbone, is ideal for infrared spectroscopy since it has an large transition dipole moment and is spectrally isolated

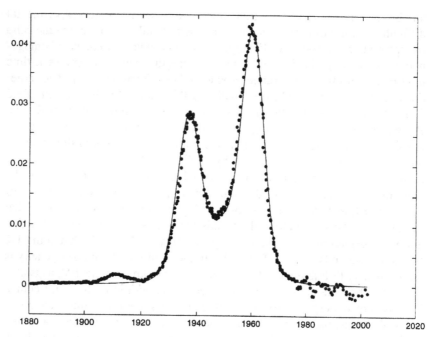

Figure 12 Frequency-resolved vibrational echo. The three-pulse echo signal from hemoglobin-CO is spectrally analyzed in the separate diagrams R_1 and R_2 (at ω_0) from R_3 (at $\omega_0 - \Delta$). (Results from Ref. 117.)

from other amide modes and from vibrational transitions of the amino acid side chains. The amide I has been widely used as a marker mode of secondary structure (see, for example, Refs. 125–128). For example, proteins containing mostly β sheets have amide I bands at lower frequency than those containing mostly α helices. These IR methods are extremely useful and have great potential. They are empirical and try to predict the presence of various secondary structures of a protein by comparisons with vibrational spectra from samples with known structure.

We will show in this section that by applying nonlinear infrared methods, such as IR-pump–IR-probe, dynamical hole burning, and IR photon echoes, one can gather significantly more detailed information on the structure and dynamics of the amide I band than is possible with conventional (linear) absorption spectroscopy. Starting with some knowledge of the underlying contributions to amide I absorption, such as obtained by the aforementioned empirical approaches, nonlinear spectroscopy could provide measures of the spatial disposition and internal structure of these spectral components. In particular, we will demonstrate a novel, two-dimensional IR spectroscopy on the amide I band, which allows the determination of the strength of coupling between pairs of peptide groups with the help of cross peaks, (30,42) in complete analogy to two-dimensional NMR spectroscopy (COSY) (1–4,32).

A. An Excitonic Model for the Amide I Band

Krimm et al. (118) investigated the relationship between the vibrational spectrum of the amide I mode and the structure of the protein very systematically and tried to deduce underlying force fields. Normal mode calculations have been performed, (118–121) based on empirical force fields deduced from spectroscopic data (118) or from ab initio calculations on model compounds such as hydrogen bonded N-methylacetamide (NMA, a model of a single peptide group) (122) or dipeptides. One very important result from these studies was the finding that the amide I modes from different peptide groups cannot be viewed as being independent, but they are strongly coupled. Using as examples highly symmetrical polypeptides, Krimm et al. showed that the dominating coupling mechanism is transition dipole coupling (TDC). The large splitting found between the symmetry-allowed transitions of these samples was explained with this potential. The strength of the coupling deduced from the splitting at spectral lines was consistent with the known strength and direction of the individual transition dipoles.

In a more recent approach by Tasumi et al. (123,124), the amide I vibrational subspace was treated as completely separable from all other vibrations. It was furthermore assumed that the force fields that couple independent amide I states originate only from transition dipole-dipole coupling (TDC) and that through-bond effects can be completely neglected. This model follows the idea of a normal mode calculation and consequently the coupling constants have the units of forces. However, the underlying ideas are similar to those used for molecular vibrational excitons (vibrons), and as described by Davidov for electronic excitations, (131) the coupling Hamiltonian has the units of an energy. In this model, the connection between the secondary structure of the peptide backbone and the absorption spectrum of the amide I band is given by the angular and distance dependence of the dipole interactions. Despite the simplicity of this approach, a qualitative agreement was found between this simple model calculation and measured FTIR spectra for some mid-size proteins such as myoglobin (123,124).

We have adopted the excitonic band model not only because it describes conventional absorption spectroscopy (linear spectroscopy), but because it enables an extremely convenient description of nonlinear experiments, such as pump-probe, dynamical hole burning, or photon echoes. In these third-order experiments one has to consider not only transitions from the ground state to the one-excitonic states but also transitions from the one-excitonic to the two-excitonic states (see Fig. 13). These additional transitions reveal the required information to deduce, at least in principle, the complete coupling scheme.

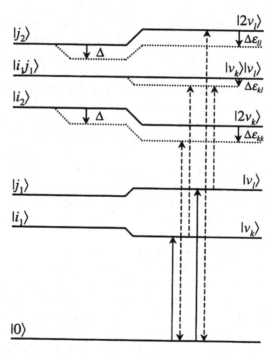

Figure 13 Energy level scheme for a system of two coupled oscillators. The isolated peptide states (left side) are coupled by some weak interaction, which mixes them to generate the excitonic states (right side). Anharmonicity, which is crucial for understanding the 2D pump probe spectra, is introduced into this model by lowering the energies of the double excited monomeric site states $|i_2\rangle$ and $|j_2\rangle$ by Δ from their harmonic energies $2\varepsilon_{ii}$. This anharmonicity mixes into all coupled states, giving rise to diagonal anharmonicity ($\Delta\varepsilon_{kk}$) and off-diagonal anharmonicity (mixed-mode anharmonicity, $\Delta\varepsilon_{kl}$) in the basis of the normal modes discussed in the text.

To describe such an excitonic system, the eigenstates of

$$H = H_0 + V = \sum_{i=1}^{n} H_i + \sum_{i<j} V_{ij} \tag{22}$$

are needed, where the H_i describes the n noninteracting (monomeric) peptide groups and the V_{ij} the interpeptide potential. Whether the coupling term V consists only of dipole-dipole coupling [as proposed by Tasumi et al. (123)] or includes higher-order multipole interactions or additional through-bond effects is not vital at this point. The only assumption made is that V_{ij} is bilinear in the coordinates of site i and j. The diagonalization is readily accomplished in the basis of site excitations involving 0,1,2 quanta: the ground state $|0\rangle$, having no vibrational quanta excited, the states $\{|i_1\rangle\}$, have one vibrational quantum on the ith site, and the two-particle states $\{|i_1 j_1\rangle\}$, $i \neq j$, having one quantum at two different sites i and j, and $\{|i_2\rangle\}$, the biphonon (overtone) states, having two quanta at a single site i, respectively. In this basis, the matrix elements of H are:

$$\langle 0|H|0\rangle \equiv 0 \tag{23a}$$
$$\langle i_1|H|i_1\rangle = \varepsilon_i \tag{23b}$$
$$\langle i_1|H|j_1\rangle = \beta_{ij} \tag{23c}$$
$$\langle i_1 j_1|H|i_1 j_1\rangle = \varepsilon_i + \varepsilon_j \tag{23d}$$
$$\langle i_2|H|i_2\rangle = 2\varepsilon_i - \Delta \tag{23e}$$
$$\langle i_1 j_1|H|i_1 k_1\rangle = \beta_{jk} \tag{23f}$$
$$\langle i_1 j_1|H|i_2\rangle = \sqrt{2}\beta_{ij} \tag{23g}$$

where the ε_i are the vibrational zero-order frequencies of the uncoupled states, Δ the anharmonicity of these states (assumed to be the same for all groups), and β_{ij} the excitonic resonance couplings. For the sake of clarity, the vacuum-to-aggregate shift terms ΔD, which are required in a fully general form of Equation (23) (131), are omitted here and are included into an effective anharmonicity. (30) The factor $\sqrt{2}$ in Equation (23g) originates from a harmonic approximation. Site i is de-excited $v = 1 \rightarrow v = 0$, while site j undergoes a $v = 1 \rightarrow v = 2$ transition so that the matrix element of Equation (23g) is proportional to $\langle i_1|q_i|i_0\rangle\langle j_1|q_j|j_2\rangle = \sqrt{2}\langle i_1|q_i|i_0\rangle\langle j_0|q_j|j_1\rangle$, where q_i and q_j are the oscillation coordinates of the amide I mode on sites i and j, respectively. The resonance term in Equation (23f), on the other hand, includes only $v = 0 \rightarrow v = 1$ transitions so that no corresponding factor appears.

If the individual amide modes were harmonic oscillators, the exciton-ically coupled states of the polypeptide would also be a set of n independent harmonic normal modes (under the assumption of a bilinear coupling term), which are completely decoupled from each other. In this case, the two-exciton states could be obtained directly without explicit diagonalization of the two-exciton matrix, for they are merely product states of the one-excitonic states. However, the third-order response of such a harmonic system would be exactly zero, since all transitions depicted in Fig. 13 would cancel. Therefore, it is essential to consider anharmonicity in order to under-stand the third-order response of the amide I band (as it is in any nonlinear spectroscopy on vibrational systems). It is nevertheless instructive to think about the problem in a picture of "almost harmonic" excitonically coupled states and consider anharmonicity as a perturbation. The anharmonicity Δ lowers the zero-order energies of the doubly excited states $|i_2\rangle$ from their harmonic values $2\varepsilon_i$ [Equation (23e) and dotted lines in Fig. 13], while the energies of states $|i_1 j_1\rangle$, having two excitations at two different sites, are not affected by anharmonicity. The anharmonicity Δ of the monomeric sites has been determined with the help of IR-pump–IR-probe experiments on NMA to be $\Delta = 16$ cm^{-1} (30). This anharmonicity is mixed into all coupled excitonic states, and the resulting normal modes now are asso-ciated with a diagonal anharmonicity $\Delta\varepsilon_{kk} = \varepsilon_{kk} - 2\varepsilon_k$ and off-diagonal (or mixed-mode) anharmonicity $\Delta\varepsilon_{kl} = \varepsilon_{kl} - \varepsilon_k - \varepsilon_l$ (see Fig. 13). In other words, the right-hand side of the level scheme in Fig. 13, and as a conse-quence all nonlinear IR experiments on the amide I band, can be readily understood in terms of ordinary anharmonicity. The crucial point is that the excitonic coupling model provides a convenient and extremely simple way to relate these anharmonicities to the coupling Hamiltonian (ultimately related to the three-dimensional structure of the protein). The anharmonic-ities would otherwise have to be determined from unrealistically computer expensive quantum chemistry methods, requiring the calculation of the 2nd, 3rd, and 4th derivatives of the ground state potential surface [the largest molecule for which this has been demonstrated, is benzene (132)].

The picture of "almost harmonic" excitonically coupled states is particularly appropriate in the localization or weak coupling limit. This limit will be valid in smaller peptides that do not have the rather strict symmetries of helices or sheets. It is very likely that the vibrational frequencies of each amide unit will be different even in the absence of any coupling. An example of this limit is found in the pentapeptide discussed below. If the frequency separations between the uncoupled modes are large compared with the individual coupling terms, $|\beta_{ij}/(\varepsilon_i - \varepsilon_j)| < 1$, the coupled states are predominantly localized on single peptide units. In weak coupling, the diagonal anharmonicities are given to first order by $\Delta\varepsilon_{kk} = -\Delta$, and the off-diagonal anharmonicities are calculated from perturbation theory (in lowest order of Δ and β_{kl}; see Appendix):

$$\Delta\varepsilon_{kl} = -4\Delta\frac{\beta_{kl}^2}{(\varepsilon_k - \varepsilon_l)^2} \tag{24}$$

In addition, one obtains for the transition dipole moments the relation $\vec{\mu}_{|\nu_k\rangle\to|\nu_k\rangle|\nu_l\rangle} = \vec{\mu}_{|0\rangle\to|\nu_l\rangle}$, $k \neq l$, and $\vec{\mu}_{|\nu_k\rangle\to|\nu_k\rangle|\nu_k\rangle} = \sqrt{2}\vec{\mu}_{|0\rangle\to|\nu_k\rangle}$, respectively, and all other transitions $|\nu_k\rangle \to |\nu_l\rangle|\nu_m\rangle$, $k \neq l$, m, are forbidden.

In the strong coupling limit, for example, for a symmetrical α- helix, where all peptide groups are essentially equivalent so that the zero-order energies ε_i are all identical, it is more complicated to find simple pictures for the oscillators. Not only is a numerical diagonalization of the two-exciton matrix then needed to find the diagonal and off-diagonal anharmonicities, but also the transition dipoles between $\vec{\mu}_{|0\rangle\to|\nu_l\rangle}$ and $\vec{\mu}_{|\nu_k\rangle\to|\nu_k\rangle|\nu_l\rangle}$ change considerably and even transitions such as $|\nu_k\rangle \to |\nu_l\rangle|\nu_m\rangle$, $k \neq l$, m, which are forbidden both in the harmonic and the localization limit, become weakly allowed. In the case when anharmonicity is large compared with excitonic couplings $\Delta > \beta_{ij}$, the two-excitonic states completely lose their identity of being products of one-excitonic states. This situation will most likely prevail for the N-H vibrations of the peptides.

B. Response for N Coupled Oscillators

An alternative perspective on third-order responses of N coupled vibrators, which will be particularly helpful to describe spectral diffusion processes in such coupled systems (see Section IV. D), can be developed by assuming that R_1 and R_2 are the same and writing the total response function as:

$$R = \sum_{i,j,k}\langle 2\mu_{0j}(t_3)\mu_{j0}(t_1)\rho_{00}\mu_{0i}(0)\mu_{i0}(t_2)$$
$$- \mu_{ik}(t_3)\mu_{jk}(t_2)\mu_{j0}(t_1)\rho_{00}\mu_{0i}(0)\rangle \tag{25}$$

After factoring out the parts involving the anharmonicity, the response becomes

$$R = 2\sum_{i,j,k}\langle(1 - \gamma_{ij,k}e^{i\Delta_{kj,i0}(t_3-t_2)+i\int_{t_2}^{t_3}\delta\Delta_{kj,i0}(\tau)\,d\tau})$$
$$\times \mu_{0j}(t_3)\mu_{j0}(t_1)\rho_{00}\mu_{0i}(0)\mu_{i0}(t_2)\rangle \tag{26}$$

where i and j each take N values corresponding to the one quantum state, the $\mu(t)$ are given by Equation (3), and k runs over all of the $\frac{1}{2}N(N+1)$ two quantum states. The anharmonicity has both diagonal and off diagonal contributions, $\Delta_{kj,i0} = \omega_{nj} - \omega_{i0}$. The transition dipole factor $\gamma_{ij,k}$ is:

$$\gamma_{ij,k} = \frac{\mu_{ik}\mu_{kj}}{2\mu_{0i}\mu_{j0}} \tag{27}$$

In the absence of coupling and electrical anharmonicity, $\gamma_{ij,k} = \frac{1}{2}$ when k is one of the $\frac{1}{2}N(N-1)$ two-particle states, and $\gamma_{ij,k} = 1$ when k is one of the N biphonon states.

C. Two-Dimensional IR Spectroscopy on the Amide I Band

We will describe in the following a novel two-dimensional IR (2D-IR) spectroscopic technique on some peptide samples. These experiments have been performed in order to verify the excitonic coupling and to establish the nature and strength of the coupling. Our intention is the use the foregoing principles to develop a novel structure analysis method with the potential of ultrahigh time resolution. In our first approach to 2D-IR spectra, the transient response of the sample was measured with the help of a broadband, ultrashort probe pulse (120 fs) as a function of the peak frequency of a narrower band pump pulse (ca. 10 cm^{-1}). The pump pulse was generated from the output of the femtosecond IR light source (see Sec. II) by means of an adjustable IR-Fabry-Perot filter (30,42). The essence of the experiment is to selectively populate individual one-excitonic levels with the pump pulse and probe the response of the sample by means of transitions back to the ground state and to the two-excitonic states. In that way, diagonal anharmonicity $\Delta\varepsilon_{kk}$ is sensed along the diagonal of the 2D spectra, where the negative bleach and stimulated emission signals are observed at the frequency ε_k of the pumped state $|v_k\rangle$ and excited state absorption involves transitions to the double excited state $|v_k\rangle|v_k\rangle$. In the off-diagonal region corresponding to the kl cross-peak, one finds a bleach of the probed transition ε_l, arising from the depopulation of the common ground state and an excited state absorption to the mixed state $|v_k\rangle|v_l\rangle$, which is red-shifted from the bleach by the off-diagonal anharmonicity $\Delta\varepsilon_{kl}$.

In order to picture the underlying principles of nonlinear 2D-IR spectra, a model calculation of an idealized system consisting of two coupled vibrators is shown in Fig. 14b together with their linear absorption spectrum (Fig. 14a). The frequencies of these transitions were chosen as 1615 cm^{-1} and 1650 cm^{-1}, the anharmonicity $\Delta = 16$ cm^{-1}, and the coupling $\beta_{12} = 7$ cm^{-1}. We used a homogeneous dephasing rate of $T_2 =$

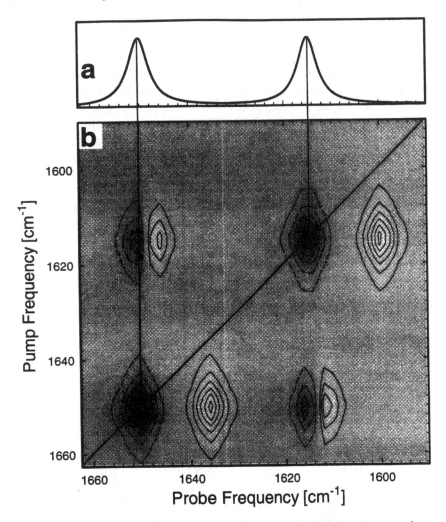

Figure 14 A model calculation of the 2D-IR spectra of a idealized system of two coupled vibrators. The frequencies of these transitions were chosen as 1615 cm^{-1} and 1650 cm^{-1}, the anharmonicity as $\Delta = 16$ cm^{-1}, the coupling as $\beta_{12} = 7$ cm^{-1}, and the homogeneous dephasing rate as T$_2$ = 2 ps. The direction of both transitions as well as the polarization of the pump and the probe pulse were set perpendicular. The spectral width of the pump pulses was assumed 5 cm^{-1}. The figure shows (a) the linear absorption spectrum and (b) the nonlinear 2D spectrum. In the 2D spectra, light gray colors and solid contour lines symbolize regions with a positive response, while negative signals are depicted in dark gray colors and with dashed contour lines.

2 ps in order to make the features as distinct as possible. In Fig. 14b, where the coupling is switched on, pairs of a positive and a negative peak appear along the diagonal of the 2D spectrum at positions corresponding to those of the related fundamental lines. In addition, off-diagonal peaks show up, each consisting of a positive and a negative contribution. When the coupling is switched off, the off-diagonal peaks would disappear, emphasizing that this technique is analogous to 2D NMR (COSY) spectroscopy. Of course in this IR experiment the pump pulse is not transferring coherence. A fixed delay time of approximately 0.6 ps was introduced and the narrow frequency band of the pump field restricted the range of coherences that are initially excited.

Experimental 2D spectra were obtained from three different peptide samples, whose structures are shown in Fig. 15. The first sample, a de novo

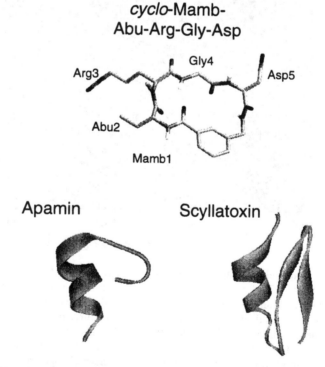

Figure 15 Structures of the three peptide samples investigated by two-dimensional IR spectroscopy: a de novo cyclic penta peptide (*cyclo*-Mamb-Abu-Arg-Gly-Asp), apamin, and scyllatoxin.

cyclic penta peptide (*cyclo*-Mamb-Abu-Arg-Gly-Asp), is small enough so that all amide I transitions are spectrally resolved in the linear absorption spectrum (see Fig. 16a), allowing the properties of the nonlinear 2D-IR spectra and the validity of the excitonic coupling model to be examined in detail. Both its NMR and x-ray structures are known (133). The peptide is stabilized by a hydrogen bond between the Mamb1-Abu2 peptide bond and the Arg3-Gly4 peptide bond and was specifically designed to form a single

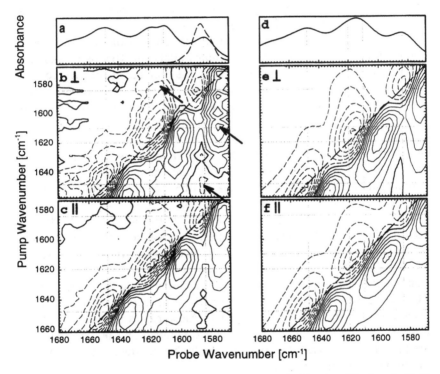

Figure 16 Absorption spectrum of the *cyclo*-Mamb-Abu-Arg-Gly-Asp in D_2O. The dashed line shows a representative spectrum of the pump pulses (width ~ 12 cm^{-1}) utilized to generate the 2D-IR spectra. (b,c) 2D pump-probe spectra of the same sample measured with the polarization of the probe pulse perpendicular and parallel to the polarization of the pump pulse, respectively. The dashed contour lines mark regions where the difference signal is negative (bleach and stimulated emission), while the solid contours lines mark regions where the response is positive (excited state absorption). The most prominent off-diagonal bands are marked by arrows. (d,e,f) A global least-squares fit of the experimental data, used to refine the coupling Hamiltonian in Equation (29c). (From Ref. 42.)

well-defined conformation in solution with an almost ideal type II' β-turn centered at the Abu-Arg residues (133). The other samples investigated are apamin (134), a small neurotoxic peptide component found in the honeybee venom and scyllatoxin (135,136), a scorpion toxin with a high affinity for apamin-sensitive potassium channels. Both peptides were chosen for their variety of structural motifs, which are stabilized in the solution phase by disulfide bridges. Apamin (18 amino acids) is one of the smallest globular peptides known and has a short α helix and a β turn (134). Scyllatoxin, with 31 amino acids, is the smallest known natural peptide containing both an α helix and a β sheet (135,136).

Figure 16 shows the nonlinear 2D-IR spectra of the cyclic penta-peptide for perpendicular (Fig. 16b) and parallel (Fig. 16c) pump and probe polarizations, respectively. In both cases, the dominating signals are found along the diagonal of the 2D-IR spectrum, where the spectra are signifi-cantly better resolved than the linear absorption spectrum (Fig. 16a). More germane for this discussion, however, is the off-diagonal region, where cross-peaks appear (see arrows in Fig. 16b), the strongest of which was for the $1610-1584$ cm^{-1} level pair. The other cross-peaks are weaker but are easily verified in cuts through the 2D-IR spectra along the probe axis for certain pump frequencies, where they can be identified by their dispersive shapes (see, for example, arrow in Fig. 17).

The intensity of the cross-peak, relative to the intensity of the diagonal contribution, is larger when the spectrum is measured with the polarization of the pumped and probed beams perpendicular (Fig. 16b). The anisotropy r_{kl} of each peak measures the angle between the pumped and the probed transitions through $r_{kl} = \frac{1}{5}(3\cos^2 \varphi_{kl} - 1)$ so that along the diagonal of the 2D spectra values close to 0.4 are observed as expected, since here the same transition is pumped as is probed. Anisotropies smaller than 0.4 are observed in the off-diagonal region since pumped and probed transitions are in general not parallel, explaining the higher contrast of the cross-peaks in the perpendicular spectrum.

The ultimate goal is to deduce from these experiments the resonance couplings β_{ij} since these are the numbers that may be directly related to the structure of the peptide, given that one has a reliable model for computing the coupling Hamiltonian from the structure. The off-diagonal anharmonic-ities $\Delta\varepsilon_{kl}$ can be deduced from the intensities of the off-diagonal peaks (42), so that the resonance couplings β_{ij} could be computed according to Equation (24) in the weak coupling limit (valid for the cyclic penta peptide) or with the help of a diagonalization of the two-excitonic matrix. However, additional ambiguity arises since the zero-order energies ε_i, on

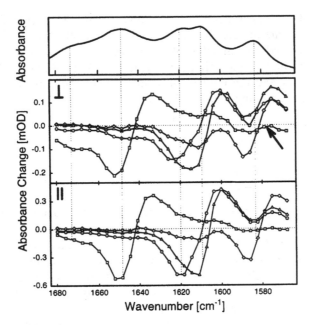

Figure 17 Cuts through the 2D-IR pump probe spectra shown in Fig. 16b and c along the direction of the probe axes for selected pump frequencies which were chosen to match the peaks in the linear absorption spectrum (squares: 1648 cm^{-1}; circles: 1620 cm^{-1}; triangles: 1610 cm^{-1}; diamonds: 1584 cm^{-1}). The frequency positions of the pump pulses are marked by the vertical dotted lines. (From Ref. 42.)

which the nonlinear response of the system depends very critically, are a priori not known. In the localization limit, where each one-excitonic state is predominantly localized on one individual molecular site, their eigenenergies (observed in the absorption spectrum, Fig. 16a) are essentially the same as the zero-order energies ε_i. Nevertheless, there are still $5! = 120$ permutations of how to distribute these frequencies to the five peptide groups. In Ref. 42 it was shown that this ambiguity can be resolved with the additional information obtained from the measured anisotropies and with empirical rules for the amide I frequencies.

In the simplest model, as proposed by Krimm et al. (118) and Tasumi et al. (123), the coupling Hamiltonian is given by a simple dipole-dipole coupling term:

$$\beta_{ij} = \frac{\vec{\mu}_i \cdot \vec{\mu}_j}{r_{ij}^3} - 3\frac{(\vec{r}_{ij} \cdot \vec{\mu}_i)(\vec{r}_{ij} \cdot \vec{\mu}_j)}{r_{ij}^5} \qquad (28)$$

where the directions of the transition dipoles $\vec{\mu}_i$ and the vectors connecting two sites $\vec{r}_{ij}$ relate the coupling Hamiltonian to the structure of the peptide. The position and the direction of each dipole vector with respect to the peptide bond have been assigned as depicted in Fig. 18a (118,123). Using this formalism and the known x-ray structure of the peptide, one would obtain for the coupling Hamiltonian (in cm^{-1})

$$\beta_{kl} = \begin{pmatrix} \cdots & -6 & -8 & -8 & -1 \\ -6 & \cdots & 8 & 4 & -1 \\ -8 & 8 & \cdots & 0 & 1 \\ -8 & 4 & 0 & \cdots & 4 \\ -1 & -1 & 1 & 4 & \cdots \end{pmatrix} \tag{29a}$$

(where numbering starts from the Mamb-Abu-peptide bond). However, the fact that the position of the dipole with respect to the peptide group has to be specified clearly stresses that the dipole approximation is not appropriate to describe the coupling (137). In other words the size of the peptide group whose distributed charges give rise to the transition dipole is of the same order as the separation between pairs of peptide groups. Nevertheless, it would be very valuable to find an effective Hamiltonian based on Coulomb interactions in order that the coupling matrix elements can be related directly to structure. In the present case we have performed density

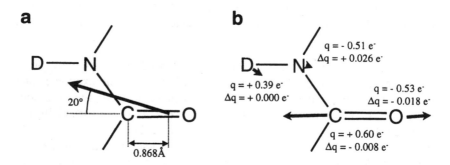

Figure 18 Models from which the excitonic coupling between pairs of peptide groups were calculated: (a) The direction and location of the transition dipole of the amide I mode (118,123) from which the coupling between two peptide groups is calculated according to a dipole-dipole interaction term [Eqaution (28)] (b) The nuclear displacements, partial charges, and charge flow of the amide I normal mode obtained from a DFT calculation on deuterated N-methylacetamide (all experiments were performed in D$_2$O) (42). With this set of transition charges, the multipole interaction is computed, avoiding the limitations of the dipole approximation.

functional theory (DFT) calculation on model compounds chosen to mimic the local electronic structure of individual peptide bonds (such as deuterated N-methylbenzamide, $C_6H_5-COND-CH_3$, and N,N-dimethylacetamide, and NMA) in order to calculate the nuclear displacements, partial charges, and charge flows during amide I vibration (see Fig. 18b). With this set of transition charges, the multipole electrostatic interaction has been calculated, yielding as coupling Hamiltonian:

$$\beta_{kl} = \begin{pmatrix} \cdots & -10 & -7 & -1 & 0 \\ -10 & \cdots & 4 & 6 & -2 \\ -7 & 4 & \cdots & -4 & 1 \\ -1 & 3 & -4 & \cdots & -11 \\ 0 & -2 & 1 & -11 & \cdots \end{pmatrix} \tag{29b}$$

whose patterns are somewhat similar to the dipole-dipole Hamiltonian in Equation (29a), but which in detail differs considerably. The Hamiltonian in Equation (29b), together with an assignment of the observed absorption frequencies to the five peptide groups, was used as a starting point of a least-square Levenberg Marquardt algorithm to globally fit all the experimental information available (i.e., parallel and perpendicular 2D-IR spectrum and linear absorption spectrum; see Fig. 16a,b,c) simultaneously. The perpendicular 2D-IR spectrum was weighted most in this global fit since it is believed that it carries the most significant structural content. The modeling included homogeneous and inhomogeneous broadening mechanisms with widths of 12 and 10 cm^{-1}, respectively (30,42). The resulting fits are shown in Fig. 16d,e,f, yielding as a refined Hamiltonian:

$$H_{kl}^{refine} = \begin{pmatrix} 1618 & -11 & -7 & -1 & 0 \\ -11 & 1588 & 4 & 6 & -2 \\ -7 & 4 & 1671 & -6 & 1 \\ -1 & 6 & -6 & 1648 & 1 \\ 0 & -2 & 1 & 1 & 1610 \end{pmatrix} \tag{29c}$$

Except for the coupling constant between the Gly4-Asp5 peptide group and the Asp5-Mambl peptide group [the 4-5 element in Equation (29c)], the refined Hamiltonian in Equation (29c) is almost identical to the Hamiltonian in Equation (29b). Apparently electrostatic interaction describes the coupling between two peptide groups reasonably well when the groups are not neighboring in the peptide chain. However, it appears as if the through-bond effect cannot be neglected for chemically bonded pairs of peptide groups. Our ab initio calculations on model compounds (NMA) clearly showed that the amide I normal modes are accompanied by a flow of charge to the methyl groups (corresponding to the C_α atoms in the peptide chain) and also a bending of the methyl C–H bonds. Both effects will couple neighboring groups by through bond interactions involving the C_α atom, and the amide I mode cannot be viewed as entirely localized on the peptide group (as assumed in Fig. 18b). Quantum chemistry calculations on dipeptides and tripeptides, which are definitely feasible with present-day computer technology, will enable a better description of the coupling between adjacent peptide bonds.

Figure 19 shows the results on apamin and scyllatoxin (30). Owing to the larger size of these peptides, the different amide I states underneath the amide I band are no longer spectrally resolved, so that the absorption spectrum appears as a broad band (width 30–40 cm^{-1}) with only very

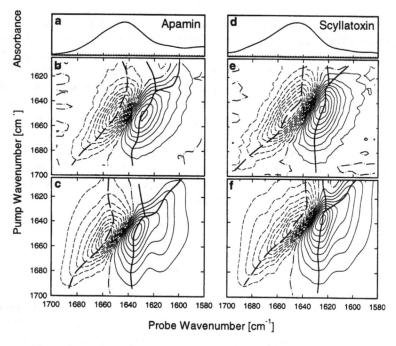

Figure 19 Absorption spectrum, measured 2D-IR spectrum, and modeled 2D-IR spectrum (from top to bottom) of apamin (a, b, c) and scyllatoxin (d, e, f) in D$_2$O. The dashed contour lines mark regions where the difference signal is negative (bleach and stimulated emission), while the solid contour lines mark regions where the response is positive (excited state absorption). The isosbestic line is marked by a dashed-dotted line. The thick lines mark the position of the local maxima (thick solid lines) and minima (thick broken line) of the probe spectra as a function of the pump frequency.

weak substructure (Fig. 19a,d). Nevertheless, in this case it can be unambiguously concluded from the 2D-IR spectra (Fig. 19b,e) that the amide I states are delocalized vibrational excitons. The variation of the response with pump frequency is enhanced by the extra lines in the contour plots in Fig. 19, which mark the positions of the local minima (thick dashed lines) and maxima (thick solid lines) of the transient probe spectra as a function of pump frequency. If the states excited by the narrowband pump pulse were localized on individual sites, one would observe only their anharmonic response so that these lines would directly follow the diagonal of the 2D graph. This is clearly is not what is observed. Model calculations of these spectra, based on the known structure of the peptides (see Fig. 17c,f), feature an excellent qualitative agreement with the experimental results. In contrast to the experiments described before, no detailed information on the zero-order frequencies ε_i is available so that they were divided into two groups: those peptide groups that are hydrogen bonded within the macro molecule and those that are not (30). The dependence of the transient response on parameters for homogeneous and inhomogeneous broadening permitted an estimate of their values (12 cm^{-1} and 24 cm^{-1}, respectively). Homogeneous broadening is essentially determined by the width of the transient holes, while inhomogeneity controls how much the transient response deviates from the diagonal (30). In addition, since the excitonic wave functions are known from the fit, the degree of delocalization could be estimated to be approximately 8 Å. This corresponds to approximately 11/2 helix turns in an α helix. An upper limit is thereby set on how far coherent transport of vibrational energy can take place, such as would be required for the propagation of the so-called Davidov soliton (138,139). Furthermore, if experiments can establish the coherence length at 300 K for excitations in the various secondary structure motifs, the computational effort needed to search for structures that match spectra might be considerably deduced.

Coherent transport of vibrational energy is further limited by vibrational energy relaxation. Experiments on the amide I band of different peptides (NMA, apamin, scyllatoxin BPTI, and the cyclic pentapeptide) revealed a vibrational relaxation rate of approximately $T_1 = 1.2$ ps, which is essentially independent of the particular peptide (30,53). A similar value has recently been reported for myoglobin at room temperature, with only a weak dependence of the relaxation rate on temperature down to cryogenic temperatures (140). In other words, vibrational relaxation of the amide I mode reflects an intrinsic property of the peptide group itself rather than a specific characteristic of the primary or secondary structural motifs of the peptide. It is significantly faster than that of other $C=O$ modes, such as in acetylbromide $(CH_3BrC=O)$ (65), heme groups (67,98,141,142), or metal-carbonyls (143). The relaxation rate of ^{15}N–NMA is essentially the same as that of ^{14}N–NMA (95), suggesting that the Fermi resonances responsible for the fast relaxation rate of the amide I mode do not involve much motion of the N atom. On the other hand, vibrational relaxation limits the maximum time window in which spectral diffusion processes can be observed by nonlinear IR techniques, so that knowledge and hence control of the mechanism of vibrational relaxation of the amide I band might help to extend this observation limit.

D. Spectral Diffusion of the Amide I Band

We have presented two types of nonlinear IR spectroscopic techniques sensitive to the structure and dynamics of peptides and proteins. While the 2D-IR spectra described in this section have been interpreted in terms of the static structure of the peptide, the first approach (i.e., the stimulated photon echo experiments of test molecules bound to enzymes) is less direct in that it measures the influence of the fluctuating surroundings (i.e., the peptide) on the vibrational frequency of a test molecule, rather than the fluctuations of the peptide backbone itself. Ultimately, one would like to combine both concepts and measure spectral diffusion processes of the amide I band directly. Since it is the geometry of the peptide groups with respect to each other that is responsible for the formation of the amide I excitation band, its spectral diffusion is directly related to structural fluctuations of the peptide backbone itself. A first step to measuring the structural dynamics of the peptide backbone is to measure stimulated photon echoes experiments on the amide I band (51).

The result of such an experiment on the de novo cyclic penta peptide, which has been introduced previously in this paragraph, is shown in Fig. 20. Qualitatively, the results are very similar to the results of the stimulated photon echo system on isolated test molecules embedded to proteins. As a function of T, the signal decays on a time scale corresponding to vibrational relaxation of the amide I states $T_1/2 = 600$ fs. As a function of τ, on the other hand, a significant peak shift is again obtained. As in the previous case, the peak shift, represented in Fig. 20 by the normalized first moment $M_1(T)$, slightly decays within the first ps, which is the time window accessible to these experiments in the moment. Similar results are obtained for apamin (51).

However, the interpretation of these results is considerably more complex since one has to deal with spectral diffusion of coupled states,

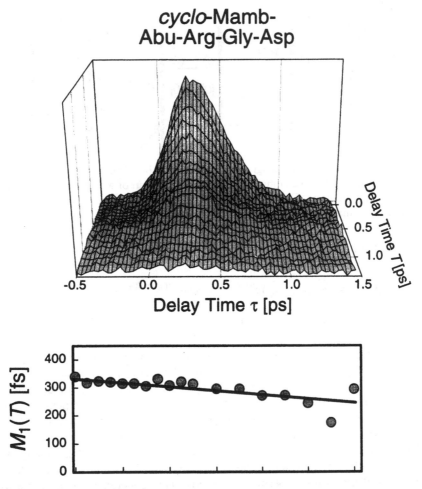

Figure 20 The stimulated (three-pulse) photon echo signal of the amide I band of *cyclo*-Mamb-Abu-Arg-Gly-Asp as function of delay time τ and T (see Fig. 3) and its normalized first moment. The first moment decays with time (T) due to conformational fluctuations of the peptide backbone.

rather than that of isolated vibrations. At this stage, we model the experiment within a simple Bloch picture, which assumes a strict separation of time scales of dephasing that implies a homogeneous bandwidth in a fixed inhomogeneous distribution of transitions. The relevant Feynman diagrams are depicted in Fig. 21. In the weak coupling limit (see Section IV.A), we

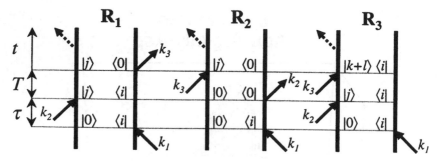

Figure 21 The Feynman diagrams, which have to be taken into account to model the rephasing part of the stimulated photon echo system of a excitonically coupled system of vibrations. The $\{|i\rangle\}$ and $\{|k+1\rangle\}$ are states of the one-exciton and two-exciton manifold, respectively.

obtain for the total response function (51):

$$\sum_{l=1}^{3} R_l = \sum_{i,j} \langle \mu_{0,i}^2 \mu_{0,j}^2 \rangle e^{i(\varepsilon_i \tau - \varepsilon_j t)} \cdot [1 + e^{i(\varepsilon_i - \varepsilon_j)T}] \cdot [1 - e^{i\Delta\varepsilon_{ij}t}]$$

$$\times e^{-(\tau+t)/T_2} e^{-T/T_1} \tag{30}$$

which, in the presence of inhomogeneity, has to be convoluted with distribution functions for the excitonic energies ε_j. The response in Equation (30) is exactly zero in the harmonic limit ($\Delta\varepsilon_{ij} = 0$), as expected. A model calculation of Equation (30), based on the known structures of these peptides, the coupling constants determined from the 2D-IR experiment [Equation (29c)] and parameters for vibrational relaxation ($T_1 = 1.2$ ps) and homogeneous broadening ($T_2 = 0.7$ ps), but neglecting inhomogeneous broadening, is shown in Fig. 22b. A sharp coherence spike at $T = \tau = 0$ is obtained, in clear contrast to the experimental results. This coherence spike is due to the interstate coherences $|j\rangle\langle i|$, that form after the second light interaction, which are expressed by the term $(1 + e^{i(\varepsilon_i - \varepsilon_j)T})$ in Equation (30) and which are all in phase only at delay zero $T = \tau = 0$. As time T increases, these terms dephase rapidly owing to their different beating frequencies. The square law detector in this particular experiment, which measures the t-integrated intensity of the third-order polarization [Equation (17)], strongly averages out the beatings in the third polarization, so that only weak structures remains in the signal (see arrow in Fig. 22b).

In order to obtain a model that indeed reproduces the experimental results, inhomogeneous broadening (modeled by a diagonal disorder of

cyclo-Mamb-Abu-Arg-Gly-Asp

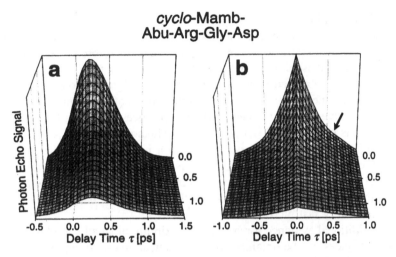

Figure 22 (a) Model calculation for the stimulated photon echo signal of the cyclic model peptide (*cyclo*-Mamb-Abu-Arg-Gly-Asp) based on its known structures. The same coupling constants were employed as in the model simulations of the 2D-IR spectrum in Fig. 16. The parameters for homogeneous broadening ($T_2 = 0.7$ ps), vibrational relaxation ($T_1 = 1.2$ ps), and inhomogeneous broadening (diagonal disorder 20 cm^{-1}) were also the same. The pulse duration of the laser pulses was set to 120 fs. (b) The same calculation as in (a), but with δ-shaped laser pulses and neglecting the inhomogeneous broadening. A sharp coherence spike now occurs at $T = \tau = 0$, which is not seen experimentally.

20 cm^{-1}) and the finite duration of the laser pulses (120 fs) had to be taken into account, giving the result shown in Fig. 22a, which features an excellent qualitative fit to the experiments. Inhomogeneity almost completely suppresses the coherence spike and gives rise to a large peak shift. Somewhat surprisingly, the main content of a stimulated photon echo experiment on a system of coupled states is essentially the same as in the case of an isolated transition, despite the apparent complication of the vibrational exciton coherence terms, namely that the peak shift is a qualitative measure of the inhomogeneity of the transitions.

In agreement with this conclusion, the Bloch picture applied here to derive Equation (30) does not predict a decay of the first moment, since the Bloch description omits spectral diffusion processes. Nevertheless, it is possible to understand the existence of a peak shift within the Bloch description, and this suggests a qualitative interpretation for its decay. As we have seen from photon echo experiments on spectroscopic probes

embedded within proteins, the peptide is fluctuating on a wide range of
time scales starting from the subpicosecond regime. The coupling scheme
in the excitonic system of the amide I band must therefore be continuously
rearranging on these time scales. The decay of the first moment found in the
experimentally obtained stimulated photon echo signal is considered to be
a direct experimental manifestation of fluctuations of the peptide backbone.
A more involved theory of spectral diffusion processes in coupled excitonic
systems has been worked out recently by Mukamel et al. (48,49) and will
help to better place these experiments on a more quantitative basis. The
photon echo experiment presented here is not mode selective and measures
spectral diffusion processes that represent an average of all peptide groups.
Nevertheless, the results clearly prove that the peptide backbone is fluctuating on a very fast picosecond time scale.

Additional evidence for structural flexibility of small peptides on very
fast time scales can be obtained from dynamical hole-burning experiments.
An example is shown in Fig. 23, (30) which reflects a cut through the 2D

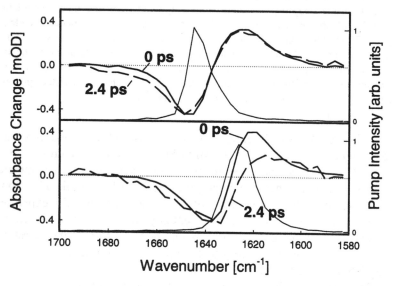

Figure 23 The response of a hole burned into the amide I band of scyllatoxin
by a narrowband pump pulse (width ~10 cm⁻¹, the spectrum of the pump pulse
shown as thin solid line) as a function of the delay time between pump and probe
pulse (thicker solid line: 0 ps; thick dashed line: 2.4). Vibrational T₁ relaxation was
compensated for in this plot by scaling the 2.4 ps spectrum in order to facilitate a
direct comparison with the time zero signal.

spectrum of Fig. 19e with the time separation of pump and probe pulse being varied between 0 and 2.4 ps. After selectively exciting some of the excitonic states of the amide I band of scyllatoxin with the narrowband pump pulse (width ca. 10 cm^{-1}), a spectral hole and an anharmonically shifted excited state absorption are observed. As time goes by, the hole broadens and the signal starts to wash out. Even though this process is slower than T_1 relaxation of 1.2 ps, it is clearly observable within the available time window. In the more conventional case of an inhomogeneously broadened band, this process is usually referred to as spectral diffusion. However, the situation can be given a more quantitative basis in this example since the energies of the different transitions underneath the amide I band are strongly correlated. Structural fluctuations of the peptide backbone continuously rearrange the excitonic coupling scheme so that population can flow between the different excitonic states. Two contributions to the spectral diffusion process might be distinguished: that of the diagonal and that of the off-diagonal elements of the coupling Hamiltonian. The former might reflect, for example, fluctuations of the structures of the hydrogen bonds formed by the peptides that have a major effect on the diagonal energies. On the other hand, spectral diffusion from fluctuation of the off-diagonal elements of the exciton matrix would directly relate to structural fluctuations of the peptide backbone since the coupling constants are a sensitive measure of the orientations and separations between neighboring peptide groups. The results discussed in the previous paragraph suggest that electrostatic Coulomb interaction can account for this interaction so that an extremely simple geometric expression, rather than sophisticated quantum chemistry calculations, would be required to calculate the correlation functions of the coupling (48,49). In addition, the technique introduced in Fig. 23 is potentially mode selective both in the diagonal and the off-diagonal region. Consequently, one might be able to deduce from 2D-IR spectroscopy a complete set of diagonal and off-diagonal fluctuation correlation functions, where the time separation of pump and probe is introduced as a third dimension (3D spectroscopy). The off-diagonal fluctuation correlation functions also might be obtainable from MD simulations so that a direct link between theory and experiment could be established.

E. 2D-IR Spectroscopy Using Semi-Impulsive Methods

The IR-2D spectroscopic technique applied in Section IV.C (30,42) utilized the frequency domain: after selectively bleaching individual one-excitonic states using a narrowband intense pump pulse, a broadband probe pulse recorded the nonlinear response of the bleached transition, as well as that of other transitions coupled to the bleached transition. By continuously tuning the frequency of the pump pulse, 2D spectra were constructed. One frequency dimension is the center frequency of the pump pulses, and the other comes from dispersing the probe pulse. It has been shown that cross peaks in those 2D spectra are related to the strength of coupling between pairs of peptide units (42).

Mukamel and others have proposed time domain 2D spectroscopic techniques on vibrational transitions that utilize impulsive excitation through a fifth-order Raman effect of low-frequency modes (33,34,48). Lately this method has been demonstrated experimentally on neat liquids (36,37,40). More recently, Mukamel and coworkers (47) have described third-order nonlinear coherent experiments on excitonically coupled two- and three-level systems, in which electronic transitions are excited with three laser pulses. The pulses are chosen short compared with relaxation and coupling mechanisms, but long compared with the transition frequency, corresponding to the so-called semi-impulsive limit. Model spectra on coupled two-level systems comparing various time orderings (photon echo, reverse photon echo, transient grating, reverse transient grating) illustrate that the excitonic coupling gives rise to cross peaks, from which the strength of coupling between individual pairs of transitions can be determined.

From the experimental viewpoint, these concepts require a measurement of the complete third-order field generated by the interaction of the sample with three incident fields. Such a measurement requires a heterodyne detection scheme using phase locked laser fields for the pump pulses and a local oscillator pulse with which to perform spectral interferometry. We have recently presented a much simpler semi-impulsive scheme (53), which, in terms of the underlying nonlinear response functions, resembles the transient grating experiment discussed by Mukamel et al. (47). Such a transient grating experiment can be thought of as a field from each of the first and second pulses (wavevectors k_1 and $-k_2$), which arrive at the sample simultaneously, forming a grating that scatters a field from the third pulse (wavevectors k_3) into the direction $k_s = k_1 - k_2 + k_3$. Our experiments also work in the time domain in the semi-impulsive limit. In the proposed simplified scheme, the first and second light field interactions corresponding to wavevectors $-k_1$ and k_1 originate from one laser pulse, while both the third field and the local oscillator field originate from the a second laser pulse with wavevector k_2, so that the scattered field has a wavevector $k_s = k_1 - k_1 + k_2 = +k_2$. In other words, the scheme is a conventional pump-probe configuration, where the third-order polarization generated by the combined pump and probe pulses generates an electric field propagating in the direction k_2 of the probe pulse. The probe field acts as the local oscillator in heterodyning the generated field. Therefore we define these as pump/probe self-heterodyne experiments, rather than pump-probe experiments, in order to emphasize its relationship to semi-impulsive methods and that interstate coherences of multilevel systems, rather than incoherent population states, are being probed.

The relevant Feynman diagrams to describe such experiments are essentially the same as those depicted in Fig. 21, with appropriately altered labeling of the wave vectors and time coordinates. In the weak coupling limit, we obtain for the total response function (53):

$$\sum_1 R_1 = 2 \sum_{i,j} \mu_{0,i}^2 \mu_{0,j}^2 \cdot \left[e^{-i\varepsilon_j t_2} - e^{-i(\varepsilon_j - \Delta\varepsilon_{ij})t_2} \right] e^{-t_2/T_2}$$

$$\times \left[1 + e^{i(\varepsilon_i - \varepsilon_j)t_1} \right] e^{-t_1/T_1} \tag{31}$$

The complex third-order generated field, which could be obtained experimentally from a phase locked heterodyne configuration, is proportional to the two-dimensional Fourier transform of Equation (31):

$$E^{(3)}(\omega_1, \omega_2) = \int_0^\infty dt_1 e^{i\omega_1 t_1} \int_0^\infty dt_2 e^{i\omega_2 t_2} \sum_{l=1}^{8} R_1(t_1, t_2)$$

$$= 2 \sum_{i,j} \mu_{0,i}^2 \mu_{0,j}^2 \cdot \left[\frac{1}{i(\varepsilon_j - \omega_2) + \dfrac{1}{T_2}} - \frac{1}{i(\varepsilon_j - \Delta\varepsilon_{ij} - \omega_2) + \dfrac{1}{T_2}} \right]$$

$$\times \left[\frac{1}{-i\omega_1 + \dfrac{1}{T_1}} + \frac{1}{i(\varepsilon_j - \varepsilon_i - \omega_1) + \dfrac{1}{T_1}} \right] \tag{32}$$

Along the $\omega_2 = 0$ axis, the 2D spectrum exhibits resonance peaks at the fundamental frequencies of the exciton states ε_j together with resonance peaks at frequencies $\varepsilon_j - \Delta\varepsilon_{ij}$ shifted by diagonal anharmonicity and having opposite phase. In the cross peak region ($\omega_1 \neq 0$), pairs of resonance peaks show up (again having opposite phase), which are now separated by off-diagonal anharmonicity $\Delta\varepsilon_{ij}$. These cross peaks can be unambiguously identified by their ω_1-frequency, which is equal to the separations between the corresponding fundamental frequencies $\varepsilon_j - \varepsilon_i$ (which can be either positive or negative). The off-diagonal anharmonicity is related to the resonance couplings β_{ij} [Equation (24)], so that this method is another means of two-dimensional spectroscopy from which structural information can be deduced.

In a self-heterodyne experiment, however, there is no independent control over the phase of the local oscillator field, so that the complete information on the complex third-order polarization of Equation (32) cannot be obtained. It is necessary to analyze in more detail the measurement process in order to determine the accessible information. In the actual experiment the spectrometer performs the Fourier transform of the generated third-order field of Equation (31) with respect to time coordinate t_2, generating the field components of $E^{(3)}(t_1; \omega_2)$ given by:

$$E^{(3)}(t_1; \omega_2) = 2\sum_{i,j} \mu_{0,i}^2 \mu_{0,j}^2 \cdot \left[\frac{1}{i(\varepsilon_j - \omega_2) + \dfrac{1}{T_2}} \right.$$

$$\left. - \frac{1}{i(\varepsilon_j - \Delta\varepsilon_{ij} - \omega_2) + \dfrac{1}{T_2}} \right] \cdot [1 + e^{i(\varepsilon_i - \varepsilon_j)t_1}] e^{-t_1/T_1} \quad (33)$$

The square-law detector then measures the total intensity at ω_2, which includes the probe electric field $E_2(\omega_2)$:

$$|E_2(\omega_2) - E^{(3)}(t_1; \omega_2)|^2 \cong |E_2(\omega_2)|^2 - 2\,\text{Re}(E_2^*(\omega_2)E^{(3)}(t_1; \omega_2)) \quad (34)$$

so for the δ-function pulses, the difference signal has the form:

$$\Delta A(t_1; \omega_2) = -4\,\text{Re}\sum_{i,j} \mu_{0,i}^2 \mu_{0,j}^2 \cdot \left[\frac{1}{i(\varepsilon_j - \omega_2) + \dfrac{1}{T_2}} \right.$$

$$\left. - \frac{1}{i(\varepsilon_j - \Delta\varepsilon_{ij} - \omega_2) + \dfrac{1}{T_2}} \right] \cdot [1 + e^{i(\varepsilon_i - \varepsilon_j)t_1}] e^{-t_1/T_1} \quad (35)$$

This expression describes the difference signal in the time domain taken in the manner of the experimental data shown in Fig. 25a. A spectral analysis of this signal is obtained utilizing a cosine-Fourier transform of the signal along the t_1 axis:

$$S^{(3)}(\omega_1, \omega_2) = -2 \int_0^\infty dt_1 \cos(\omega_1 t_1) \operatorname{Re} E^{(3)}(t_1; \omega_2) \qquad (36)$$

An analytical form for Equation (36) can be obtained in a straightforward manner, but it does not make the important features of the signal transparent. Instead, the signal is illustrated in Fig. 24 from a model

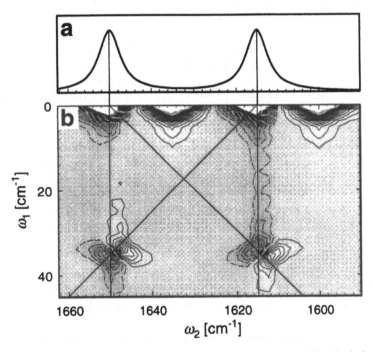

Figure 24 Self-heterodyned 2D-IR signal simulation: A model calculation for an idealized system of two coupled vibrators with frequencies 1615 cm^{-1} and 1650 cm^{-1}, diagonal anharmonicity $\Delta\varepsilon_{11} = \Delta\varepsilon_{22} = 16 \text{ cm}^{-1}$, off-diagonal anharmonicity $\Delta\varepsilon_{12} = 2.5 \text{ cm}^{-1}$, homogeneous dephasing rate $T_2 = 2$ ps, and a population relaxation time $T_1 = 4$ ps. The figure shows (a) the linear absorption spectrum (b) the semi-impulsive self-heterodyne 2D spectrum. For the sake of clarity, δ-shaped laser pulses are assumed, in which case the response function $\sum R_1$ directly corresponds to the third-order polarization $P^{(3)}(t_1, t_2)$. In the 2D spectra, light gray colors and solid contour lines symbolize regions with a positive difference IR signal, while negative signals are depicted in dark gray colors and with dashed contour lines. The cross peaks in (b) have different shapes, allowing them to be assigned unambiguously.

calculation for an idealized system of two coupled vibrators. The same parameters as in Fig. 14 were chosen: frequencies 1615 cm^{-1} and 1650 cm^{-1}, diagonal anharmonicity $\Delta = 16$ cm^{-1}, coupling $\beta_{12} = 7$ cm^{-1} resulting in a off-diagonal anharmonicity of $\Delta\varepsilon_{12} = 2.5$ cm^{-1} [Equation (24)]. We used a homogeneous dephasing rate of $T_2 = 2$ ps and a population relaxation time $T_1 = 4$ ps in order to make the features as distinct as possible. The 2D spectrum is shown in Fig. 14b together with the linear absorption spectrum in Fig. 14a. Two cross peaks emerge at positions where the vertical lines, originating from the peaks of the linear absorption spectrum, cross the unit slope diagonals originating from the corresponding other absorption peak. Since only the real part of $E^{(3)}(t_1; \omega_2)$ participates in the self-heterodyne experiment, positive and negative ω_1 frequencies cannot be distinguished. As a consequence, the cosine Fourier transform of Equation (36) exhibits cross peaks in the $\omega_1 > 0$ region, which originate both from $|\varepsilon_i - \varepsilon_j|$ and from $|\varepsilon_j - \varepsilon_i|$. This is the reason why diagonals in both directions are drawn in the 2D spectra of Fig. 24b. In effect the loss of information about the imaginary part of the third-order polarization causes the negative half side of the ω_1-axis to be flipped onto the positive side, and each cross peak term then appears twice in a 2D spectrum, which is defined only for the positive half of the ω_1-axis. This superposition of cross peaks will introduce a loss of resolution and make it more difficult to analyze congested spectra. However, as also seen from Fig. 24, the duplicated cross peaks have different shapes and therefore may be distinguished by analyzing both their positions and shapes.

Figure 25a shows the response of the cyclic penta peptide introduced in Section IV.C after semi-impulsive excitation with an intense, ultrashort pump pulse, in the time domain. Pronounced beatings, originating from

---

Figure 25 Self-heterodyned 2D-IR semi-impulsive 2D-IR spectra: (a) The frequency resolved signal of the cyclic pentapeptide as a function of probe frequency ω_2 and delay time t_1 between pump and probe pulse. Pronounced beatings are observed in the signal, marked, for example, by the arrow. (b) The cosine-Fourier transform with respect to delay time t_1 of the data in (a). The data along $\omega_1 = 0$ are suppressed in the 2D contour plot for a better distinctness of the cross peak region. Light gray colors and solid contour lines symbolize regions with positive response, while negative signals are depicted in dark gray colors and with dashed contour lines. Cross peaks appear in the 2D spectrum where vertical lines, which mark the peak position in the linear absorption spectrum, cross diagonal lines originating from a coupled band. The most prominent cross peak is related to the absorption lines at 1584 cm^{-1} and 1620 cm^{-1} (peak I and I').

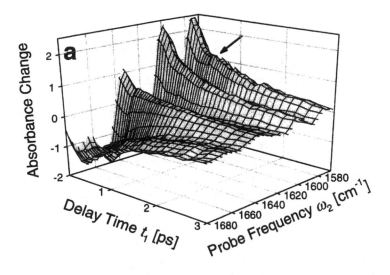

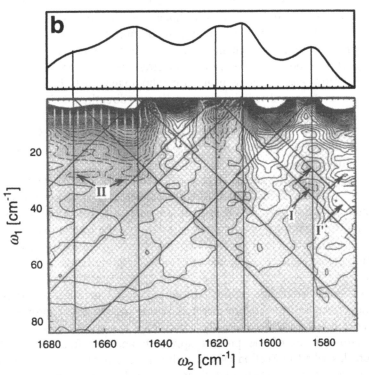

interstate coherences between excitonic states, are observed in the signal as a function of delay time t_1 of the separation between pump and probe pulse (marked, for example, by the arrow). The spectral analysis of these data according to Equation (36) is shown in Fig. 25b together with the linear absorption spectrum of the peptide. Prominent positive and negative peaks are observed along the ω_2-axis for $\omega_1 = 0$, which essentially reflect the bleach, stimulated emission, and excited state absorption of each individual linear absorption bands. Furthermore, cross peaks appear in the $\omega_1 > 0$ region (some of which marked by arrows) at positions where the vertical lines marking the peak positions in the linear absorption spectrum cross-diagonal lines originating from related absorption lines. The most prominent cross peak is related to the 1584 cm^{-1} and 1620 cm^{-1} transition (peak I and I'), in perfect agreement with the frequency domain 2D-IR spectra shown in Fig. 16b and Fig. 16c, where the strongest cross peak was observed between the same set of transitions. A few other distinct cross peaks are marked as well. The elongated shape of the cross peak connecting the absorption line at 1673 cm^{-1} with that at 1648 cm^{-1} (peaks II) emphasizes the higher spectral resolution of this method in the ω_1 direction, which is limited by T_1 vibrational relaxation. In addition, there is inhomogeneous broadening because the high flexibility of the corresponding Arg3-Gly4-Asp5 peptide groups permits a range of structures to exist in equilibrium (42,133). However, the cross peaks are inhomogeneously broadened only along the ω_2-axis and remain sharp in the ω_1 direction. This is a consequence of the inhomogeneous distributions being correlated so that they affect neither the level spacings nor the interstate coherences. Clearly this may not always be the case, and methods to evaluate the degree of correlation will be needed.

For uncoupled systems, where all groups are independent of each other, off-diagonal anharmonicity would vanish, $\Delta\varepsilon_{ij} = 0$, $i \neq j$, and only diagonal anharmonicity would remain. In that case, Equation (35) would not exhibit oscillations as a function of t_1 so that the oscillations observed in the experimental signal are an independent proof of the existence of off-diagonal anharmonicity and hence of exciton coupling between the amide I states.

In conclusion, the rather straightforward one-color pump-probe scheme of the self-heterodyne method seems particularly appropriate for smaller coupled systems where the vibrational spectra are less congested. The main content of the present approach and the frequency domain method described in Section IV.C are the same, namely the existence and magnitudes of cross peaks and their relationship to couplings between sites. However, they are complementary in terms of spectral resolution and broadening mechanisms. Fully heterodyned echo experiments in which all the Bohr frequencies are examined separately were recently reported (146).

Can Peptide Structures Be Determined by Nonlinear 2D-IR Spectroscopy?

We shall conclude this chapter with a few speculative remarks on possible future developments of nonlinear IR spectroscopy on peptides and proteins. Up to now, we have demonstrated a detailed relationship between the known structure of a few model peptides and the excitonic system of coupled amide I vibrations and have proven the correctness of the excitonic coupling model (at least in principle). We have demonstrated two realizations of 2D-IR spectroscopy: a frequency domain (incoherent) technique (Section IV.C) and a form of semi-impulsive method (Section IV.E), which from the experimental viewpoint is extremely simple. Other 2D methods, proposed recently by Mukamel and coworkers (47), would not pose any additional experimental difficulty. In the case of NMR, time domain Fourier transform (FT) methods have proven to be more sensitive by far as a result of the multiplex advantage, which compensates for the small population differences of spin transitions at room temperature. It was recently demonstrated that FT methods are just as advantageous in the infrared regime, although one has to measure electric fields rather than intensities, which cannot be done directly by an electric field detector but requires heterodyned echoes or spectral interferometry (146). Future work will have to explore which experimental technique is most powerful and reliable.

We have investigated peptides whose structures were known beforehand from NMR or x-ray spectroscopy and related these structures to 2D-IR spectroscopy. Ultimately, one would like to deduce the structure of an unknown sample from a 2D-IR spectrum. In the case of 2D NMR spectroscopy, two different phenomena are actually needed to determine peptide structures. Essentially, correlation spectroscopy (COSY) is utilized in a first step to assign protons that are adjacent in the chemical structure of the peptide so that J coupling gives rise to cross peaks in these 2D spectra. However, this through-bond effect cannot be directly related to the three-dimensional structure of the sample, since that would require quantum chemistry calculations, which presently cannot be performed with sufficient accuracy. The nuclear Overhauser effect (NOE), which is an incoherent population transfer process and has a simple distance dependence, is used as an additional piece of information in order to measure the distance in space between assigned protons. These constraints, together with molecular mechanics (MM) calculations, are generally sufficient to unambiguously determine the 3D structure of a peptide.

Similar procedures might turn out to be necessary in the case of 2D vibrational spectroscopy. The spectroscopy demonstrated so far essentially reflects the first step of the procedure in NMR spectroscopy (COSY). So far, we have investigated only the amide I subspace. However, all amide vibrations (N–H, amide II, etc.) might turn out to be equally important in revealing the required information, each of them giving different and hopefully complementary pieces of information. Couplings between different amide subspaces need to be explored. Incoherent population transfer out of the amide I transition appears to be very efficient ($T_1 =$ 1.2 ps), and the mechanism and the course of this transfer is completely unknown.

One reason for our hesitancy in predicting the structure of the cyclic pentapeptide from the 2D spectra was the inaccuracy of the coupling Hamiltonian. If one wants to calculate couplings from a structure, one will have to separate the problem into as small as possible units, since a complete and sufficiently accurate quantum chemistry calculation of the entire peptide will not be feasible. We used, as a first step, the –CO–NH– group as such a unit, which turned out to be too small, at least for peptide groups adjacent in the amino acid sequence. However, even larger units such as di- and tripeptides are still within the scope of state-of-the-art computer technology. Quantum chemistry calculations may be sufficiently accurate to calculate couplings (i.e. the off-diagonal elements of the coupling Hamiltonian), even though they might fail to predict the transition frequencies (i.e., the diagonal elements). However, since both quantities are measured independently in 2D-IR spectroscopy, in contrast to conventional 1D spectroscopy, the former information might be sufficient.

In conclusion, what we have summarized in this article can be viewed as a very first step in the direction of a novel structural analysis method, which opens a vast new field of experimental research where much work remains to be done. Any 2D-IR spectroscopy certainly will be limited to molecular sizes smaller than are tractable by 2D NMR and x-ray spectroscopy because of the intrinsically poor spectral resolution of vibrational transitions. However, the one tremendous prospect that makes this research worthwhile is the inherent high time resolution of IR spectroscopy, which covers all biologically relevant time scales, starting from the subpicosecond time regime. The IR photon echo experiments have proven experimentally for the first time that the peptide backbone itself indeed fluctuates on these fast time scales. If one succeeds to predict structures from 2D-IR spectroscopy one will also be able to follow structural modifications on the same ultrafast 1 ps time scale. These could be either equilibrium fluctuation of the peptide backbone or concerted structural rearrangements after triggering, for example, with the help of photochemical reactions (144,145) or temperature jumps.

APPENDIX: DIAGONAL AND OFF-DIAGONAL ANHARMONICITY IN THE WEAK COUPLING LIMIT

The one-exciton Hamiltonian H_1 mixes the monomeric site states $|i\rangle$ to create the one-excitonic states $|\nu_k\rangle = \sum_k q_{ki}|i\rangle$ (see Fig. 13). When the system is harmonic ($\Delta = 0$), the two-excitonic eigenstates of the two-excitonic Hamiltonian $H_2^{(0)}$ are simply Boson product states of the one excitonic states: $\{a_{ij}(|\nu_k\rangle|\nu_l\rangle + |\nu_l\rangle|\nu_k\rangle)\}$, $1 \le k$, where the normalization factors are $a_{ii} = \frac{1}{2}$ for $l = k$ and $a_{ij} = 1/\sqrt{2}$ for $l \ne k$, respectively. The corresponding eigenvalues are $\varepsilon_k + \varepsilon_l$. The transformation matrix between the two-exciton basis and the site basis $\{\alpha_{ij}(|i\rangle|j\rangle + |j\rangle|i\rangle)\}$, $i \le j$, is

$$Q_{kl,ij} = 2a_{ij}a_{kl}(q_{ki}q_{lj} + q_{li}q_{ki}) \tag{37}$$

Anharmonicity is introduced by reducing the site energies of only the double excited monomeric site states $|i\rangle|i\rangle$ by an energy Δ (see Fig. 13). The perturbed Hamiltonian $H_2 = H_2^{(0)} + V$ consists of a harmonic part $H_2^{(0)}$ and an anharmonicity term V, which mixes the harmonic two excitonic states. The matrix elements of V in the site basis are:

$$V_{ij,i'j'} = -\Delta \cdot \delta_{ij}\delta_{i'j'}\delta_{ii'} \tag{38}$$

The matrix elements of V in the excitonic basis are:

$$V_{kl,mn} = -\Delta \sum_{ij} Q_{kl,ij}\delta_{ij}Q_{ij,mn}^{-1} = -\Delta \sum_{i} Q_{kl,ii}Q_{mn,ii} \tag{39}$$

Up to this point, this expression is exact. The problem can be evaluated assuming that the coupling is weak so that each one-excitonic state is predominantly localized on an individual monomeric site:

$$q_{ij} = \delta_{ij} + q'_{ij} = \delta_{ij} + \frac{\beta_{ij}}{\varepsilon_j - \varepsilon_i} \quad \text{with} \quad \left|\frac{\beta_{ij}}{\varepsilon_j - \varepsilon_i}\right| < 1 \tag{40}$$

where β_{ij} are the off-diagonal terms of the one-exciton Hamiltonian H_1 in the site basis. Then we obtain for the diagonal anharmonicities in first order of V and lowest order in β_{ij}:

$$\Delta\varepsilon_{kk} = V_{kk,kk} = -\Delta \tag{41}$$

and for the off-diagonal anharmonicities

$$\Delta\varepsilon_{kl} = V_{kl,kl} = -2\Delta\sum_i q_{ki}^2 q_{li}^2 = -4\Delta\frac{\beta_{kl}^2}{(\varepsilon_k - \varepsilon_l)^2} \tag{42}$$

ACKNOWLEDGMENTS

We would like to thank Manho Lim for the significant contribution in setting up the instrumentation and taking the data presented here and William DeGrado for introducing us to the cyclic model penta-peptide. We are indebted to Dr. Nien-Hui Ge for her careful contributions regarding some of the key formulas in this paper. Dr. Matthew Asplund gratefully contributed the spectrally resolved echoes prior to their publication. The research was supported by NIH and NSF with instrumentation developed under NIH RR13456. Peter Hamm is grateful to the DFG for a postdoctoral fellowship.

REFERENCES

1. Wüthrich K. NMR in Biological Research: Peptides and Proteins. New York: American Elsevier, 1976.
2. Bax A. Two-Dimensional Nuclear Magnetic Resonance in Liquids. Boston: Kluwer, 1982.
3. Ernst RR, Bodenhausen G, Wokaun A. Principles of Nuclear Magnetic Resonance in One and Two Dimensions. Oxford: Oxford University Press, 1987.
4. Sanders JK, Hunter BK. Modern NMR Spectroscopy. New York: Oxford University Press, 1993.
5. Karplus M, Petsko GA. Molecular dynamics simulations in biology. Nature 1990; 347:631–639.
6. Elber R, Karplus M. Enhanced sampling in molecular dynamics: use of the time-dependent hartree approximation for a simulation of carbon monoxide diffusion through myoglobin. J Am Chem Soc 1990; 112:9161–9175.
7. Zhou HX, Wlodek S, McCommon JA. Conformation gating as a mechanism for enzyme specificity. Proc Natl Acad Sci USA 1998; 95:9280–9283.

8. Soman KV, Karimi A, Case DA. Unfolding of an α-helix in water. Biopolymers 1991; 31:1351–1361.

9. Daggett V, Levitt M. Molecular dynamics simulations of helix denaturation. J Mol Biol 1992; 223:1121–1138.

10. Sung SS, Wu XW. Molecular dynamics simulation of synthetic peptide folding. Proteins Struct Funct Genet 1996; 25:202–214.

11. Srajer V, Teng T, Ursby T, Pradervand C, Ren Z, Adachi S, Schildkamp W, Bourgeois D, Wulff M, Moffat K. Photolysis of the carbon monoxide complex of myoglobin: nanosecond time-resolved crystallography. Science 1996; 274:1726–1729.

12. Helliwell JR, Rentzepis PM, eds. Time Resolved Diffraction. Oxford: Clarendon Press, 1997.

13. Silvestri SD, Weiner AM, Fujimoto JG, Ippen EP. Femtosecond dephasing studies of dye molecules in a polymer host. Chem Phys Lett 1984; 112:195–199.

14. Joo T, Albrecht AC. Electronic dephasing of molecules in solution at room temperature by femtosecond degenerate four wave mixing. Chem Phys 1993; 176:233–247.

15. Fidder H, Wiersma DA. Exciton dynamics in disordered molecular aggregates: dispersive dephasing probed by photon echo and Rayleigh scattering. J Phys Chem 1993; 97:11603–11610.

16. Vöhringer P, Arnett D, Westervelt R, Feldstein M, Scherer NF. Optical dephasing on femtosecond timescales: direct measurement and calculation from solvent spectral densities. J Chem Phys 1995; 102:4027–4036.

17. Fleming GR, Cho M. Chromophore-solvent dynamics. Annu Rev Phys Chem 1996; 47:109–134.

18. Emde MF, Baltuska A, Kummrow A, Pshenichnikov MS, Wiersma DA. Ultrafast librational dynamics of the hydrated electron. Phys Rev Lett 1998; 80:4645–4648.

19. Joo T, Jia Y, Yu JY, Lang MJ, Fleming GR. Third-order nonlinear time domain probes of solvation dynamics. J Chem Phys 1996; 104:6089–6108.

20. Mercer IP, Abend S, Gould IR, Klug DR. A quantum mechanical/molecular mechanical approach to solvation dynamics tested by three pulse photon echo measurements. In: Elsaesser T, Fujimoto JG, Wiersma DA, Zinth W, eds. Ultrafast Phenomena, Berlin: Springer-Verlag, 1998:532–534.

21. DeBoeij WP, Pshenichnikov MS, Wiersma DA. Ultrafast solvation dynamics explored by femtosecond photon echo spectroscopy. Annu Rev Phys Chem 1998; 49:99–123.

22. Homoedelle BJ, Edington MD, Diffey WM, Beck WF. Stimulated photon echo and transient grating studies of protein-matrix solvation dynamics and interexciton radiationless decay in α-phycocyanin and allphycocyanin. J Phys Chem 1998; 102:3044–3052.

23. Joo T, Jia Y, Yu JY, Jonas DM, Fleming GR. Dynamics in isolated bacterial light harvesting antenna (LH2) of *Rhodobacter sphaeroides* at room temperature. J Phys Chem 1996; 100:2399–2409.

24. Groot ML, Yu JY, Agarwal R, Norris JR, Fleming GR. Three-pulse photon echo experiments on the accessory pigment in the reaction center of *Rhodobacter sphaeroides*. J Phys Chem B 1998; 102:5923–5931.

25. Jimenez R, Mourik F, Yu JY, Fleming GR. Three-pulse photon echo measurements on LH1 and LH2 complexes of *Rhodobacter sphaeroides*: nonlinear spectroscopic probe of energy transfer. J Phys Chem B 1997; 101:7350–7359.

26. Austin RH, Beeson KW, Eisenstein L, Frauenfelder H, Gunsalus IC. Dynamics of ligand binding to myoglobin. Biochemistry 1975; 14:5355–5373.

27. Leeson DT, Wiersma DA, Fritsch K, Friedrich J. The energy landscape of myoglobin: an optical study. J Chem Phys 1997; 101:6331–6340.

28. Johnson JB, Lamb DC, Frauenfelder H, Muller JD, McMahon B, Nienhaus GU, Young RD. Ligand binding to heme proteins IV. Interconversion of taxonomic substates in carbonmonoxymyoglobin. Biochemistry 1996; 71:1563–1573.

29. Mercer IP, Gould IR, Klug DR. Optical properties of solvated molecules calculated by a QMMM method. Faraday Discuss 1997; 108:51–62.

30. Hamm P, Lim M, Hochstrasser RM. The structure of the amide I band of peptides measured by femtosecond nonlinear IR spectroscopy. J Phys Chem B 1998; 102:6123–6138.

31. Lim M, Hamm P, Hochstrasser RM. Protein fluctuations are sensed by stimulated infrared echoes of the vibrations of carbon monoxide and azide probes. Proc Natl Acad Sci USA 1998; 95:15315–15320.

32. Aue WP, Bartoldi E, Ernst RR. Two-dimensional spectroscopy: applications to nuclear magnetic resonance. J Chem Phys 1976; 64:2229–2246.

33. Tanimura Y, Mukamel S. Two-dimensional femtosecond vibrational spectroscopy of liquids. J Chem Phys 1993; 99:9496–9511.

34. Okumura K, Tanimura Y. Two-dimensional THz spectroscopy of liquids: non-linear vibrational response to a series of THz laser pulses. Chem Phys Lett 1998; 295:298–304.

35. Okumura K, Tanimura Y. The $(2n + 1)$th-order off-resonant spectroscopy from the $(n + 1)$th-order anharmonicities of molecular vibrational modes in the condensed phase. J Chem Phys 1997; 106:1687–1698.

36. Tominaga K, Yoshihara K. Fifth order optical response of liquid CS_2 observed by ultrafast nonresonant six-wave mixing. Phys Rev Lett 1995; 74:3061–3064.

37. Steffen T, Duppen K. Femtosecond two-dimensional spectroscopy of molecular motion in liquids. Phys Rev Lett 1996; 76:1224–1227.

38. Steffen T, Fourkas JT, Duppen K. Time resolved four and six-wave mixing in liquids. I. Theory. J Chem Phys 1997; 105:7364–7382.

39. Steffen T, Duppen K. Time resolved four and six-wave mixing in liquids. II. Experiments. J Chem Phys 1997; 106:3854–3864.

40. Tokmakoff A, Lang MJ, Larsen DS, Fleming GR, Chernyak V, Mukamel S. Two-dimensional Raman spectroscopy of vibrational interactions in liquids. Phys Rev Lett 1997; 79:2702–2705.

41. Hamm P, Lim M, Hochstrasser RM. Non-Markovian dynamics of the vibrations of ions in water from femtosecond infrared three pulse photon echoes. Phys Rev Lett 1998; 81:5326–5329.

42. Hamm P, Lim M, DeGrado WF, Hochstrasser RM. The two-dimensional IR nonlinear spectroscopy of a cyclic penta-peptide in relation to its three-dimensional structure. Proc Natl Acad Sci USA 1999; 96:2036–2041.

43. Hybl JD, Albrecht AW, Sarah M, Gallagher F, Jonas DM. Two-dimensional electronic spectroscopy. Chem Phys Lett 1998; 297:307–313.

44. Mukamel S, Piryatinski A, Chernyak V. Two-dimensional Raman echoes: femtosecond view of molecular structure and vibrational coherence. Acc Chem Res 1999; 32:145–154.

45. Noda I. Two-dimensional infrared spectroscopy. J Am Chem Soc 1989; 111:8116–8118.

46. Mukamel S. Principles of Nonlinear Optical Spectroscopy. New York: Oxford University, 1995.

47. Zang WM, Chernyak V, Mukamel S. Multidimensional femtosecond correlation spectroscopy of electronic and vibrational excitons. J Chem Phys 1999; 110:5011–5028.

48. Chernyak V, Zhang WM, Mukamel S. Multidimensional femtosecond spectroscopies of molecular aggregates and semiconductor nanostructures: The nonlinear exciton equation. J Chem Phys 1998; 109:9587–9601.

49. Meier T, Chernyak V, Mukamel S. Femtosecond photon echoes in molecular aggregates. J Chem Phys 1997; 107:8759–8780.

50. Hamm P. Lim M, Hochstrasser RM. Vibrational relaxation and dephasing of small molecules strongly interacting with water. In: Elsaesser T, Fujimoto JG, Wiersma DA, Zinth W, eds. Ultrafast Phenomena. Berlin: Springer-Verlag, 1998:514–516.

51. Hamm P, Lim M, DeGrado WF, Hochstrasser RM. Stimulated photon echoes from amide I vibrations. J Phys Chem 1999; 103:10049–10053.

52. Hamm P, Lim M, Asplund M, Hochstrasser RM. The fifth-order contribution to the oscillations in photon echoes of anharmonic vibrators. Chem Phys Lett 1997; 301:167–174.

53. Hamm P, Lim M, DeGrado WF, Hochstrasser RM. Two dimensional self-heterodyned spectroscopy of vibrational transitions of a small globular peptide. J Chem Phys 2000; 112:1907–1916.

54. Nisoli M, De Silvestri S, Magni V, Svelto O, Danielius R, Piskarskas A, Valiulis G, Varanavicius A. Highly efficient parametric conversion of femtosecond Ti:sapphire laser pulses at 1 kHz. Opt Lett 1994; 19:1973–1995.

55. Yakovlev VV, Kohler B, Wilson KR. Broadly tunable 30-fs pulses produced by optical parametrical amplification. Opt Lett 1994; 19:2000–2002.

56. Wilson KR, Yakovlev VV, Ultrafast rainbow: tunable ultrashort pulses from a solid state kilohertz system. J Opt Soc Am B 1997; 14:444–448.

57. Seifert F, Petrov V, Woerner M. Solid-state laser system for the generation of mid-IR femtosecond pulses tunable from 3.3 to 10 μm. Opt Lett 1994; 19:2009–2011.

58. Gale GM, Gallot G, Hache F, Sander R. Generation of intense highly coherent femtosecond pulses in the mid infrared. Opt Lett 1997; 22:1253–1255.

59. Woutersen S, Emmerichs U, Bakker HJ. Femtosecond mid-IR pump-probe spectroscopy of liquid water: evidence for a two-component structure. Science 1997; 278:658–660.

60. Woutersen S, Emmerichs U, Bakker HJ. A femtosecond midinfrared pump-probe study of hydrogen bonding in ethanol. J Chem Phys 1997; 107:1483–1490.

61. Woutersen S, Emmerichs U, Nienhuys HK, Bakker HJ. Anomalous temperature dependence of vibrational lifetimes in water and ice. Phys Rev Lett 1998; 81:1106–1109.

62. Laenen R. Rauscher C, Laubereau A. Vibrational energy redistribution of ethanol oligomers and dissociation of hydrogen bonds after ultrafast infrared excitation. Chem Phys Lett 1998; 283:7–14.

63. Laenen R. Rauscher C, Laubereau A. Dynamics of local substructures in water observed by ultrafast infrared hole burning. Phys Rev Lett 1998; 80:2622–2625.

64. Gale GM, Gallot G, Hache F, Lascoux N, Bratos S, Leicknam JCL, Femtosecond dynamics of hydrogen bonds in liquid water: a real time study. Phys Rev Lett 1999; 82:1068–1071.

65. Hochstrasser RM, Anfinrud PA, Diller R, Han C, Iannone M, Lian T, Locke B. Femtosecond Infrared Spectroscopy of complex molecules using cw IR lasers. In: Harris CB, Ippen EP, Mourou GA, Zewail AH, eds. Ultrafast Phenomena VII. Berlin: Springer-Verlag, 1990:429–433.

66. Li M, Owrutsky J, Sarisky M, Culver JP, Yodh A, Hochstrasser RM. Vibrational and rotational relaxation times of solvated molecular ions. J Chem Phys 1993; 98:5499–5507.

67. Owrutsky JC, Ki M, Locke B, Hochstrasser RM. Vibrational relaxation of the CO stretch vibration in hemoglobin-CO, myoglobin-CO and protoheme-CO. J Phys Chem 1995; 99:4842–4845.

68. Kleiman VD, Arrivo SM, Melinger JS, Heilweil EJ. Controlling condensed-phase vibrational excitation with tailored infrared pulses. Chem Phys 1998; 223:207–216.

69. Kaindl RA, Smith DC, Joschko M, Hasselbeck MP, Woerner M, Elsaesser T. Femtosecond infrared pulses tunable from 9 to 18 μm at an 88 MHz repetition rate. Opt Lett 1998; 23:861–863.

70. Hamm P, Lauterwasser C, Zinth W. Generation of tunable subpicosecond light pulses in the midinfrared between 4.5 and 11.5 μm. Opt Lett 1993; 18:1943–1945.

71. Hamm P, Lim M, Hochstrasser RM. Vibrational energy relaxation of the cyanide ion in water. J Chem Phys 1997; 107:10523–10531.

72. Brewer RG, Shoemaker RL. Photon echo and optical mutation in molecules. Phys Rev Lett 1971; 27:631–634.

73. Zimdars D, Tokmakoff A, Chen S, Greenfield SR, Fayer MD, Smith TL, Schwettman HA. Picosecond infrared vibrational photon echoes in a liquid and glass using free electron laser. Phys Rev Lett 1993; 70:2718–2721.

74. (a) Tokmakoff A, Urdahl RS, Zimdars D, Francis RS, Kwok AS, Fayer MD. Vibrational spectral diffusion and population dynamics in a glass-forming liquid: variable bandwidth picosecond infrared spectroscopy. J Chem Phys 1995; 102:3919–3931; (b) Tokmakoff A, Fayer MD. Homogeneous vibrational dynamics and inhomogeneous broadening in glass-forming liquids: infrared photon echo experiments from room temperature to 10 K. J Chem Phys 1995; 103:2810–2826.

75. Rector KD, Kwok AS, Ferrante C, Tokmakoff A, Rella CW, Fayer MD. Vibrational anharmonicity and multilevel vibrational dephasing from vibrational echo beats. J Chem Phys 1997; 106:10027–10036.

76. Oxtoby DW. A molecular dynamics simulation of dephasing in liquid nitrogen. J Chem Phys 1978; 68:5528–5533.

77. Kubo R. A stochastic theory of line shape. Adv Chem Phys 1969; 15:101–127.

78. Yee TK, Guststafson. Diagrammatic analysis of the density operator for nonlinear optical calculations. Phys Rev A 1978; 18:1597.

79. Berne BJ, Pecora R. Dynamical Light Scattering. New York: Wiley, 1996.

80. Wang CC, Pecora R. Time-correlation function for restricted rotational diffusion. J Chem Phys 1980; 72:5333–5340.

81. Horng ML, Gardecki JA, Papzyan A, Maroncelli M. Subpicosecond measurement of polar solvent dynamics, coumarin 153 revisited. J Phys Chem 1995; 99:17311–17337.

82. Perchard C, Perchard JP. Liaison hydrogène en phase liquide et spectrométrie Raman. J Raman Spec 1977; 6:74–79.

83. Shelley VM, Yarwood J. The noncoincidence effect in N,N-dimethyl-formamide: a comparison of theoretical predictions and experimental results. Chem Phys 1984; 137:277–280.

84. Thomas HD, Jonas J. High pressure Raman study of the CO stretching mode in liquid N,N-dimethylacetamide. J Chem Phys 1989; 90:4144–4149.

85. Schindler W, Sharko PT, Jonas J. Raman study of pressure effects on frequencies and isotropic line shapes in liquid acetone. J Chem Phys 1982; 76:3493–3504.

86. Torii H, Tasumi M. Local order and transition dipole coupling in liquid methanol and acetone as the origin of the Raman noncoincidence effect. J Chem Phys 1993; 99:8459–8465.

87. Czeslik C, Jonas J. Effect of pressure on local order in liquid dimethyl sulfoxide. J Phys Chem A 1999; 103:3222–3227.

88. Torii H, Tasumi M. Raman noncoincidence effect and intermolecular interactions in liquid dimethyl sulfoxide: simulations based on the transition dipole coupling mechanism and liquid structures derived by Monte Carlo method. Bull Chem Soc Jpn 1995; 68:128–134.

89. Logan D. The non-coincidence effect in the Raman spectra of polar liquids. Chem Phys 1986; 103:215–225.

90. Logan D. The Raman non-coincidence effect in dipolar binary mixtures. Chem Phys 1989; 131:199–207.

91. Logan D. On the isotropic Raman spectra of isotopic binary mixtures. Mol Phys 1986; 58:97–129.

92. Gamba Z, Klein ML. Short-range structure of liquid pyrrole. J Chem Phys 1990; 92:6973–6974.

93. Ambroseo JR, Hochstrasser RM. Pathways of relaxation of the N-H stretching vibration of pyrrole in liquids. J Chem Phys 1988; 89:5956–5957.

94. Fourkas JT, Kawashima H, Nelson KA. Theory of nonlinear optical experiments with harmonic oscillators. J Chem Phys 1995; 103:4393–4407.

95. Hamm P, Lim M, Hochstrasser RM. Unpublished results.

96. Hesselink WH, Wiersma DA. Theory and experimental aspects of photon echoes in molecular solids. In: Agranowich VM, Hochstrasser RM, eds. Spectroscopy and Excitation Dynamics of Condensed Molecular Systems. North Holland: Elsevier Science, 1983:249–299.

97. Becker PC, Fragnito HL, Bigot JY, Brito Cruz CH, Fork RL, Shank CV. Femtosecond photon echoes from molecules in solution. Phys Rev Lett 1989; 63:505–507.

98. Hill JR, Tokmakoff A, Peterson KA, Sauter B, Zimdars D, Dlott DD, Fayer MD. Vibrational dynamics of carbon monoxide at the active side of myoglobin: picosecond infrared free-electron laser pump-probe experiments. J Phys Chem 1994; 98:11213–11219.

99. Cho M, Yu JY, Joo T, Nagasawa Y, Passino SA, Fleming GR. The integrated photon echo and solvation dynamics. J Phys Chem 1996; 100:11944–11953.

100. DeBoeij WP, Pshenichnikov MS, Wiersma DA. On the relation between the echo-peak shift and Brownian oscillator correlation function. Chem Phys Lett 1996; 253:53–60.

101. Levenson MD, Kano SS. Introduction to Nonlinear Laser Spectroscopy. Boston: Academic Press, 1988.

102. Ferrario M, Klein ML, McDonald IR. Dynamical behavior of the azide ion in protic solvents. Chem Phys Lett 1993; 213:537–540.

103. Ferrario M, McDonald IR, Symons MCR. Solvent-solute hydrogen bonding in dilute solutions of CN^- and CH_3CN in water and methanol. Mol Phys 1992; 77:617–627.

104. Rey R, Hynes JT. Vibrational phase and energy relaxation of CN^- in water. J Chem Phys 1988; 108:142–153.

105. Polak M, Gruebele M, Saykally RJ. Velocity modulation laser spectroscopy of negative ions: the v_3 band of azide (N_3^-). J Am Chem Soc 1987; 109:2884–2887.

106. Jimenez R, Fleming GR, Kumar PV, Maroncelli M. Femtosecond solvation dynamics of water. Nature 1994; 369:471–473.

107. Jönsson BM, Hakansson K, Liljas A. The structure of human carbonic anhydrase II in complex with bromide and azide. FEBS Lett 1993; 322:186–190.

108. Merz KM, Banci L. Binding of azide to human carbonic anhydrase II: the role electrostatic complementary plays in selecting the preferred resonance structure of azide. J Phys Chem 1996; 100:17414–17420.

109. Derewenda Z, Dodson G, Emsley P, Harris D, Nagai K, Perutz M, Reynaud JP. Stereochemistry of caron monoxide binding to normal human adult and cowtown hemoglobins. J Mol Biol 1990; 211:515–519.

110. Maurus R, Bogumil R, Nguyen NT, Mauk AG, Brayer G. Structural and spectroscopic studies of azide complexes of horse heart myoglobin and the His-64-Thr variant. Biochem J 1998; 332:67–74.

111. Locke B, Lian T, Hochstrasser RM. Determination of Fe-CO geometry and heme rigidity in carbonmonoxyhemoglobin using femtosecond IR spectroscopy. Chem Phys 1991; 158:409–419.

112. Lian L, Locke B, Kitagawa T, Nagai M, Hochstrasser RM. Determination of Fe-CO geometry in the subunits of carbonmonoxy hemoglobin using femtosecond infrared spectroscopy. Biochemistry 1993; 32:5809–5814.

113. Lim M, Jackson TA, Anfinrud PA. Binding of CO of myoglobin from a heme pocket docking site to form nearly linear Fe–C–O. Science 1995 269:962–966.

114. Lin SH, Yu NT, Tame J, Shih D, Renaud JP, Pagnier J, Nagai K. Effect of the distal residues on the vibrational modes of the Fe–CO bond in hemoglobin studies by protein engineering. Biochemistry 1990; 29:5562–5566.

115. Li T, Quillin TL, Phillips GN, Olson JS. Structural determinants of the stretching frequency of CO bound to myoglobin. Biochemistry 1994; 33:1433–1446.

116. Braunstein DP, Chu K, Egeberg KD, Frauenfelder H, Mourant JR, Nienhaus GU, Ormos P, Sligar SG, Springer BA, Young RD. Ligand binding to heme proteins III: FTIR studies of His-E7 and Val-E11 mutants of carbonmonoxymyoglobin. Biophys J 1993; 65:2447–2454.

117. Asplund M, Lim M, Hochstrasser RM. Spectrally resolved three pulse photon echoes in the vibrational infrared. Chem Phys Lett 2000; 323:269–277.

118. Krimm S, Bandekar J. Vibrational spectroscopy and conformation of peptides, polypeptides and proteins. Adv Protein Chem 1986; 38:181–365.
119. Krimm S, Reisdorf WC. Understanding normal modes of proteins. Faraday Discuss 1994; 99:181–197.
120. Reisdorf WC, Krimm S. Infrared Dichroism of amide I and amide II modes of α_I- and α_{II}-helix segments in membrane proteins. Biophys J 1995; 69:271–273.
121. Reisdorf WC, Krimm S. Infrared amide I band of the coiled coil. Biochemistry 1996; 35:1383–1386.
122. Mirkin NG, Krimm S. Ab initio vibrational analysis of hydrogen bonded *trans* and *cis* N-methylacetamide. J. Am. Chem. Soc. 1991; 113:9742–9747.
123. Torii H, Tasumi M. Model calculations on the amide I infrared bands of globular proteins. J Chem Phys 1992; 96:3379–3387.
124. Torii H, Tasumi M. Theoretical analysis of the amide I infrared bands of globular proteins. In: Mantsch HH, Chapman D, eds. Infrared Spectroscopy of Biomolecules. New York: Wiley-Liss, Inc., 1996:1–18.
125. Baumruk V. Pancoska P. Keiderling TA. Prediction of secondary structure using statistical analysis of electronic and vibrational circular dichroism and Fourier transform infrared spectra of proteins in H_2O. J Mol Biol 1996; 259:774–791.
126. Dong A, Huang P, Caughey WS. Protein Secondary structure in water from second derivative amide I infrared spectra. Biochemistry 1990; 29:3303–3308.
127. Surewicz WK, Mantsch HH, Chapman D, Determination of protein secondary structure by FTIR spectroscopy: a critical assessment. Biochemistry 1993; 32:389–394.
128. Prestrelski SJ, Byler DM, Liebman MN. Generation of a substructure library for the description and classification of protein secondary structure. II. Application to spectrastructure correlations in Fourier transform infrared spectroscopy. Proteins 1992; 14:440–450.
129. Wilson G, Hecht L, Barron LD. Evidence for a new cooperative transition in native lysozyme from temperature-dependent Raman optical activity. J Phys Chem B 1997; 101:694–698.
130. Teraoka J, Bell AF, Hecht L, Barron LD. Loop structure in human serum albumin from Raman optical activity. J Raman Spectrosc 1998; 29:67–71.
131. Davydov AS. Theory of Molecular Excitons. New York: Plenum, 1971.
132. Maslen PE, Handy NC, Amos RD, Jayatilaka D. Higher analytic derivates IV: anharmonic effects in the benzene spectrum. J Chem Phys 1992; 97:4233–4254.
133. Bach AC, Eyermann CJ, Gross JD, Bower MJ, Harlow RL, Weber PC, DeGrado WF. Structural studies of a family of high affinity ligands for GPIIb/IIIa. J Am Chem Soc 1994; 116:3207–3219.

134. Pease JHB, Wemmer DE. Solution structure of apamin determined by nuclear magnetic resonance and distance geometry. Biochemistry 1988; 27:8491–8498.

135. Pagel MD, Wemmer DE. Solution structure of a core peptide derived from scyllatoxin. Proteins 1994; 18:205–215.

136. Martins JC, van de Ven FJM, Borremans FAM. Determination of the three-dimensional solution structure of scyllatoxin by ^{1}H nuclear magnetic resonance. J Mol Biol 1995; 253:590–603.

137. Cheam TC, Krimm S. Infrared intensities of amide modes in N-methylacetamide and poly(glycine I) from ab initio calculations of dipole moment derivatives of N-methylacetamide. J Chem Phys 1985; 82:1631–1641.

138. Davydov AS. Solitons and energy transfer along protein molecules. J Theor Biol 1977; 66:379–387.

139. Davydov AS. Solitons in molecular systems. Phys Scr 1997; 20:387–394.

140. Peterson KA, Rella CW, Engholm JR, Schwettman HA. Ultrafast vibrational dynamics of the myoglobin amide I band. J Phys Chem B 1999; 103:557–561.

141. Hochstrasser RM. Femtosecond infrared probing of biomolecules. Proc SPIE 1992; 16:1921.

142. Owrutsky JC, Li M, Culver JP, Sarisky MJ, Yodh AG, Hochstrasser RM. Vibrational dynamics of condensed phase molecules studied by ultrafast infrared spectroscopy. In: Lau A, Siebert F, Werncke W, eds. Time Resolved Vibrational Spectroscopy IV. Berlin: Springer-Verlag, 1993:63–67.

143. Heilweil EJ, Casassa MP, Cavanagh RR, Stephenson JC. Picosecond vibrational energy transfer studies of surface adsorbates. Annu Rev Phys Chem 1989; 40:143–171.

144. Volk M, Kholodenko Y, Lu HSM, Gooding EA, DeGrado WF, Hochstrasser RM. Peptide conformational dynamics and vibrational Stark effects following photoinitiated disulfide cleavage. J Phys Chem B 1997; 101:8607–8616.

145. Behrendt R, Renner C, Schenk M, Wang F, Wachtveitl J, Oesterhelt D, Moroder L. Photomodulation of conformational states of cyclic peptides with a backbone-azobenzene moiety. Angew Chem Int Ed Engl 1999; 38:2771–2774.

146. Asplund MC, Zanni MT, Hochstrasser RM. Two-dimensional infrared spectroscopy of peptides by phase-controlled femtosecond vibrational photon echoes. Proc Natl Acad Sci USA 2000; 97:8219–8224.

8

Two-Dimensional Coherent Infrared Spectroscopy of Vibrational Excitons in Polypeptides

Andrei Piryatinski, Vladimir Chernyak, and Shaul Mukamel
University of Rochester, Rochester, New York

I. INTRODUCTION

The infrared absorption of proteins and polypeptides in the amide I $(1600-1700 \text{ cm}^{-1})$ spectral region originates from the stretching motion of the peptide CO bond coupled to N–H bending and C–H stretching. This mode has a strong ($\sim$0.4 D) transition dipole moment and is clearly distinguishable from other vibrational modes of the amino acid side chains. Early study of symmetrical model polypeptides conducted by Krimm and Bandeker (1) demonstrated that the dipole-dipole interaction between the CO-stretching modes results in the delocalization of amide I states, which can be modeled as Frenkel vibrational excitons. Assuming dipole-dipole coupling between peptide groups, Torii and Tasumi (2) performed model calculations of the absorption lineshape for a few mid-size ($\sim$100 peptide) globular proteins with known structures and obtained good agreement with experiment. The dependence of the coupling on relative orientations and distances of the interacting dipoles results in a unique amide I band signature of the particular secondary structure motif. This is widely utilized in studying protein and polypeptide structures in the native state (3) and for monitoring the early events in protein folding (4–8).

349

In linear spectroscopy, the signal depends on a single frequency (e.g, absorption) or time (e.g., free induction decay) variable and can therefore be considered one-dimensional (1D). The amount of information it carries is limited and, except for very simple systems, does not lend itself to a unique modeling. The information extracted from 1D infrared (IR) spectra is limited since proteins usually fold into a complex three-dimensional structure, consisting of several polypeptide segments forming different types of secondary structures. The amide I band thus consists of a number of unresolved spectral lines associated with vibrational motions of different structural elements. Conformational fluctuations within a particular three-dimensional protein structure and local interaction with solvent induce inhomogeneous broadening, and the spectrum is typically highly congested. Fourier-transform infrared (FTIR) spectroscopy has been employed to somewhat improve the resolution of these spectra (3).

Nonlinear visible and infrared techniques offer a broad range of novel spectroscopies, which can probe the structure and dynamics of complex molecules, aggregates, solvent-solute interactions, and liquids. Time-resolved nonlinear multiple-pulse techniques are multidimensional and commonly used in nuclear magnetic resonance (9,10), electron spin resonance (11), and in the microwave. Extremely valuable microscopic information can be revealed by varying time delays between pulses, carrier frequencies, phases, and envelopes. The multidimensional nature of the signals has not been fully appreciated in the early days of optical spectroscopy, and experiments have been interpreted in terms of simple few-level systems with phenomenological homogeneous or inhomogeneous broadening. Only with recent experimental and theoretical advances has this multidimensionality been realized, and a variety of multidimensional off-resonant Raman (12,13) and resonant optical and IR techniques has been developed (14–18).

In impulsive multidimensional (ND) Raman spectroscopy a sample is excited by a train of N pairs of optical pulses, which prepare a wavepacket of quantum states. This wavepacket is probed by the scattering of the probe pulse. The electronically off-resonant pulses interact with the electronic polarizability, which depends parametrically on the vibrational coordinates (19), and the signal is related to the $2N + 1$ order nonlinear response (18). Seventh-order three-dimensional (3D) coherent Raman scattering, technique has been proposed by Loring and Mukamel (20) and reported in Refs. 12 and 21. Fifth-order two-dimensional (2D) Raman spectroscopy, proposed later by Tanimura and Mukamel (22), had triggered extensive experimental (23–28) and theoretical (13,25,29–38) activity. Raman techniques have been reviewed recently (12,13) and will not be discussed here.

In resonant infrared multidimensional spectroscopies the excitation pulses couple directly to the transition dipoles. The lowest order possible technique in noncentrosymmetrical media involves three-pulses, and is, in general, three dimensional (Fig. 1A). Simulating the signal requires calculation of the third-order response function. In a small molecule this can be done by applying the sum-over-states expressions (see Appendix A), taking into account all possible Liouville space pathways described by the Feynman diagrams shown in Fig. 1B. The third-order response of coupled anharmonic vibrations depends on the complete set of one- and two-exciton states coupled to thermal bath (18), and the sum-over-states approach rapidly becomes computationally more expensive as the molecule size is increased.

An alternative quasiparticle description of the optical response is possible using the nonlinear exciton equations (NEE) (39). The response function is then represented in terms of one-exciton Green functions and exciton-exciton scattering matrix. Four coherent ultrafast 2D techniques have been proposed (16,17), and computer simulations of the 2D response were performed for model aggregates made out of a few two-level chromophores.

Experimental 2D IR PE study of stretching motion of carbon monoxide in myoglobin-CO and rare-earth carbonyls were carried out by Fayer and coworkers (14). The echo beats were observed and the relaxation parameters were determined. The vibrational motion of each CO group was modeled as an anharmonic three-level system. The fifth-order contribution to the PE signal from CO modes in hemoglobin-CO has been studied as well (40). Incoherent 2D IR studies of the amide I spectral region of small proteins such as apamin, scyllatoxin, and bovine pancreatic trypsin inhibitor (BPTI) has been reported by Hamm et al., (15) who had employed femtosecond pump-probe and dynamic hole-burning techniques. The anharmonicity, relaxation, and energy equilibration times have been measured, and the disorder-induced delocalization length of vibrational excitons has been estimated. Similar to 2D NMR spectroscopy (9,10), the sensitivity of the 2D IR signal to protein geometry can be used for structure determination. This has been demonstrated experimentally by dynamic hole-burning measurements on a model pentapeptide *cyclo*-Abu-Arg-Gly-Asp-Mamb molecule (41), whose structure is known from x-ray and nuclear magnetic resonance (NMR) study (42). The cross-peak positions and intensities were detected and used to determine the coupling energies between the amide groups. Computer simulations of the 2D IR PE signal from glycine dipeptide, based on the NEE approach, were carried out (43). Different models of spectral

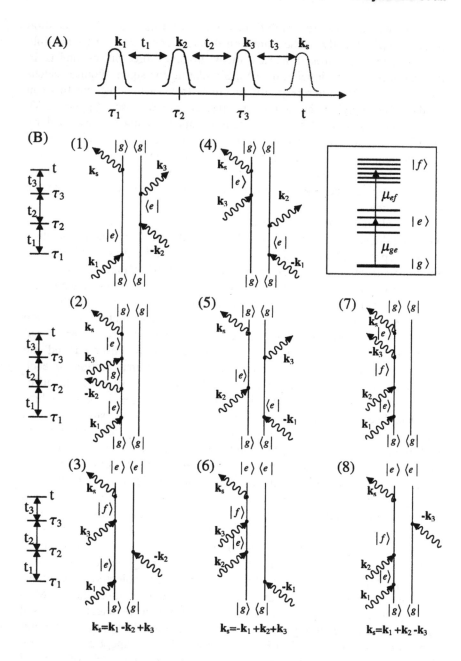

(A)

(B)

broadening show completely different 2D signals, even when the 1D infrared spectra are very similar. We have further demonstrated how the phase of the 2D signal may be used to distinguish between different types of the two-exciton states.

In this chapter we survey the basic ideas behind the theoretical interpretation of resonant multidimensional spectroscopies. For specificity and clarity we focus on three-pulse infrared techniques that probe vibrational motions. However, the same ideas apply in the visible for electronic excitations. In Section II we survey different detection modes of time domain and three-pulse femtosecond spectroscopies and represent different signals in terms of the third-order response function. The response function is expressed using the single exciton Green function and the exciton-exciton scattering matrix in Section III. The structure of the one- and two-vibrational-exciton manifolds of cyclic pentapeptide and different models for line broadening are described in Section IV. Section V presents simulations of the 2D PE signal. By considering the absolute value as well as the real and imeginary parts of the signal in the heterodyne detection mode, the signatures of one- and two-exciton dynamics and the various line broadening models are studied. Final discussion and conclusions are given in Section VI.

II. THREE-PULSE MULTIDIMENSIONAL FEMTOSECOND OPTICAL SPECTROSCOPIES

In a four-wave mixing experiment, the induced polarization, to third order in the driving field, is given by (18):

$$P^{(3)}(t) = \int_{-\infty}^{t} d\tau_3 \int_{-\infty}^{\tau_3} d\tau_2 \int_{-\infty}^{\tau_2} d\tau_1 R(t, \tau_3, \tau_2, \tau_1)$$
$$\times \mathcal{E}(\mathbf{r}, \tau_3)\mathcal{E}(\mathbf{r}, \tau_2)\mathcal{E}(\mathbf{r}, \tau_1) \tag{1}$$

Figure 1 (A) Pulse sequence in three-pulse experiment. (B) Double-sided Feynman diagrams representing the Liouville space pathways contributing to the third-order response. In the rotating wave approximation each diagram contributes to a four-wave-mixing signal in a distinct direction (combination of wave vectors), as indicated. Pathways (1), (2), (4), and (5) run over one-exciton states $|e\rangle$ only and (3), (6)–(8) over one-($|e\rangle$) and two-exciton $|f\rangle$ states shown in the insert. The various 2D techniques (17) are determined by the following diagrams: photon echo $(t_2 = 0)$ by (4)–(6) (see also Fig. 5), reverse photon echo $(t_1 = 0)$ by (7), (8), transient grating $(t_1 = 0)$ by (1), (3), (5), (6), and reverse transient grating $(t_2 = 0)$ by (1), (2), (6), (7), (8).

where the total field $\mathcal{E}(\mathbf{r}, t)$ consists of three applied fields:

$$\mathcal{E}(\mathbf{r}, \tau) = \sum_{j=1}^{3} [E_j(\tau) \exp(i\mathbf{k}_j\mathbf{r} - i\omega_j\tau) + E_j^*(\tau) \exp(-i\mathbf{k}_j\mathbf{r} + i\omega_j\tau)] \quad (2)$$

where ω_j, $\mathbf{k}_j$, and E_j is the frequency, wave vector, and envelope of the jth field, respectively. Even though a polypeptide is typically small compared to the optical wavelength, the sample is larger than the wavelength, and the optical signal is subjected to the phase-matching condition (18). This implies that the four-wave mixing signal may only be generated in the directions defined by the wave vectors $\mathbf{k}_s = \pm\mathbf{k}_1 \pm \mathbf{k}_2 \pm \mathbf{k}_3$. For resonant techniques the signal is dominated by contributions that satisfy the rotating wave approximation (RWA). All other terms are highly oscillatory and may be neglected. The three columns in Fig. 1(B) show the Feynman diagrams corresponding to the three wavevectors which survive the RWA for the present model.

We next survey the possible detection modes in a three pulse experiment. The excitation pulses come at times $\tau_1 = t - t_1 - t_2 - t_3$, $\tau_2 = t - t_2 - t_3$, and $\tau_3 = t - t_3$, where t_1 and t_2 are delay times between the pulses and t_3 is the delay time between the last excitation pulse and actual time when the polarization is measured (Fig. 1A). Equation (1) can be recast in the following form:

$$P^{(3)}(t) = \int_0^\infty dt_3 \int_0^\infty dt_2 \int_0^\infty dt_1 \mathcal{R}(t_3, t_2, t_1)\mathcal{E}(t - t_1 - t_2 - t_3)$$
$$\times \mathcal{E}(t - t_2 - t_3)\mathcal{E}(t - t_3) \quad (3)$$

where $\mathcal{R}(t_3, t_2, t_1) \equiv 6R(t, t - t_3, t - t_2 - t_3, t - t_1 - t_2 - t_3)$. The third-order response function is given by

$$R(t, \tau_3, \tau_2, \tau_1) \equiv i^3\text{Tr}\{\hat{P}(t)[\hat{P}(\tau_3), [\hat{P}(\tau_2), [\hat{P}(\tau_1), \overline{\rho}]]]\}, \quad (4)$$

and

$$\mathcal{R}(t_3, t_2, t_1) \equiv i^3\text{Tr}\{\hat{P}(t_3 + t_2 + t_1)[\hat{P}(t_2 + t_1), [\hat{P}(t_1), [\hat{P}(0), \overline{\rho}]]]\} \quad (5)$$

where $\hat{P}(\tau)$ are the polarization operators in the Heisenberg representation and $\overline{\rho}$ is the equilibrium density matrix.

Eq. (5) has eight terms coming from the three commutators. For each choice of a wavevector, only some of these terms survive the RWA. Let us consider an impulsive technique involving short pulses and denote the relevant terms of $\mathcal{R}$ by $\mathcal{R}'$. Various detection modes that probe different projections of the third-order response function $\mathcal{R}(t_3, t_2, t_1)$ may be employed. The time-integrated homodyne signal is given by

$$S_{\text{hom}}(t_2, t_1) = \int dt_3 |\mathcal{R}'(t_3, t_2, t_1)|^2 \tag{6}$$

The time-gated homodyne signal is given by

$$S_{\text{TG}}(t_3, t_2, t_1) = |\mathcal{R}'(t_3, t_2, t_1)|^2 \tag{7}$$

Frequency dispersed homodyne detection gives

$$S_{\text{FD}}(\omega_3, t_2, t_1) = |R(\omega_3, t_2, t_1)|^2 \tag{8}$$

where

$$R(\omega_3, t_2, t_1) = \int dt_3 \mathcal{R}'(t_3, t_2, t_1) \exp(i\omega_3 t_3) \tag{9}$$

The time-gated heterodyne signal obtained in a phase-locked measurement gives the real and the imaginary parts of the response function:

$$S_R(t_3, t_2, t_1) = \text{Re}\,\mathcal{R}'(t_3, t_2, t_1), \tag{10}$$
$$S_I(t_3, t_2, t_1) = \text{Im}\,\mathcal{R}'(t_3, t_2, t_1).$$

The most detailed detection is given by the Wigner spectrogram, which is bilinear in the response function (45–49):

$$S_{\text{RAM}}(\omega_3; t_3, t_2, t_1) = \int d\tau \mathcal{R}'(t_3 + \tau, t_2, t_1) \mathcal{R}'(t_3 - \tau, t_2, t_1) \exp(2i\omega_3\tau) \tag{11}$$

The time-gated (frequency-dispersed) signals are obtained by integrating the spectrogram over frequency (time), i.e.

$$S_{\text{TG}}(t_3, t_2, t_1) = \int dt_3 S_{\text{RAM}}(\omega_3; t_3, t_2, t_1) \tag{12}$$

and

$$S_{\text{FD}}(\omega_3, t_2, t_1) = \int dt_3 S_{\text{RAM}}(\omega_3; t_3, t_2, t_1) \tag{13}$$

In the next section we present a closed form expression for the vibrational response function using the Frenkel-exciton model.

III. THE THIRD-ORDER RESPONSE OF VIBRATIONAL EXCITONS

We model the amide band as a system of N interacting localized vibrations. For the sake of third-order spectroscopies, we only need to consider the lowest three levels of each peptide group with energies 0, Ω_m, Ω'_m ($m = 1, \ldots, N$). The matrix elements of the dipole operator corresponding to the 0-1 and 1-2 transitions are denoted μ_m and μ'_m, respectively, and their ratio is $\kappa_m \equiv \mu'_m/\mu_m$. To introduce the vibrational Frenkel exciton model, we define the exciton-oscillator operators (17,39,50):

$$\hat{B}_n^\dagger = |1\rangle_{nn}\langle 0| + \kappa_n|2\rangle_{nn}\langle 1| \tag{14}$$

with the commutation relations

$$[\hat{B}_m, \hat{B}_n^\dagger] = \delta_{mn}[1 - (2 - \kappa_m^2)\hat{B}_m^\dagger\hat{B}_m] \tag{15}$$

The polarization operator that describes the coupling to the driving field $\mathcal{E}(t)$ is then given by

$$\hat{P} = \sum_n \mu_n(\hat{B}_n + \hat{B}_n^\dagger) \tag{16}$$

The Hamiltonian can be represented in the form (17,39,50)

$$H = \sum_{mn} h_{mn}\hat{B}_m^\dagger\hat{B}_n + \sum_n \frac{g_n}{2}(\hat{B}_n^\dagger)^2(\hat{B}_n)^2 + H_b - \mathcal{E}(t)\hat{P} \tag{17}$$

where $h_{mn} = \delta_{mn}\Omega_m + J_{mn}$, with J_{mn} being the hopping matrix, and $g_m \equiv 2\kappa_m^{-2}[\Delta_m + (2 - \kappa_m^2)\Omega_m]$ is an exciton-exciton interaction energy, where $\Delta_m \equiv (\Omega'_m - 2\Omega_m)$ is the vibrational anharmonicity. This Hamiltonian describes excitons as oscillator (quasiparticle) degrees of freedom.* H_b represents a bath Hamiltonian. We shall not specify it and merely require that it conserves the number of excitons. The bath induces relaxation kernels. The structure of the final expression is independent of the specific properties of the bath; the latter only affects the microscopic expression for the relaxation kernels (51).

The response function has been calculated in Ref. 39. It depends on the following three Green functions:

$$G_{mn}(\tau) \equiv \theta(\tau)\text{Tr}\{\hat{B}_m(\tau)\hat{B}_n^\dagger(0)\bar{\rho}\} \tag{18}$$

---

* Both the commutation relations [Equation (15)] and the Hamiltonian [Equation (17)] contain higher order products of $\hat{B}_n^\dagger$ and $\hat{B}_n$. These higher terms do not contribute to the third-order optical response and were neglected.

$$G^{(2)}_{mn,kl} \equiv \theta(\tau)\text{Tr}\{\hat{B}_m(\tau)\hat{B}_n(\tau)\hat{B}^\dagger_k(0)\hat{B}^\dagger_l(0)\bar{\rho}\} \tag{19}$$

$$G^{(p)}_{mk,ln} \equiv \theta(\tau)\text{Tr}\{\hat{B}_m(\tau)\hat{B}^\dagger_k(0)\bar{\rho}\hat{B}_l(0)\hat{B}^\dagger_n(\tau)\} \tag{20}$$

where $\bar{\rho}$ is the equilibrium density matrix, $\hat{B}(\tau)[\hat{B}^\dagger(\tau)]$ are the annihilation [creation] operators in the Heisenberg representation, and $\theta(\tau)$ is the Heavyside step function ($\theta(\tau) = 0$ for $\tau < 0$ and $\theta(\tau) = 1$ for $\tau > 0$).

$G_{mn}(\tau)$ is the one-exciton Green function. $G^{(2)}$ represents the time evolution of the two exciton states and satisfies the Bethe-Salpeter equation:

$$G^{(2)}_{mn,kl} = G_{mk}(\tau)G_{nl}(\tau) + G_{ml}(\tau)G_{nk}(\tau) + \sum_{pqp'q'}\int_0^\tau d\tau'' \int_0^{\tau''} d\tau'$$

$$G_{mp}(\tau - \tau'')G_{nq}(\tau - \tau'')\bar{\Gamma}_{pq,p'q'}(\tau'' - \tau')G_{p'k}(\tau')G_{q'l}(\tau')$$

where $\bar{\Gamma}$ is the exciton-exciton scattering matrix. The response function will be expressed in terms of $\bar{\Gamma}$ rather than $G^{(2)}$. The third Green function $G^{(p)}$ represents the evolution of the exciton density matrix $\langle B^+_n B_m\rangle$ (populations and coherences). We have recast it in the form:

$$G^{(p)}_{mk,ln} = G_{mk}(\tau)G^\dagger_{nl}(\tau) + \bar{G}_{mn,kl}(\tau)$$

The first term represents the factorized evolution in terms of exciton amplitude $\langle B^+B\rangle = \langle B^+\rangle\langle B\rangle$. $\bar{G}$ is the irreducible (unfactorized) part of the Green function. The response function will be expressed in terms of $\bar{G}$ rather than $G^{(p)}$.

In summary, the response function depends on the exciton Green function G, the exciton-exciton scattering matrix Γ, and the irreducible part $\bar{G}$ of the density matrix Green function. It has two components:

$$R(t, \tau_3, \tau_2, \tau_1) = R_c(t, \tau_3, \tau_2, \tau_1) + R_i(t, \tau_3, \tau_2, \tau_1) \tag{21}$$

The coherent part is

$$R_c(t; \tau_3, \tau_2, \tau_1) = (-i^3)\sum_{\text{perm}}\sum \mu_n\mu_{m_1}\mu_{m_2}\mu_{m_3}\int_{-\infty}^\infty d\tau''$$

$$\times \int_{-\infty}^\infty d\tau' G_{n'm_1}(\tau' - \tau_1)G_{m'm_2}(\tau' - \tau_2)G^\dagger_{m_3m''}(\tau'' - \tau_3)G_{nn''}(t - \tau'')$$

$$\times \bar{\Gamma}_{n''m'',n'm'}(\tau'' - \tau') + \text{c.c.} \tag{22}$$

where $\sum_{\text{perm}}$ denotes summation over the six permutations of the times τ_1, τ_2, and τ_3.* The incoherent part is

$$R_i(t; \tau_3, \tau_2, \tau_1) = 2 \sum_{\text{perm}} \sum \mu_n \mu_{m_1} \mu_{m_2} \mu_{m_3} \int_{-\infty}^{\infty} d\tau''' \int_{-\infty}^{\infty} d\tau'' \int_{-\infty}^{\infty} d\tau'$$

$$G_{nn''}(t - \tau'') G_{n'm_3}(\tau' - \tau_3) G_{im''}^{\dagger}(\tau'' - \tau') \overline{\Gamma}_{n''m'',n'm'}(\tau'' - \tau')$$

$$\times \overline{G}_{m'i,kj}(\tau' - \tau''') [G_{km_1}(\tau''' - \tau_1) \delta_{m_2 j} \delta(\tau''' - \tau_2)$$

$$\times G_{m_2 j}^{\dagger}(\tau''' - \tau_2) \delta_{m_1 k} \delta(\tau''' - \tau_1)] + \text{c.c.} \tag{23}$$

The advantage of the present NEE Green function approach is that modeling of the two-exciton dynamics requires calculation of $N \times N$ exciton scattering matrix, where N is the number of peptide units. Computational time of the coherent component of the 2D signal scales as $\sim N^4$, allowing application to larger polypeptides and averaging over a sufficient number of Monte Carlo runs to account for static disorder. Moreover, interference effects are naturally built in, making it particularly suitable for inverting 2D signals to yield the structure and dynamic parameters. The sum-over-state expressions for the third-order response are given in Appendix A. In this approach the calculation of the response function [Equations (34)–(36)] requires the diagonalization of full two-exciton Hamiltonian, presented by a $N^2 \times N^2$ matrix. This corresponds to the total computational time of the 2D signal to scale as $\sim N^6$. The NEE, which take into account the coupling with a thermal bath, makes it possible to model the exciton relaxation dynamics as well. In particular, the calculations based on the exciton scattering matrix account for the renormalization of the two-exciton dephasing rate (provided $2\Gamma \neq \gamma^{(2)}$) determined by the relative contribution of the overtone and collective doubly-excited states to a given two exciton state, as defined in the next section.

In the next section we specify the parameters of the vibrational exciton Hamiltonian for cyclic pentapeptide and discuss the structure of one- and two-vibrational-exciton manifold.

IV. VIBRATIONAL EXCITONS IN CYCLIC PENTAPEPTIDE

The structure of several cyclic pentapeptides has been investigated in solution using 2D NMR spectroscopy and compared with the crystal x-ray

---

* The τ' and τ'' integrations run from $-\infty$ to ∞ since the Green functions $G(\tau)$ and the scattering matrix $\overline{\Gamma}(\tau)$ are nonzero for $\tau > 0$ only.

structures (42). The structure is not unique, and several conformations are compatible with the NMR and x-ray measurements. In our study we used the crystallographic structure to obtain the atomic coordinates of *cyclo*(Abu-Arg-Gly-Asp-Mamb) shown in Fig. 2. The backbone conformation traces out a rectangular shape with a β-turn centered at the Abu-Arg bond.

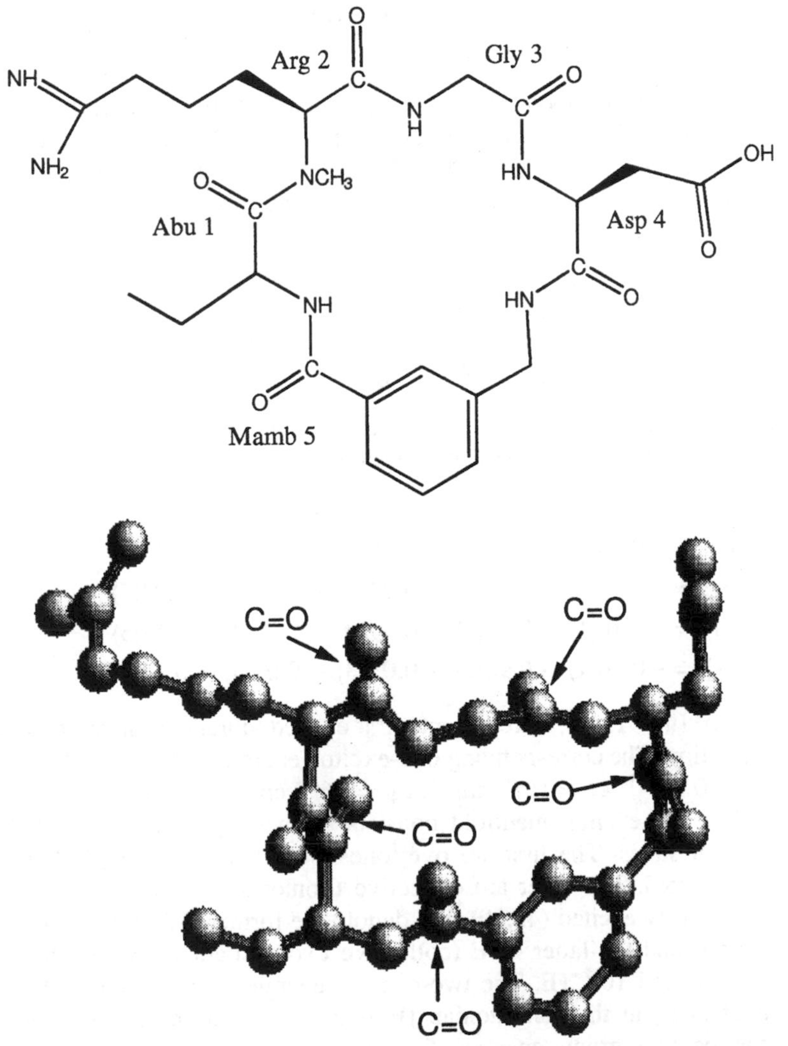

Figure 2 3D structure of the pentapeptide. (From Ref. 42.)

To obtain the one-exciton Hamiltonian we used the central frequencies for the peptide CO vibrations reported in Ref. 41 and assigned $\Omega_1 = 1588$ cm^{-1} to Abu-Arg, $\Omega_2 = 1671$ cm^{-1} to Arg-Gly, $\Omega_3 = 1648$ cm^{-1} to Gly-Asp, $\Omega_4 = 1610$ cm^{-1} to Asp-Mamb, and $\Omega_5 = 1618$ cm^{-1} to Mamb-Abu. The dipole-dipole couplings among CO vibrations were calculated from

$$J_{mn} = \frac{(\mu_m \cdot \mu_n) - 3(\hat{m} \cdot \mu_m)(\hat{m} \cdot \mu_n)}{|R_{mn}|^3} \qquad m, n = 1, \ldots, 5 \qquad (24)$$

by assigning each dipole on a CO bond 0.868 Å from the carbon atom and forming 25° angle with respect to the bond. The absolute value of each dipole moment is 0.37 D (1,2,15,41). Using these parameters, the one-exciton Hamiltonian (in cm^{-1}) assumes the form

$$h = \begin{pmatrix} 1588 & 7.2 & 5.7 & -1.7 & -7.0 \\ 7.2 & 1671 & -0.7 & 0.6 & -7.6 \\ 5.7 & -0.7 & 1648 & 2.2 & -6.2 \\ -1.7 & 0.6 & 2.2 & 1610 & 0.3 \\ -7.0 & -7.6 & -6.2 & 0.3 & 1618 \end{pmatrix} \qquad (25)$$

Since $|\Omega_n - \Omega_m| > J_{nm}$, each one-exciton state is a weakly perturbed localized CO vibration. The one-exciton eigenstates are

$$|e_1\rangle = 0.98|1\rangle - 0.07|2\rangle - 0.07|3\rangle + 0.08|4\rangle + 0.18|5\rangle \qquad (26)$$
$$|e_2\rangle = -0.05|1\rangle - 0.03|2\rangle - 0.08|3\rangle + 0.98|4\rangle - 0.17|5\rangle$$
$$|e_3\rangle = 0.16|1\rangle - 0.15|2\rangle - 0.21|3\rangle - 0.17|4\rangle - 0.94|5\rangle$$
$$|e_4\rangle = -0.10|1\rangle + 0.07|2\rangle - 0.97|3\rangle - 0.05|4\rangle + 0.19|5\rangle$$
$$|e_5\rangle = -0.10|1\rangle - 0.98|2\rangle - 0.03|3\rangle - 0.01|4\rangle + 0.15|5\rangle$$

where $|n\rangle$, $(n = 1, \ldots, 5)$ represents first excited vibrational state of the nth CO vibration. The corresponding one-exciton energies are $\varepsilon_1 = 1586$ cm^{-1}, $\varepsilon_2 = 1610$ cm^{-1}, $\varepsilon_3 = 1617$ cm^{-1}, $\varepsilon_4 = 1650$ cm^{-1}, and $\varepsilon_5 = 1673$ cm^{-1}.

The two-exciton manifold consists of two types of doubly excited vibrational states. The first are overtones (local), where a single bond is doubly excited. The other are collective (nonlocal), where two bonds are simultaneously excited (43,50). We denote the former OTE (overtone two-excitation) and the latter CTE (collective two-excitation). A pentapeptide has 5 OTE and 10 CTE. The two-exciton energies are determined by the parameters g_n in the Hamiltonian [Equation (17)], which in turn depend on the peptide group energies Ω_n, the anharmonicity Δ_n, and dipole moment ratio κ_n, $n = 1, \ldots, 5$. We set them equal for all CO units and $\Delta = -16$ cm^{-1} adopted from the experiment (15,41) and $\kappa = \sqrt{2}$. The two-exciton energies are obtained from the poles of the exciton scattering matrix (rather than a direct diagnolization of the 15×15 two-exciton Hamiltonian, which is much more expensive). They are $\bar{\varepsilon}_1 = 3157$ cm^{-1}, $\bar{\varepsilon}_2 = 3195$ cm^{-1}, $\bar{\varepsilon}_3 = 3200$ cm^{-1}, $\bar{\varepsilon}_4 = 3205$ cm^{-1}, $\bar{\varepsilon}_5 = 3220$ cm^{-1}, $\bar{\varepsilon}_6 = 3227$ cm^{-1}, $\bar{\varepsilon}_7 = 3235$ cm^{-1}, $\bar{\varepsilon}_8 = 3258$ cm^{-1}, $\bar{\varepsilon}_9 = 3259$ cm^{-1}, $\bar{\varepsilon}_{10} = 3264$ cm^{-1}, $\bar{\varepsilon}_{11} = 3283$ cm^{-1}, $\bar{\varepsilon}_{12} = 3287$ cm^{-1}, $\bar{\varepsilon}_{13} = 3288$ cm^{-1}, $\bar{\varepsilon}_{14} = 3322$ cm^{-1}, and $\bar{\varepsilon}_{15} = 3331$ cm^{-1}. In general the two-exciton eigenstates are linear combinations of the OTE and the CTE. However, since $|\Omega'_m - 2\Omega_n| > \kappa^2 J_{nm}$, most two-exciton states can be classified as weakly perturbed OTE or CTE type.

Having introduced the one- and two-exciton states, we next turn to the line broadening. We denote the dephasing rate of the first vibrational transition by Γ and the overtone by $\gamma^{(2)}$. In all calculations Γ and $\gamma^{(2)}$ are set identical for all peptide groups. The anharmonicity $\Delta = -16$ cm^{-1} is fixed and independent of disorder. We have employed six models:

A. Small homogeneous dephasing rates $\Gamma = 0.2$ cm^{-1} and $\gamma^{(2)} = 0.4$ cm^{-1}.

B. Large homogeneous dephasing rates $\Gamma = 5$ cm^{-1} and $\gamma^{(2)} = 10$ cm^{-1}, which correspond typically to experimental values (15,41).

C. Static diagonal disorder. The n'th peptide energy is represented as

$$\Omega_n = \overline{\Omega}_n + \xi_n \qquad n = 1, \ldots, 5 \tag{27}$$

where $\overline{\Omega}_n$ is average energy of the n'th peptide group set to the central frequencies in the one-exciton Hamiltonian [Equation (25)]. The random variables ξ_n, representing energy disorder, are assumed to be uncorrelated random Gaussian variables with variance $\sigma_d = 12$ cm^{-1} equal for all peptide groups: $\Gamma = 0.2$ cm^{-1}, $\gamma^{(2)} = 0.4$ cm^{-1}.

D. Same as model 3 except that the homogeneous dephasing rates of each peptide are adopted from experiment and set to $\Gamma = 5$ cm^{-1}, $\gamma^{(2)} = 10$ cm^{-1}.

E. Static off-diagonal disorder. The exciton coupling is given by

$$J_{mn} = \bar{J}_{mn} + \zeta_{mn} \qquad n \neq m; \quad n, m = 1, \ldots, 5 \tag{28}$$

where $\bar{J}_{mn}$ is average coupling energy between the m'th and the nth peptide groups, whose values are given in Equation (25). ζ_{mn} are uncorrelated Gaussian random variables with equal variances

$\sigma_{od} = 12$ cm^{-1} for all ζ_{mn}, (n $\neq$ m; n, m = 1, ..., 5). The homogeneous dephasing rates are $\Gamma = 0.2$ cm^{-1}, $\gamma^{(2)} = 0.4$ cm^{-1}.

F.   Same as model 5 except that the homogeneous dephasing rates are taken from experiment and set to $\Gamma = 5$ cm^{-1}, $\gamma^{(2)} = 10$ cm^{-1}.

Different peptide groups in the pentapeptide have inhomogeneous broadening, which varies in the range of $\sigma \sim 3$–12 cm^{-1} (41). Some of the experimentally observed lines are therefore dominated by homogeneous and others by inhomogeneous broadening. We introduced these different models in order to study the signature homogeneous and inhomogeneous broadening on the 2D PE spectra. In models (A), (C), and (E) we used small homogeneous broadening in order to resolve all resonances. Some of these resonances are not resolved in models (B), (D), and (F) which use larger, more realistic, homogeneous broadening.

The degree of one-exciton state localization can be described by the inverse participation ratio (52–54):

$$P(\varepsilon) = \left\langle \sum_{n=1}^{5} |\psi_\varepsilon(n)|^4 \right\rangle^{-1} \tag{29}$$

where $\psi_\varepsilon(n)$ is the n'th component of one-exciton wavefunction with energy in the interval $[\varepsilon, \varepsilon + d\varepsilon]$. For our model $P(\varepsilon)$ may vary between $P = 1$ (localized state) and $P = 5$ (delocalized state). The participation ratio distribution as well as the density of states are shown in Fig. 3. For models A and B the exciton states are well localized, since $|\Omega_n - \Omega_m| < J_{nm}$. In models C and D diagonal disorder slightly increases the one-exciton state delocalization, resulting in the participation ratio ~ 1.3 in the maximum of the density of states (dotted line in the plot), and in models E and F the off-diagonal disorder corresponds to the state delocalization within ~ 2 peptide groups near the density of states (dotted line) maxima.

The linear (1D) absorption spectra of all models are presented in Fig. 4. Model A shows five well-resolved one-exciton lines. In model B the lines ε_2 and ε_3 are poorly resolved due to the increased homogeneous broadening. Diagonal disorder in models C and D further broadens the spectra. Since off-diagonal disorder induces state delocalization, the one-exciton resonances shift for models E and F and become $\varepsilon'_1 = 1578$ cm^{-1}, $\varepsilon'_2 = 1605$ cm^{-1}, $\varepsilon'_3 = 1618$ cm^{-1}, $\varepsilon'_4 = 1652$ cm^{-1}, and $\varepsilon'_5 = 1679$ cm^{-1}.

Additional information related to the one- and two-exciton dynamics can be obtained from the 2D spectra, as will be shown in the following section.

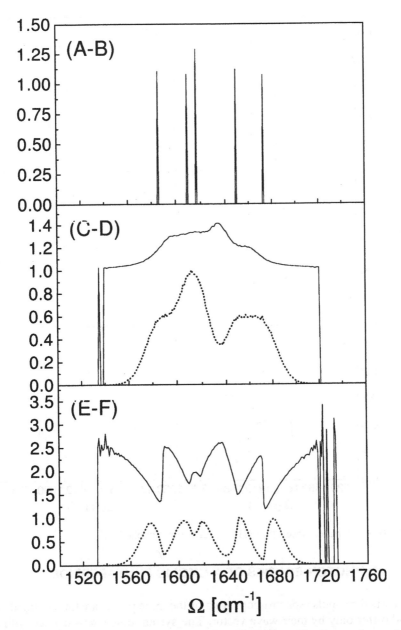

Figure 3 Solid line: Inverse participation ratio of one-exciton states for models A–F. Dotted line: Density of one-exciton states for models C–F.

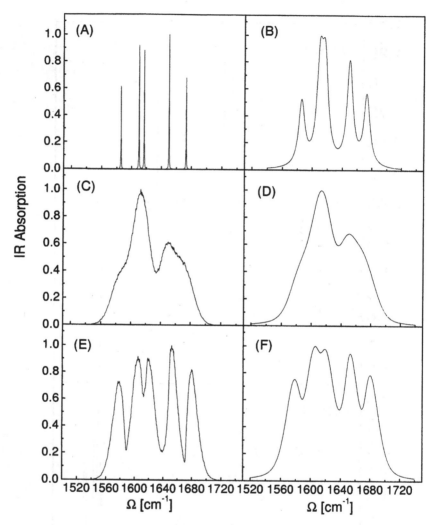

Figure 4　Infrared absorption (1D) spectra for models A–F.

V.　2D PHOTON ECHOES OF A CYCLIC PENTAPEPTIDE

In a 2D three-pulse spectroscopy, two of the three pulses are time-coincident and differ only by their wave vector. The system thus interacts once with a single pulse and twice with a pulse pair. In the 2D photon echo technique we set $t_2 = 0$. We consider the heterodyne signal [Equation (10)]. This signal can be measured by mixing the third-order signal with the heterodyne pulse arriving with delay time t_3 in the direction determined by the phase matching conditions $\mathbf{k}_s = \mathbf{k}_3 + \mathbf{k}_2 - \mathbf{k}_1$. We shall display the 2D PE signal in the frequency domain by performing a double Fourier transform:

$$S(\Omega_2, \Omega_1) = \int_0^\infty dt_3 \int_0^\infty dt_1 \exp(i\Omega_2 t_3 + i\Omega_1 t_1) S(t_3, 0, t_1). \qquad (30)$$

The 2D PE signal is computed using the response function given by Equations (21)–(23). Since we consider the 2D response on the time scale smaller than the dephasing times, only the coherent component of the response function [Equation (21)] contributes to the signal. The 2D Fourier transform PE signal determined by Equations (22) and (30) has the following form (17):

$$S(\Omega_2, \Omega_1) = S^{(1)}(\Omega_2, \Omega_1) + S^{(2)}(\Omega_2, \Omega_1) \qquad (31)$$

The first component,

$$S^{(1)}(\Omega_2, \Omega_1) = \frac{i}{2} \sum_{abcd} \mu_a \mu_b \mu_c \mu_d \frac{1}{\Omega_2 - \varepsilon_d + i\Gamma} \frac{1}{\Omega_1 + \varepsilon_c + i\Gamma}$$

$$\times \frac{\overline{\Gamma}_{cd,ab}(\varepsilon_c + \varepsilon_d)}{\varepsilon_c + \varepsilon_d - (\varepsilon_a + \varepsilon_b) + 2i\Gamma} \qquad (32)$$

represents correlations between one-exciton states shown by the Feynman diagram (Fig. 5(1)). The second component,

$$S^{(2)}(\Omega_2, \Omega_1) = \sum_{abcd} \mu_a \mu_b \mu_c \mu_d \frac{1}{\Omega_1 + \varepsilon_c + i\Gamma} \frac{1}{2\pi} \int d\omega \overline{\Gamma}_{cd,ab}(\omega - i\gamma_0)$$

$$\times \frac{1}{\omega - \Omega_2 - \varepsilon_c - i(\Gamma + \gamma_0)} \frac{1}{\omega - (\varepsilon_a + \varepsilon_b) + i(2\Gamma - \gamma_0)}$$

$$\times \frac{1}{\omega - (\varepsilon_c + \varepsilon_d) - i\gamma_0} \qquad (33)$$

is induced by correlations between one- and two-exciton states and is represented by the Feynman diagram in Fig. 5(2).*

Below we first analyze the absolute value of the 2D PE signal and then consider its phase by looking at the real and the imaginary parts separately.

---

* The components of the signal calculated according to these diagrams, using the sum-over-state approach presented in Appendix A, coincide with Equations (32) and (33) in the narrow line limit $\Gamma \ll (\Delta', J)$

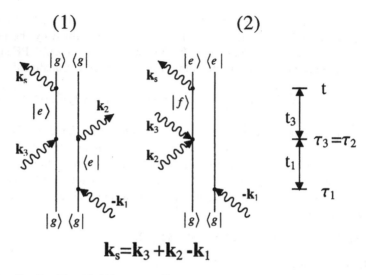

Figure 5 Double-sided Feynman diagrams representing the two Liouville space pathways contributing to photon echo representing (1) correlations between one-exciton states, and (2) correlations between one- and two-exciton states.

Numerical work involved the calculation of the exciton-exciton scattering matrix $\bar{\Gamma}$ and evaluating the integral in Equation (33). The 2D PE signal for models C–F was averaged over 10^5 disorder realizations.

A. Absolute Value of the 2D Signal

Figure 6 shows the calculated absolute value of the 2D PE signal $|S(\Omega_2, \Omega_1)|$. The resonances representing various correlations between one-exciton states $(-\varepsilon_a, \varepsilon_a)$ (diagonal peaks) and $(-\varepsilon_a, \varepsilon_b)$ a $\neq$ b and a, b = 1, ..., 5 (off-diagonal peaks) as well as between one- and two-exciton states $(-\varepsilon_a, \bar{\varepsilon}_b - \varepsilon_a)$, a, b = 1, ..., 5 (cross-peaks) are well resolved in model A. Model B has the same resonant energies, but since the homogeneous broadening is comparable with the anharmonicity and coupling energies, many resonances are unresolved. Five slices of the 2D spectra along the Ω_2 direction at $\Omega_1 = -\varepsilon_a$, a = 1, ..., 5, for models A and B are displayed in Fig. 7. The well-resolved resonances in model A are marked in the plot and identified in Table 1; the broader curves represent model B.

The 2D PE signal for models C and D (Fig. 6) shows inhomogeneously broadened resonances stretched along the $-\Omega_1 = \Omega_2$ direction. They represent self-correlations of the one-exciton states and

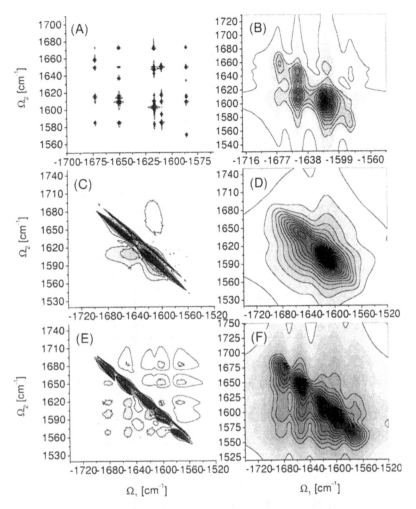

Figure 6 Absolute value $|S(\Omega_2, \Omega_1)|$ of 2D infrared photon echo signal for models A–F. (The different panels have different intensity codes. Light gray is zero and maximum value is dark gray.)

correlations between one-exciton and two-exciton (OTE) states on the same sites. The resonances representing correlations between different one-exciton states (off-diagonal peaks) and correlations between the one-exciton and two-exciton states make a negligible contribution to the spectra of models C and D. This is a consequence of the uncorrelated disorder

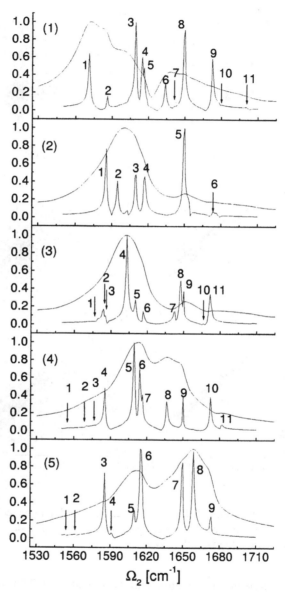

Figure 7 Slices of the signal of Fig. 6 for models A and B for (1) $\Omega_1 = -\varepsilon_1$, (2) $\Omega_1 = -\varepsilon_2$, (3) $\Omega_1 = -\varepsilon_1$, (4) $\Omega_1 = -\varepsilon_4$, and (5) $\Omega_1 = -\varepsilon_5$. Model A shows detailed structure; the broader curve corresponds to model B. All resonances of model A are identified in Table 1.

Table 1 Energy of the Resonances Identified in Fig. 7 for Model A

Ω_2 resonances	$\Omega_1 = -\varepsilon_1$	$\Omega_1 = -\varepsilon_2$	$\Omega_1 = -\varepsilon_3$	$\Omega_1 = -\varepsilon_4$	$\Omega_1 = -\varepsilon_5$
1.	$\bar{\varepsilon}_1 - \varepsilon_1$	$\bar{\varepsilon}_2 - \varepsilon_2$	$\bar{\varepsilon}_2 - \varepsilon_3$	$\bar{\varepsilon}_4 - \varepsilon_4$	$\bar{\varepsilon}_6 - \varepsilon_5$
2.	ε_1	$\bar{\varepsilon}_4 - \varepsilon_2$	$\bar{\varepsilon}_3 - \varepsilon_3$	$\bar{\varepsilon}_5 - \varepsilon_4$	$\bar{\varepsilon}_7 - \varepsilon_5$
3.	$\bar{\varepsilon}_2 - \varepsilon_1$	$\bar{\varepsilon}_5 - \varepsilon_2$	ε_1	$\bar{\varepsilon}_6 - \varepsilon_4$	$\bar{\varepsilon}_8 - \varepsilon_5$
4.	$\bar{\varepsilon}_3 - \varepsilon_1$	$\bar{\varepsilon}_6 - \varepsilon_2$	$\bar{\varepsilon}_5 - \varepsilon_3$	$\bar{\varepsilon}_7 - \varepsilon_4$	$\bar{\varepsilon}_{10} - \varepsilon_5$
5.	ε_3	ε_4	$\bar{\varepsilon}_6 - \varepsilon_3$	$\bar{\varepsilon}_9 - \varepsilon_4$	$\bar{\varepsilon}_{11} - \varepsilon_5$
6.	$\bar{\varepsilon}_5 - \varepsilon_1$	$\bar{\varepsilon}_{11} - \varepsilon_2$	ε_3	$\bar{\varepsilon}_{10} - \varepsilon_4$	$\bar{\varepsilon}_{13} - \varepsilon_5$
7.	$\bar{\varepsilon}_6 - \varepsilon_1$	—	$\bar{\varepsilon}_9 - \varepsilon_3$	ε_3	$\bar{\varepsilon}_{14} - \varepsilon_5$
8.	$\bar{\varepsilon}_7 - \varepsilon_1$	—	$\bar{\varepsilon}_{10} - \varepsilon_3$	$\bar{\varepsilon}_{11} - \varepsilon_4$	$\bar{\varepsilon}_{15} - \varepsilon_5$
9.	$\bar{\varepsilon}_8 - \varepsilon_1$	—	ε_4	ε_4	ε_5
10.	$\bar{\varepsilon}_{10} - \varepsilon_1$	—	$\bar{\varepsilon}_{11} - \varepsilon_3$	$\bar{\varepsilon}_{14} - \varepsilon_4$	—
11.	$\bar{\varepsilon}_{12} - \varepsilon_1$	—	$\bar{\varepsilon}_{13} - \varepsilon_3$	$\bar{\varepsilon}_{15} - \varepsilon_4$	—

distribution for different peptide groups, which does not allow the rephasing of the PE signal during the t_3 time delay (Fig. 5). We assumed that the anharmonicity is independent of disorder. Otherwise, for uncorrelated disorder, the OTE cross peaks would not be resolved.

Slices of the 2D signal of models C and D along Ω_2 at $\Omega_1 = -\varepsilon_a$, $a = 1, \ldots, 5$, are shown in Fig. 8. The lines in this plot are homogeneously broadened, since the photon echo technique removes the inhomogeneous broadening. Each slice of model C contains two distinct peaks. The narrow one (width Γ) represents self-correlation of the one-exciton states and the broader one (width $\gamma^{(2)} + \Gamma$) is due to the OTE state. Since the resonances present in each panel represent self-correlation of single excitons and of single exciton with the OTE excited on the same peptide group, the energy splitting of their maxima should provide an anharmonicity close to the anharmonicity of a single peptide unit Δ. In panels (1), (2), and (5) the anharmonicity is -15 cm^{-1} and in panel D it is -14 cm^{-1}. These are very close to the anharmonicity of a single peptide unit $\Delta = -16$ cm^{-1}, indicating that the corresponding OTE states are weakly perturbed by the weak coupling to the other two-exciton states. The anharmonicity -16 cm^{-1} in panel (3) indicates that the OTE with energy $\bar{\varepsilon}_5$ is decoupled from the other doubly excited states. The structure of spectra of model C suggests that fitting the spectrum of model D by two Lorentzian lines, one of each representing a diagonal peak, should provide the anharmonicity and dephasing rate of the two-exciton states.

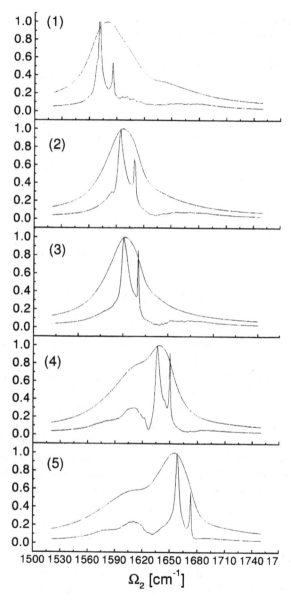

Figure 8 Slices of the signal of Fig. 6 for models C and D for (1) $\Omega_1 = -\varepsilon_1$, (2) $\Omega_1 = -\varepsilon_2$, (3) $\Omega_1 = -\varepsilon_1$, (4) $\Omega_1 = -\varepsilon_4$, and (5) $\Omega_1 = -\varepsilon_5$. Model C shows detailed structure; the broader curve corresponds to model D.

The 2D PE signal (Fig. 6) of models E and F is dominated by inhomogeneous broadening along the $-\Omega_1 = \Omega_2$ direction of the diagonal peaks and the cross peaks associated with the OTE states. In addition, weak off-diagonal and the cross-peaks are clearly seen in the plot. Slices of the 2D PE signal along Ω_2 at $\Omega_1 = -\varepsilon'_a$, $a = 1, \ldots, 5$, are shown in Fig. 9. As in Fig. 8, these spectra are homogeneously broadened; the broader curves represent model F and the underlying structure is clearly seen for model E. Self-correlations of one-exciton states are represented by the narrow lines and correlations between the one- and two-exciton states are given by the broader lines. In contrast to models C and D, the energy differences between the maxima of the one-exciton and strong two-exciton resonances are not equal to Δ, indicating that the relevant two-exciton states are not necessarily the localized OTE type. Panels (2) and (4) show two equally strong two-exciton lines close to the one-exciton line. They are clearly seen in the 2D PE plot of model E as two inhomogeneous resonances stretched in different directions which are close to $-\Omega_1 = \Omega_2$. The strong two-exciton lines seen in panels (1) and (4) of Fig. 9 correspond to the 2D resonances stretched in the direction close (but again not exactly equal to) $-\Omega_1 = \Omega_2$. This indicates that the two-exciton states in models E and F depend on the coupling J_{nm} and are delocalized. The weak signal observed in the 2D plot represents the off-diagonal peaks and the cross peaks inhomogeneously broadened in directions $\Omega_1 \sim \Omega_2$.

So far we have discussed the absolute value of the 2D PE signal. The phase of the signal (which can be observed as well) carries additional information (43). In particular, the phase (or equivalently the real and the imaginary parts of the signal) can explain the weak intensity of the cross peaks in models E and F. Below we examine the real and imaginary parts of the 2D PE signal.

B. Real and Imaginary Parts of the 2D Signal

The real and imaginary parts of the 2D PE signal are displayed in the left and the right columns, respectively, of Figs. 10 and 11. To clearly show the 2D resonance structure in model A, we display the signal on an expanded scale, showing Ω_2 resonances only at $\Omega_1 = -\varepsilon_4$. The diagonal peak 1 $(-\varepsilon_4, \varepsilon_4)$, and the cross peak 2 $(-\varepsilon_4, \bar{\varepsilon}_{12})$ due to the OTE are marked in the plot. The real part of the signal is dispersive in the $\Omega_1 = \Omega_2$ direction across the resonances. It approaches zero and changes sign (edge between the light gray and dark gray regions) along the line connecting the resonances in the $-\Omega_1 \sim \Omega_2$ direction. The imaginary part of the signal has positive (dark gray) and negative (light gray) maxima at the

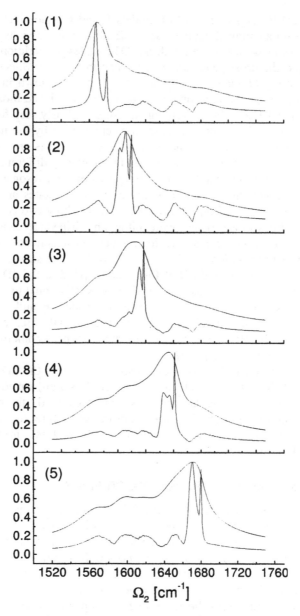

Figure 9 Slices of the signal of Fig. 6 for models E and F for (1) $\Omega_1 = -\varepsilon_1'$, (2) $\Omega_1 = -\varepsilon_2'$, (3) $\Omega_1 = -\varepsilon_3'$, (4) $\Omega_1 = -\varepsilon_4'$, and (5) $\Omega_1 = -\varepsilon_5'$. Model E shows detailed structure; the broader curve corresponds to model F.

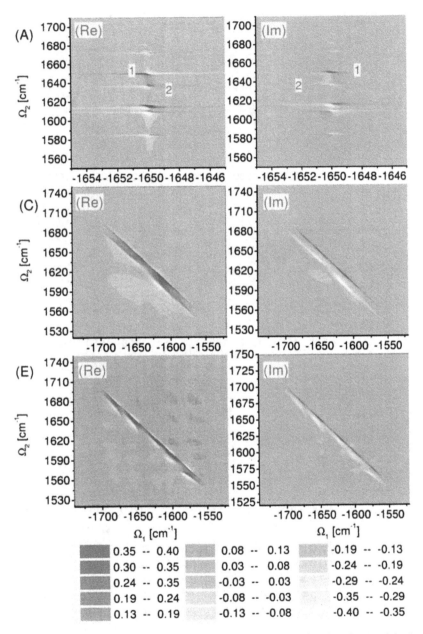

Figure 10   Real and imaginary parts of the 2D signal $S(\Omega_2, \Omega_1)$ for models A, C, and E (small homogeneous dephasing).

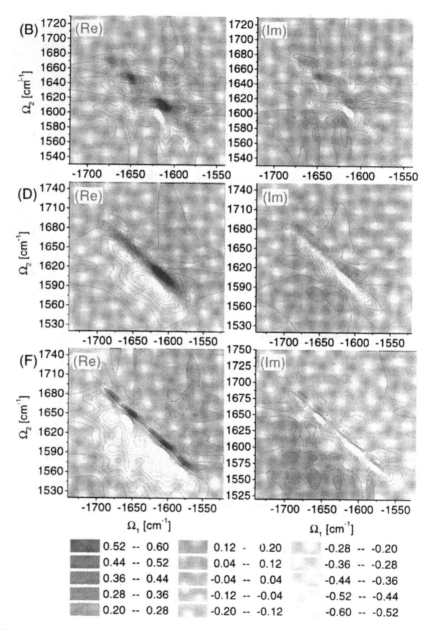

Figure 11 Real and imaginary parts of 2D signal $S(\Omega_2, \Omega_1)$ for models B, D, and F (large homogeneous dephasing).

resonances 1 and 2, respectively. It changes sign twice along the $\Omega_1 = \Omega_2$ direction and does not change sign in the $-\Omega_1 = \Omega_2$ direction. According to Equations (32) and (33), this indicates that the relevant components of the exciton scattering matrix are real. This is the case as long as $|2\varepsilon - \bar{\varepsilon}_a| \gg 3\Gamma$ ($a = 1, \ldots, 15$), which is possible provided $\Gamma \ll (\Delta, J)$. If $|2\varepsilon - \bar{\varepsilon}_a| \sim 3\Gamma$, the scattering matrix acquires an imaginary part. The real and imaginary parts of the signal are then superpositions of the real and imaginary parts marked 1 (or 2), whose relative contributions are determined by the real and imaginary parts of the exciton scattering matrix. For $\Gamma \ll (\Delta, J)$ the exciton scattering matrix has weak imaginary parts contributing to the OTE resonances and off-diagonal peaks, shown in Fig. 10 A. For $\Gamma \sim (\Delta, J)$ the scattering matrix has comparable real and imaginary parts corresponding to the homogeneously broadened signal (model B) shown in Fig. 11.

Since the real and the imaginary parts of the signal in models A and B change sign across the resonances in the direction $\Omega_1 = \Omega_2$ and remain positive or negative along $-\Omega_1 = \Omega_2$, for close disorder realizations the resonances should add up constructively in the directions $-\Omega_1 \sim \Omega_2$, and destructively in the perpendicular direction. Diagonal disorder (models C and D) corresponds to the inhomogeneous broadening of the diagonal peaks and the OTE cross peaks only in the direction $-\Omega_1 = \Omega_2$ corresponding to the signal shown in Figs. 10 and 11. By examining the real and imaginary parts of the signal of model C, we note that they have similar behaviour in the direction $\Omega_1 = \Omega_2$ to peaks 1 and 2 in model A and that the Ω_2 difference between the lines along which the signal changes sign (edges between the dark gray and light gray regions) reflects the magnitude of the anharmonicity Δ. For off-diagonal disorder, the inhomogeneously broadened resonances in the direction $-\Omega_1 = \Omega_2$ are clearly seen in models E and F. However, the off-diagonal and cross peaks induced by exciton delocalization are very weak, since they are inhomogeneously broadened in the vicinity of the $\Omega_1 = \Omega_2$ direction.

VI. DISCUSSION

In this chapter we surveyed the theoretical analysis of resonant multidimensional spectroscopies generated by the interaction of 3 fs pulses with a Frenkel exciton system. Closed expressions for the time-domain third-order response function derived by solving the NEE are given in terms of various exciton Green functions. Alternatively, the multidimensional time-domain signal can be calculated starting from the frequency domain the third-order optical susceptibility presented in Appendix B. Equations (43)–(45) only contain products of the exciton Green functions and the scattering matrix. The time domain response [Equations (21)–(23)] is more complex and contains convolutions of time domain Green functions and the scattering matrix. The third-order susceptibility can also be employed to derive the frequency domain counterpart of the 2D three pulse techniques proposed in Ref. 17.

We have applied these expressions to calculate the 2D PE signal from cyclic pentapeptide in the amide I spectral region. The one- and two-exciton states are well localized. Different models of homogeneous and inhomogeneous line broadening show distinct 2D patterns. From the absolute value of the signal we have identified all resonances associated with one-exciton correlations (diagonal and off-diagonal peaks) as well as those associated with one- and two-exciton state correlations (cross-peaks) in models A and B. Uncorrelated diagonal disorder (models C and D) induces inhomogeneous broadening of the 2D PE spectra along $-\Omega_1 = \Omega_2$ directions. In this case the signal has diagonal peaks (due to one-exciton self correlations) and cross peaks due to correlations of the one-exciton states with the OTE excited on the same peptide group. Off-diagonal disorder (models E and F) induces exciton delocalization over two peptides. In this case the signal contains inhomogeneously broadened diagonal, off-diagonal, and cross peaks. Examination of the real and the imaginary parts of the 2D PE signal shows that in the vicinity of the 2D resonances its profile is determined by the the ratio of the real and the imaginary parts of the exciton scattering matrix (44). Static disorder may lead to the formation or destruction of echo components in certain directions.

Determination of polypeptide structure requires measurements of the coupling energy J_{nm}. In the weak coupling limit, the absolute value of an off-diagonal peak $(-\varepsilon_n, \varepsilon_m)$ is proportional to $J_{nm}\Delta$. The anharmonicity Δ can be determined from the Ω_2 energy difference between the diagonal peak and the OTE cross peak, for fixed Ω_1. However, as we have shown, uncorrelated diagonal energy disorder destroys coherences between exciton states on different peptide groups and consequently the components of the 2D PE signal that carry structural information. Dynamical hole-burning measurements in pentapeptides (41) demonstrated that some of the peptide units have weak inhomogeneous broadening, so that even in the presence of diagonal disorder the 2D PE experiment can be helpful. In contrast, off-diagonal disorder results in the delocalization of exciton states, leading to correlations between vibrational excitations on different peptides, which show up as off-diagonal peaks and cross peaks due to CTE.

Since 2D PE spectroscopy is a femtosecond impulsive technique, it can be used for real-time study of early events in protein folding, which are the focus of extensive effort (55). 2D NMR spectroscopies (10) are limited to much slower time scales (ms), thus 1D IR (4–8) and luminescence (56,57) spectroscopies are employed for following faster conformational changes. 2D IR spectroscopy should provide more detailed information.

We have demonstrated that one- and two-exciton homogeneous dephasing rates can be obtained from PE signal as the Ω_2 HWFM of the diagonal and off-diagonal peaks. The sum of one- and two-exciton homogeneous dephasing rates is the Ω_2 HWFM of the cross peaks. Dynamical hole-burning experiments (15,41) were employed to measure the one-exciton state relaxation parameters. However, the two-exciton dephasing rate in the amide I region, which may reveal new information related to CO group coupling with intraprotein vibrational modes, has not been reported. This was measured using the PE technique for a single CO group attached to the hemo pocket of hemoglobin protein and several model molecules (14).

Finally, we note that the time scale for the PE experiment is determined by the dephasing times, which are very short in proteins ($\sim$300 fs) (41). Other complementary 2D techniques were proposed in Ref. 17. In particular, energy relaxation, which occurs in proteins and polypeptides on a large time scale ($\sim$2–15 ps) (15,41), can be studied by utilizing the transient grating and the three pulse PE techniques. These can be calculated as well using the third-order response function presented here.

ACKNOWLEDGMENTS

We gratefully acknowledge the support of the National Science Foundation, the United States Air Force Office of Scientific Research, and the Petroleum Research Fund sponsored by the American Chemical Society.

APPENDIX A: SUM-OVER-STATE REPRESENTATION OF THE THIRD-ORDER RESPONSE

In this appendix we present the sum-over-one- and two-exciton state expressions for the third-order response function. Double-sided Feynman diagrams representing the Liouville space pathways contributing to the four wave mixing in the RWA are given in Fig. 1B. The response function is

$$R(t, \tau_1, \tau_2, \tau_3) = R_1(t, \tau_1, \tau_2, \tau_3) + R_2(t, \tau_1, \tau_2, \tau_3) \qquad (34)$$

The first term contains a summation over one-exciton states $|e\rangle$ [Feynman diagrams (1), (2), (4), −(5)] only

$$R_1(t, \tau_1, \tau_2, \tau_3) \tag{35}$$
$$= i^3 \sum_{e,e'} [\mu_{ge}\mu_{eg}\mu_{e'g}\mu_{ge'}G_{eg}(t - \tau_3)G_{gg}(\tau_3 - \tau_2)G_{ge'}(\tau_2 - \tau_1)$$
$$+ \mu_{ge}\mu_{e'g}\mu_{eg}\mu_{ge'}G_{eg}(t - \tau_3)G_{ee'}(\tau_3 - \tau_2)G_{ge'}(\tau_2 - \tau_1)$$
$$+ \mu_{ge}\mu_{e'g}\mu_{ge'}\mu_{eg}G_{eg}(t - \tau_3)G_{ee'}(\tau_3 - \tau_2)G_{eg}(\tau_2 - \tau_1)$$
$$+ \mu_{ge'}\mu_{e'g}\mu_{ge}\mu_{eg}G_{e'g}(t - \tau_3)G_{gg}(\tau_3 - \tau_2)G_{eg}(\tau_2 - \tau_1)]$$

The second term represents the sum over one- and two-exciton $|f\rangle$ [Feynman diagrams (3), (6)–(8)] states

$$R_2(t, \tau_1, \tau_2, \tau_3) \tag{36}$$
$$= -i^3 \sum_{e,e',f} [\mu_{e'f}\mu_{fe}\mu_{eg}\mu_{ge'}G_{fe'}(t - \tau_3)G_{ee'}(\tau_3 - \tau_2)G_{ge'}(\tau_2 - \tau_1)$$
$$+ \mu_{e'f}\mu_{fe}\mu_{eg}\mu_{ge'}G_{fe'}(t - \tau_3)G_{ee'}(\tau_3 - \tau_2)G_{eg}(\tau_2 - \tau_1)$$
$$+ \mu_{e'f}\mu_{ge'}\mu_{fe}\mu_{eg}G_{fe'}(t - \tau_3)G_{fg}(\tau_3 - \tau_2)G_{eg}(\tau_2 - \tau_1)$$
$$+ \mu_{ge'}\mu_{e'f}\mu_{fe}\mu_{eg}G_{e'g}(t - \tau_3)G_{fg}(\tau_3 - \tau_2)G_{eg}(\tau_2 - \tau_1)]$$

In these expressions the Liouville space Green function matrix elements representing one-exciton coherence are

$$G_{e,g}(\tau) = \theta(\tau)\exp[-i(\varepsilon_e - i\Gamma_e)t] \tag{37}$$

where Γ_e is one-exciton dephasing rate. The two-exciton coherence Green function has the form

$$G_{f,g}(\tau) = \theta(\tau)\exp[-i(\bar{\varepsilon}_f - i\Gamma_f^{(2)})t] \tag{38}$$

where $\Gamma_f^{(2)}$ is the two-exciton dephasing rate. In the absence of pure dephasing between one- and two-exciton states it can be factorized as

$$G_{f,e}(\tau) = G_{f,g}(\tau)G_{g,e}(\tau) \tag{39}$$

The ground state population Green function is $G_{gg} = 1$, and the one-exciton population Green function $G_{e,e'}$ can be partitioned into the coherent and the incoherent parts, respectively:

$$G_{e,e'}(\tau) = G_{e,g}(\tau)G_{g,e'}(\tau) + \bar{G}_{e,e'}(\tau) \tag{40}$$

Substitution of Equation (40) into Equations (35) and (36) leads to partitioning of the response function given by Equation (34) into its coherent and incoherent parts.

APPENDIX B: GREEN-FUNCTION REPRESENTATION OF THE THIRD-ORDER SUSCEPTIBILITY

In four-wave mixing spectroscopy, the polarization $P(\mathbf{r}, t)$ can be represented using the third-order optical susceptibility $\chi^{(3)}(-\omega_s; \omega_a, \omega_b, \omega_c)$:

$$P(\mathbf{r}, t) = \frac{1}{(2\pi)^3} \int d\omega_a \int d\omega_b \int d\omega_c \exp(-i\omega_s t) \chi^{(3)}$$
$$\times (-\omega_s; \omega_a, \omega_b, \omega_c) \bar{\mathcal{E}}(\mathbf{r}, \omega_a) \bar{\mathcal{E}}(\mathbf{r}, \omega_b) \bar{\mathcal{E}}(\mathbf{r}, \omega_c) \qquad (41)$$

where

$$\bar{\mathcal{E}}(\mathbf{r}, \omega) \equiv \int_{-\infty}^{\infty} \mathcal{E}(\mathbf{r}, t) \exp(i\omega t)\, dt \qquad (42)$$

is the driving field in the frequency domain and $\mathbf{r}$ denotes the position of the molecule whose size is smaller than the optical wavelength (Latin indices ω_a etc. denote fields with no particular time ordering). The signal frequency ω_s is $\omega_s = \omega_a + \omega_b + \omega_c$.

The expression for the optical susceptibility can be partitioned into a coherent and an incoherent part:

$$\chi^{(3)}(-\omega_s; \omega_a, \omega_b, \omega_c) = \chi_c^{(3)}(-\omega_s; \omega_a, \omega_b, \omega_c) + \chi_i^{(3)}(-\omega_s; \omega_a, \omega_b, \omega_c) \qquad (43)$$

The coherent contribution in terms of the one exciton Green functions and the scattering matrix has the following form (18,58):

$$\chi_c^{(3)}(-\omega_s; \omega_a, \omega_b, \omega_c)$$
$$= \frac{1}{3!} \sum_{p(\omega_a, \omega_b, \omega_c)} \sum_{nm_1m_2m_3} \sum_{n'n''} \mu_n \mu_{m_1} \mu_{m_2} \mu_{m_3} G_{n'm_1}(\omega_a) G_{n'm_2}(\omega_b)$$
$$\times G_{m_3n''}^{\dagger}(-\omega_c) G_{nn''}(\omega_s) \bar{\Gamma}_{n''n'}(\omega_a + \omega_b) + \text{c.c.} \qquad (44)$$

where the summation $p(\omega_a, \omega_b, \omega_c)$ is overall $3! = 6$ permutations of frequencies ω_a, ω_b, and ω_c.

The incoherent contribution adopts the form:

$$\chi_i^{(3)}(-\omega_s; \omega_a, \omega_b, \omega_c)$$
$$= \frac{1}{3!} \sum_{p(\omega_a, \omega_b, \omega_c)} \sum \mu_n \mu_{m_1} \mu_{m_2} \mu_{m_3} \int_{-\infty}^{\infty} \frac{d\varepsilon}{2\pi} G_{nn''}(\omega_s) G_{n'm_3}(\omega_b)$$
$$\times G_{im''}^{\dagger}(-\varepsilon) \bar{\Gamma}_{n''m'',n'm'}(\omega_a + \omega_b + \omega_c - \varepsilon) \bar{G}_{m'i,kj}(\omega_a + \omega_c)$$
$$\times [G_{km_1}(\omega_a)\delta_{m_2j} + G_{m_2j}^{\dagger}(-\omega_c)\delta_{km_1}] + \text{c.c.} \qquad (45)$$

The one-exciton Green functions, the scattering matrix, and the irreducible part of the two-exciton Green function entering these equations are given by a Fourier transformation of Equations (18)–(21) to the frequency domain.

REFERENCES

1. Krimm S, Bandeker J. J Adv Protein Chem 38:181, 1986.
2. Torii H, Tasumi M. J Chem Phys 96:3379, 1992.
3. (a) Surewicz WK, Mantsch HH. Biochem Biophys Acta 952:115, 1988.
   (b) Surewicz WK, Mantch HH, Chapman D. Biochemistry 32:389, 1993.
   (c) Jeckson M, Mantsch H. Crit Rev Biochem Mol Biol 30:95, 1995.
4. Phillips CM, Mizutani Y, Hochstrasser RM. Proc Natl Acad Sci USA 92:7292, 1995.
5. Williams S, Causgrove TP, Gilmanish R, Fang KS, Callender RH, Woodruff WH, Dyer RB. Biochemistry 35:691, 1996.
6. Reinstadler D, Fabian H, Backmann J, Naumannl D. Biochemistry 35:15822, 1996.
7. Gilmanshin R, Williams S, Callender RH, Woodruff WH, Dyer RB. Proc Natl Acad Sci USA 94:3709, 1997.
8. Dyer RB, Gai F, Woodruff WH, Gilmanshin R, Callender RH. Acc Chem Res 31:709, 1998.
9. (a) Ernst RR, Bodenhausen G, Wokaun A. Principals of Nuclear Magnetic Resonance in One and Two Dimensions. Oxford: Clarendon Press, 1987.
   (b) JKM Sanders, BH Hunter. Modern NMR Spectroscopy. New York: Oxford, 1993.
10. van Nuland NAJ, Forge V, Balbach J, Dobson CM. Acc Chem Res 31:773, 1998.
11. Lee S, Budil DE, Freed JH. J Chem Phys 101:5529, 1994.
12. (a) Bout DV, Muller LJ, Berg M. Phys Rev Lett 67:3700, 1991. (b) Berg M, Bout DAV. Acc Chem Res 30:65, 1997.
13. Mukamel S, Piryatinski A, Chernyak V. Acc Chem Res 32:145, 1999.
14. Rector KD, Kwok AS, Ferrante C, Tokmakoff A, Rella CW, Fayer MD. J Chem Phys 106:10027, 1997.
15. Hamm P, Lim M, Hochstrasser RM. J Phys Chem B 102:6123, 1998.
16. (a) Zhang WM, Chernyak V, Mukamel S. In: Elsaesser T, Fujimoto JG, Wiersma DA, Zinth W, eds. Ultrafast Phenomena XI. New York: Springer, 1998:663. (b) Mukamel S, Zhang WM, Chernyak V. In: Garab G, ed. Photosynthesis: Mechanisms and Effects. 1:3 Dordrecht: Kluwer Academic, 1998.
17. Zhang WM, Chernyak V, Mukamel S. J Chem Phys 110:5011, 1999.
18. Mukamel S. Principles of Nonlinear Optical Spectroscopy. New York: Oxford University Press, 1995.

19. Hellwarth RW. Prog Quantum Electron 5:2, 1977.
20. (a) Loring RF, Mukamel S. J Chem Phys 83:2119, 1985. (b) Mukamel S, RF Loring. J Opt Soc Am B 3:595, 1986.
21. (a) Inaba R, Tominaga K, Tasumi K, Nelson M. Chem Phys Lett 211:183, 1993. (b) Tominaga K, Inaba R, Kang TJ, Naitoh Y, Nelson KA, Tasumi M, Yoshihara K. Proceedings of the Raman XIV International Conference on Raman Spectroscopy. New York: Wiley, 1994. (c) Yoshihara K, Inaba R, Okamoto H, Tasumi M, Tominaga K, Nelson KA. In: Wiersma D, ed. Femtosecond Reaction Dynamics. Amsterdam: North-Holland, 100, 1994.
22. Tanimura Y, Mukamel S. J Chem Phys 99:9496, 1993.
23. Tokmakoff A, Lang MJ, Larsen DS, Fleming GR, Chernyak V, Mukamel S. Phys Rev Lett 79:2702, 1997.
24. (a) Steffen T, Duppen K. Phys Rev Lett 76:1224, 1996. (b) J Chem Phys 106:3854, 1997.
25. Steffen T, Fourkas JT, Duppen K. J Chem Phys 105:7364, 1996.
26. (a) Tominaga K, Yoshihara K. Phys Rev Lett 74:3061, 1995. (b) Tominaga K, Keogh GP, Naitoh Y, Yoshihara K. J Raman Spectrosc 26:495, 1995. (c) Tominaga K, Yoshihara K. J Chem Phys 104:1159,4419, 1996; Phys Rev Lett 76:987, 1996; Phys Rev A 55:831, 1997.
27. (a) Tokmakoff A, Fleming GR. J Chem Phys 106:2569, 1997. (b) Tokma A-koff, Lang MJ, Larsen DS, Fleming GR. Chem Phys Lett 272:48, 1997.
28. Tokmakoff A. J Chem Phys 105:13, 1996.
29. Khidekel V, Mukamel S. Chem Phys Lett 240:304, 1995; 263:350, 1996 (E).
30. Palese SP, Buontempo JT, Schilling L, Lotshaw WT, Tanimura Y, Mukamel S, Miller RJD. J Phys Chem 98:12466, 1994.
31. Okumura K, Tanimura Y. Phys Rev E 53:214, 1996; J Chem Phys 105:7294, 1996; J Chem Phys 106:1687, 1997.
32. Okumura K, Tanimura Y. Chem Phys Lett 277:159, 1997.
33. (a) Okumura K, Tanimura Y. Chem Phys Lett 278:175, 1997. (b) Cho M, Okumura K, Tanimura Y. J Chem Phys 108:1326, 1998.
34. Saito S, Ohmine I. J Chem Phys 108:240, 1998.
35. Chernyak V, Mukamel S. J Chem Phys 108:5812, 1998.
36. (a) Piryatinski A, Chernyak V, Mukamel S. In: Elsaesser T, Fujimoto JG, Wiersma DA, Zinth W, eds. Ultrafast Phenomena XI. New York: Springer, 1998:541. (b) Mukamel S, Piryatinski A, Chernyak V. J Chem Phys 110:1711, 1999.
37. Chernyak V, Piryatinski A, Mukamel S. Laser Chem 19:109, 1999.
38. Okumura K, Tokmakoff A, Tanimura Y. J Chem Phys 111:492, 1999.
39. Chernyak V, Zhang WM, Mukamel S. J Chem Phys 109:9587, 1998.
40. Hamm P, Lim M, Asplund M, Hochstrasser RM. Chem Phys Lett 301:167, 1999.
41. Hamm P, Lim M, DeGrado WF, Hochstrasser R. Proc Natl Acad Sci 96:2036, 1999.

42. Bach AC, Eyermann CJ, Gross JD, Bower MJ, Harlow RL, Weber PC, DeGrado WF. J Am Chem Soc 116:3207, 1994.
43. Piryatinski A, Tretiak S, Chernyak V, Mukamel S. J. Raman Spectrosc 31:125, 2000.
44. Chernyak V, Wang N, Mukamel S. Phys Rep 263:213, 1995.
45. (a) Lepetit L, Cheriaux G, Joffre M. J Opt Soc Am B 12:2467, 1995. (b) Lepetit L, Joffre M. Opt Lett 21:564, 1996. (c) Likforman J-P, Joffre M, Theirry-Mieg V. Opt Lett 22:1104, 1997.
46. (a) Chemla DS, Bigot JY, Mycek M-A, Weiss S, Schäfer W. Phys Rev B 69: 3631, 1992. (b) Bigot JY, Mycek M-A, Weiss S, Ulbrich RG, Chemla DS. Phys Rev Lett 70:3307, 1993. (c) Chemla DS, Bigot JY, Mycek M-A, Weiss S, Schäfer W. Phys Rev A 50:8449, 1994.
47. Mukamel S, Ciordas-Ciurdariu C, Khidekel V. IEEE J Quantum Electron 32:1278, 1996.
48. Meier T, Chernyak V, Mukamel S. J Chem Phys 107:8759, 1997.
49. (a) Mukamel S. J Chem Phys 107:4165, 1997. (b) Yokojima S, Meier T, Chernyak V, Mukamel S. Phys Rev B 59:12584, 1999.
50. Kühn O, Chernyak V, Mukamel S. J Chem Phys 105:8586, 1996.
51. Zwanzig R. Lect Theor Phys 3:106, 1961; Physica 30:1109, 1964.
52. Economou E. Green's Function in Quantum Physics. New York: Springer, 1994.
53. Thouless D. Phys Rep Phys Lett 13C:93, 1974.
54. Spano FC, Kuklinski JR, Mukamel S. Phys Rev Lett 65:211, 1990.
55. Valentine J, ed. Acc Chem Res 31:697–780, 1998.
56. Ballew RM, Sabelko J, Gruebele M. Nature Struct Biol 3:923, 1996.
57. Thompson PA, Eaton WA, Hofrichter J. Biochemistry 36:9200, 1997.
58. Chernyak V, Mukamel S. Phys Rev B 48:2470, 1993.

9

Vibrational Dephasing in Liquids: Raman Echo and Raman Free-Induction Decay Studies

Mark A. Berg
University of South Carolina, Columbia, South Carolina

I. INTRODUCTION

Dephasing is one of the basic results of the interaction of a molecular vibration with its surroundings. Since the early 1970s, vibrational dephasing has been under continuous study through its affect on Raman line shapes and Raman free-induction decays (FID) (1,2). In the early 1990s, advances in ultrafast lasers led to the introduction of a fundamentally new type of dephasing measurement, the Raman echo, (3). Results from this new experiment, in combination with the more familiar Raman FID, are building a detailed picture of how vibrations interact with their surroundings. This chapter reviews recent dephasing studies based on Raman echo experiments (3–11). It looks at how the Raman echo is performed, how it is interpreted, and how it is changing our picture of vibrational dephasing.

Dephasing is defined as the loss of coherence or, equivalently, the loss of memory of an initial phase. In a precise description, this phase is the quantum phase between two eigenstates. For molecular vibrations, a classical description is often qualitatively useful. In such a description, the phase is the phase of the classical oscillation of the molecule. Picture an ensemble of vibrators in which all oscillations are started with the same phase at $t = 0$. In the absence of any external interactions, these oscillations will remain in phase with each other forever. In real materials, the molecular vibrators are perturbed by other molecules in their surroundings. Because these perturbations differ from one vibrator to the next, the molecular oscillations will lose their phase relationship with each other. The time required for this loss is the vibrational dephasing time.

Why is vibrational dephasing interesting? Most obviously, dephasing holds fundamental interest as a basic molecular process. In addition, chemical reactions can produce coherently excited products (12–15) and subsequent fast processes can be altered by the persistence or decay of that coherence. However, the most important reason may be that vibrational dephasing provides a simple context within which we can learn about the interactions of solvents and chemical systems. A strong effect of the solvent on a chemical process is the rule, rather than the exception (16). Solvents change rates of isomerization, bond formation, electronic-state relaxation, and so on. These effects are caused by static and dynamic perturbations of both electronic and nuclear coordinates of the reacting solutes. Vibrational dephasing probes the dynamic nuclear portion of the problem. Solvent interactions with electronic coordinates are discussed elsewhere (17–19).

Understanding solute-solvent interactions is a twofold problem. It involves both how the solvent moves and how that solvent motion couples into the nuclear motions of the solute. Treating the solvent dynamics is an important part of the problem but is not sufficient by itself. The coupling is also critically important. Many solvent motions are irrelevant because they cannot effectively alter solute nuclear motion. Other, apparently small, solvent motions may have a large effect if they couple strongly. Different approximations are appropriate for treating different solvent motions. By selecting the relevant solvent motions, the coupling determines the dynamical models appropriate for treating the solvent dynamics.

Vibrational dephasing is an excellent model for the general problem of solvent-solute interactions because it probes both essential elements: solvent dynamics and coupling to solute nuclear motion. Although dephasing refers only to solute vibrational motion, the lessons learned for this nuclear motion will transfer to the more general nuclear motions in reacting systems. This inherent mixing of dynamics and coupling is also a major challenge in interpreting dephasing experiments. Disentangling the effects of solvent dynamics from the effects of solvent coupling is a major aim of this work.

The majority of studies of vibrational dephasing have looked at the width or shape of the isotropic Raman line. The Raman line shape is the Fourier transform of the coherence decay function that characterizes dephasing (20,21). The coherence decay can also be measured directly in the time domain (22,23). Reported under a variety of names, including time-resolved CARS, time-resolved CSRS, and time-resolved coherent Raman scattering, we will call this experiment the Raman free-induction decay (FID) to emphasize its relationship to the corresponding FID experiment in NMR (24). Despite some early controversy (25), Loring and Mukamel clearly established that the Raman FID is exactly the Fourier transform of the Raman line shape (26). Early Raman line shape and FID work was reviewed by Oxtoby (1); a review by Morresi et al. covers more recent studies (2).

The concept of the Raman echo extends back to Hartmann in 1968 (27), and a few early experiments were performed on gas-phase electronic (28,29) and vibrational (30) transitions. However, it was the paper by Loring and Mukamel in 1985 that pointed out the importance of the Raman echo for studying condensed-phase vibrational dephasing (26). Initial attempts to perform the Raman echo in liquids failed (31), but technical improvements allowed the first successful Raman echo experiment in a liquid in 1991 (3).

The Raman FID is a third-order spectroscopy with one experimental time variable. The Raman echo is the straightforward extension to higher order; it is a seventh-order spectroscopy and has two time variables. In this way, the Raman echo and Raman FID are analogous to the spin echo and FID in NMR spectroscopy (24) or to the photon echo (18,32–34) and absorption line shape in electronic spectroscopy. In general, high-order spectroscopies contain qualitatively new dynamic information. The number of time variable reflects the level of detail retained in the experiment. In the cases just cited, the echo spectroscopies distinguish "homogeneous" and "inhomogeneous" contributions to the line shape. But whereas the inhomogeneous broadening in NMR or electronic spectroscopy is a trivial experimental artifact, in the case of vibrational transitions, the "inhomogeneous" dephasing processes are due to real molecular interactions; they are just slower than the "homogeneous" dephasing processes. Thus the goal of the Raman echo is to provide information on the rate of solvent-solute interactions that is not available from the FID or line shape alone.

A similar idea is exploited in recent work with the infrared (IR) echo, although with IR-active rather than Raman-active vibrations (35–42). Although the basic concepts in the Raman and IR echoes are the same, they each work best on different systems. The infrared echo is best for vibrations with strong IR transitions (and therefore potential for resonant energy transfer and sensitivity to local electric fields), for dilute solutes, and for systems with slow rotation. In contrast, the Raman echo is best for vibrations with strong Raman transitions (and generally weak IR transitions), for neat or concentrated solutions, and can be applied to quickly rotating solutes. Thus the two echo techniques are complementary, rather than competitive, routes to studying vibrational dephasing. The infrared echo is reviewed by Rector and Fayer in this volume (Chapter 6).

A number of fifth-order coherent Raman spectroscopies have appeared in recent years (11,43–55), and they can show echo-like behavior (56–60). However, the interpretation of these "echoes" is more complex than the seventh-order echoes. Moreover, many of these experiments are aimed at low-frequency intermolecular motions; only a few look at intramolecular vibrational dephasing (47,48). These experiments are not considered in detail in this chapter but are discussed by Blank et al. (Chapter 10) and Fourkas (Chapter 11).

This chapter is organized into three main sections. Section II is an overview of basic concepts. It discusses both the Raman FID and the Raman echo with the aim of illustrating the different information content of the two experiments. It also reviews the existing theoretical ideas on vibrational dephasing mechanisms that motivate the later experiments. Section III describes how the experiments are actually performed. It pays special attention to the unique requirements and problems of performing a high-order spectroscopy like the Raman echo. New results are discussed in Section IV. There are experimental results on four different systems and a new theory of dephasing prompted by these results. Each of the experimental systems displays unique dephasing properties. Nonetheless, a coherent unifying framework that can accommodate all these results is starting to emerge. A final summary (Section V) outlines our current understanding of this overall scheme and its implications for understanding solute-solvent interactions in general.

II. OVERVIEW OF VIBRATIONAL DEPHASING AND COHERENT RAMAN SPECTROSCOPY

A. One-Dimensional Measurements: Raman Line Shape and Free Induction Decays

The experimental dephasing time is defined by the decay of the correlation function of the vibrational coordinate q:

$$C_{FID}(\tau) = \langle q(\tau)q(0) \rangle \tag{1}$$

This function is measured directly in the Raman-FID experiment. It can also be derived from the shape of the isotropic Raman line, which is the

Fourier transform of C_{FID}:

$$I_{iso}(\Omega - \Omega_0) = \int_0^\infty C_{FID}(\tau)e^{-i(\Omega - \Omega_0)\tau}\,d\tau \tag{2}$$

Although experiments are most directly expressed in terms of the oscillator coordinate $q(\tau)$, the quantity of physical relevance is the instantaneous oscillator frequency $\omega(t)$, or, expressed more conveniently, the instantaneous deviation from the mean frequency, $\delta\omega(t) = \omega(t) - \langle\omega\rangle$. This frequency deviation is the direct consequence of time-dependent forces exerted on the oscillator by its environment. At a semiclassical level, the FID correlation function is related to the time-dependent frequency by

$$C_{FID}(\tau) = \left\langle \exp\left[-i\int_0^\tau \delta\omega(t)\,dt\right]\right\rangle \tag{3}$$

Unfortunately, the conversion from $\delta\omega(t)$ to $C_{FID}(t)$ involves a significant averaging and loss of information.

This loss of information is easier to see after applying a cumulant approximation (61). This approximation is equivalent to assuming that the distribution of $\delta\omega$ is always Gaussian and simplifies Equation (3) to

$$C_{FID}(\tau) = \exp\left\{-\left[\int_0^\tau dt'\right] \otimes \int_0^{t'} C_\omega(t'')dt''\right\} \tag{4}$$

$$C_\omega(t) = \langle\delta\omega(t)\delta\omega(0)\rangle \tag{5}$$

where $\otimes$ indicates operator multiplication. The FID correlation function $C_{FID}(\tau)$ is derived from the frequency correlation function $C_\omega(t)$ by averaging in time.

If we also assume that the vibrational frequency is modulated by a single process, $\delta\omega(t)$ is fully characterized by two parameters: Δ_ω, the rms magnitude of $\delta\omega$, and τ_ω, the correlation time of $\delta\omega(t)$. In the Kubo-Anderson model (61–63), the frequency correlation decay is taken to be exponential:

$$C_\omega(t) = \Delta_\omega^2 e^{-t/\tau_\omega} \tag{6}$$

and the FID correlation function becomes

$$C_{FID}(\tau) = \exp\left[-\Delta_\omega^2\tau_\omega^2\left(e^{-\tau/\tau_\omega} - 1 + \frac{\tau}{\tau_\omega}\right)\right] \tag{7}$$

In the limit $\Delta_\omega\tau_\omega \ll 1$, called the fast modulation or homogeneous limit, the FID correlation function reduces to

$$C_{FID}(\tau) = e^{-\tau/T_2} \tag{8}$$

where $T_2 = \Delta_\omega^2 \tau_\omega$, whereas in the limit $\Delta_\omega \tau_\omega \gg 1$, called the slow modulation or inhomogeneous limit, it becomes

$$C_{FID}(\tau) = e^{-\tau^2/2\Delta_\omega^2} \tag{9}$$

In each limit, the experimental decay is characterized by a single parameter, whereas the underlying frequency correlation function is characterized by at least two parameters. This problem exists for other models as well. In general, the problem of inverting $C_{FID}(\tau)$ to find $C_\omega(t)$ is underdetermined and requires model-dependent assumptions to complete an analysis. This problem reflects the loss of information caused by the time averaging in Equation (4).

If we restrict ourselves to the Kubo-Anderson model, the problem does not seem entirely intractable. The shape of the decay distinguishes the limits: the fast limit has an exponential decay; the slow limit has a Gaussian decay. In the intermediate regime [Equation (7)], the details of the decay shape can be used to find both Δ_ω and τ_ω.

However, these results are highly model dependent (64–66). For example, a model based on a random distribution of dipole-dipole interactions gives an exponential decay in the slow modulation limit (64). As another example, a nonexponential frequency correlation decay gives different intermediate results than the Kubo-Anderson model (67,68).

The situation becomes even more complicated when multiple dephasing processes are considered. Figure 2c shows FID curves from two distinctly different modulation processes. In case #1 the are two processes modulating the vibrational frequency: one fast and one slow. In case #2, there is a single process at an intermediate rate. Real experimental data contain both random noise and systematic errors from tails of other transitions, weak overtones, hot bands, and isotope lines and from background emissions from the sample. In the face of these factors, the subtle differences between $C_{FID}(\tau)$ in case #1 and in case #2 are almost impossible to distinguish, despite the significant differences in the underlying physics.

In the fast modulation limit [Equation (8)], the loss of information is fundamental. The lifetime of the frequency perturbations is short compared to their magnitude, and the uncertainty principle precludes a full characterization of $\delta\omega(t)$ by any experimental technique. However, in the slow modulation limit and in the intermediate regimes, the loss of information in the FID is not fundamental. The next section shows that the Raman echo contains additional information about the rate of the frequency fluctuations that is not present in the FID. By using a combination of Raman echo and

FID measurements, a much more robust and more nearly unique model of $C_\omega(t)$ can be extracted.

B. A Two-Dimensional Measurement: The Raman Echo

Loring and Mukamel pointed out that additional information can be obtained by a higher order spectroscopy, the Raman echo (26). The corresponding observed correlation function is [see Equation (3)]

$$C_{RE}(\tau_1, \tau_3) = \left\langle \exp\left[i \int_{\tau_1}^{\tau_1+\tau_3} \delta\omega(t)dt - i \int_0^{\tau_1} \delta\omega(t)dt\right] \right\rangle. \qquad (10)$$

The Raman echo correlation function has two time variables and provides direct information on the rate of frequency modulation (26,69,70). In essence, the echo tests whether $\delta\omega(t)$ differs between two different time intervals, whereas the FID only examines $\delta\omega(t)$ during a single time interval [Equation (3)]. If $\delta\omega(t)$ changes slowly, i.e., $\tau_\omega \gg \tau_1, \tau_3$, the two integrals in Equation (10) cancel, with complete cancellation at $\tau_1 = \tau_3$. More specifically,

$$C_{RE}(\tau_1, \tau_3) = C_{FID}(|\tau_1 - \tau_3|) \qquad (11)$$

in the slow modulation limit. On the other hand, if $\delta\omega(t)$ randomizes quickly, the change in sign in the integrals is unimportant, no cancellation occurs, and

$$C_{RE}(\tau_1, \tau_3) = C_{FID}(\tau_1 + \tau_3) \qquad (12)$$

in the fast modulation limit. Thus, a comparison of the Raman echo to the FID provides qualitative and model-independent discrimination between fast and slow modulation dephasing.

For more complex systems with intermediate modulation or multiple modulation time scales, a cumulant approximation is useful [see Equation (4)] (34):

$$C_{RE}(\tau_1, \tau_3) = \exp\left\{-2\left[2\int_0^{\tau_1} dt' - \int_0^{\tau_1+\tau_3} dt' + 2\int_0^{\tau_3} dt'\right]\right.$$
$$\left. \otimes \int_0^{t'} dt'' C_\omega'(t'')\right\}. \qquad (13)$$

The echo and FID decays derive from the same frequency correlation function, but the echo dissects $C_\omega(t)$ with a more complex and informative operator.

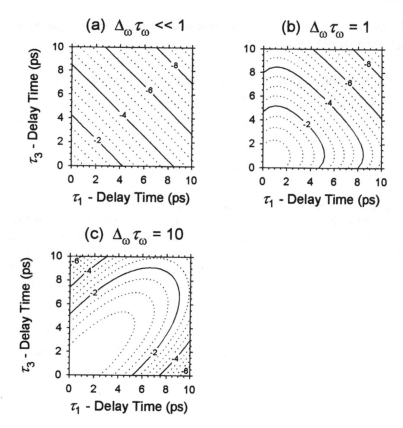

Figure 1 Two-dimensional contour plots of the log of the Raman-echo correlation function, ln $C_{RE}(\tau_1, \tau_3)$, showing the effect of changing the rate of solvent-induced perturbations. All cases give a Raman line with the same FWHM (5 cm^{-1}) and FIDs with similar decay times but give very different Raman echo results. (a) Fast modulation; (b) intermediate modulation ($\Delta_\omega = 3.32$ cm^{-1}, $\tau_\omega = 1.60$ ps); (c) slow modulation. Calculations are based on a single Kubo-Anderson process [Equations (7)–(9)].

This idea is illustrated in Fig. 1, which shows the behavior of the Raman echo signal for fast, intermediate, and slow modulation dephasing. In each of the cases shown, the Raman linewidth is the same, and the FID decays times are very similar. Nonetheless, the Raman echo decays are qualitatively different, depending on the rate of frequency modulation.

The additional information in the Raman echo is even more important when multiple dephasing mechanisms might be operating simultaneously.

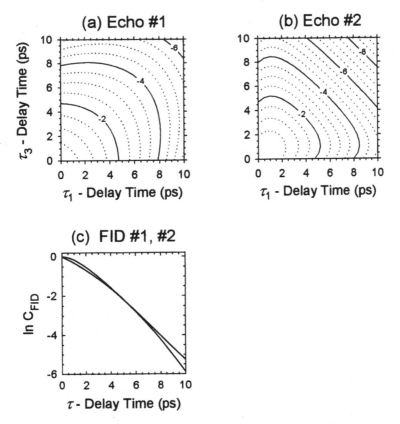

Figure 2 The Raman echo has the ability to distinguish single from multiple solvent interactions. (a) & (b) Two-dimensional contour plots of the log of the Raman-echo correlation function, ln $C_{RE}(\tau_1, \tau_3)$. Case #1: A combination of two interactions: one in the fast modulation limit [$T_2 = 3.47$ ps; Equation (8)], and one in the slow modulation limit [$\Delta_\omega = 1.30$ cm^{-1}, $\tau_\omega = 40.9$ ps; Equation (9). Case #2: A single interaction with intermediate modulation (same as Fig. 1b). The echo signals clearly differ for these two cases. However, the Raman lines have identical FWHM's (5 cm^{-1}), and the Raman FID signals (c) are difficult to distinguish experimentally.

An example is illustrated in Fig. 2. In case #1, there are two mechanisms operating: one in the fast modulation limit, and one in the slow modulation limit. This is the classic case of combined homogeneous and inhomogeneous line broadening. In case #2, there is a single mechanism, but it is in the intermediate modulation region. The Raman linewidths are identical for the two cases, and the Raman FID decays (and line shapes) are very similar (Fig. 2c). However, the Raman echo signals show clear and easily distinguishable differences between the two cases (Fig. 2a and b).

Thus the Raman echo adds vital new information on the dephasing process. This fact does not obviate Raman line shape or FID measurements. Because of their relative simplicity and experimental ease, these techniques remain valuable tools. However, using them in comparison with the Raman echo [Equations (11) and (12)] provides a much more complete and less model-dependent picture of vibrational dephasing than is possible with line shape or FID measurements alone.

C. Vibrational Dephasing Mechanisms

Although the combination of echo and FID techniques can more precisely define the vibrational frequency perturbations $\delta\omega(t)$, connecting $\delta\omega(t)$ with the dynamics of the solvent and the solvent-vibration coupling requires a correct model of the dephasing mechanism. This section briefly reviews the potential dephasing mechanisms relevant to interpreting the experiments presented in Section IV. More comprehensive reviews can be found elsewhere (2,71).

1. Connecting Frequency Fluctuations to Solvent Motion

In pure dephasing, the loss of coherence is caused by time-dependent perturbations of the vibrational frequency. Pure dephasing has been assumed in the earlier portions of this chapter and, in general, dominates vibrations in liquids. Within the cumulant approximation, both the Raman echo and FID decays can be calculated from $C_\omega(t)$ [Equations (4) and (13)]. Relating these measurement to properties of the solvent and solute requires a molecular model for $C_\omega(t)$.

To calculate this correlation function, the system Hamiltonian is divided into components for the vibration H_{vib}, solvent H_{sol}, and solvent-vibration coupling V, which depend on the vibrational coordinate q and the set of solvent coordinates Q:

$$H = H_{vib}(q) + V(q,Q) + H_{sol}(Q) \tag{14}$$

Following Oxtoby (72), the vibrational Hamiltonian is expanded to include 3rd-order anharmonicity:

$$H_{vib} = \tfrac{1}{2}\omega_0^2 q^2 + K_{111}q^3 + \cdots \tag{15}$$

where K_{111} is the anharmonicity constant. The coupling is expanded up to the quadratic term

$$V = V(0, Q) + \frac{\partial V(0, Q)}{\partial q}q + \frac{\partial^2 V(0, Q)}{\partial q^2}q^2 + \cdots \tag{16}$$

The perturbation of the $v = 0$ to $v = 1$ transition frequency by the solvent is

$$\delta\omega(t) = -\frac{3\hbar K_{111}}{\omega_0^3}\frac{\partial V(0, Q(t))}{\partial q} + \frac{\hbar}{2\omega_0}\frac{\partial^2 V(0, Q(t))}{\partial q^2} \tag{17}$$

Note that a harmonic vibration and linear coupling is insufficient to shift the frequency or cause dephasing; nonlinearity must be present in either the vibration or in the coupling.

For the term based on vibrational anharmonicity, the frequency perturbation is proportional to the force exerted by the solvent along the vibrational coordinate

$$\delta\omega(t) \propto F(t) \equiv -\frac{\partial V(0, Q(t))}{\partial q} \tag{18}$$

There is a clear and direct connection between dephasing measurements, the solvent-solute interaction V and the solvent dynamics Q(t). Interest then focuses on finding a suitable model for the solvent dynamics and calculating the resulting solvent force on the vibration.

The first vibrational-anharmonicity term of Equation 18 has been found to dominate over nonlinear coupling in the specific case of N_2, [73] and Oxtoby has argued that this term will dominate in general. [72] However, several theories have looked at dephasing due to the second, nonlinear coupling term of Equation 18. [71,74–76] Under reasonable approximations and with a steeply repulsive V, the anharmonic term also produces frequency shifts proportional to the solvent force on the vibrator, just as in Equation 19. [72] Again the issue returns to an accurate treatment of the solvent dynamics and the nature of the solvent-solute coupling.

2. Fast Modulation Pure Dephasing Theories

The isolated binary collision (IBC) theory is one of the earliest dephasing theories developed for liquids (77). Despite criticism of its assumptions (78), it remains popular for the interpretation of experimental data, because it relates the dephasing to easily measurable solvent properties. It assumes that the solvent dynamics are essentially gas-like, consisting of uncorrelated collisions of the vibrating molecule with solvent molecules. The solvent-solute interaction is taken as purely repulsive. The dependence of the linewidth Γ (FWHM) on temperature T and viscosity η reduces to

$$\Gamma = \frac{1}{\pi c T_2} \propto T\eta \qquad (19)$$

Oxtoby took a different approach and treated the solvent dynamics using hydrodynamic equations (72). On first impression, the IBC and hydrodynamic approaches seem very different: the IBC theory includes solvent molecular structure explicitly but ignores solvent correlation; the hydrodynamic theory ignores molecular structure in the solvent but includes strong hydrodynamic correlations. Nevertheless, the final predictions of the two theories are nearly identical. In particular, Equation (19) is also derived from hydrodynamic theory.

Several other papers have proposed other dephasing mechanisms that result in fast modulation (79–83). Despite the differences in the details of these theories, they all predict the same direct proportionality of the linewidth to the viscosity (or, equivalently, to the diffusion constant) that is predicted by IBC or hydrodynamic theory [Equation (19)]. Because this result was obtained by very different methods of treating solvent correlations, Oxtoby concluded that solvent correlations are not important in vibrational dephasing (72). However, Equation (19) is clearly inappropriate for supercooled liquids, where the viscosity diverges at a finite temperature. In Section IV.D we will show that this problem is resolved by including the distinction between the inertial diffusive components of the solvent dynamics, a distinction absent in earlier theories. This change in the treatment of the solvent dynamics also revises Oxtoby's conclusion — the dephasing from inertial dynamics will be shown to depend critically on the degree of correlation in the solvent bath.

The IBC model can be modified by using the Enskog time to estimate the collision frequency (84). This modification gives a solid-like character to the dynamics, in that reasonable collision times are given for a solid material. This modification also removes any explicit dependence of the dephasing on the viscosity. Other solid-like treatments of the solvent give the same result (71,75). Although the lack of a viscosity dependence is not unphysical a priori, experiments presented in Section IV.C will show that a strong viscosity dependence does exist in supercooled liquids.

3. Slow Modulation Pure Dephasing Theories

In many cases, the dynamics in low-viscosity liquids are sufficiently rapid to put vibrational dephasing in the fast modulation limit (73,85). However, several plausible mechanisms for producing slow modulation dephasing have been proposed. The Raman echo is especially useful in characterizing these mechanisms and in distinguishing them from fast modulation mechanisms.

Bondarev and Mardaeva were the first to observe that the Raman linewidth in mixtures could have an extra broadening, which reaches a maximum at intermediate concentrations (86). This broadening was attributed to fluctuations in the local concentration around the vibrator. This idea was elaborated theoretically by Knapp and Fisher (87). In a binary mixture of solvents, the first solvation shell of a vibration will have substantial fluctuations in composition. If the vibrational frequency is sensitive to the local composition, dephasing will result. Because it takes on the order of 10 ps for molecules to diffuse in and out of the first solvation shell, the frequency perturbations will last much longer than collisonal perturbations, and the resulting dephasing will be either in the slow modulation limit or in the intermediate range.

In an influential paper, Schweizer and Chandler proposed another potential source of slow modulation dephasing (88). If the vibration's frequency is sensitive to the total number of solvent molecules in the first solvation shell, long-lived density fluctuations will cause slow modulation dephasing. They also pointed out the important distinction between coupling by short-range repulsive intermolecular forces and long-range attractive forces. Because only small amplitude motion (~ 0.1 molecular diameter) is needed to modulate the very rapidly varying repulsive forces, they associated these coupling forces with collision-like solvent dynamics and fast modulation dephasing. On the other hand, large amplitude solvent motion (~ 1 molecular diameter) is needed to modulate slowly varying attractive forces. They associated attractive coupling forces with diffusive solvent dynamics and slow modulation dephasing, whereas repulsive coupling forces were associated with collision-like dynamics and fast modulation dephasing.

The pure dephasing theories discussed in the last two subsections pose the basic questions we have explored with the Raman echo experiment. Is dephasing due to fast or slow modulation processes? What is the relative role of collisional versus diffusive dynamics in the solvent? What are the relative roles of repulsive versus attractive coupling forces? Can dephasing over the full range of viscosities through the supercooled liquid region and into the solid glass be correctly be described by a single theory?

4. "Impure" Dephasing

Up to this point, this chapter has assumed that dephasing is due to elastic modulations of the vibrational transition frequency, i.e., pure dephasing.

Pure dephasing has proven to be the dominant line-broadening mechanism in most liquids. However, the diversity among systems is great enough that potential contributions from other mechanisms must be considered for each system studied.

The population lifetime T_1 adds to the pure dephasing time T_2^* to give the total dephasing time T_2 (1):

$$\pi c \Gamma = \frac{1}{T_2} = \frac{1}{T_2^*} + \frac{1}{2T_1} \qquad (20)$$

The contribution of T_1 to the total linewidth Γ (FWHM) is highly variable. Table 1 shows the contributions for the molecules studied here. The lifetime is a significant factor in pure CH_3I, a small correction in CH_3CN, and negligible in ethanol. The lifetime has not been measured in toluene, but a similar frequency mode in benzene has a long lifetime (98).

Resonant energy transfer between molecules can also cause line broadening and the associated Raman noncoincidence effect (1,2,103,104). The definitive test for this mechanism is an isotopic dilution, which removes the resonance without changing any other properties of the liquid. Most verified cases of energy-transfer line broadening involve molecules with large IR transition moments (103). Because the Raman echo experiments focus on molecules with small IR moments, energy transfer is expected to be weak (104). Isotopic dilution experiments summarized in Table 1 verify that this line-broadening mechanism is small or negligible for the molecules studied here.

Table 1 Properties of the Vibrations Studied[a]

	$\bar{\nu}$ (cm^{-1})	Γ (cm^{-1})	Γ(dilute) (cm^{-1})	$\Gamma(T_1)$ (cm^{-1})	ρ	T_2 (ps)	τ_{rot} (ps)
CH_3I	2951	5.7	5.2[89]	3.5–5.7[92]	0.004[99]	1.9	1.5[101]
CH_3CN	2945	6.5	6.0[90,91]	1.1[93]	0.009[90]	1.65	1.5[101]
CH_3CD_2OH	2936	15	14[5]	0.24[94–97]	—	0.725	—
$C_6H_5CH_3$	1003	1.8[b]	—	(0.13)[c] [98]	0.02[100]	6.0[b]	1.2/3.0/7.9[d] [102]

[a] Center frequency $\bar{\nu}$, linewidth Γ (FWHM), linewidth upon isotopic dilution Γ (dilute), contribution to the linewidth from population relaxation $\Gamma(T_1)$, Raman depolarization ratio ρ, dephasing time T_2 and rotation time τ_{rot}.

[b] 250 K.

[c] 992 cm^{-1} mode of benzene.

[d] x-,y-, and z-axes.

III. IMPLEMENTING COHERENT RAMAN EXPERIMENTS

A. Raman FID

The Raman FID was pioneered by the Kaiser group (22,23) and has been refined by a number of workers over the years (105–116). Because a large amount of detail is available in the literature, only the basic features are reviewed here.

The light-matter interactions of the Raman FID experiment are illustrated in Fig. 3a. Light pulses are needed at two frequencies: Laser (L) and Stokes (S), with their frequency difference adjusted to the vibrational transition energy. An initial pair of Laser and Stokes pulses (pair I) excites the vibration through a Raman interaction. The density matrix of the vibration is transferred from the pure ground state (ρ_{00}) to a coherent superposition of the $\nu = 0$ and $\nu = 1$ states (ρ_{01}).

As soon as the coherent superposition is created, it begins to decay due to dephasing processes. The amount of coherence remaining after a period τ is measured with a probe (II) Laser pulse. This pulse stimulates the coherent superposition to emit an anti-Stokes (AS) pulse as it returns to the ground state. The decay of the coherence is recorded by measuring the intensity of the AS pulse as the time period τ is increased.

The coherent probe is only effective if the angles of the pulses are adjusted to satisfy the phase-matching condition

$$\mathbf{k}_{AS} = \mathbf{k}_{LI} - \mathbf{k}_{SI} + \mathbf{k}_{LII} \tag{21}$$

In the absence of dispersion in the index-of-refraction, this condition is satisfied for collinear pulses. In real liquids, noncollinear pulses are needed. If we choose to work with coplanar pulses for simplicity, there are three angles to choose to satisfy one phase-matching condition [Equation (9)]. The relevant angles (ϕ_1, ϕ_2, α) are defined in Fig. 4b. For a given choice of output angel α, the input angles are (6)

$$\phi_1^2 = \frac{2(\nu_L - \nu_S)}{\nu_L \nu_S}\delta + \frac{\nu_{AS}}{\nu_S}\alpha^2 \tag{22}$$

$$\phi_2 = \frac{\nu_S}{\nu_L - \nu_S}\phi_1 + \left(\frac{\nu_L}{\nu_L - \nu_S} + 1\right)\alpha \tag{23}$$

where the index-of-refraction mismatch δ is defined by

$$\delta = 2\frac{n_{AS} - n_L}{\bar{n}}\nu_L - \frac{n_{AS} - n_S}{\bar{n}}\nu_S \tag{24}$$

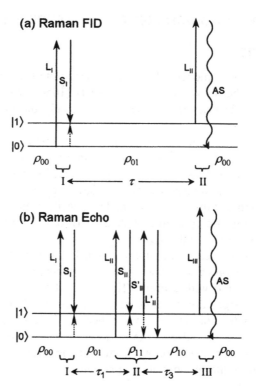

Figure 3　A ladder diagram illustrating the sequence of laser pulses in time and their interactions with the ground ($|0\rangle$) and first excited ($|1\rangle$) vibrational states. Light pulse frequencies: L = laser; S = Strokes; AS = anti-Stokes. The vibrational density matrix element (ρ) during each time interval (τ) is indicated below the energy levels. Several light-matter interactions (arrows) occur during each laser pulse (I, II, or III). (a) In the FID, the vibration evolves in the same coherence (ρ_{01}) for the entire experiment. (b) In the Raman echo, an additional set of interactions (II) splits the evolution between coherences of opposite phase (ρ_{01} and ρ_{10}), causing cancellation of the dephasing caused by long-lived perturbations.

and small angles are assumed. The frequencies of the pulses are given by the ν_i's and the corresponding indices-of-refraction by the n_i's and the mean index-of-refraction by $\bar{n}$. The index-of-refraction mismatch is most easily found by optimizing a two-pulse CARS experiment, where there is only a single angle to scan (Fig. 4c). In terms of the optimum CARS angle ψ,

$$\delta = \frac{\nu_L \nu_S}{2\nu_L - \nu_S}\psi^2 \tag{25}$$

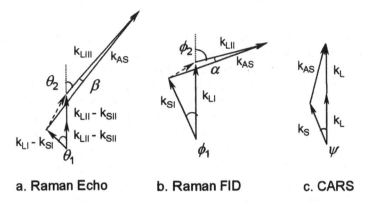

Figure 4 Phase-matching configurations used for several coherent Raman experiments. All angles are positive as shown. (Adapted from Ref. 6.)

Competing processes are another concern in real experiments. These processes result from interactions with different time orderings of the pulses and with perturbation-theory pathways proceeding through nonresonant states. They correspond to the constant nonresonant background seen in CARS and other frequency-domain spectroscopies. These nonresonant interactions are only possible when the excitation and probe pulses are overlapped in time, so they add an instantaneous component to the total material response function

$$M_{FID}(t) = \chi_R C_{FID}(t) + \chi_{NR}\delta(t) \tag{26}$$

where χ_R and χ_{NR} are the amplitudes of the resonant and nonresonant contributions, respectively, and $\delta(t)$ is the Dirac delta function. When the effect of nonzero pulse widths is added, the FID signal is

$$I_{FID}(\tau) \propto \int_{-\infty}^{\infty} dt \left| E_{LII}(t - \tau) \int_{0}^{\infty} dt_1 M_{FID}(t_1) F_I(t - t_1) \right|^2 \tag{27}$$

$$F(t) = E_L(t)E_S(t). \tag{28}$$

The nonresonant term leads to an extra peak at $t = 0$ with a width determined by the pulse widths (109,111). Because the early time data are essential to determining the shape of $C_{FID}(t)$, this nonresonant term is important in fitting real experimental data taken with finite pulse widths.

B. Raman Echo

The light-matter interactions of the Raman echo are illustrated in Fig. 3b. The first excitation pulse (I) and the coherent probing (III) are essentially identical to the corresponding processes in the Raman FID. The Raman echo differs by the introduction of an intermediate set of pulses (II), which divide the total time interval into two portions τ_1 and τ_3. The net effect of the interactions caused by these pulses is to reverse the sign of the coherence, i.e., to transfer density from ρ_{01} to ρ_{10}. The change in sign of the coherence leads to the change in the sign of the integral over time in Equation (10). As a result, the dephasing during τ_1 can be canceled during τ_3 if the dephasing results from perturbations that remain constant during the entire time interval $\tau_1 + \tau_3$. As τ_1 and τ_3 are varied, the extent of cancellation changes. The characteristic patterns of cancellation provide information on the vibrational frequency-modulation rates (Figs. 1 and 2).

A total of seven interactions with input fields are required, so the Raman echo is a $\chi^{(7)}$ process. However, in practice, a single laser pulse is used to generate both the L_{II} and $L_{II'}$ interactions, and similarly, a single pulse generates both S_{II} and $S_{II'}$. Thus, five real pulses are used.

In principle, the II and II' interactions could be performed by separate pulses, in which case the time separation of these pulses would become a third time variable τ_2. This experiment would be the stimulated Raman echo (SRE). During τ_2, the vibrator would be in a pure state (ρ_{11} in Fig. 3, ρ_{00} in other pathways), and the dynamics would be governed by the population relaxation decay time T_1. Thus the SRE is a three-dimensional experiment that measures the correlation function

$$C_{SRE}(\tau_1, \tau_2, \tau_3) = \left\langle \exp\left[i \int_{\tau_1+\tau_2}^{\tau_1+\tau_2+\tau_3} \delta\omega(t)\, dt - \frac{\tau_2}{T_1} - i \int_0^{\tau_1} \delta\omega(t)\, dt \right] \right\rangle$$

$$(29)$$

Although there is no fundamental barrier to performing the SRE, no one has yet attempted this experiment.

An essential point in the quantitative interpretation of the Raman echo is that the two-dimensional Raman echo is really a reduction of the SRE experiment, which is fundamentally three dimensional. In the idealized Raman echo correlation function presented earlier [Equation (10)], the reduction occurs by fixing $\tau_2 = 0$ in Equation (29). However, in real experiments using pulses with nonzero duration, interactions II and II' can occur at different times within the pulses. Although the observed Raman signal intensity I_{RE} approximates the ideal correlation function C_{RE} in the limit of infinitely short pulses, I_{RE} should be derived from C_{SRE} when the pulses have finite duration,

$$
\begin{aligned}
I_{RE}(\tau_1, \tau_3) \propto \int_{-\infty}^{\infty} dt \bigg| E_{LIII}(t - \tau_1 - \tau_3) \int_0^{\infty} dt_1 \int_0^{\infty} dt_2 \\
\times \int_0^{\infty} dt_3 C_{SRE}(t_1, t_2, t_3) \times F_I(t - t_3 - t_2 - t_1) \\
\times F_{II}(t - t_3 - t_2 - \tau_1) F_{II}(t - t_3 - \tau_1) \bigg|^2
\end{aligned}
\tag{30}
$$

where E_L and E_S are the electric field envelopes of the Laser and Stokes pulses, respectively. In linear spectroscopy, pulse-width effects are included by using a one-dimensional convolution integral. For nonlinear experiments like the Raman echo, fourfold integrals like Equation (30) are typically needed to include the pulse-width broadening.

This is a convenient point in the discussion to compare the Raman echo to other "echo" spectroscopies. All echo spectroscopies originate from the same three-dimensional stimulated-echo correlation function defined in Equation (29). However, the full three-dimensional experiments needed to completely map out $C_{SRE}(t_1, t_2, t_3)$ are often too complex to be practical. Thus, the dimensionality of the experiment needs to be reduced and a strategy for efficient sampling the remaining dimensions needs to be defined to simplify both the experiment and its analysis. A variety of different strategies have been implemented, resulting in a bewildering set of variations on echo spectroscopy (18,117–124). In the Raman echo, the second time dimension is fixed near zero to reduce the dimensionality. The remaining two dimensions are sampled by scanning τ_3 at various fixed values of τ_1 (see Section IV).

The infrared echo is also used to measure vibrational dynamics but in the standard implementation involves a further reduction in dimension (35,36,41,42). The excitation interactions I and II are strictly analogous to those in the Raman echo; the Raman interaction is simply replaced by a direct absorption (Fig. 3, dashed arrows). However, whereas the Raman echo time resolves the signal during τ_3, the infrared echo integrates the signal during this time period. In this way, the infrared echo reduces the correlation function to one dimension. The standard, two-pulse photon echo is reduced to one dimension in much the same way. Because the infrared echo derives from the same basic correlation function as the Raman echo, the infrared echo has the same fundamental sensitivity to the modulation rate of the vibrational frequency. However, the reduction in dimension does reduce its ability to distinguish between some models, especially when intermediate modulation rates are involved (6).

The three-pulse-echo peak shift is another two-dimensional echo technique, so far applied only to electronic transitions (122,123). It integrates over τ_3 and keeps τ_1 and τ_2 as the time variables. The data are reduced by tracing the maximum in τ_1 as a function of τ_2, resulting in a one-dimensional decay curve. Although the implementation of this type of echo spectroscopy is quite different, the essential information content is much the same as in the Raman echo approach.

Another type of echo that is receiving much attention are fifth-order echoes, which are also based on Raman interactions (56–60). The majority of studies have looked at intermolecular interactions (11,43–54), but a few studies have looked at an intramolecular vibrational overtone (47,48). Compared to the Raman echo, the fifth-order echo replaces a pair of interactions by a single, double-quantum interaction. Although the fifth-order experiments are formally of lower order than the Raman echo, the double-quantum interaction is forbidden in the harmonic approximation. As a result, it is not clear that the signal from a fifth-order echo will be stronger than that from a seventh-order Raman echo.

C. Special Problems of Seventh-Order Spectroscopy

Along with the potential for providing important information, high-order spectroscopies such as the Raman echo entail a number of unique problems. The first and most practical of these problems is finding the desired signal. Using the high-intensity pulses needed to drive a $\chi^{(7)}$ process, as many as 50–100 other signals can be identified coming from the sample resulting from lower-order $\chi^{(3)}$ and $\chi^{(5)}$ processes. $\chi^{(3)}$ Processes are so strong that beams representing fourth and fifth harmonics of the basic phase-matching condition are detectable. Cascaded processes, in which signals from one nonlinear process act to drive another, also produce readily visible beams. Fortunately, the majority of these processes produce light at frequencies well separated from the Raman echo anti-Stokes frequency, so a simple bandpass filter provides a great simplification of the observed signals.

The other major mechanism for identifying the Raman echo signal is the phase-matching requirement:

$$\mathbf{k}_{AS} = \mathbf{k}_{LI} - \mathbf{k}_{SI} - 2\mathbf{k}_{LII} + 2\mathbf{k}_{SII} - \mathbf{k}_{LIII} \tag{31}$$

Although the phase-matching requirement imposes an experimental burden in arranging the angles of the excitation beams, it also ensures that the Raman echo signal will occur in a distinct and predictable direction, separated from almost all other competing signals. Spatial filtering of the Raman echo signal provides the maximum discrimination against scattered light from competing processes.

With five pulses, there are in principle seven adjustable angles available to satisfy this single phase-matching equation [Equation (31)]. In practice, the problem is simplified by fixing each pair of pulses, L_I/S_I and L_{II}/S_{II}, to be collinear and by keeping all pulses coplanar. With these restrictions, there are only two adjustable input angles remaining (Fig. 4a). Once a value of β is chosen, the values of θ_1 and θ_2 are determined by the same index-of-refraction parameter δ [Equation (24)] that appears in the phase-matching conditions for the Raman FID and CARS [Equations (22), (25)] (6):

$$\theta_1^2 = \frac{\delta}{\nu_L - \nu_S} + \frac{\nu_L(2\nu_L - \nu_S)}{2(\nu_L - \nu_S)^2}\beta^2 \tag{32}$$

$$\theta_2 = \theta_1 + \left(\frac{2\nu_L - \nu_S}{\nu_L - \nu_S}\right)\beta \tag{33}$$

Thus, it is possible to optimize the phase-matching for the easier Raman-FID and CARS experiments and use the resulting δ to calculate the correct excitation geometry for the Raman echo.

Once a putative Raman echo signal is found, its identity must be confirmed. The most straightforward test is to make sure that the signal disappears when any of the five input pulses is blocked. All interfering processes are generated by a subset of the excitation pulses and will remain when one or more of the excitation pulses is blocked.

A more sophisticated test is to measure the dependence of the signal size on the input Laser intensity. An example is shown in Fig. 5. The Raman echo signal has an overall seventh-order dependence on the excitation energies: fourth order on the sum of the Laser intensities and third order on the sum of the Stokes intensities. Lower-order processes have a weaker dependence on excitation energy. For example the FID is second order in total Laser intensity and first order in the Stokes intensity (Fig. 5).

A natural concern with a $\chi^{(7)}$ process is the size of the signal to be measured. However, under realistic conditions, the absolute intensity of the Raman echo signal is easy to detect. The problem is scattered light from unintended FID processes. As Fig. 5 shows, the Raman echo signal grows

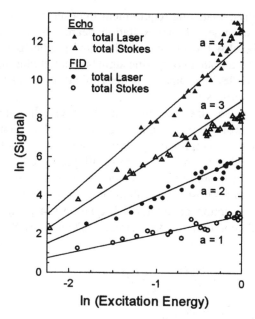

Figure 5 Log-log plot of the anti-Stokes signal intensity versus the excitation pulse energy (*sym*-methyl stretch of CH_3CN at $\tau_1 = \tau_3 = 1$ ps). In each case, the energy of all pulses of the same frequency was varied simultaneously. The lines show the expected power law dependence with exponent a. These measurements help to confirm that a true 7th-order Raman echo is being measured. (Adapted from Ref. 5.)

faster than the FID signals as the laser intensity is increased. The echo-to-FID ratio improves as the square of the total Laser intensity and the square of the total Stokes intensity. Thus, applying the maximum intensity to the sample improves the signal-to-background ratio.

The intensity that can be used is ultimately limited by self-focusing and self-phase modulation in the sample. An early study concluded that these problems would preclude Raman echo experiments (31), but careful experimental design has pushed these limits back. However, the self-focusing problem remains the primary practical limitation in expanding the application of the Raman echo.

Another unique complication of high-order experiments like the Raman echo is the presence of combined resonant/nonresonant interactions. In a third-order experiment, like the Raman FID, there is only one time interval. The desired signal occurs when a resonant interaction causes the system to occupy a real state during that period. An "artifact" arises during pulse overlap because nonresonant interactions put the system into a "virtual" state during that time. In an experiment with multiple time intervals, new artifacts arise from mixing real states during some time intervals with virtual states in other intervals (4). As a result, the full material response function for the Raman echo is

$$M_{RE}(t_1, t_2, t_3) = \chi_1 C_{SRE}(t_1, t_2, t_3) + \chi_2 C_{FID}(t_1)e^{-t_2/T_1}\delta(t_3)$$
$$+ \chi_3\delta(t_1)\delta(t_2)C_{FID}(t_3) + \chi_4\delta(t_1)\delta(t_2)\delta(t_3) \qquad (34)$$

[see Equation (26)]. This more complete response function should be substituted for the more idealized $C_{SRE}(t_1, t_2, t_3)$ in Equation (30). The final term in Equation (34) is a fully nonresonant interaction and causes an extra peak at $\tau_1 = \tau_3 = 0$. The second and third terms are partially nonresonant and contribute peaks near, but slightly after, $\tau_3 = 0$ and $\tau_1 = 0$, respectively.

These nonresonant interactions significantly complicate the analysis at short times. As a result, independent measurements the of the $\tau_1 = 0$ and $\tau_3 = 0$ points on the time scale are important ingredients in the analysis. If these parameters are removed from the fitting, the convolution analysis of the Raman echo signals is stable and unique, even at short times.

The earlier discussion focused on the $\nu = 0$ and $\nu = 1$ vibrational states. However, in a high-order experiment, higher vibrational states must be considered as well. In fact, if the vibration is purely harmonic, the Raman echo signal is exactly canceled by additional processes involving the overtone states (56–60). Thus, anharmonicity in the vibration is essential. As the difference between the $0 \rightarrow 1$ and $1 \rightarrow 2$ transition frequencies increases beyond the combined linewidths of the two transitions, the echo signal regains its full strength. However, if the two transitions still lie within the bandwidth of the excitation pulses, quantum beats arise. This effect has been most clearly documented by Tokmakoff et al. in the context of the infrared echo (40). These beats contain information on overtone transitions, but they also complicate the interpretation of the dephasing of the fundamental. In all the experimental systems discussed below, the anharmonicity is sufficient to move the overtone completely out of the excitation bandwidth, so the simple two-state analysis of Fig. 3 is sufficient.

Recently, there has been a good deal of concern about cascaded low-order processes contaminating fifth-order echo experiments (52–55,125). In a cascaded process, the signal from a third-order process acts as an excitation for a second third-order process. Certain cascaded signals mimic fifth-order echo signals in frequency, direction, and excitation power dependence. There is a natural concern that analogous processes might interfere with the seventh-order Raman echo as well. However, all known cascaded processes involve a switch between an overtone transition and a fundamental transition, relative to the direct process. In a fifth-order echo, a weak overtone transition can be replaced by a strong fundamental, resulting in a strong cascaded process (51). A cascaded analog to the Raman echo must replace a strong fundamental transition with a weak overtone, resulting in an extremely weak cascaded signal.

The lack of a known cascaded process strong enough to compete with the Raman echo is reassuring, but the possibility of as-yet-unrecognized cascaded processes still exists. However, a direct experimental test for cascaded processes exists. All cascaded processes have a signal intensity that depends on the fourth power of the concentration of the vibrators, whereas direct processes depend on the second power (51,125). Interference between a direct and a cascaded process gives a third-power dependence on concentration. Measurements of Raman echo signal strength from the methyl stretch of CH_3CN as it is diluted in CD_3CN are shown in Fig. 6. The results show the expected second-power dependence, providing direct experimental confirmation that cascaded processes are not an issue for the Raman echo.

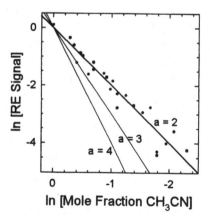

Figure 6 The concentration dependence of the Raman-echo signal (*sym*-methyl stretch of CH_3CN diluted in CD_3CN). The expected quadratic (a = 2) dependence is found. The quartic dependence (a = 4) of a cascaded nonlinear process alone or the cubic dependence (a = 3) of a cascade signal interfering with the direct signal is not found. (Adapted from Ref. 6.)

A final question in any dephasing measurement is the potential involvement of rotational dynamics. It has long been known that infrared and anisotropic Raman line shapes contain contributions from both rotational and dephasing dynamics, whereas the isotropic Raman line shape is immune to rotational dynamics and reflects only dephasing dynamics (21). The polarization conditions needed to extract analogous rotation-free, isotropic Raman FIDs have been well studied (106,113,126). In systems where rotation is slow relative to dephasing, the sensitivity of infrared measurements to rotation presents no problem. However, for small-molecule liquids, rotation is often as fast or faster than dephasing (Table 1). For this reason, Raman spectroscopy, rather than infrared, has dominated the study of dephasing (1,2).

Moving to higher-order echo experiments, it has been shown that the infrared echo is inescapably affected by rotation as well as dephasing (38). The full polarization and rotation dependence of the Raman echo has not been explored. However, if the vibrator has a purely isotropic Raman cross section, i.e., the depolarization ratio $\rho = 0$, the analysis is simple. The Raman echo is completely independent of rotational dynamics. This fact gives the Raman echo a significant advantage over the infrared echo for many systems.

All the experiments reported here are on vibrations with near-zero depolarization ratios (Table 1). In this case, the excitation pulse pairs must have the same polarization, but the relative polarization of different inter-actions is unimportant. In practice, we take all excitation polarizations perpendicular to the plane of the excitation beams and the L_{III} polar-ization parallel. In this configuration, the signals with parallel polariza-tion are only generated by scattering from L_{III}. A polarizer is placed in the signal beam to provide additional discrimination against competing nonlinear processes.

D. Experimental Equipment

The choice of laser system is dictated by the considerations outlined above. High-energy pulses of two different frequencies are needed, and good temporal synchronization between the pulses is essential. A pulse width of $0.3-1.0$ ps is adequate to resolve most dephasing processes. Good mode quality is important to minimize self-focusing.

We have performed Raman echo experiments using synchronously pumped dye lasers, followed by dye amplifiers pumped by a Q-switched Nd:YAG laser (Fig. 7). These systems are standard (127–129), but several

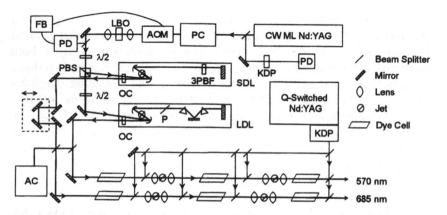

Figure 7 Schematic of the laser system used in the Raman FID and echo experiments. PC = Pulse compressor; AOM = acousto-optic modulator; PD = photodiode; FB = feedback electronics; PBS = polarizing beam-splitter; 3PBF = 3-plate birefringent filter; SDL/LDL = Stokes/Laser dye laser; P = pellicle; AC = autocorrelator; OC = output coupler; LBO/KDP = doubling crystals. Final pulses have widths of 0.5–1 ps and energies of 0.3–1 mJ (From Ref. 6.)

modifications have been made for Raman-echo experiments. Most obviously, the dye laser and amplifier chain have been duplicated to provide two independently tunable wavelengths. Instead of using the 80–100 ps pulse from a CW mode-locked Nd:YAG to pump the dye lasers directly, the Nd:YAG pulse is first compressed to ~ 2 ps in a fiber-grating compressor (130). The short pump pulse both shortens the dye laser pulses to the required 0.5–1 ps duration and markedly improves the synchronization of the two dye lasers. The synchronization is also improved by stabilizing the pump power with an acousto-optic attenuator driven by a feedback loop. Longitudinally pumped amplifiers are chosen for their good mode quality (128,129). Low-aberration optics, e.g., achromatic lenses, also improve the beam quality.

Recent advances in solid-state laser systems offer several potential improvements. In particular, tunable optical parametric amplifiers provide tunable pulses with excellent synchronization and beam quality. However, most reported systems produce pulses of <100 fs. Pulses this short are not needed to measure vibrational dephasing in most systems, and they unnecessarily exacerbate the problems of self-focusing and excitation of overtone transitions. A solid-state system optimized for longer pulses has not been used for Raman-echo experiments yet, but it has the potential to make these experiments considerably easier.

Once appropriate pulses are generated, the experimental setup is straightforward, if somewhat complex (Fig. 8). The original pulses are split and recombined to provide the required five pulses with approximately equal energy at the sample. Polarizer/half-waveplate attenuators allow fine optimization of the pulse energy. The beams are combined such that the I and II pulse pairs can be scanned in time as pairs. Each beam is separately focused, so the spot size at the sample can be controlled by the distance the sample is placed in front of the focal point. Translation stages containing the final mirror/lens combination allow the angles of the beams to be adjusted without changing the spot sizes. A chopper blocks alternate S_I pulses, and software subtracts the resulting measurements to correct for scattered light. A pinhole at the beam focus and interference filters separate the Raman echo signal from other beams. A PMT and 16-bit A/D converter have been found to provide the sensitivity, linearity, and dynamic range needed.

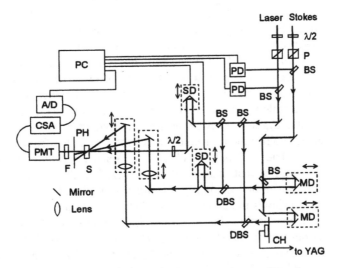

Figure 8 Schematic of the optical system used to perform the Raman FID and echo experiments. P = Polarizer; (D)BS = (dichroic) beamsplitter; MD = manual delay line; SD = computer-scanned delay line; CSA = charge sensitive amplifier; CH = chopper; PH = pinhole; S = sample; F = bandpass and neutral density filters; PD = photodiode; A/D = analog-to-digital converter; PC = computer; PMT = photomultiplier; $\lambda/2$ = half-wave plate. (From Ref. 6.)

IV. RECENT VIBRATIONAL DEPHASING RESULTS

A. Concentration Fluctuations in $CH_3I:CDCl_3$

As discussed in Section II.C, dephasing by local concentration fluctuations has been considered many times in the literature. However, the Raman echo has proven to be important in fully unraveling this process.

Although Schweizer and Chandler did not explicitly discuss concentration fluctuations (88), their central arguments on the differing effects of attractive and repulsive interactions can be applied to concentration fluctuations. If a vibration interacts with the solvent through the long-range attractive portion of the intermolecular potential, it will be sensitive to the average composition of the first solvation shell. The composition changes due to diffusion in and out of this shell, which takes 5–10 ps in typical liquids. This time is characteristically slow compared to the ~100 fs lifetime expected for short-range repulsive interactions. Schweizer and Chandler's analysis predicts that concentration fluctuations should lead to a slow dephasing process, if and only if the fluctuations are coupled to the vibration by a long-range attractive interaction. By using the Raman echo to determine the lifetime of the vibrational frequency fluctuations in liquid mixture, the validity of this scenario can be tested.

We decided to use this approach on the *sym*-methyl stretch of methyl iodide in chloroform-*d* (4). This system has a large concentration effect and has been previously studied by Döge et al. (131) and by Knapp and Fischer (87) on the basis of Raman linewidth measurements. We began by making more detailed measurements of the Raman line shape as a function of concentration. Figure 9 shows that the peak frequency shifts significantly as the average environment around the vibrator changes from pure CH_3I to pure $CDCl_3$. The shift is large (10.1 cm^{-1}) and linear in concentration. The linearity implies that there is ideal mixing, i.e., the CH_3I and $CDCl_3$ molecules are positioned randomly throughout the liquid.

Figure 10 shows the corresponding behavior of the Raman linewidth. The widths of the methyl line in a pure CH_3I environment and in a pure $CDCl_3$ environment are nearly identical. However, in a mixed environment, the line is wider and is widest for a 50:50 mixture. This behavior matches the expected effects of dephasing by local concentration fluctuations.

The FID from the 50:50 mixture is shown in Fig. 11. The decay is clearly nonexponential, and thus the line shape is not a pure Lorentzian. The decomposition of the line shape into Lorentzian and Gaussian components (132) is shown as a function of concentration in Fig. 10. Although there is some uncertainty in such a decomposition, the major features are clear.

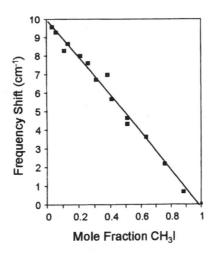

Figure 9 Shift in the peak frequency of the isotropic Raman line of the *sym*-methyl stretch in CH_3I as a function of concentration in $CDCl_3$. The shift is relatively and linear with concentration. The peak frequency in pure CH_3I is 2951 cm^{-1}. (From Ref. 4.)

There is a Lorentzian component of about 5.5 cm^{-1} FWHM, which is concentration independent. All of the concentration variation is due to a Gaussian component, which is nearly absent at the ends of the concentration range but reaches a maximum size of 4.25 cm^{-1} in the 50:50 mixture. The Lorentzian shape of the concentration independent component is suggestive of a fast modulation dephasing. The Gaussian shape of the concentration dependent component is suggestive of a slow modulation process.

Results of the Raman echo experiment on this system are shown in Fig. 12. The interpretation of these data proceeds by comparison with a series of models, which are constrained to be consistent with the linewidth and FID data. The simplest model assumes that the entire linewidth/FID is due to fast modulation processes, including the concentration-dependent process. The predictions are shown as solid curves in Fig. 12. At long τ_1 and τ_3, the predicted signal is substantially smaller than the observed signal. The enhanced signal results from a rephasing induced by the echo sequence and indicates that a slow modulation process must be present. On the other hand, the rephasing is not complete, so there must be a significant fast-modulation process as well.

The next obvious model is a combination of one fast process corresponding to the concentration-independent Lorentzian component of

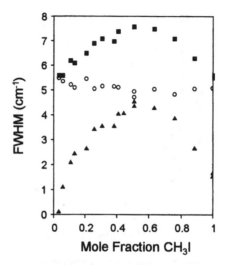

Figure 10 Width (FWHM) of the isotropic Raman line of the *sym*-methyl stretch in CH_3I as a function of concentration in $CDCl_3$ (■). Voight fits give Lorentzian (O) and Gaussian (▲) contributions to the line shape. The Lorentzian component is consistent with a concentration independent fast-modulation process. The Gaussian component suggests an additional contribution from slow concentration fluctuations. (From Ref. 4.)

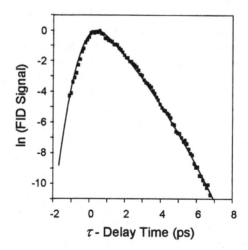

Figure 11 Raman FID of the *sym*-methyl stretch in 50:50 CH_3I:$CDCl_3$ showing a nonexponential decay (points). The fit (solid curve) is a combination of fast ($T_2 = 2.0$ ps) and slow ($\Delta_\omega = 4.25$ cm^{-1}) dephasing processes. (From Ref. 4.)

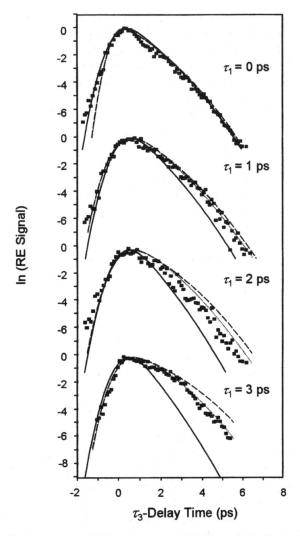

Figure 12 Raman-echo data from the *sym*-methyl stretch in 50:50 $CH_3I:CDCl_3$ (points). Three models for the dephasing are shown: solid, fast modulation only; dashed, fast and static modulation; dotted, fast and intermediate (5 ps) modulation. The best fit (dotted) is consistent with interaction of the vibration with local-concentration fluctuations. The fits are constrained to be consistent with FID and line shape data. (From Ref. 6.)

the line shape and one static process corresponding to the concentration-dependent Gaussian component. This model predicts the dashed curves in Fig. 12. The predicted intensity at large delays is larger than in the fast-modulation model in qualitative agreement with the data. But at a quantitative level this model overestimates the magnitude of the rephasing, and the predicted signal is too large at long times.

The third model assumes that the concentration fluctuations are long lived but not static. The dotted line in Fig. 12 shows good agreement with the data assuming a lifetime of 5 ps for the concentration fluctuations. A range of lifetimes from 4 to 7 ps is compatible with the data. This model not only agrees with the Raman echo data, it is also matches the FID and Raman line shape and peak position data as well. The lifetime found in the Raman echo implies that the Gaussian component of the line shape (4.25 cm^{-1}) is actually motionally narrowed from the full distribution of frequencies (5.15 cm^{-1}).

This type of lifetime data is almost impossible to extract accurately and unambiguously from line shape or FID data alone. Figure 11 shows that the second, infinite-lifetime model fits the FID very well. The Raman echo gives the lifetime information that is key to testing Schweizer and Chandler's analysis.

By extending Schweizer and Chandler's theory (88) to concentration fluctuations (see next section), we can relate the experimental dephasing parameters to molecular properties. From the size of the frequency shift upon dilution and the distribution of frequencies, the average number of solvent molecules perturbing the vibration is found to be 5.4. This result is close to the number of solvent molecules in the first solvation shell of the vibrating molecule. Thus, the relevant solute-solvent interaction has a range of approximately one molecular diameter. From the long range, we infer that the coupling is due to the attractive portion of the intermolecular potential. From the diffusion constant, the time for exchange of molecules in and out of the first solvation shell is estimated to be 6–10 ps, a range that overlaps with the experimental range of perturbation lifetimes. This system is a good example of a long-range attractive interaction producing a slow modulation of the vibration, as anticipated by the Schweizer-Chandler model.

B. Density Fluctuations in Acetonitrile

The experiments in $CH_3I:CDCl_3$ mixtures show that a long-range attractive interaction can make a vibration sensitive to fluctuations in the local composition. Do similar fluctuations in local density also perturb vibrations, even in a pure liquid? This question is the main topic of Schwiezer and Chandler's original theory (88). In their original paper, they not only point out the possibility of dephasing from density fluctuations, they also propose a method of estimating its magnitude from gas-to-liquid frequency shifts. This method predicts that in many common liquids, the magnitude of dephasing from density fluctuations will be similar to the magnitude from collisional dynamics.

Because density fluctuations produce a long-lived perturbation, whereas collisional dynamics are fast, the Raman echo is a definitive experiment for testing this prediction. An excellent system for this test is the *sym*-methyl stretch of acetonitrile (3). There have been many Raman line shape studies of this mode, which concluded that the IBC theory of collisional dynamics alone can account for the linewidth (84,90,91,133,134). On the other hand, calculations by George and Harris using Schweizer and Chandler's theory predicted that a substantial fraction of the linewidth is due to density fluctuations (135).

The Raman FID for this mode (Fig. 13) is a single exponential with $T_2 = 0.82$ ps, as expected from previous line shape measurements. Corresponding Raman echo results are shown in Fig. 14. The fit curves are derived from the FID by assuming that all the frequency modulations are fast. Simulations show that a slow component with a width of $\Delta_\omega \geq 1$ cm^{-1}

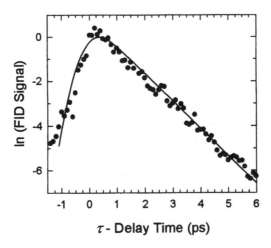

Figure 13 Raman FID of the *sym*-methyl stretch in CH_3CN (points). The exponential fit (solid curve) is consistent with the isotropic Raman line shape. (Adapted from Ref. 3.)

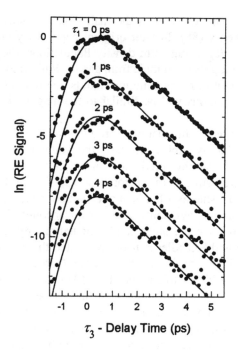

Figure 14 Raman-echo data (points) from the *sym*-methyl stretch in CH_3CN show no change as τ_1 is increased. The solid curves are predicted from the FID data assuming fast modulation. Only fast dephasing mechanisms are operating; the effects of slow density fluctuations are too small to observe. (Adapted from Ref. 3.)

would be detectable (3), whereas $\Delta_\omega = 5$ cm^{-1} was the predicted contribution from density fluctuations (135).

A similar conclusion comes from the results of Inaba et al. (10,11), who performed a Raman echo experiment on the CN stretch of benzonitrile. They saw no evidence for a slow modulation process in this liquid and, therefore, no evidence for dephasing by density fluctuations. It appears that the importance of density fluctuations is overemphasized by the Schweizer-Chandler theory and that they do not contribute significantly to dephasing in simple liquids.

Although the experimental finding of negligible dephasing from density fluctuations is clear, it needs to be reconciled with the equally clear finding of a significant role for concentration fluctuations. We extended Schweizer and Chandler's treatment of density fluctuations to a liquid experiencing both density and concentration fluctuations in order to compare the magnitudes of the two processes (4). The total magnitude of the vibrational frequency perturbations is

$$\Delta_\omega^2 = \chi \frac{\bar{\omega}^2}{N} + x_A x_B \frac{\delta\omega_{AB}^2}{N} \tag{35}$$

The first term is the contribution from density fluctuations; the second is the contribution from concentration fluctuations. The number of molecules in the solvation sphere is N, the mean gas-to-liquid shift is $\bar{\omega}$, the difference in gas-to-liquid shifts between neat and dilute solution is $\delta\omega_{AB}$, and the mole fractions of the components of the mixture are x_A and x_B. None of these factors favors the concentration term. The difference of shifts $\delta\omega_{AB}$ is generally smaller than the mean shift $\bar{\omega}$, and the maximum value of $x_A x_B = 0.5$. The largest difference between these two terms is the factor of the liquid compressibility χ. In typical liquids, $\chi = 0.02-0.05$ (88), so this factor greatly suppresses the effect of density fluctuations. Even in systems with strong coupling to concentrations fluctuations, like $CH_3I:CDCl_3$, the fluctuations are responsible for only about half of the total linewidth (see Section IV.A). If the contribution of density fluctuations is 20–50 times smaller than the contribution from concentration fluctuations, as Equation (35) predicts, then density fluctuations will be too weak to contribute significantly to the total linewidth in most systems.

A physical interpretation of the role of the compressibility can be extracted from the well-known compressibility equation (136):

$$\chi = 1 + \int dr_{12}[g_2(r_{12}) - 1] \tag{36}$$

which relates the compressibility to the radial distribution function $g_2(r_{12})$. The degree of correlation between the positions of solvent molecules is given by $g_2(r_{12}) - 1$. In liquids, these correlations are strong and the compressibility is low. If a solvent molecule tries to move to create a local density fluctuation, other solvent molecules will move in a correlated fashion to fill any resulting voids and to relieve the strain of any tightly packed regions. As a result of these correlations, density fluctuations are hard to create.

For concentration fluctuations, there is no term corresponding to χ. In an ideal solution, there are no correlations in the composition of neighboring molecules. The composition of a solvating molecule can change freely without forcing a compensating change in other solvent molecules. Concentration fluctuations are easy to create.

These ideas can be generalized to describe the effect of attractive interactions on vibrations. If the interaction couples to a highly correlated coordinate, like molecular positions, typical interaction strengths are too low to create substantial fluctuations in the vibrational frequency. However, if the interaction couples to a weakly correlated coordinate, like molecular composition or orientation, attractive interactions can compete with repulsive interactions in perturbing a vibration. Another example of this principle is the dephasing of CO bound to myoglobin, which is caused by movement of protein groups with partial charges (137–139). The charged groups are scattered throughout the protein, and as a result their movements are poorly correlated. As a result, the long-range electric fields of these charges can effectively dephasing the CO.

C. Stress Fluctuations in Toluene

Given that attractive interactions do not play an important role in pure liquids, what is the nature of the perturbations induced by repulsive interactions? In recent years there has been a growing appreciation of the fact that liquid dynamics have at least two major components: a fast inertial component and a slower diffusive component. Although these components are often distinguished by the magnitudes of their rates, a more definitive distinction is the viscosity dependence of their rates (140). The rate of inertial dynamics is independent of or weakly dependent on the viscosity. The rate of diffusive dynamics is inversely proportional to the viscosity. For "normal" low-viscosity liquids ($\sim$1 cP), the rate of diffusive and inertial dynamics can be similar. However at high viscosity, the rates become well distinguished. Thus, dephasing measurements in high-viscosity liquids are critical for separating the roles of inertial and diffusive solvent dynamics in perturbing vibrations.

Unfortunately, very little is known about dephasing in high-viscosity liquids. Aside from some recent work by Fayer's group (37,141,142), there are almost no experimental studies available. As discussed in Section II.C, the available dephasing theories do not explicitly consider this regime or distinguish between inertial and diffusive dynamics.

To explore the high-viscosity region, we performed FID measurements on the 1002 cm^{-1} of toluene over a range of temperatures (9). Toluene has a broad liquid range (384–178 K) and can be easily supercooled to $\sim$155 K. As a result, its viscosity can be varied from a low value (0.5 cP) near room temperature to a high value (100 cP) at low temperature (143).

Raman FID measurement in toluene over this range are shown in Fig. 15. In the high-temperature/low-viscosity region (250–230 K/1–2 cP),

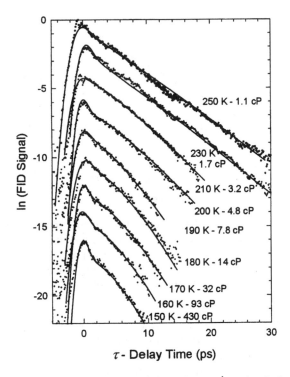

Figure 15 Raman FID measurements of the 1002 cm^{-1} mode of toluene at various temperatures and viscosities (points). The decays undergo a transition from slow and exponential in the high-temperature/low-viscosity range (250–230 K) to rapid and Gaussian in the low-temperature/high-viscosity range (180–150 K). The fit curves (solid) are a Kubo analysis of the shapes [see Equation (7) and Fig. 16].

the coherence decay is slow and exponential. In the low-temperature/high-viscosity region (180–150 K/14–430 cP), the decay is faster and close to Gaussian in shape. A transition region occurs at 210–190 K/3–10 cP. This pattern does not match the temperature/viscosity behavior predicted by any of the existing dephasing theories (Section II.C).

A Kubo analysis was attempted as a first approach to interpreting these data (61,63). The dephasing is attributed to a single frequency modulation process characterized by a magnitude Δ_ω and a correlation time τ_ω. The decay at each temperature was fit to Equation (7) to find these parameters. The results are plotted in Fig. 16. As expected, the magnitude of the frequency perturbations is independent of temperature. However, the correlation time shows a linear correlation to the viscosity of the solution. This

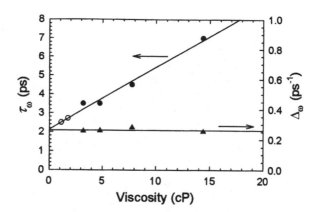

Figure 16 Parameters from the Kubo analysis of FID measurements in toluene (see Fig. 15): the frequency correlation time τ_ω(●) and the rms magnitude of the frequency perturbations Δ_ω(▲). When the FID is exponential, τ_ω(○) is estimated from T_2 by assuming that Δ_ω is a constant with temperature [Equation (8)]. The frequency correlation time is directly correlated with the viscosity.

result indicates that the frequency perturbations are related to viscous flow within the liquid. In other words, the relevant solvent dynamics includes the diffusive component, not just the inertial component. These ideas are not present in existing theories of dephasing.

D. A Viscoelastic Theory of Vibrational Dephasing

The FID results in toluene show the need for a theory of vibrational dephasing that works throughout the range from low-viscosity liquid to solid glass. It is also important that such a theory address the different roles of inertial and diffusive solvent dynamics explicitly. We developed the viscoelastic (VE) theory of vibrational dephasing to address these issues (8).

The concept behind this theory is illustrated in Fig. 17. The vibrating molecule is approximated as a spherical cavity within a continuum solvent, and the vibrational motion is approximated as a spherical "breathing" of the cavity. The radius of the cavity is determined by a balancing of forces: the tendency of the solvent to collapse an empty cavity, the intermolecular van der Waals attraction of the vibrator for the solvent molecules, and the intermolecular repulsion between the solvent molecules and the core of the vibrator. When the vibrator is in $\nu = 1$, the mean bond length of the vibrating bonds is longer due to anharmonicity. The increased bond length

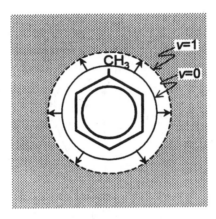

Figure 17 Schematic illustration of the viscoelastic (VE) model of dephasing. The vibrating molecule (toluene here) occupies a cavity within the solvent with a certain size in $v = 0$. In $v = 1$, the radius of the cavity is slightly larger, because of vibrational anharmonicity. This effect couples shear fluctuations of the solvent to the vibrational frequency. See also Fig. 18.

displaces the repulsive wall of the vibrator outward, forcing the solute cavity to expand.

The energetics of this model are shown in Fig. 18. The solute-cavity radius has a minimum energy at a radius r_0 when the vibrator is in $v = 0$ and a minimum at a larger radius $r_0 + dr$ when it is in $v = 1$. When the vibrator is in $v = 0$, there will be a distribution of cavity sizes at equilibrium due to thermal excitations in the solvent. These different cavities have different vertical transition energies, and the width of the size distribution maps into the width of transitions Δ_ω.

The frequency correlation time τ_ω corresponds to the time it takes for a single vibrator to sample all different cavity sizes. The fluctuation-dissipation theorem (144) shows that this time can be found by calculating the time for a vertically excited $v = 0$ vibrator to reach the minimum in $v = 1$. This calculation is carried out by assuming that the solvent responds as a viscoelastic continuum to the outward push of the vibrator. At early times, the solvent behaves elastically with a modulus G_∞. The push of the vibrator launches sound waves (acoustic phonons) into the solvent, allowing partial expansion of the cavity. This process corresponds to a rapid, inertial solvent motion. At later times, viscous flow of the solvent allows the remaining expansion to occur. The time for this diffusive motion is related to the viscosity η by G_∞ and the net force constant at the cavity

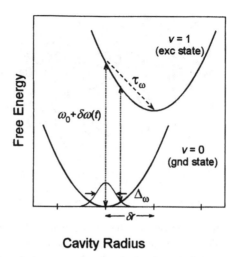

Cavity Radius

Figure 18   Free-energy potentials corresponding to the VE theory of dephasing. The equilibrium cavity radius is larger in $v = 1$ than in $v = 0$ by and amount δr (see Fig. 17). As a result, the energy of the vibrational transition depends on the cavity radius. At thermal equilibrium, shear fluctuations in the liquid will cause fluctuations in the radius, which in turn cause fluctuations in the transition frequency ω. (Adapted from Ref. 8.)

boundary K_S:

$$\tau_\omega = \left[\frac{1}{G_\infty} + \frac{4}{3K_S} \right] \eta \tag{37}$$

With this information, we can calculate the separate contributions of inertial and diffusive solvent motion to the linewidth.

The first result of this calculation is that the inertial motion causes almost no dephasing. This result is a direct contrast to models like the IBC theory, which attribute all the dephasing to collisional, i.e., inertial, dynamics. The difference between these theories lies in their assumptions about correlations in the solvent motion. The IBC explicitly assumes that the collisions are independent, i.e., the solvent motion has no correlations. As a result, the collisions are an effective sink for phase memory from the vibration. On the other hand, within the VE model the solvent motions appear as sound waves. Their effect on the vibrational frequency decays as they propagate away from the vibrator, but they remain fully coherent at all times. Because they remain coherent, they cannot destroy the phase memory of the vibration. The degree to which inertial solvent motion is coherent is an important and unresolved issue. The IBC and VE models represent extreme viewpoints on this issue, and the truth may lie somewhere in between. However, measuring the extent of vibrational dephasing due to inertial motion can provide incisive information on the degree of coherence in inertial solvent motion.

For the moment, assume that the VE picture is correct and inertial solvent motion causes negligible dephasing. Diffusive motion must be the primary cause of coherence decay. In the VE theory, the diffusive motion is the relaxation of stress fluctuations in the solvent by viscous flow. The VE theory calculates both the magnitude Δ_ω and lifetime τ_ω of the resulting vibrational frequency perturbations. A Kubo-like treatment then predicts the coherence decay as a function of the viscosity of the solvent. Figure 19 shows results for typical solvent parameters. At low viscosity, the modulation is in the fast limit, so the decay is slow and nearly exponential. Under these conditions, the dephasing time is inversely proportional to the viscosity, as in previous theories [Equation (19)]. As the viscosity increases, the modulation rate slows. The decay becomes faster and approaches a

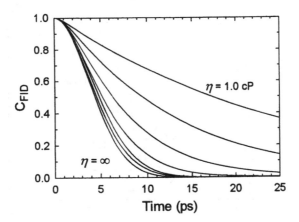

Figure 19 The VE prediction of the diffusive component of the vibrational coherence decay C_{FID} as a function of the solvent viscosity ($\eta = 1, 2, 4, 8, 16, 32$ and ∞ cP) for typical parameters. At low viscosity, the decay is exponential, its rate is inversely proportional to the viscosity, and the corresponding Raman line is homogeneously broadened. At high viscosity, the decay becomes Gaussian, its decay time reaches a limiting value, and the Raman line is inhomogeneously broadened. (From Ref. 8.)

Gaussian shape. In the limit of infinite viscosity, e.g., in a glass, the line shape approaches a well-behaved limit. These qualitative features are the same as were seen experimentally in supercooled toluene (Section IV.C).

A quantitative fit of the VE theory to the toluene data is shown in Fig. 20. In addition to the VE dephasing, an additional temperature-independent, fast-modulation dephasing process had to be included. Several reasonable mechanisms exist for this additional dephasing: the population lifetime and inertial dynamics acting through phonon scattering or through imperfect correlations in the solvent. The theory reproduces

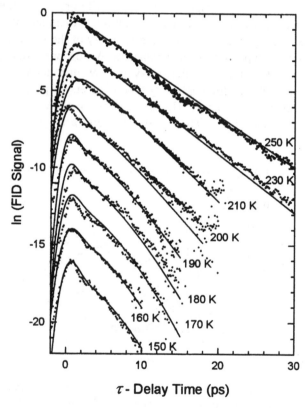

τ- Delay Time (ps)

Figure 20 The VE theory of dephasing fit to Raman-FID data on toluene (see Fig. 15). The entire range of data is fit with only three temperature-independent parameters: the solvent's high-frequency elastic modulus, a solvent-solute coupling constant, and a homogeneous dephasing time. A temperature-independent homogeneous dephasing process is assumed in addition to the VE mechanism.

the temperature-dependent changes of the decay shape very well. Only three parameters are used to fit the entire temperature range: the solvent modulus G_∞, the coupling strength between the solvent and vibration, and the dephasing time of the viscosity independent process T_2. (In contrast, the Kubo analysis (Fig. 15) uses two fitting parameters for each temperature.) The temperature-dependent changes in the coherence decay are determined by the independently measured viscosity.

The ability to correctly reproduce the viscosity dependence of the dephasing is a major accomplishment for the viscoelastic theory. Its significance can be judged by comparison to the viscosity predictions of other theories. As already pointed out (Section II.C 22), existing theories invoking repulsive interactions severely misrepresent the viscosity dependence at high viscosity. In Schweizer-Chandler theory, there is an implicit viscosity dependence that is not unreasonable on first impression. The frequency correlation time is determined by the diffusion constant D, which can be estimated from the viscosity and molecular diameter σ by the Stokes-Einstein relation:

$$\tau_\omega^{SC} = \frac{\sigma^2}{6D} = \frac{\pi\sigma^3}{2kT}\eta \tag{38}$$

In comparison, the correlation time in the viscoelastic theory is given by Equation (37). For typical parameters, $\tau_\omega^{SC} \sim 50\tau_\omega^{VE}$. Thus, the Schweizer-Chandler predicts frequency perturbations in the slow-modulation limit at low viscosity. In the VE theory, the slow-modulation limit is only reached at high viscosity. The fundamental difference is in the interaction assumed in each theory. Schweizer-Chandler theory assumes coupling through long-range attractive forces; VE theory assumes that short-range repulsive forces dominate. Thus, the toluene results are in accord with our earlier conclusion that repulsive interactions are primarily responsible for dephasing in pure liquids.

E. Solvent-Assisted IVR in Ethanol

The bandwidth of the symmetric methyl stretch in ethanol-1,1-d_2 is unusually wide (15 cm^{-1}) compared to other methyl-containing liquids, e.g., acetonitrile (6.5 cm^{-1}) and methyl iodide (5.7 cm^{-1}) (Table 1). (The deuteration prevents mixing of the methyl and methylene hydrogen vibrations.) Within the context of the dephasing ideas already presented, a number of possible reasons for this difference present themselves. However, a closer experimental examination of the system shows that none of these are correct, and a new dephasing mechanism must be considered (5).

The first experiment to show that this system is unusual is the Raman FID as a function of temperature (Fig. 21). Ethanol supercools easily and forms a glass at 97 K. But unlike the experiments in supercooled toluene (Fig. 15), the decay in ethanol at 80 K remains exponential, and the dephasing rate is changed only slightly at low temperature. This result eliminates shear fluctuations, density fluctuations, and other mechanisms with an explicit viscosity dependence.

Theories based on the Enskog collision time (84) or other solid-like approaches do not have a strongly temperature-dependent frequency correlation time. But they do have a temperature-dependent factor resulting from the need to create the solvent fluctuations in the first place. Thus, all fast-modulation theories predict that the dephasing rate will go to zero at 0 K.

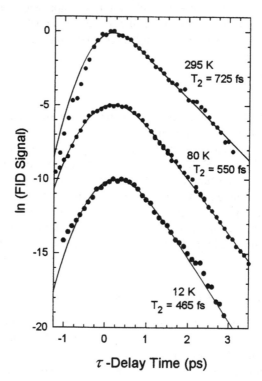

Figure 21 Raman FID (points) of the *sym*-methyl stretch in CH_3CD_2OH in the liquid (295 K), in the high-temperature glass (80 K), and in the low-temperature glass (12 K). Fits (solid curves) are based on exponential decays. The increase in dephasing rate at low temperatures is unexpected. (Adapted from Ref. 5.)

The Raman FID at 12 K shows that this prediction fails (Fig. 21). The rate is no slower at 12 K than at 80 K. In fact, it is slightly faster.

The fact that the dephasing rate is almost constant with temperature within the glass (80–12 K) suggests that the linewidth is due to a static distribution of structures frozen into the glass. This idea can be tested with the Raman echo. A line dominated by static broadening will show strong rephasing effects when τ_1 is increased. As Fig. 22 shows, this rephasing does not occur in ethanol, either in the liquid, high-temperature glass, or low-temperature glass.

The combination of FID and echo experiments has eliminated all possibilities for pure dephasing mechanisms. Slow-modulation mechanisms are inconsistent with the echo results; fast-modulation mechanisms are inconsistent with a broad line persisting at low temperature. Resonant

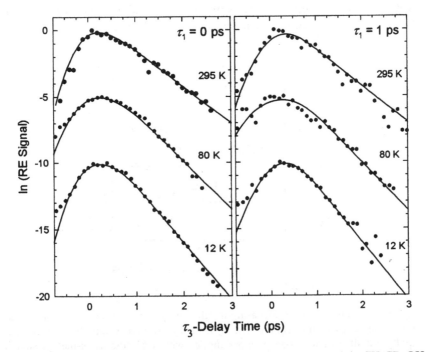

Figure 22 Raman-echo data (points) from the *sym*-methyl stretch in CH_3CD_2OH in the liquid (295 K), in the high-temperature glass (80 K), and in the low-temperature glass (12 K). In all cases, there is no change in the decays between $\tau_1 = 0$ and at $\tau_1 = 1$ ps, showing that there is no slow dephasing mechanism. (Adapted from Ref. 5.)

energy transfer can be considered, but isotope dilution experiments have eliminated this possibility as well (Table 1) (5). Population relaxation can cause line broadening [Equation (20)] but the measured time of 22 ps is much to long to contribute to the linewidth in ethanol (Table 1) (94–97).

At this point it is clear that none of the recognized dephasing mechanisms can account for the coherence decay in ethanol. We have proposed that solvent-induced intramolecular vibrational redistribution (IVR) is responsible. A number of other vibrational levels of the methyl group lie near the energy of the *sym*-methyl stretch, primarily the asymmetric stretch and the overtones and combinations of the bending modes (Fig. 23). These levels are separated by more than their respective linewidths, so they do not undergo conventional IVR. However, they do lie within kT of each other. It is possible that coupling to solvent motions causes rapid transfers between these state, i.e., solvent-induced IVR. Because the number of states involved is small, significant population will remain in the *sym*-stretch, even after the IVR process is over. The

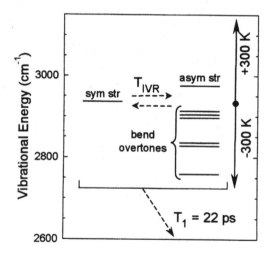

Figure 23 A proposal for dephasing in ethanol by solvent-assisted intramolecular vibrational redistribution (IVR). The *sym*-methyl stretch is initially excited, but rapidly equilibrates with one or more modes within kT (the *asym*-methyl stretch and/or CH bend overtones). Dephasing occurs with this rapid equilibration time T_{IVR}. However, significant population remains in the *sym*-methyl stretch after equilibration. Relaxation from this group of state to lower states causes the final relaxation of the population to zero, which is measured as T_1 in energy relaxation experiments. (Adapted from Ref. 7.)

measured T_1 reflects loss out of these nearly isoenergetic states to lower levels and can be much slower than the IVR time T_{IVR}.

Although T_{IVR} may not be reflected in a T_1 measurement, it will cause dephasing in the same way that T_1 would. The idea that solvent-induced IVR may be active in methyl groups has been suggested before (92,96,97), but there has not been a good estimate of its rate. If our proposal that the dephasing time is primarily due to IVR is true, the time is very fast, $T_{IVR} = $ 225 and 365 fs at 12 K and 295 K, respectively. Within this scenario, the fast T_{IVR} and broad linewidth can be attributed to the hydrogen-bonding network of ethanol, which provides an efficient bath to promote IVR.

V. SUMMARY

Before the introduction of the Raman echo, many dephasing mechanisms had been discussed, but most experimental work had been interpreted in terms of IBC or related theories. With the increased information available from the Raman echo and from measurements across a wider range of temperature, a much richer picture of vibrational dephasing is emerging. Two features of the system are important: the intermolecular coupling and the solvent dynamics. Each of these factors has two major components. The coupling has a long-range attractive part and a short-range repulsive part. The dynamics has a viscosity-independent inertial part and a viscosity-dependent diffusive part. The possible combinations between these components are summarized in Table 2.

Because the amplitude of inertial motion is small, it causes only small modulations of long-range interactions. There is no evidence that these weak interactions are important in dephasing.

Table 2 Classification of Dephasing Mechanisms by Type of Intermolecular Potential, Type of Solvent Dynamics, and Speed of the Resulting Frequency Modulation

	Potential	
Dynamics	Long range (attractive)	Short range (repulsive)
Inertial/Collisional	Weak (fast)	IBC/phonon (fast)
Diffusive	S-C (slow)	VE (fast-slow)

Long-range attractive interactions can be effectively modulated by diffusive dynamics, which have a much larger amplitude. Schweizer-Chandler theory treats dephasing from these effects. They produce slow-modulation dephasing, even at low viscosity. However, the diffusive motion must be along a coordinate with weak correlations. In the case of density fluctuations, the strong correlations between molecular positions suppress fluctuations to such an extent that dephasing by density fluctuations is not important in most simple liquids. This case is illustrated by acetonitrile (3) and benzonitrile (10,11) (Section IV.B). However, if a coordinate with weak correlations is introduced, dephasing by long-range forces can become significant. This case was illustrated by the mixture containing methyl iodide in which the number of methyl iodide neighbors is the weakly correlated coordinate (Section IV.A).

In pure liquids, short-range repulsive forces are responsible for most of the dephasing. The viscoelastic theory describes the interaction of these forces with the diffusive dynamics of the liquid (Section IV.D). The resulting frequency modulation is in the fast limit in low-viscosity liquids but can reach the slow-modulation limit at higher viscosities. This type of dephasing was seen in supercooled toluene (Section IV.C).

Short-range forces interacting with inertial dynamics are at the heart of IBC and related theories (Section II.C). These processes always produce fast-modulation dephasing. The existing studies are still unclear on the role of this kind of dephasing. Reasonable candidates exist, for example, the temperature-independent component of dephasing in toluene (Section IV.D). Additional studies are needed to make a conclusive assignment.

The enduring conclusions from dephasing studies are that there are a plethora of potential dephasing mechanisms, and the dominant mechanisms can change depending on the system examined. The framework just outlined is a useful organization of current results but probably does not exhaust the possibilities that might be found in a wider variety of systems. The ethanol study is a case in point, where none of the standard dephasing processes applies, and a new IVR process had to be postulated (Section IV.E).

Several fundamental questions have been raised by these investigations and hopefully will be answered in the near future. One set of questions concerns the role of inertial dynamics coupled by short-range forces. A clear example of this type of process has not yet been unambiguously identified. An important question is whether this type of dephasing is dominated by vibrational anharmonicity or by nonlinear coupling. In the case of vibrational anharmonicity, the degree of coherence in the inertial dynamics becomes a central issue. The IBC and VE theories make opposite and extreme assumptions about this coherence and make diametrically opposed predictions regarding the resulting dephasing.

A long-range goal of studying vibrational dephasing is to gain insight into solvent effects in general. An important step in attaining this goal is to make quantitative connections between the vibrational dephasing in a given liquid and other dynamical processes. The VE theory holds promise in this regard. An analogous VE theory of electronic solvation already exists (140,145), and an analogous theory of molecular rotation is easy to envision. Future studies will tell whether the VE framework can successfully unify these different dynamical processes.

The existing techniques of Raman FID and Raman echo in combination with studies extending across wide temperature and viscosity ranges have the potential to answer all these questions. With the ever-increasing sophistication of ultrafast laser technology, these experiments are becoming easier and more accessible. Armed with these techniques and the understanding obtained in simple liquids, vibrational dephasing also promises to be a route to deciphering the dynamics of more complex systems, such as polymers and biological systems.

ACKNOWLEDGMENTS

I thank the coworkers who helped with the original work discussed here: Prof. David A. Vanden Bout, Dr. Laura J. Muller, Dr. John E. Freitas, Dr. Xiaotian Zhang, and Hugh H. Hubble. I also thank Dr. Xun Pan and Prof. Richard MacPhail of Duke University for providing the Raman line shape data on $CH_3I:CDCl_3$. This work was supported by the National Science Foundation.

REFERENCES

1. Oxtoby DW. Adv Chem Phys 40:1, 1979.
2. Morresi A, Mariani L, Distefano MR. J Raman Spectrosc 26:179, 1995.
3. Vanden Bout D, Muller LJ, Berg M. Phys Rev Lett 67:3700, 1991.
4. Muller LJ, Vanden Bout DA, Berg M. J Chem Phys 99:810, 1993.
5. Vanden Bout D, Freitas JE, Berg M. Chem Phys Lett 229:87, 1994.
6. Vanden Bout D, Berg M. J Raman Spectrosc 26:503, 1995.
7. Berg M, Vanden Bout DA. Acc Chem Res 30:65, 1997.
8. Berg M. Chem Phys 233:257, 1998.
9. Hubble HW, Lai T, Berg MA. J Chem Phys (submitted).

10. Inaba R, Tominaga K, Tasumi M, Nelson KA, Yoshihara K. Chem Phys Lett 211:183, 1993.
11. Tominaga K, Yoshihara K. Prog Cryst Growth Charact Mat 33:371, 1996.
12. Banin U, Bartana A, Ruhman S, Kosloff R. J Chem Phys 101:8461, 1994.
13. Waldman A, Ruhman S, Shaik S, Sastry GN. Chem Phys Lett 230:110, 1994.
14. Wang Q, Schoenlein RW, Peteanu LA, Mathies RA, Shank CV. Science 266:422, 1994.
15. Vos MH, Jones MR, Martin J-L. Chem Phys 233:179, 1998.
16. Reichardt C. Solvents and Solvent Effects in Organic Chemistry. 2nd ed. Weinheim: VCH Publishers, Weinheim, 1988.
17. Stratt RM, Maroncelli M. J Phys Chem 100:12981, 1996.
18. Fleming GR, Cho MH. Ann Rev Phys Chem 47:109, 1996.
19. Ma J, Fourkas JT, Vanden Bout DA, Berg M. In: Fourkas JT, Kivelson D, Mohanty U, eds. Supercooled Liquids: Advances and Novel Applications. Vol. 676. Washington, DC: American Chemical Society, 1997, p 199.
20. Gordon RG. Advances in Magnetic Resonance. Vol. 3. New York: Academic Press, 1968:1–42.
21. Nafie LA, Peticolas WL. J Chem Phys 57:3145, 1972.
22. Laubereau A, Kaiser W. Rev Mod Phys 50:607, 1978.
23. Penzkofer A, Laubereau A, Kaiser W. Prog Quantum Electron 6:55, 1979.
24. Slichter CP. Principles of Magnetic Resonance. Berlin: Springer-Verlag, 1989.
25. Zinth W, Polland HJ, Laubereau A, Kaiser W. App Phys B 26:77, 1981.
26. Loring RF, Mukamel S. J Chem Phys 83:2116, 1985.
27. Hartmann SR. IEEE J Quantum Electron QE–4:802, 1968.
28. Leung KP, Mossberg TW, Hartmann SR. Opt Commun 43:145, 1982.
29. Leung KP, Mossberg TW, Hartmann SR. Phys Rev A 25:3097, 1982.
30. Brückner V, Bente EAJM, Langelaar J, Bebelaar D. Opt Commun 51:49, 1984.
31. Müller M, Wynne K, Van Voorst JDW. Chem Phys 128:549, 1988.
32. Levenson MD. Introduction to Nonlinear Laser Spectroscopy. San Jose, CA: Academic Press, 1982.
33. Walsh CA, Berg M, Narasimhan LR, Fayer MD. Acc Chem Res 20:120, 1987.
34. Mukamel S. Principles of Nonlinear Optical Spectroscopy. New York: Oxford University Press, 1995.
35. Zimdars D, Tokmakoff A, Chen S, Greenfield SR, Fayer MD, Smith TI, Schwettman HA. Phys Rev Lett 70:2718, 1993.
36. Tokmakoff A, Zimdars D, Sauter B, Francis RS, Kwok AS, Fayer MD. J Chem Phys 101:1741, 1994.
37. Tokmakoff A, Zimdars D, Urdahl RS, Francis RS, Kwok AS, Fayer MD. J Phys Chem 99:13310, 1995.
38. Tokmakoff A, Fayer MD. J Chem Phys 103:2810, 1995.

39. Tokmakoff A, Fayer MD. Acc Chem Res 28:437, 1995.
40. Tokmakoff A, Kwok AS, Urdahl RS, Francis RS, Fayer MD. Chem Phys Lett 234:289, 1995.
41. Hamm P, Lim M, Asplund M, Hochstrasser RM. Chem Phys Lett 301:167, 1999.
42. Hamm P, Lim M, Hochstrasser RM. Phys Rev Lett 81:5326, 1998.
43. Steffen T, Duppen K. Phys Rev Lett 76:1224, 1996.
44. Steffen T, Fourkas JT, Duppen K. J Chem Phys 105:7364, 1996.
45. Tominaga K, Keogh GP, Naitoh Y, Yoshihara K. J Raman Spectrosc 26:445, 1995.
46. Tominaga K, Yoshihara K. J Chem Phys 104:4419, 1995.
47. Tominaga K, Yoshihara K. Phys Rev Lett 76:987, 1996.
48. Tominaga K, Yoshihara K. J Phys Chem A 102:4222, 1998.
49. Tokmakoff A, Lang MJ, Larsen DS, Fleming GR. Chem Phys Lett 272:48, 1997.
50. Pfeiffer M, Lau A. J Chem Phys 108:4159, 1998.
51. Ulness DJ, Kirkwood JC, Albrecht AC. J Chem Phys 108:3897, 1998.
52. Kirkwood JC, Ulness DJ, Albrecht AC, Stimson MJ. Chem Phys Lett 293:417, 1998.
53. Kirkwood JC, Albrecht AC. J Chem Phys 111:1, 1999.
54. Kirkwood JC, Albrecht AC, Ulness DJ, Stimson MJ. J Chem Phys 111:1, 1999.
55. Blank DA, Kaufman LJ, Fleming GR. J Chem Phys 111:3105, 1999.
56. Tanimura Y, Mukamel S. J Chem Phys 99:9496, 1993.
57. Khidekel V, Mukamel S. Chem Phys Lett 240:304, 1995.
58. Leegwater JA, Mukamel S. J Chem Phys 102:2365, 1995.
59. Mukamel S, Piryatinski A, Chernyak V. Acc Chem Res 32:145, 1999.
60. Fourkas JT, Kawashima H, Nelson KA. J Chem Phys 103:4393, 1995.
61. Kubo R. Fluctuation, Relaxation and Resonance in Magnetic Systems. London: Oliver and Boyd, 1961:23–68.
62. Anderson PW. J Phys Soc Jpn 9:316, 1954.
63. Rothschild WG. J Chem Phys 65:455, 1976.
64. Klauder JR, Anderson PW. Phys Rev 125:912, 1962.
65. Mims WB. Phys Rev 168:370, 1968.
66. Hu P, Hartmann SR. Phys Rev B 9:1, 1974.
67. Oxtoby DW. J Chem Phys 74:1503, 1981.
68. Rothschild WG, Perrot M, Guillaume F. Chem Phys Lett 128:591, 1986.
69. Wynne K, Müller M, Brandt D, Van Voorst JDW. Chem Phys 125:211, 1988.
70. Müller M, Wynne K, Van Voorst JDW. Chem Phys 125:225, 1988.
71. Oxtoby DW, Rice SA. Chem Phys Lett 42:1, 1976.
72. Oxtoby DW. J Phys Chem 70:2605, 1979.
73. Oxtoby DW. J Chem Phys 68:5528, 1978.
74. Madden PA, Lynden-Bell RM. Chem Phys Lett 38:163, 1976.

75. Diestler DJ. Chem Phys Lett 39:39, 1976.
76. Knauss DC. Mol Phys 36:413, 1978.
77. Fischer SF, Laubereau A. Chem Phys Lett 35:6, 1975.
78. Dardi PS, Cukier RI. J Chem Phys 89:4145, 1988.
79. Lynden-Bell RM. Mol Phys 33:907, 1977.
80. Wertheimer RK. Mol Phys 35:257, 1978.
81. Ouillon R, Sergiescu V, Tascón D'León. Mol Phys 49:151, 1983.
82. Levine AM, Shapiro M, Pollak E. J Chem Phys 88:1959, 1988.
83. Fukuda T, Ikawa S, Kimura M. Chem Phys 133:137, 1989.
84. Schroeder J, Schiemann VH, Sharko PT, Jonas J. J Chem Phys 66:3215, 1977.
85. Schvaneveldt SJ, Loring RF. J Chem Phys 104:4736, 1996.
86. Bondarev AF, Mardaeva AI. Opt Spectrosc 35:167, 1973.
87. Knapp EW, Fischer SF. J Chem Phys 76:4730, 1982.
88. Schweizer KS, Chandler D. J Chem Phys 76:2296, 1982.
89. Keutel D, Seifert F, Oehme K-L. J Chem Phys 99:7463, 1993.
90. Yarwood J, Arndt R, Doge G. Chem Phys 25:387, 1977.
91. Tanabe K. Chem Phys 38:125, 1979.
92. Spanner K, Laubereau A, Kaiser W. Chem Phys Lett 44:88, 1976.
93. Deak JC, Iwaki LK, Dlott DD. J Phys Chem A 102:8193, 1998.
94. Laubereau A, von der Linde D, Kaiser W. Phys Rev Lett 28:1162, 1972.
95. Alfano RR, Shapiro SL. Phys Rev Lett 29:1655, 1972.
96. Laubereau A, Kehl G, Kaiser W. Opt Commun 11:74, 1974.
97. Fendt A, Fischer SF, Kaiser W. Chem Phys 57:55, 1981.
98. Iwaki LK, Deak JC, Rhea ST, Dlott DD. Chem Phys Lett 303:176, 1999.
99. Döge G, Arndt R, Khuen A. Chem Phys 21:53, 1977.
100. Amorim da Costa AM, Norman MA, Clarke JHR. Mol Phys 29:191, 1975.
101. Bartoli FJ, Litovitz TA. J Chem Phys 56:413, 1972.
102. Bauer DR, Alms GR, Brauman JI, Pecora R. J Chem Phys 61:2255, 1974.
103. Logan DE. Chem Phys 103:215, 1986.
104. Döge G, Schneider D. Mol Phys 80:525, 1993.
105. Zinth W, Leonhardt R, Holzapfel W, Kaiser W. IEEE J Quantum Electron 24:455, 1988.
106. Okamoto H, Yoshihara K. J Opt Soc Am B 7:1702, 1990.
107. Inaba R, Okamato H, Yoshihara K, Tasumi M. J Phys Chem 97:7815, 1993.
108. Gale GM, Guyot-Sionnest P, Zheng WQ. Opt Commun 58:395, 1986.
109. Zinth W, Laubereau A, Kaiser W. Opt Commun 26:457, 1978.
110. Kohles N, Laubereau A. App Phys B 39:141, 1986.
111. Kohles N, Aechtner P, Laubereau A. Opt Commun 65:391, 1988.
112. Fickenscher M, Laubereau A. J Raman Spectrosc 21:857, 1990.
113. Purucker H-G, Tunkin V, Laubereau A. J Raman Spectrosc 24:453, 1993.
114. Schaertel SA, Lee D, Albrecht AC. J Raman Spectrosc 26:889, 1995.
115. Joo T, Dugan MA, Albrecht AC. Chem Phys Lett. 177:4, 1991.
116. Joo T, Albrecht AC. J Chem Phys 99:3244, 1993.

117. Bigot JY, Portella MT, Schoenlein RW, Bardeen CJ, Migus A, Shank CV. Phys Rev Lett 66:1138, 1991.
118. Bardeen CJ, Shank CV. Chem Phys Lett 203:535, 1993.
119. Joo T, Jia Y, Fleming GR. J Chem Phys 102:4063, 1995.
120. Vöhringer P, Arnett DC, Yang TS, Scherer NF. Chem Phys Lett 237:387, 1995.
121. de Boeij WP, Pshenichnikov MS, Wiersma DA. Chem Phys Lett 238:1, 1995.
122. Cho M, Yu J-Y, Joo T, Nagasawa Y, Passino SA, Fleming GR. J Phys Chem 100:11944, 1996.
123. Passino SA, Nagasawa Y, Joo T, Fleming GR. J Phys Chem A 101:725, 1997.
124. Gallagher SM, Albrecht AW, Hybl JD, Landin BL, Rajaram B, Jonas DM. J Opt Soc Am B 15:2338, 1998.
125. Ivanecky JE, Wright JC. Chem Phys Lett 206:437, 1993.
126. Dick B. Chem Phys 113:131, 1987.
127. Migus A, Shank CV, Ippen EP, Fork RL. IEEE J Quantum Electron QE-18:101, 1982.
128. Koch TL, Chiu LC, Yariv A. Opt Commun 40:364, 1982.
129. Koch TL, Chiu LC, Yariv A. J Appl Phys 53:6047, 1982.
130. Kafka JD, Kolner BH, Baer T, Bloom DM. Opt Lett 9:505, 1984.
131. Döge G, Arndt R, Buhl H, Bettermann G. Z Naturforsch A 35:468, 1980.
132. Davies JT, Vaughan JM. Astrophys J 137:1302, 1963.
133. Yarwood J, Ackroyd R, Arnold KE, Doge G, Arndt R. Chem Phys Lett 77:239, 1981.
134. Tanabe K. Chem Phys Lett 84:519, 1981.
135. George SM, Harris CB. J Chem Phys 77:4781, 1982.
136. McQuarrie DA. Statistical Mechanics. New York: Harper Collins, 1976.
137. Rella CW, Rector KD, Kwok A, Hill JR, Schwettman HA, Dlott DD, Fayer MD. J Phys Chem 100:15620, 1996.
138. Rector KD, Rella CW, Hill JR, Kwok AS, Sligar SG, Chein EYT, Dlott DD, Fayer MD. J Phys Chem B 101:1468, 1997.
139. Rector KD, Engholm JR, Hill JR, Myers DJ, Hu R, Boxer SG, Dlott DD, Fayer MD. J Phys Chem B 102:331, 1998.
140. Berg M. J Phys Chem A 102:17, 1998.
141. Rector KD, Fayer MD. J Chem Phys 108:1794, 1998.
142. Rector KD, Kwok AS, Ferrante C, Francis RS, Fayer MD. Chem Phys Lett 276:217, 1997.
143. Barlow AJ, Lamb J, Matheson AJ. Proc Roy Soc A 292:322, 1966.
144. D Chandler. Introduction to Modern Statistical Mechanics. New York: Oxford University Press, 1987.
145. Berg M. Chem Phys Lett 228:317, 1994.

10

Fifth-Order Two-Dimensional Raman Spectroscopy of the Intermolecular and Vibrational Dynamics in Liquids

Graham R. Fleming and David A. Blank*
University of California at Berkeley, and Lawrence Berkeley National Laboratory, Berkeley, California

Minhaeng Cho
Korea University, Seoul, South Korea

Andrei Tokmakoff
Massachusetts Institute of Technology, Cambridge, Massachusetts

I. INTRODUCTION

In recent years there has been significant interest in the extension of nonlinear optical spectroscopy to higher orders involving multiple time and/or frequency variables. The development of these multidimensional techniques is motivated by the desire to probe the microscopic details of a system that are obscured by the ensemble averaging inherent in linear spectroscopy. Much of the recent work to extend time domain vibrational spectroscopy to higher dimensionality has involved the use of nonresonant Raman-based techniques. The use of Raman techniques has followed directly from the rapid advancements in ultrafast laser technology for the visible and near-IR portions of the spectrum. Time domain nonresonant Raman spectroscopy provides access to an extremely

---

* *Current affiliation*: University of Minnesota, Minneapolis, Minnesota

broad range of frequencies, from intermolecular motions (far-IR) to intramolecular vibrations (mid-IR). The entire frequency range can be probed simultaneously with the lower limit on the frequency determined by the maximum delay between laser pulses and the upper limit on the frequency determined by the bandwidth of the laser pulses.

In time domain nonresonant Raman spectroscopy of isotropic media, the third-order response is the lowest-order nonzero response. The measured response is governed by the third-order susceptibility, $\chi^3(\omega)$, or equivalently in the time domain the third-order response function, $R^{(3)}(\tau)$ (1). Experimental examples include impulsive stimulated scattering (ISS) (2), optical Kerr-effect spectroscopy (OKE), and optical heterodyne detected Raman-induced Kerr-effect spectrocopy (OHD-RIKES) (3,4). In these experiments, two light field interactions, initially overlapped in time, drive a vibrational coherence, the evolution of which is probed by a final Raman interaction occurring at some adjustable time later. There is a single time period under experimental control, and the measured response carries exactly the same information that is contained in an incoherent light scattering experiment involving a single frequency variable (1). Thus, the third-order time domain nonresonant Raman experiment is considered a one-dimensional measurement that probes a macroscopic ensemble averaged response.

Fifth-order Raman spectroscopy, originally proposed by Tanimura and Mukamel, was initially directed towards investigating the relative contributions of homogeneous and inhomogeneous line broadening (5). The experiment involves preparation and probing steps that are very similar to those of the third-order Raman experiment. However, between the preparation of the initial vibrational coherence and the probing step, there is a second interaction with an additional pair of light fields that transfers the initial vibrational coherence to a second vibrational coherence. The second coherence is then probed in the final step. Thus, fifth-order time domain Raman spectroscopy is a two-dimensional technique in that it has two controllable time periods, and the response is governed by the fifth-order response function, $R^{(5)}(\tau_2, \tau_4)$. The sequence of pulses and a representative ladder diagram and Feynman diagram for an $R^{(5)}(\tau_2, \tau_4)$ response are shown in Fig. 1. The salient feature of these experiments is the direct transfer of the first vibrational coherence to the second vibrational coherence. For example, transferring an initial coherence to a second coherence of nearly equal frequency and opposite phase results in a rephasing process during the second time interval: this is analogous to spin echo (6,7) and photon echo (8–10) experiments (Fig. 1b). In this example, heterogeneous line broadening can then be removed via the rephasing events.

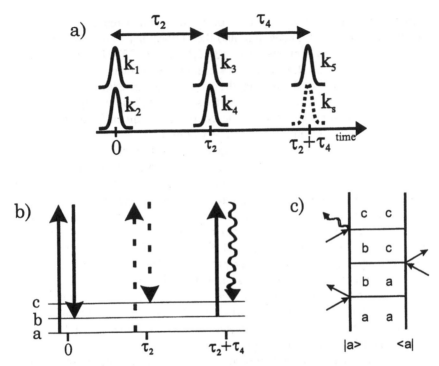

Figure 1 (a) Diagram of the temporal relationship between the five incoming laser pulses and the signal. (b) Ladder diagram for the fifth order nonresonant Raman response. Solid arrows represent a field-matter interaction on the ket, dotted arrows are field-matter interaction on the bra, and the wavy line represents the emitted signal field. (c) A two-sided Feynman diagram depicting the same pathway as (b). Time is running vertically up the figure.

Following the initial proposal, fifth-order time domain Raman spectroscopy received considerable attention both theoretically (11–20) and experimentally (21–32). For the case of intermolecular motions, the majority of the experimental efforts have involved probing the intermolecular modes of liquid CS_2, a standard system in nonresonant Raman spectroscopy due to its very large polarizability and the wealth of available experimental results. Experimental efforts to probe intramolecular vibrations are fewer in number, with the only published example probing modes of liquid CH_3Cl and CCl_4 (24). It was quickly realized that, owing to the direct transfer of the first vibrational coherence to the second, the experiment offered substantially more information than had initially been considered. There is not only the ability to generate rephasing events in individual vibrational modes, but also the opportunity to directly probe the coupling between vibrations (5,11,13,16,17,22–26,33). The existence of coupling between the Raman active vibrations is intrinsic to the generation of the fifth-order signal, making the experiment an exquisitely sensitive probe of these microscopic interactions. The fifth-order signal has the potential to identify the nature of the coupling and quantify the magnitude of the coupling between Raman active vibrations.

While being one of the techniques greatest strengths, the direct dependence on the nonlinearity of the system, along with the fact that the technique involves a high-order nonresonant response, results in an extremely small relative magnitude for the signal. As a direct consequence of the small signal magnitude, it is important to consider, and difficult to avoid, interference from a large potential number of unwanted signals. Very recently it has become clear that one of the most difficult interfering signals to avoid is generated by cascading third-order processes. These lower-order cascades generate signals in the same phase matched direction and have the same overall dependence on the laser fluence as the direct fifth-order process. As a result, interference from these third-order cascaded signals currently poses the greatest obstacle to the broad success of two-dimensional fifth-order nonresonant Raman spectroscopy.

In this chapter we will provide a theoretical description of the two-dimensional fifth-order nonresonant Raman response, and we will present the current state of experimental progress in this area. Our discussions will include a detailed analysis of the cascading third-order processes, which, as will be demonstrated, currently play a leading role in the experimental pursuit of the unadulterated true fifth-order response.

II. THEORETICAL BACKGROUND

For a more detailed description of the topics in this section, see Ref. 34.

A. General: Nonresonant Nonlinear Optical Response

In the most general case of the nonresonant fifth-order light scattering measurement, five off-resonant optical pulses are injected, as shown in Fig. 1. Two pairs of pulses are used to create two consecutive vibrational coherence states, and the fifth pulse is scattered by the temporal and spatial grating thus created. The controlled delay times between the first two pairs of pulses and between the second pair of pulses and the final pulse are denoted as τ_2 and τ_4, respectively (see Fig. 1). The external field is thus given by a sum of the five fields:

$$E_{ex}(\mathbf{r}, t) = \sum_{j=1}^{5} E_j(\mathbf{r}, t) \tag{1}$$

where

$$E_j(\mathbf{r},t) = u_j \varepsilon_j(t) \exp(i\mathbf{k}_j \cdot \mathbf{r} - i\omega_0 t) + c.c. \tag{2}$$

Here c.c. refers to the complex conjugate and $\varepsilon_j(t)$ and u_j are the temporal envelope function and unit vector of the polarization of the jth electric field. The frequencies of the five fields are assumed to be identical, i.e., $\omega_j = \omega_0$ for $j = 1, 2, 3, 4$, and 5. This can be experimentally achieved by using time-delayed pulses generated from a common laser oscillator. Although the femtosecond pulses generated in the laboratory have a finite width, for the sake of simplicity the laser pulses are assumed to be impulsive in this section, i.e.:

$$\varepsilon_1(t) = \varepsilon_2(t) = \varepsilon\delta(t)$$
$$\varepsilon_3(t) = \varepsilon_4(t) = \varepsilon\delta(t - \tau_2)$$
$$\varepsilon_5(t) = \varepsilon\delta(t - \tau_2 - \tau_4) \tag{3}$$

Even though the laser pulses are approximated as delta functions, the slowly varying amplitude approximation can still be applied to pulses as wide as tens of femtoseconds, where the time scales for the nuclear degrees of freedom remain much slower than the pulse width (1).

When the external field frequency is far off-resonant with respect to the optical transition, the field-matter interaction can be described by the effective interaction Hamiltonian given by

$$H_I^{eff} = -\hat{\alpha}{:}\mathbf{E}^2(\mathbf{r}, t) \tag{4}$$

Then, the Liouville equation for the ground state density matrix, $\rho_g(\mathbf{r},t)$, is given as

$$\frac{\delta}{\delta t}\rho_g(\mathbf{r}, t) = -\frac{i}{\hbar}[H_0 + H_I^{eff}, \rho_g(\mathbf{r}, t)] \tag{5}$$

where H_0 is the material Hamiltonian determining the nuclear dynamics in the electronic ground state. The second-order density matrix, expanded with respect to the field-matter interactions, can be obtained as

$$\rho_g^{(2)}(\mathbf{r}, t) = \frac{i}{\hbar}\int_0^{\infty} dt_1 \mathbf{E}^2(\mathbf{r}, t - t_1) e^{-iH_0 t_1/\hbar}[\hat{\alpha}, \rho_{eq}]e^{iH_0 t_1/\hbar} \tag{6}$$

where ρ_{eq} denotes the equilibrium density operator in the electronic ground state. It should be noted that the effective interaction Hamiltonian defined in Equation (6) represents two field-matter interactions. Thus, the first-order term with respect to the effective interaction Hamiltonian is essentially identical to the second-order perturbation expansion term with respect to the field-matter interaction when the hyperpolarizability contribution is ignored.

The $2n$-order expanded density matrix, which is obtained by the n-order perturbation expansion with respect to the effective interaction Hamiltonian, is then given as

$$\rho_g^{(2n)}(\mathbf{r}, t) = \left(\frac{i}{\hbar}\right)^n \int_0^\infty dt_n \cdots$$

$$\int_0^\infty dt_1 \mathbf{E}^2(\mathbf{r}, t - t_n) \cdots \mathbf{E}^2(\mathbf{r}, t - \cdots t_2 - t_1)$$

$$\times e^{-iH_0 t_n / \hbar} \lfloor \hat{\alpha}, \cdots e^{-iH_0 t_2 / \hbar} \lfloor \hat{\alpha}, e^{-iH_0 t_1 / \hbar} [\hat{\alpha}, \rho_{eq}] e^{iH_0 t_2 / \hbar} \rfloor e^{iH_0 t_2 / \hbar} \cdots \rfloor e^{iH_0 t_n / \hbar}$$

$$(7)$$

By definition, the polarization is the expectation value of the dipole operator,

$$\mathbf{P}^{(n)}(\mathbf{r}, t) = N \, \mathrm{Tr}[\hat{\mu}_{ind}(\mathbf{r}, t)\rho_g^{(n-1)}(\mathbf{r}, t)]$$

$$= N\mathbf{E}(\mathbf{r}, t) \cdot \mathrm{Tr}[\hat{\alpha}\rho_g^{(n-1)}(\mathbf{r}, t)] \qquad (8)$$

where the induced dipole operator was defined as $\hat{\mu}_{ind}(\mathbf{r}, t) = \mathbf{E}(\mathbf{r}, t) \cdot \hat{\alpha}$, and N denotes the number density.

The spatial amplitude of the electric field generated by one of the n-order induced polarizations should satisfy the following Maxwell equation:

$$\nabla \times \nabla \times \{\varepsilon_s^{(n)}(\mathbf{r}, t) \exp(i\mathbf{k}_s' \cdot \mathbf{r})\} - \frac{n_s^2 \omega_s^2}{c^2} \varepsilon_s^{(n)}(\mathbf{r}, t) \exp(i\mathbf{k}_s' \cdot \mathbf{r})$$

$$= \frac{4\pi \omega_s^2}{c^2} \overline{\mathbf{P}}_s^{(n)}(t) \exp(i\mathbf{k}_s \cdot \mathbf{r}) \qquad (9)$$

Here we look for a solution of the form, $\mathbf{E}^{(n)}(\mathbf{r}, t) = \varepsilon^{(n)}(\mathbf{r}, t) \exp(i\mathbf{k}_s' \cdot \mathbf{r} - i\omega_s t)$, where $\mathbf{k}_s'$ is different from $\mathbf{k}_s$, which is given by a combination of the incoming wave vectors, due to the frequency dispersion of the refractive index of the optical sample.

Within the slowly varying amplitude approximation, the generated electric field amplitude grows linearly with respect to the distance from the front boundary, z, of the optical sample. Direct integration over z gives the generated electric field amplitudes of the n-order NLO processes, approximately given as

$$\varepsilon^{(n)}(t) = iA\ F^{(n)}\overline{P}^{(n)}(t) \tag{10}$$

where the constant A is defined as

$$A_{sj} \equiv \frac{2\pi|\omega_{sj}|}{n_s c}\left(\frac{1}{2}\right) \tag{11}$$

and the phase matching factor, $F^{(n)}$, is defined as

$$F^{(n)} = \frac{\sin[\Delta kl/2]}{\Delta kl/2}\exp(i\Delta kl/2) \tag{12}$$

with $\Delta k = k - k'$. Here n_s is the refractive index of the sample at the frequency of ω_s, and l is the sample thickness. The factor of 2 in the denominator inside the parenthesis of Equation (1) is introduced to take the average amplitude of the generated field within the sample — note that the generated field amplitude increases linearly from 0 at the front boundary to a maximum value proportional to l at the rear boundary so that the average field amplitude within the optical sample is proportional to 1/2.

B. Direct Fifth-Order Electrically Nonresonant Scattering

The fifth-order nonlinear polarization can be obtained by inserting the electric field given in Equation (1) into Equation (8). In particular, we look for a specific polarization component whose wave vector and frequency are given as, for example, $k_s = k_1 - k_2 - k_3 + k_4 + k_5$ and $\omega_s = \omega_1 - \omega_2 - \omega_3 + \omega_4 + \omega_5 = \omega_0$, respectively:

$$P_{dir}^{(5)}(r, t) = \overline{P}_{dir}^{(5)}(t)\exp(ik_s \cdot r - i\omega_s t) \tag{13}$$

Then, within the impulsive limit, carrying out the double integration involved in the calculation of $\rho_g^{(4)}(r, t)$ and using Equation (4), we find the amplitude of the direct fifth-order off-resonant scattering polarization:

$$\overline{P}_{dir}^{(5)}(t) = N\varepsilon^5 R^{(5)}(\tau_2, \tau_4)\delta(t - \tau_2 - \tau_4) \tag{14}$$

Using Equation 10, the electric field generated from the direct fifth-order polarization can be expressed as

$$\varepsilon_{dir}^{(5)}(t) = iAF^{(5)}\overline{P}_{dir}^{(5)}(t) = iAF^{(5)}N\varepsilon^5 R^{(5)}(\tau_2, \tau_4)\delta(t - \tau_2 - \tau_4) \tag{15}$$

where the fifth-order nonlinear response function is defined as

$$R^{(5)}(\tau_2, \tau_4) = -\frac{1}{\hbar^2} \langle [[\hat{\alpha}(\tau_2 + \tau_4), \hat{\alpha}(\tau_2)], \hat{\alpha}(0)] \rho_{eq} \rangle \tag{16}$$

This is the standard result originally obtained by Tanimura and Mukamel (5). It should be noted that α is a second rank tensor with respect to the relative polarizations of the interacting fields, and thus $R^{(5)}$ is a sixth rank tensor. However, for the sake of simplifying the notation, we have dropped the tensoral indices.

The 2D response [Equation (16)] can be expressed in terms of the coordinates via a Taylor expansion of the polarizability operator (12,16,27):

$$\alpha(t) = \alpha_0 + \sum_j \alpha_j^{(1)} Q_j(t) + \frac{1}{2} \sum_{j,k} \alpha_{j,k}^{(2)} Q_j(t) Q_k(t) + \cdots \tag{17}$$

where $\alpha^{(n)}$ is the nth derivative of the polarizability with respect to the coordinate(s). For third-order nonresonant Raman spectroscopy, it is common to use the Placzek approximation and assume that all higher-order terms in the Taylor expansion are negligible, thus truncating the expansion at the linear term. However, the assumption of a linear dependence of the polarizability on the coordinates within a harmonic ground state potential leads to a zero value for the three-point correlation function in Equation (16). This demonstrates the intrinsic dependence of the fifth-order signal on the microscopic coupling in the system. There are two types of coupling that can generate the fifth-order signal: anharmonicity in the vibrational potential, AN, and nonlinearity in the dependence of the polarizability on the vibrational coordinate, NP (5,12,28). The resulting response can be expressed as the sum of the two individual contributions:

$$R^{(5)}(\tau_2, \tau_4) = R^{(5),AN}(\tau_2, \tau_4) + R^{(5),NP}(\tau_2, \tau_4) \tag{18}$$

The anharmonic response can be obtained by inclusion of the cubic anharmonicity in the vibrational potential,

$$V \cong \frac{1}{2} \sum_j k_j Q_j^2 + \frac{1}{3!} \sum_{ijk} g_{ijk}^{(3)} Q_i Q_j Q_k \tag{19}$$

where $g^{(3)}$ is the third derivative of the ground state potential with respect to the coordinates. The magnitude of the individual AN and NP responses reflects the magnitude of the potential anharmonicity, $g^{(3)}$, and the magnitude of the nonlinearity of the polarizability, $\alpha^{(2)}$, respectively. The individual responses based on AN and NP coupling can be expressed in terms of single time, two point coordinate correlation functions (12,16):

$$G_j(t) \equiv -\tfrac{i}{2}\langle[Q_j(t), Q_j(0)]\rangle$$

$$R^{(5),AN}(\tau_2, \tau_4) \propto \sum_{ijk} g^{(3)}_{ijk}\alpha^{(1)}_i\alpha^{(1)}_j\alpha^{(1)}_k$$

$$\int_0^\infty d\tau G_i(\tau_4 - \tau)G_j(\tau)G_k(\tau_2 - \tau) \tag{20}$$

$$R^{(5),NP}(\tau_2, \tau_4) \propto \sum_{ij} \alpha^{(2)}_{ij}\alpha^{(1)}_i\alpha^{(1)}_j G_i(\tau_4)G_j(\tau_2 + \tau_4)$$

$$+ \alpha^{(1)}_i\alpha^{(2)}_{ij}\alpha^{(1)}_j G_i(\tau_4)G_j(\tau_2) \tag{21}$$

C. Cascaded Fifth-Order Electronically Nonresonant Scattering

In the above derivation of the direct fifth-order response, the five field-matter interactions used to create the macroscopic induced polarization, [Equation (14)] were restricted to the five external fields. In other words, internal fields generated by lower-order induced polarizations were completely ignored in the field-matter interactions. This approximation is very good for the lowest-order NLO processes, such as four-wave mixing spectroscopies. However, the internal field produced by the lower-order induced nonlinear polarization can be of crucial importance in higher-order NLO processes such as the fifth-order processes considered here.

There are two types of third-order cascades that can lead to an overall fifth-order signal. Ladder diagrams for the two types of cascades are shown in Fig. 2a and b, and we label them sequential and parallel. A sequential cascade (Fig. 2a) involves the emission of a field from the first vibrational coherence that then participates in driving a vibrational coherence on a different chromophore. A parallel cascade (Fig. 2b) begins with the preparation of vibrational coherences on separate chromophores as a result of interactions with the first and second pairs of laser fields. The probing interaction on one of the chromophores then results in a field that is involved in a probing interaction on the other chromophore.

We consider here only those cascading processes that produce a signal field satisfying the same example phase matching condition considered above for the direct fifth-order process, $\mathbf{k}_s = \mathbf{k}_1 - \mathbf{k}_2 - \mathbf{k}_3 + \mathbf{k}_4 + \mathbf{k}_5$. For this phase matching condition we must consider a total of four third-order intermediate polarizations: two for each of the two types of cascades. The two symmetrical cascade intermediates are (a) $\mathbf{k}_{s1} = \mathbf{k}_1 -$

a) Sequential Cascade

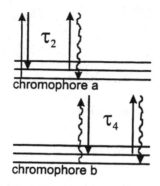

b) Parallel Cascade

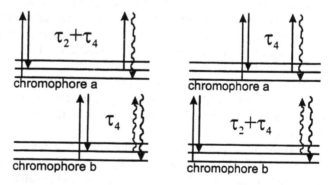

Figure 2 Representative ladder diagrams for the possible cascading fifth-order pathways. (a) The sequential cascade pathway. (b) The parallel cascade pathways.

$k_2 - k_3$, $\omega_{s1} = \omega_1 - \omega_2 - \omega_3 = -\omega_0$, (b) $k_{s2} = k_1 - k_2 + k_4$, $\omega_{s2} = \omega_1 - \omega_2 + \omega_4 = \omega_0$; the two parallel cascade intermediates are (a) $k_{p1} = k_1 - k_2 + k_5$, $\omega_{p1} = \omega_1 - \omega_2 + \omega_5 = \omega_0$, and (b) $k_{p2} = -k_3 + k_4 + k_5$, $\omega_{p2} = -\omega_3 + \omega_4 + \omega_5 = \omega_0$. Using Equation (8), the corresponding third-order induced polarizations are expressed as

$$\overline{P}_{s1}^{(3)}(t) = \overline{P}_{s2}^{(3)}(t) = N\varepsilon^3 R^{(3)}(t)\delta(t - \tau_2)$$

$$\overline{P}_{p1}^{(3)}(t) = N\varepsilon^3 R^{(3)}(t)\delta(t - \tau_2 - \tau_4)$$

$$\overline{P}_{p2}^{(3)}(t) = N\varepsilon^3 R^{(3)}(t - \tau_2)\delta(t - \tau_2 - \tau_4) \tag{22}$$

where the third-order Raman response function is

$$R^{(3)}(t) = \frac{i}{\hbar} \langle [\hat{\alpha}(t), \hat{\alpha}(0)] \rho_{eq} \rangle \tag{23}$$

The spatial amplitude of the electric fields generated by these third-order induced polarizations can be obtained using Equation (10). The four electric fields are then expressed as

$$\varepsilon_{s1}^{(3)}(t) = iA_{s1}F_{s1}^{(3)}\overline{P}_{s1}^{(3)}(t) = iA_{s1}F_{s1}^{(3)}N\varepsilon^3 R^{(3)}(t)\delta(t - \tau_2)$$

$$\varepsilon_{s2}^{(3)}(t) = iA_{s2}F_{s2}^{(3)}\overline{P}_{s2}^{(3)}(t) = iA_{s2}F_{s2}^{(3)}N\varepsilon^3 R^{(3)}(t)\delta(t - \tau_2)$$

$$\varepsilon_{p1}^{(3)}(t) = iA_{p1}F_{p1}^{(3)}\overline{P}_{p1}^{(3)}(t) = iA_{p1}F_{p1}^{(3)}N\varepsilon^3 R^{(3)}(t)\delta(t - \tau_2 - \tau_4)$$

$$\varepsilon_{p2}^{(3)}(t) = iA_{p2}F_{p2}^{(3)}\overline{P}_{p2}^{(3)}(t) = iA_{p2}F_{p2}^{(3)}N\varepsilon^3 R^{(3)}(t - \tau_2)\delta(t - \tau_2 - \tau_4) \tag{24}$$

with A_{xj} and F_{xj} defined in Equations (11) and (12).

In addition to the five external fields, the molecular system can interact with these four internal fields generated by the third-order NLO scattering processes. The total electric field is therefore

$$E(r,t) = E_{ex}(r,t) + E^{(3)}(r,t) + \cdots \tag{25}$$

where the external field, $E_{ex}(r,t)$, was given in Equation (1), and the third-order internal field, which is involved in the cascading processes contributing to the fifth-order signal, is given as

$$E^{(3)}(r, t) = \sum_{j=1}^{4} \varepsilon_{sj}^{(3)}(t) \exp(ik_{sj} \cdot r - i\omega_{sj}t) + c.c. \tag{26}$$

If one of these internal fields, in place of one of the external fields, participates in one of the three field-matter interactions of another third-order NLO process, the cascading coherent field will be generated. The corresponding four fifth-order cascaded polarization amplitudes are then expressed as

$$\overline{P}_{seq1}^{(5)}(t) = iA_{s1}F_{s1}^{(3)}N^2\varepsilon^5 R^{(3)}(t - \tau_2)R^{(3)}(\tau_2)\delta(t - \tau_2 - \tau_4)$$

$$\overline{P}_{seq2}^{(5)}(t) = iA_{s2}F_{s2}^{(3)}N^2\varepsilon^5 R^{(3)}(t - \tau_2)R^{(3)}(\tau_2)\delta(t - \tau_2 - \tau_4)$$

$$\overline{P}_{par1}^{(5)}(t) = iA_{p1}F_{p1}^{(3)}N^2\varepsilon^5 R^{(3)}(t - \tau_2)R^{(3)}(t)\delta(t - \tau_2 - \tau_4)$$

$$\overline{P}_{par2}^{(5)}(t) = iA_{p2}F_{p2}^{(3)}N^2\varepsilon^5 R^{(3)}(t - \tau_2)R^{(3)}(t)\delta(t - \tau_2 - \tau_4) \tag{27}$$

As can be seen from Equation (27), the sequential cascades result in responses that are symmetrical along the two adjustable time variables, while the parallel cascades result in responses that are asymmetrical along the two time variables. Adding together all four cascaded contributions to obtain the total fifth-order cascaded polarization,

$$\overline{\mathbf{P}}_{\text{cas}}^{(5)}(t) = iAN^2\varepsilon^5(F_{s1}^{(3)} + F_{s2}^{(3)})R^{(3)}(t - \tau_2)R^{(3)}(\tau_2)\delta(t - \tau_2 - \tau_4)$$
$$+ iAN^2\varepsilon^5(F_{p1}^{(3)} + F_{p2}^{(3)})R^{(3)}(t - \tau_2)R^{(3)}(t)\delta(t - \tau_2 - \tau_4) \quad (28)$$

where $A = A_{s1} = A_{s2} = A_{p1} = A_{p2}$. Using Equation (10) the field associated with the total cascaded response is expressed as

$$\varepsilon_{\text{cas}}^{(5)}(t) = iAF^{(5)}\overline{\mathbf{P}}_{\text{cas}}^{(5)}(t) = -A^2F^{(5)}N^2\varepsilon^5(F_{s1}^{(3)} + F_{s2}^{(3)})$$
$$\times R^{(3)}(t - \tau_2)R^{(3)}(\tau_2)\delta(t - \tau_2 - \tau_4)$$
$$- A^2F^{(5)}N^2\varepsilon^5(F_{p1}^{(3)} + F_{p2}^{(3)})R^{(3)}(t - \tau_2)R^{(3)}(t)\delta(t - \tau_2 - \tau_4) \quad (29)$$

D. The Total Nonresonant Fifth-Order Raman Signal

Since the direct and cascaded responses satisfy the same phase matching condition, the total nonresonant fifth-order Raman signal will contain both contributions. Using Equations (11), (14), and (28), the ratio between the absolute values of the direct and cascaded contributions is

$$\frac{|\overline{\mathbf{P}}_{\text{cas}}^{(5)}(t)|}{|\overline{\mathbf{P}}_{\text{dir}}^{(5)}(t)|} = \left(\frac{\pi\omega_0 lN}{nc}\right)\frac{|\{(F_{s1}^{(3)} + F_{s2}^{(3)})R^{(3)}(\tau_2) + (F_{p1}^{(3)} + F_{p2}^{(3)})R^{(3)}(\tau_2 + \tau_4)\}R^{(3)}(\tau_4)|}{|R^{(5)}(\tau_2, \tau_4)|} \quad (30)$$

Thus, the ratio is dependent on experimental parameters such as the optical path length, sample number density, and the phase matching conditions for the intermediate third-order processes, as well as the ratio of the third- and fifth-order response functions. The ratio of the response functions is directly related to the magnitude of the nonlinearity in the system, which is reflected by the magnitude of the potential anharmonicity, $g^{(3)}$, and the nonlinearity in the polarizability, $\alpha^{(2)}$. For example, let us consider only the NP contribution to the direct fifth-order response [Equation (21)]. For simplicity we will consider a system represented by a single mode, in other words the response is isotropic. If we express the third-order response functions in term of the coordinate [Equation (17)] and ignore all higher order terms,

$$R^{(3)}(t) \propto \alpha^{(1)}\alpha^{(1)}C(t) \quad (31)$$

Then the ratio between the cascaded and direct response functions becomes

$$\frac{R_{cascade}^{(5)}}{R_{NP}^{(5)}} \propto \frac{[\alpha^{(1)}]^4}{[\alpha^{(1)}]^2 \alpha^{(2)}} = \frac{[\alpha^{(1)}]^2}{\alpha^{(2)}} \tag{32}$$

Thus, in the NP case, the ratio on the right-hand side of Equation (30) is proportional to the ratio of the square of the linear coefficient of the polarizability to the quadratic anharmonicity in the polarizability with respect to the coordinate.

In the experimental section of this chapter we will present spectra obtained using both homodyne and heterodyne detection methods. The total homodyne detected signal is expressed as

$$\begin{aligned}
S_{homo}(\tau_2, \tau_4) &= \frac{n_s c}{8\pi} \int_{-\infty}^{\infty} dt \ |\varepsilon_{cas}^{(5)}(t) + \varepsilon_{dir}^{(5)}(t)|^2 \\
&= \frac{n_s c}{8\pi} \int_{-\infty}^{\infty} dt \ |iAF^{(5)}\overline{P}_{cas}^{(5)}(t) + iAF^{(5)}\overline{P}_{dir}^{(5)}(t)|^2
\end{aligned} \tag{33}$$

and will inherently contain both the cascaded and direct contributions to the fifth-order signal. In contrast to homodyne detection, heterodyne detection is sensitive to the phase of the electric fields. Bearing this aspect in mind, consider the two fifth-order fields associated with the cascaded and direct processes [Equations (29) and (15)]. If the phase-matching conditions for both third- and fifth-order processes are assumed to be perfect so that $F_{sj}^{(3)} = F^{(5)} \cong 1$, the amplitude of the cascaded fifth-order scattering field, $\varepsilon_{cas}^{(5)}(t)$, is purely real, whereas that of the direct process, $\varepsilon_{dir}^{(5)}(t)$, is purely imaginary. Therefore, if a heterodyne-detection technique based on injecting a phase-controlled local oscillator field is used, one may be able to separately detect the cascaded and direct contributions. That is (35),

$$\begin{aligned}
S_{hetero}(\tau_2, \tau_4) &= \frac{n_s c}{4\pi} \int_{-\infty}^{\infty} dt \ Re\{\varepsilon_{LO}^*(t, \phi) \cdot \varepsilon_{cas}^{(5)}(t)\} \\
&\quad + Re\{\varepsilon_{LO}^*(t, \phi) \cdot \varepsilon_{dir}^{(5)}(t)\}
\end{aligned} \tag{34}$$

This result suggests that controlling the phase factor, ϕ, to make $\varepsilon_{LO}^*(t, \phi)$ imaginary, one can selectively measure the direct component alone. In this case, the optical phase of the local oscillator field should be controlled with respect to that of the final laser field. However, experimentally this is quite challenging to implement. The inability to have perfect phase matching in an experiment with finite laser pulse dimensions results in phase shifting of the signals, thus lessening the discrimination against the cascaded components. An example of this approach is presented in the experimental section using an intrinsic heterodyne technique where the local oscillator is provided by a scattered third-order signal.

III. SIMULATIONS: THE BROWNIAN OSCILLATOR MODEL

To demonstrate the potential of two-dimensional nonresonant Raman spectroscopy to elucidate microscopic details that are lost in the ensemble averaging inherent in one-dimensional spectroscopy, we will use the Brownian oscillator model and simulate the one- and two-dimensional responses. The Brownian oscillator model provides a qualitative description for vibrational modes coupled to a harmonic bath. With the oscillators ranging continuously from overdamped to underdamped, the model has the flexibility to describe both collective intermolecular motions and well-defined intramolecular vibrations (1). The response function of a single Brownian oscillator is given as,

$$G_i(t) = \eta_i \sin(\Omega_i t) \exp(-\Lambda_i t) \tag{35}$$

where $\eta_i = -\hbar/2m_i\Omega_i$, $\Omega_i = \sqrt{\omega_i - \Lambda_i^2}$ is the reduced frequency, and $\Lambda_i = \gamma_i/2$ is the damping constant. Within this model the nonlinear polarizability contribution to the two-dimensional response is directly apparent from Equation (21). The anharmonic contribution can be obtained from Equation (20) by inserting the Brownian oscillator model and evaluating the integral

$$R^{(5)}_{AN,ijk}(\tau_2, \tau_4) = 12g^{(3)}_{ijk}\alpha^{(1)}_i\alpha^{(1)}_j\alpha^{(1)}_k\eta_i\eta_j\eta_k$$

$$\times \sum_{a,b=\pm 1} \frac{\exp(-\Lambda_k\tau_2)}{\Gamma^2_{ijk}}\{ab\exp(-(\Lambda_k + \Lambda_j)\tau_4)$$

$$\times [\xi_{ijk,ab}\cos(\Omega_k(\tau_2 + \tau_4) + b\Omega_j\tau_4)$$

$$+ \Gamma_{ijk}\sin(\Omega_k(\tau_2 + \tau_4) + b\Omega_j\tau_4)$$

$$- b\exp(-\Lambda_i\tau_4)(\xi_{ijk,ab}\cos(a\Omega_k\tau_2 + \Omega_i\tau_4))$$

$$+ \Gamma_{ijk}\sin(a\Omega_k\tau_2 + \Omega_i\tau_4)]\} \tag{36}$$

with $\xi_{ijk,ab} = (\Omega_i - ab\Omega_j - a\Omega_k)$ and $\Gamma_{ijk} = (\Lambda_i - \Lambda_j - \Lambda_k)$.

A. Intermolecular Motions in CS_2

We first consider the intermolecular modes of liquid CS_2. One of the details that two-dimensional Raman spectroscopy has the potential to reveal is the coupling between intermolecular motions on different time scales. We start with the one-dimensional Raman spectrum. The best linear spectra are based on time domain third-order Raman data, and these spectra demonstrate the existence of three dynamic time scales in the intermolecular response. In Fig. 3 we have modeled the one-dimensional time domain spectrum of CS_2 for 3 cases: (A) a single mode represented by the sum of three Brownian oscillators, (B) three Brownian oscillators, and (C) a distribution of 20 arbitrary Brownian oscillators. Case (A) represents the fully coupled, or isotropic case where the liquid is completely homogeneous on the time scales of the simulation. Case (B) deconvolutes the linear response into the three time scales that are directly evident in the measured response and is in the limit that the motions associated with each of the three timescales are uncoupled. Case (C) is an example where the liquid is represented by a large distribution of uncoupled motions.

As can be seen in Fig. 3, the simulations in all three cases result in the same one-dimensional time domain Raman response, thus providing a direct example of the insensitivity of a linear measurement to the microscopic details of the system. However, the two-dimensional responses demonstrate striking differences between the three cases. For the case of nonlinear polarizability coupling, we can simulate the direct two-dimensional fifth-order nonresonant Raman response using Equation (20). The two-dimensional simulations for cases (A)–(C) are shown in Fig. 3. All three simulations show a peak around $\tau_2 = \tau_4 = 180$ fsc that corresponds to the peak in the time domain third-order signal. While having finite values along the $\tau_2 = 0$ axis, all three cases have responses that rise from zero at the $\tau_4 = 0$ axis as a result of causality. Aside from these two similarities in the simulated two-dimensional spectra, the two-dimensional responses vary substantially among the three cases. For the isotropic case, (A), there is a clear asymmetry to the response along the two adjustable time variables, with the decay along τ_4 falling away faster than the decay along τ_2. There is no evidence for any rephasing events, such as illustrated in the ladder diagram in Fig. 1b, which would lead to a ridge along the time diagonal, $\tau_2 = \tau_4$. Case (B) is nearly symmetrical in the two time variables and shows some signs of a developing ridge along the diagonal, and in case (C) there is a clear ridge along the time diagonal. The diagonal ridge reflects rephasing within the large distribution of uncoupled oscillators analogous to photon echoes (9).

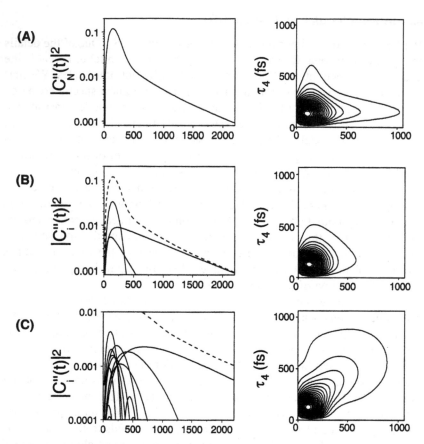

Figure 3 Brownian oscillator simulations for the 1D (left) and 2D (right) Raman response for liquid CS_2 at room temperature. (A) The response is modeled as a single oscillator represented by a sum of the three Brownian oscillators in (B). (B) The response is modeled by three independent Brownian oscillators. (C) The response is modeled by 20 randomly distributed Brownian oscillators. Note that all three cases reproduce the same 1D time dependent response but exhibit clear differences in the 2D responses.

Figure 3 is a clear example of the ability of the two-dimensional fifth-order Raman response to deconvolve the intermolecular spectral density based on the degree of coupling between motions on different time scales. Although the association of the generalized time scales represented by Brownian osciallators with specific molecular motions is certainly imperfect, it is possible to make some broad assignments. For example, it is well accepted that the fastest components in the spectral density can be assigned to inertial motions and the slowest components can be assigned to diffusive reoganizational motions (36,37). Thus, we can associate these two types of molecular motions with the highest-and lowest-frequency Brownian oscillators in Fig. 3B respectively. The remaining intermediate time scale is less well defined and could be associated with translations that generate an interaction-induced polarizability (38,39). Figures 3A and 3B then represent the two limiting cases where these three types of motions are (A) fully coupled or (B) fully uncoupled on the time scales of the simulations. Of course there is a full range of available coupling between each of the assigned types of motion (assigned time scales), and each given set of couplings will produce a unique two-dimensional response. One can then use these coupling magnitudes as adjustable parameters when simulating experimental spectra (27). Although we have given an example for the NP coupling model, and thus the couplings refer to the nonlinearity in the polarizability, $\alpha_{ij}^{(2)}$, the AN coupling model will also generate unique two-dimensional responses that are different from the NP case and strongly dependent on the anharmonic couplings, $g_{ijk}^{(3)}$ (27). These simulations demonstrate that the fifth-order two-dimensional Raman response is sensitive to both the nature and magnitude of coupling between intermolecular motions on different time scales.

B. Intramolecular Vibrations in Carbon Tetrachloride

To provide an example of the two-dimensional response from a system containing well-defined intramolecular vibrations, we will use simulations based on the polarized one-dimensional Raman spectrum of CCl_4. Due to the continuous distribution of frequencies in the intermolecular region of the spectrum, there was no obvious advantage to presenting the simulated responses of the previous section in the frequency domain. However, for well-defined intramolecular vibrations the frequency domain tends to provide a clearer presentation of the responses. Therefore, in this section we will present the simulations as Fourier transformations of the time domain responses. Figure 4 shows the Fourier transformed one-dimensional Raman spectrum of CCl_4. The spectrum contains three intramolecular vibrational modes — ν_2 at 218 cm^{-1}, ν_4 at 314 cm^{-1}, and ν_1 at 460 cm^{-1} — and a broad contribution from intermolecular motions peaked around 40 cm^{-1}. We have simulated these modes with three underdamped and one overdamped Brownian oscillators, and the simulation is shown over the data in Fig. 4.

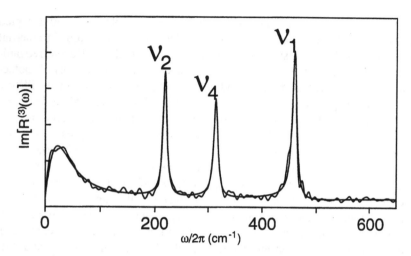

Figure 4 The 1D polarized Raman spectrum from liquid CCl_4 at room temperature fit with a sum of four Brownian oscillators.

For the two-dimensional response we will first consider the case of nonlinear polarizability coupling and simulate the response using Equation (21) and the Brownian oscillators used to fit the one-dimensional spectrum in Fig. 4. Figure 5 shows the simulations for the limiting cases where the system is (a) fully uncoupled, $\alpha_{ij}^{(2)} = \delta_{ij}$, and (b) fully coupled, $\alpha_{ij}^{(2)} = 1$. For the uncoupled case (Fig. 5a), the response is simply additive in the four modes. The peaks along the frequency diagonals ($\omega_2 = \omega_4 = \pm\Omega_i$) are the peaks observed in the one-dimensional spectrum. The only other features observed are the sum and difference frequency peaks that arise from the interaction of a vibrational coordinate with itself: the overtones ($2\omega_2 = \omega_4 = 2\Omega_i$) and zero-frequency peaks ($\omega_2 = \pm\Omega_i; \omega_4 = 0$). Figure 5b shows the simulation, including the off-diagonal coupling. The ability of the radiation field to create coherences between each of the modes leads to the appearance of various sum and difference frequency peaks, in addition to the features previously exhibited in the decoupled spectrum. The placement of these combination peaks in ω_4 is correlated with the coupled fundamentals in ω_2 in a manner that makes assignment of the origin simple.

We can also consider the case of anharmonic coupling between the modes. This will lead to peaks in the same positions as shown in Fig. 5b, but the relative amplitudes of the peaks will now be dictated by the magnitude

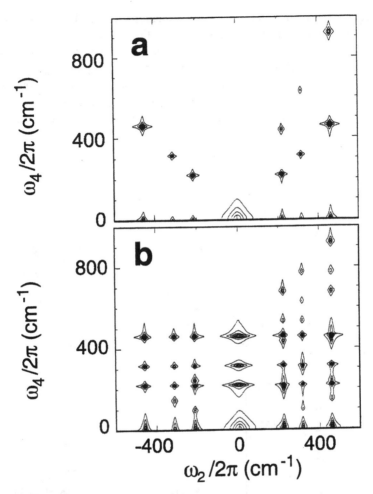

Figure 5 Simulations of the 2D Raman spectrum for the Brownian oscillators in Fig. 4, CCl_4, for nonlinear polarizability coupling [Equation (21)] in the (a) fully uncoupled, $\alpha_{ij}^{(2)} = \delta_{ij}$, and (b) fully coupled, $\alpha_{ij}^{(2)} = 1$, limits.

of the cubic anharmonicity, $g_{ijk}^{(3)}$. Thus, the different patterns for the different relative magnitudes between the diagonal, cross-, and difference-frequency peaks provide a direct indication of the nature and magnitude of the underlying nonlinearity. The different patterns will also depend on the interaction ordering and phase that lead to the response. We can simulate examples of different anharmonic couplings using Equations (21) and (36) and the two highest frequency Brownian oscillators in Fig. 4. Equation (21) is a sum over the modes with each permutation of the modes generating a separate response weighted by the magnitude of the coupling, $g_{ijk}^{(3)}$. Figure 6 shows an example of two different responses labeled by their respective weighting factors, $g_{411}^{(3)}$ and $g_{441}^{(3)}$, where the subscripts refers to the ν_1 and ν_4 vibrational modes. The differences between these two examples, each

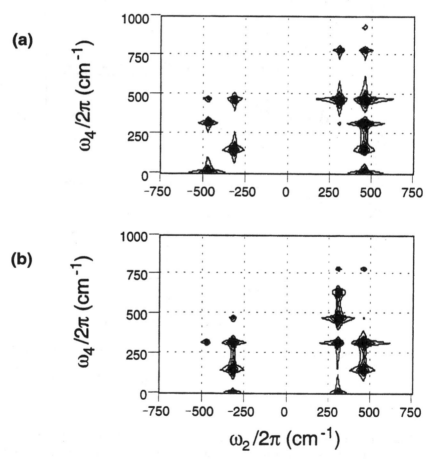

Figure 6 Simulations of the 2D Raman spectrum for the anharmonically (fully) coupled ν_1 and ν_4 modes [Equation (20)]. (a) The $g_{411}^{(3)}$ coupled response. (b) The $g_{441}^{(3)}$ coupled response.

involving a response generated by coupling between the same two vibrational modes, demonstrates the direct sensitivity of the two-dimensional response to different orderings of the individual field-matter interactions. We should note that, although there will be different patterns generated for different interaction time orderings even when there are the same number of interactions for each vibrational mode, for example, $g_{411}^{(3)}$ and $g_{141}^{(3)}$, if the coupling is considered time independent the responses must be weighted equally and therefore they cannot be considered on an individual basis (27).

IV. EXPERIMENTS

There have been a number of recent experimental investigations aimed at measuring the two-dimensional nonresonant Raman response (21–32). One of the greatest challenges in measuring the direct two-dimensional Raman spectrum comes from contamination by the third-order cascaded responses (see Section II.C). At the time of this writing, the potential magnitude of this contamination in current experimental results had only just become evident (32). In this section we will present the current state of experimental progress, and we will discuss the potential for future investigations designed to measure the unadulterated direct two-dimensional response.

A. Experimental Setup

The results presented here were obtained using a standard Ti:sapphire-based regeneratively amplified 3 kHz laser system that provided 45–65 fs pulses (FWHM Gaussian) centered at 800 nm with ~30 μJ/pulse. A diagram for the generation of 5 individual pulses is shown in Fig. 7. The initial beam was first split into three separate pulses with computer-controlled stages providing the primary time delays, τ_2 and τ_4. Two of the beams were then split again, resulting in five beams of approximately equal intensity. The power was attenuated using a $\lambda/2$-plate and cube polarizer, and the polarization of each beam at the sample was individually adjustable with $\lambda/2$-plates. For all of the spectra presented, the polarization of all the incoming beams was set parallel. The individual beams were focused into a room-temperature, 1.0 mm thick liquid sample cell using a 30 cm singlet lens. After the sample, the signal along a chosen phase matched direction was selected with an iris and imaged onto a silicon photodiode. Modulating one of the incoming beams with a mechanical chopper wheel, the signal was collected using a lockin amplifier.

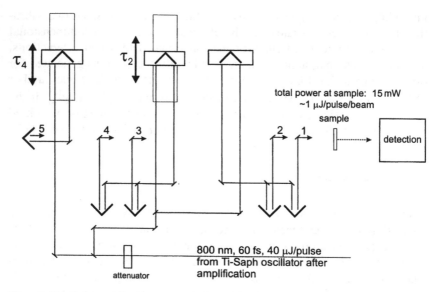

Figure 7 Schematic diagram of the experimental setup to split the incoming laser pulse into 5 separate pulses with the two adjustable time delays.

B. Intermolecular Motions in CS$_2$

Experiments on liquid CS$_2$ were conducted using a homodyne detection configuration. The expression for the overall homodyne detected signal is given in Equation (33). To ascertain the relative contribution from direct and cascaded responses in the fifth-order experiments on liquid CS$_2$, we employed three different phase matching geometries. These geometries are shown in Fig. 8a–c. The signals were collected along the phase matched direction $\mathbf{k}_1 - \mathbf{k}_2 - \mathbf{k}_3 + \mathbf{k}_4 + \mathbf{k}_5$. In addition to the overall fifth-order phase matched direction, we are also concerned with the intermediate steps involved in both the sequential and parallel cascade processes. There are two possible intermediate steps for the sequential cascade, along $\mathbf{k}_1 - \mathbf{k}_2 - \mathbf{k}_3$ and $\mathbf{k}_1 - \mathbf{k}_2 + \mathbf{k}_4$, and there are two possible intermediate steps for the parallel cascade along $\mathbf{k}_1 - \mathbf{k}_2 + \mathbf{k}_5$ and $-\mathbf{k}_3 + \mathbf{k}_4 + \mathbf{k}_5$. The overall signal intensity will be weighted by the magnitude of the phase matching factor for each individual process [Equation (33)]. Using Equation (12) we can express the magnitude of the phase matching factor, F, as sinc(Δkl/2), where Δk represents the difference between the incoming wavevectors and the signal wavevector and l is the path length.

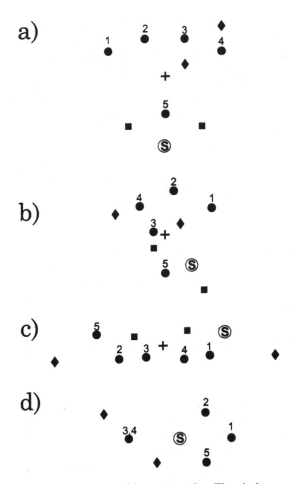

Figure 8 Experimental phase matching geometries. The circles represent the five incoming laser pulse positions. The open circles with an S inside are the fifth-order signal positions. The squares are the parallel cascade intermediate positions, and the diamonds are the sequential cascade intermediates. The phase matching magnitudes and effective path lengths are listed in Table 1. (a)–(c) Homodyne detected geometries described in the text. (d) Heterodyne detection geometry. Note that the $-\mathbf{k}_1 + \mathbf{k}_2 + \mathbf{k}_5$ parallel intermediate is located at the same position as the signal, thus serving as a local oscillator for the heterodyne detected signal.

Although our sample cell has a path length of 1 mm, the actual path length is determined by the crossing of the incoming beams in the sample. To estimate the effective path length as determined by the beam crossing, we calculate the overlap volume of all five beams as a function of position in the sample. Our beams had diameters of 3 mm at the focusing lens. The crossing volume was weighted by a Gaussian transverse beam amplitude profile with a FWHM of 40% of the beam diameter. We assign the FWHM of the Gaussian fit to the crossing volume distributions as the effective path length. The resulting effective path lengths are listed in Table 1 for each of the phase matching geometries employed along with the resulting values of the magnitude of the phase matching factors for the overall fifth-order process and the possible third-order cascade intermediates. The positions of the cascade intermediates and overall fifth-order signals are shown in Fig. 8.

The spectra obtained in phase matching geometries 8a and 8b are shown in Fig. 9a and 9b. The spectrum in Fig. 9a shows a high degree of asymmetry, while the spectrum in Fig. 9b appears quite symmetrical in the behavior along the two adjustable time variables. The symmetrical

Table 1 The Magnitude of the Wavevector Matching Factor for the Overall Fifth-Order Response and the Cascade Intermediates, $\text{sinc}(\Delta kl/2)$

		$\text{sinc}(\Delta kl/2)$			
	Phase matching	8a	8b	8c	8d
Overall fifth-order signal	$k_1 - k_2 - k_3 + k_4 - k_5$ $(-k_1 + k_2 - k_3 + k_4 - k_5)$	0.89	0.46	0.88	1.0
Sequential intermediates	$k_1 - k_2 - k_3$ $(-k_1 + k_2 - k_3)$	−0.20	0.96	−0.09	0.77
	$k_1 - k_2 + k_4$ $(-k_1 + k_2 + k_4)$	−0.19	0.96	−0.09	0.27
Parallel intermediates	$k_1 - k_2 + k_5$ $(-k_1 + k_2 + k_5)$	0.75	−0.09	−0.15	1.0
	$-k_3 + k_4 + k_5$ $(-k_3 + k_4 + k_5)$	0.75	−0.14	−0.15	1.0
Effective path length (mm)		0.61	0.73	0.73	1.0[a]

The wavevector matching conditions in parenthesis refer to geometry 8(d); others refer to geometries 8a–8c. The beam geometries are shown in Fig. 8. The effective path lengths are the FWHM of the Gaussian fits to the crossing volume distributions.
[a] The FWHM of the crossing volume distribution for geometry 8d was 1.2 mm, therefore, the effective path length was set to the smaller sample cell path length of 1.0 mm.

spectrum, Fig. 9b, has the same time-dependent behavior along each of the two time variables as the one dimension time domain spectrum shown in Fig. 3. The response at early times rises to a maximum at $\tau \sim 180$ fs and then evolve into a single exponential decay at long times with a time constant of $t_c \sim 0.8$ ps. In the asymmetrical spectrum (Fig. 9a), there is a rapid decay along τ_2 from a maximum at $\tau_2 = 0$ that at long times becomes the 0.8 ps decay seen along both dimensions in Fig. 9b. Along τ_4 the initial behavior is similar to the inertial behavior at early times in the third-order response, but the signal then decays much more rapidly, $t_c \sim 0.25$ ps, than along τ_2. There is a prominent ridge along both time axes in and along

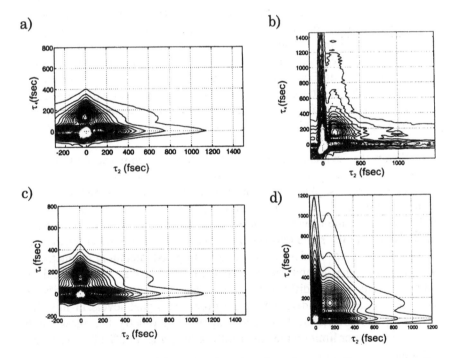

Figure 9 Measured 2D fifth-order time domain nonresonant Raman spectra for CS_2 at room temperature: (a) using the phase matching geometry shown in Fig. 8a and (b) using the phase matching geometry shown in Fig. 8b. Figures (c) and (d) are the cascading fifth-order 2D Raman spectra simulated directly from the measured 1D Raman spectrum using the phase matching magnitude values listed in Table 1 for the cascade intermediates and Equation (28) for the same phase matching geometries used in (a) and (b), respectively.

the $\tau_4 = 0$ axis in Fig. 9a that is pulse width limited in one dimension and shows nearly identical time-dependent behavior to the third-order response along the other. Both spectra also contain a strong peak at the time origin that is pulse width limited in both time dimensions, similar to the electronic response at the time origin in the third-order spectrum.

While both phase matching geometries 8a and 8b are reasonably well phase matched for the overall fifth-order response, they are very differently phase matched for the cascade intermediates (Table 1). Geometry 8a is well phase matched for the parallel cascade intermediates and poorly phase matched for the sequential cascade intermediates, and geometry 8b presents the opposite case. If the measured signals were predominantly generated by the direct fifth-order process, we would expect the response to be identical between the two experiments. However, the clear difference between the two spectra provides a strong indication that the signals are dominated by the fifth-order cascading processes. We can simulate the fifth-order cascaded spectra directly from the measured third-order spectrum in Fig. 3 using Equation (27). Figure 9c shows the simulation for the parallel cascaded response, and Fig. 9d shows the simulation for the sequential cascaded response. There is excellent agreement between Figs. 9a and 9c and Figs. 9b and 9d consistent with the relative values of the phase matching factors in Fig. 8.

With the evidence from phase matching geometries 8a and 8b demonstrating the measured fifth-order signals to be dominated by cascaded processes, phase matching geometry 8c was designed to provide a maximum level of discrimination against all four of the cascade intermediates while leaving the direct overall fifth-order process well phase matched. Using the same experimental conditions as employed for the data in Fig. 9, there was no detectable signal with phase matching geometry 8c. This confirms the domination by cascaded responses in Figs. 9a and 9b. Although it is always difficult to apply a negative result and quantify the lack of a signal, consideration of the observed signal levels in figs. 9(a) and 9(b), and the failure to detect a signal in geometry 8c, allows us to estimate an upper limit on the direct fifth-order signal at 2% of the cascaded response.

In the case of nonlinear polarizability coupling, the ratio of the cascaded to direct fifth-order response can be expressed in terms of the ratio of the first and second derivative of the polarizability [Equation (32)]. Using instantaneous normal mode simulations, Murry et al. (40) have calculated the relative ratios of $\alpha^{(1)}$ and $\alpha^{(2)}$ for 5000 intermolecular modes in CS_2. Although their results show this ratio to be somewhat randomly distributed among the 5000 modes, the majority of the modes have a ratio of $\alpha^{(1)}:\alpha^{(2)} > 100$. This result implies that for the homodyne measurement, where the signal is proportional to $|R^{(5)}|^2$, the cascaded signal will be 10^2–10^4 times larger than the fifth-order signal from NP coupling. This is a value well above our lower limit for this ratio of 50. For anharmonic coupling, even though one might expect a high degree of anharmonicity for intermolecular motions at room temperature, it is the magnitude, and not degree, of the cubic anharmonicity that will be reflected in the magnitude of the direct fifth-order signal. The magnitude of the anharmonicity will be quite small for intermolecular vibrations simply due to the low frequencies of the modes involved. In addition, Murry and Fourkas (18) have suggested that a cancellation among the continuous distribution of frequencies for the anharmonic coupling will result in a much lower contribution to the fifth-order response than NP coupling.

C. Intramolecular Vibrations in Carbon Tetrachloride and Chloroform

To provide the additional sensitivity needed to probe the intramolecular vibrations in liquid CCl_4 and $CHCl_3$, time domain Raman spectra were taken using an intrinsic heterodyne technique. As a heterodyne detection technique, the measured signal is the cross-term between a local oscillator field and a signal field such as the fifth-order signal [Equation (34)]. Intrinsic refers to the fact that the local oscillator field is derived from one of the third-order responses using a phase matching geometry designed to generate both the third- and fifth-order signals along the same phase matched direction. The phase matching geometry is shown in Fig. 8d. In this geometry the overall signal is phase matched along $k_{s5} = -k_1 + k_2 - k_3 + k_4 + k_5$. In addition to the fifth-order signal, the third-order signal $k_{s3} = -k_1 + k_2 + k_5$ is also generated along the same direction and serves as the local oscillator. Assuming $R^{(3)} \gg R^{(5)}$ and using Equation (34), the resulting signal intensity will be proportional to the cross-term between the third- and fifth-order responses,

$$I(\tau_2, \tau_4) \propto 2\,\mathrm{Re}\lfloor R^{(3)}(\tau_2 + \tau_4) \cdot R^{(5)}(\tau_2, \tau_4)\rfloor \qquad (37)$$

Figure 10 shows the Fourier transformations of intrinsic heterodyne detected time domain two-dimensional Raman spectra of (a) CCl_4, (b) $CHCl_3$, and (c) a 50:50 molar ratio CCl_4:$CHCl_3$ mixture. Only quadrants I and II are shown, since the other two quadrants are equal by inversion symmetry. Aside from the diagonal and axial peaks, there

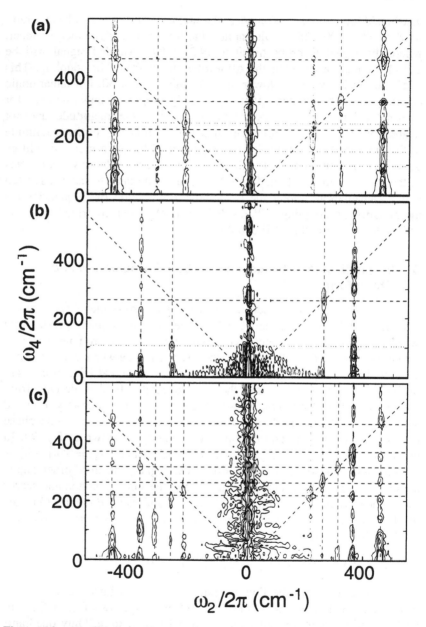

Figure 10 2D fifth-order Raman spectra of (a) CCl_4, (b) $CHCl_3$, and (c) a 1:1 mole ratio CCl_4:$CHCl_3$ mixture.

are a number of cross (off-diagonal) peaks in the spectra that could originate from the nonlinearities between the intramolecular vibrational modes in these samples. For example, there is a strong cross peak in the CCl_4 spectrum between the ν_2 and ν_1 vibrational modes ($\omega_2 = 460$ cm^{-1}, $\omega_4 = 219$ cm^{-1}). One might expect a relatively strong coupling between these two modes as a result of the Fermi resonance between $2\nu_2$ and ν_1. There are also examples of cross peaks in quadrant II such as the peak at [$\omega_2 = -460$ cm^{-1}, $\omega_4 = 219$ cm^{-1}] and the difference peak at [$\omega_2 = 219$ cm^{-1}, $\omega_4 = (460-219)$ cm^{-1}]. Off-diagonal peaks are also evident in the $CHCl_3$ spectrum and the CCl_4:$CHCl_3$ mixture spectrum. In addition to the cross peaks found in the spectra for neat CCl_4 and $CHCl_3$, the CCl_4:$CHCl_3$ mixture spectrum also contains cross peaks between modes from each of the constituents. An example is the peak at [$\omega_2 = 368$ cm^{-1}, $\omega_4 = 219$ cm^{-1}] between the ν_3 mode of $CHCl_3$ and the ν_2 mode of CCl_4.

While the off-diagonal peaks in the CCl_4 and $CHCl_3$ spectra could represent intramolecular coupling between vibrations, they may also contain contributions from intermolecular coupling. Within the direct fifth-order response, intermolecular coupling would be required to generate cross peaks between CCl_4 and $CHCl_3$ in the mixture spectrum (Fig. 10c). The cross peaks between vibrational modes of CCl_4 and $CHCl_3$ in the CCl_4:$CHCl_3$ mixture spectrum could be the result of anharmonic coupling of the involved coordinates on the collective liquid potential or intermolecular interaction-induced effects that couple the fluctuations in the electronic states between molecules in the polarizability. However, assigning the off-diagonal features in any of the two-dimensional spectra to coupling between the vibrational modes requires that the observed signal in question originate from direct fifth-order scattering. As shown in Equation (34), and clearly demonstrated for the intermolecular motions in CS_2, we must consider both the direct and cascaded contributions to the measured response. Figure 11 shows simulated sequential (a) and parallel (b) cascaded two-dimensional spectra for CCl_4. To simulate the heterodyne detected data, the spectra represent the calculated $|R^{(3)}(\tau_2 + \tau_4) \cdot R^{(5)}_{cascade}(\tau_2, \tau_4)|$ cross-term and were simulated directly from the measured third-order spectrum in Fig. 4. In Fig. 11 it is clear that cascaded responses will produce off-diagonal peaks between all of the detected vibrational modes. Note that these off-diagonal peaks carry no information concerning microscopic coupling between the different vibrational modes. The challenge becomes determining the relative contributions in the measured two-dimensional spectra from the direct and cascaded responses.

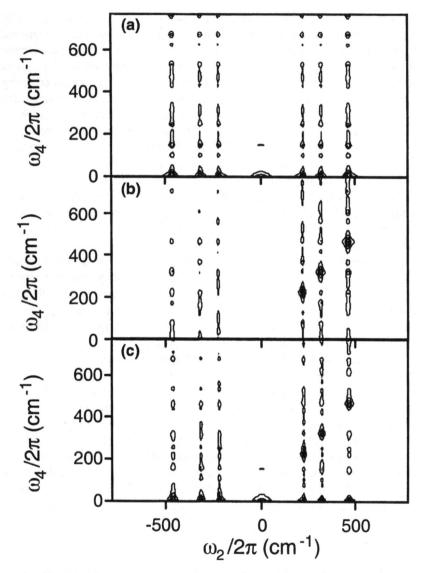

Figure 11 Cascaded 2D fifth-order Raman spectra for CCl_4 simulated directly from the 1D spectrum shown in Fig. 4. (a) The sequential cascade spectrum. (b) The parallel cascade spectrum. (c) A best fit to the 2D spectrum in Fig. 10a using a linear combination of the sequential and parallel cascaded spectra. The ratio is 0.8:1 sequential:parallel.

We can simulate the total cascaded response by a linear combination of the sequential and parallel cascaded spectra in Fig. 11, weighting each by the wavevector matching factors for the cascade intermediates shown in Table 1. However, unlike the homodyne detection employed for CS_2, in this case we must also account for the phase sensitivity inherent in a heterodyne detected signal, [Equation (34)]. To demonstrate the consequences of phase-sensitive detection, we first consider an ideal case where the direct response and all four cascade intermediates are perfectly phase matched. In this case, keeping in mind that $R^{(3)}$ and $R^{(5)}$ are purely real, the local oscillator field ($\mathbf{k}_{LO} = \mathbf{k}_1 + \mathbf{k}_2 + \mathbf{k}_5$) and the direct fifth-order signal field will be purely imaginary [Equation (15)]. In other words, the local oscillator will be perfectly in phase with the direct fifth-order signal. On the other hand, as shown in Equation (29), the cascaded signal field will be purely real, thus $\pi/2$ phase shifted from the local oscillator and perfectly out of phase. In this ideal example, the heterodyne signal contains only the desired direct fifth-order signal.

Unfortunately, real experiments seldom, if ever, present the ideal case. In these experiments the phase matching is not perfect. Equation (12) shows how a deviation from perfect phase matching will result in a complex value of the phase matching factor, F, and a phase shift in the emitted signal field. The magnitudes of the phase matching factors in Table 1 shows that both of the sequential intermediates deviate from perfect phase matching under the assumption of delta-function beam diameters. In addition, due to the finite dimensions of the incoming laser pulses, there will be a range of phase matching conditions for all of the considered processes associated with the range of incoming beam angles. To see how the phase matching condition can effect the phase of the emitted signal field, consider the sequential intermediate $\mathbf{k}_{seq} = \mathbf{k}_1 - \mathbf{k}_2 - \mathbf{k}_3$. The magnitude of the phase matching factor is shown in Table 1 as 0.77. This value is the magnitude of the complex value for the phase matching factor, $F = 0.27 + i^*0.72$. The large imaginary component in F will result in a large imaginary component in the emitted field, Equation (29), which would be in-phase with the ideal local oscillator discussed in the previous paragraph. With our experiments containing a range of phases for the potential signals and the local oscillator, it becomes quite complicated to directly simulate the cascaded response.

Although the complexity of the experiment makes it very difficult to generate an exact simulation of the expected cascaded response, it is possible to demonstrate that the measured spectra cannot be the result of cascaded processes alone. A linear least-squares fit to the measured CCl_4

spectrum using a linear combination of the sequential and parallel cascade simulations, Figs. 11a and 11b, is shown in Fig. 11(c). A comparison of Fig. 11c and Fig. 10a shows that the measured CCl_4 spectrum cannot be quantitatively simulated considering only the cascaded processes. Ab initio calculation on $CHCl_3$ have found that for the lower-frequency intramolecular modes in $CHCl_3$, such as those shown in Fig. 10b, the direct fifth-order signal is within the same order of magnitude as the cascaded signal (30,34). The direct response becomes the dominant response for the highest frequency intramolecular modes ($\omega > 1000$ cm^{-1}) reflecting the increase in magnitude of the cubic anharmonicity with increasing frequency. With the magnitude of the direct fifth-order signal within an order of magnitude of the cascaded signal, and considering the potential discrimination against the cascades provided by the heterodyne detection, it is reasonable to conclude that Spectra in Fig. 10 contain significant contributions from the direct fifth-order Raman response. This result provides clear evidence of the ability to measure the direct fifth-order signal. Unfortunately, our current inability to satisfactorily deconvolve the direct and cascaded contributions to the measured spectra significantly compromises our ability to obtain conclusions concerning the microscopic coupling between the observed vibrational modes.

D. Future Experimental Directions

Based on the current experimental evidence, it is clear that the key to the successful application of two-dimensional nonresonant Raman spectroscopy lies in finding experimental solutions to the problem of contamination from cascading third-order signals. From Equation 30, there are a few obvious experimental adjustments that can be made to improve the ratio of direct to cascaded fifth-order signals, such as using a shorter optical path length and a longer wavelength radiation source. One can also consider lowering the number density, N, to improve the ratio. This idea can readily be applied to the case of well-defined intramolecular vibrations, where the polarizability can be considered molecularly additive, by simply lowering the concentration of the chromophores of interest. Note that the number density dependence also provides an indication of whether the signal originates from the direct or cascaded processes since the signal will have an N^2 or N dependence (N^4 or N^2 for homodyne detection), respectively (Equation (15) and (29)]. However, variation of the number density cannot easily be applied to the case of intermolecular motions. The main problem lies in how to experimentally vary the concentration of intermolecular chromophores.

The typical approach taken when attempting to vary the concentration of intermolecular modes is the use of binary mixtures. When one considers the many body nature of the intermolecular modes and the complexity of binary mixtures, it is not directly evident that there is any proportionality between the ill-defined concept of concentration for intermolecular modes and the binary mixture fraction. An additional complication in the use of binary mixtures comes from the significant changes in the polarizability weighted density of states as a function of binary mixture fraction. In other words, the intermolecular spectrum is changing with binary mixture fraction. These types of effects are clearly evident in third-order measurements of CS_2 in binary mixtures (3).

Other experimental considerations that can aid in improving the ratio of direct to cascaded signals have already been introduced above; they are the exploitation of phase matching and the use of phase-sensitive detection to remove the cascaded third-order processes. For example, it should be possible to improve the discrimination provided by phase matching geometry 8c with small alternations in the geometry and the use of smaller diameter input fields. This would have to be accompanied by improvements in detection sensitivity to allow detection of the much smaller direct fifth-order signal. However, phase-sensitive detection currently appears to hold the greatest promise in the pursuit of the uncontaminated direct fifth-order signal. The intrinsic heterodyne experiments presented above give a clear indication of the ability of phase-sensitive detection to aid in discriminating against the cascaded signals while providing the additional sensitivity necessary to measure spectra from systems with polarizabilities that are less than that of CS_2. Many of the problems associated with the intrinsic heterodyne experiment, such as the time dependence and spatial phase dependence of the local oscillator, can be remedied using an external local oscillator. One possibility would be to use one of the five input beams as the local oscillator, spatially and temporally overlapping the signal and LO in an interferometer after the sample. Another approach would be the use of diffractive optics to generate, in addition to the five input fields, a local oscillator field along the signal direction. Examples of this approach have been demonstrated in four wave-mixing experiments (41,42).

V. CONCLUDING REMARKS

In this chapter we have demonstrated the great promise of two-dimensional Raman spectroscopy to go beyond the ensemble average of linear spectroscopy and reveal the microscopic details that underlie the vibrational spectral density. This technique can be applied to a wide frequency range; from intermolecular motions on time scales as long as many nanoseconds to well-defined intramolecular vibrations on femtosecond time scales. The technique has the potential to determine both the nature of coupling between motions on different time scales, i.e., anharmonicity in the vibrational potential or nonlinearity in the polarizability, and to quantify these couplings. We have also demonstrated that the experimental implementation of this technique poses many challenges. Currently the greatest hindrance to measuring the unadulterated direct fifth-order two-dimensional Raman response comes from interfering cascaded third-order signals. Using an intrinsic heterodyne detection geometry, we showed that signals from the intramolecular vibrations in CCl_4 and $CHCl_3$ that contain contributions from the direct fifth-order response can be measured. From this result we can confidently conclude that experimental progress in the near future will bring measurements of the direct fifth-order two-dimensional Raman response that will realize the potential held by this technique.

ACKNOWLEDGMENTS

The authors would like to acknowledge the funding agencies that made possible the work presented — NSF and CRM-KOSEF. The authors would like to acknowledge Laura Kaufman, who made important contributions to the experiments demonstrating third-order cascading in CS_2, and Jaeyoung Sung, who made important contributions to the theoretical background section of this chapter.

REFERENCES

1. Mukamel S. Principles of Nonlinear Optical Spectroscopy. New York: Oxford: University Press, 1995.
2. Ruhman S, Joly AG, Nelson KA. IEEE J Quantum Electron 24:460, 1988.
3. McMorrow D, Thantu N, Melinger JS, Kim SK, Lotshaw WT. J Phys Chem 100:10389–10399, 1996.
4. Lotshaw WT, McMorrow D, Thantu N, Melinger JS, Kitchenham R. J Raman Spectrosc 26:571–583, 1995.
5. Tanimura Y, Mukamel S. J Chem Phys. 99:9496–9511, 1993.
6. Ernst R, Bodenhausen G, Wokaun A. Principles of Nuclear Magnetic Resonance in One and Two Dimensions. Oxford: Clarendon Press, 1987.
7. Slichter CP. Principles of Magnetic Resonance. 3rd ed. New York: Springer, 1978.

8. Abella ID, Kurnit NA, Hartmann SR. Phys Rev 141:391–141, 1966.
9. Hartmann SR. Sci Am 218:32–40, 1968.
10. Mossberg T, Flusberg A, Kachru R, Hartmann SR. Phys Rev Lett 42:1665–1669, 1978.
11. Steffen T, Fourkas JT, Duppen K. J Chem Phys 105:7364–7382, 1996.
12. Cho M. in Lin SH, Villaeys AA, Fujimura Y, eds. Advances in Multi-Photon Processes and Spectroscopy. Vol. 12. Singapore: World Scientific, 1999, pp 229.
13. Tanimura Y. Chem Phys 233:217–229, 1998.
14. Okumura K, Tanimura Y. J Chem Phys 106:1687–1698, 1997.
15. Okumura K, Tanimura Y. J Chem Phys 107:2267–2283, 1997.
16. Mukamel S, Piryatinski A, Chernyak V. Acc Chem Res 32:145–154, 1999.
17. Saito S, Ohmine I. J Chem Phys 108:240–251, 1998.
18. Murry RL, Fourkas JT. J Chem Phys 107:9726–9740, 1997.
19. Murry RL, Fourkas JT, Keyes T. J Chem Phys 109:7913–7922 1998.
20. Hahn S, Park K, Cho M. J Chem Phys, in press.
21. Steffen T, Duppen K. Phys Rev Lett 76:1224–7 1996.
22. Steffen T, Duppen K. Chem Phys Lett 273:47–54, 1997.
23. Steffen T, Duppen K. J Chem Phys 106:3854–3864, 1997.
24. Tokmakoff A, Lang MJ, Larsen DS, Fleming GR, Chernyak V, Mukamel S. Phys Rev Lett 79:2702–2705, 1997.
25. Tokmakoff A, Lang MJ, Larsen DS, Fleming GR. Chem Phys Lett 272:48–54, 1997.
26. Tokmakoff A, Fleming GR. J Chem Phys 106:2569–2582, 1997.
27. Tokmakoff A, Lang MJ, Jordanides XJ, Fleming GR. Chem Phys 233:231–242, 1998.
28. Tominaga K, Keogh GP, Naitoh Y, Yoshihara K. J Raman Spectrosc 26:495–501, 1995.
29. Tominaga K, Yoshihara K. Phys Rev Lett 74:3061–3064, 1995.
30. Tominaga K, Yoshihara K. Phys Rev Lett 76:987–990, 1996.
31. Tominaga K, Yoshihara K. J Chem Phys 104:4419–4426, 1996.
32. Blank D, Kaufman L, Fleming GR. J Chem Phys, in press.
33. Steffen T, Duppen K. Chem Phys Lett 290:229–236, 1998.
34. Cho M, Blank D Sung J, Park K, Hahn S, Fleming GR. In preparation.
35. Eesley GL, Levenson MD, Tolles WM. JEEE J Quantum Electron QE-14:45–49, 1978.
36. Fleming GR, Cho M. Ann Rev Phys Chem 47:109–134, 1996.
37. Stratt RM, Marconcelli M. J Phys Chem 100:12981–12996, 1996.
38. Ladanyi BM, Lian YQ. J Chem Phys 103:6325–6332, 1995.
39. Geiger LC, Ladanyi BM. J Chem Phys 87:191–202, 1987.
40. Murry RL, Fourkas JT, Keyes T. J Chem Phys 109:2814–2825, 1998.
41. Maznev AA, Nelson KA, Rogers JA. Opt Lett 23:1319–1321, 1998.
42. Goodno GD, Astinov V, Miller RJD. J Phys Chem B 103:603–607, 1999.

11

Nonresonant Intermolecular Spectroscopy of Liquids

John T. Fourkas
Eugene F. Merkert Chemistry Center, Boston College, Chestnut Hill, Massachusetts

1. INTRODUCTION

The dynamics of a chemical process can change considerably in going from the gas phase to the liquid phase. One fundamental reason for such differences is that liquids are able to solvate chemical species. For example, solvation might stabilize the transition state in a chemical reaction to a greater extent than it stabilizes the reactants, thereby accelerating the reaction rate. Of course, solvation itself is a dynamic process, which has important implications for chemical processes in solution. If the lifetime of a transition state is shorter than the inherent dynamic time scale of the solvent, for instance, solvation will not be able to stabilize the transition state to the fullest possible extent. The above example illustrates the importance of gaining a molecular-level understanding of the dynamics of solvents.

Although the microscopic motions in a liquid occur on a continuum of time scales, one can still partition this continuum into two relatively distinct portions. The short-time behavior in a liquid is characterized by frustrated inertial motions of the molecules. While an isolated molecule in the gas phase can translate and rotate freely, in a liquid these same motions are interrupted by collisions with other molecules. Liquids are dense enough media that collisions occur very frequently, so that molecules undergo pseudo-oscillatory motion in the local potentials defined by their nearest neighbors. Given the heterogeneity of local environments in a liquid, these intermolecular vibrations are characterized by a broad distribution of inherent frequencies, ranging from a few to perhaps 200 cm^{-1}. However, since the structure of a liquid evolves with time, an intermolecular vibrational mode cannot retain its identity indefinitely. In practice, the local mode picture breaks down on time scales longer than a picosecond or so. On these longer time scales, the microscopic motions in a liquid are nonoscillatory and irreversible, as the positions and orientations of the molecules in a liquid evolve diffusively.

Ideally, one would like to be able to study the specific solvent dynamics associated with a chemical process of interest. While a number of techniques exist for achieving this end (1–5), they also tend to give rather indirect information about the relevant molecular motions of the solvent. However, the molecular dynamics of a solvent around a solute can be described to a first approximation by the dynamics of the pure solvent. Thus, there is considerable interest in trying to understand the inherent microscopic dynamics of pure solvents.

Three types of techniques are widely used to probe the microscopic dynamics of pure liquids: inelastic neutron scattering (6), far-infrared (FIR) spectroscopies (7), and low-frequency Raman spectroscopies (8). Of these, inelastic neutron scattering is the most general, in that it is capable of probing all of the microscopic motions of a liquid. On the other hand, it is also an expensive technique that can only be performed at a limited number of sites in the world. Conversely, FIR and Raman techniques are subject to spectroscopic selection rules, and therefore cannot probe all of the microscopic dynamics of a liquid. As is the case for intramolecular vibrational spectroscopies, due to different selection rules intermolecular IR and Raman techniques yield complementary information. Technical limitations have in the past made good FIR sources and detectors expensive, which has limited the application of FIR spectroscopies to liquids [although this situation may be changing with the advent of pulsed THz radiation sources based on ultrafast laser systems (9)]. On the other hand, Raman techniques, by virtue of being nonresonant, can be implemented with relatively inexpensive visible sources and detectors. Accordingly, low-frequency Raman spectroscopies have been an extremely popular means of studying liquid dynamics and will be the subject of this chapter.

The earliest and simplest nonresonant technique for studying intermolecular dynamics was spontaneous low-frequency Raman spectroscopy (also known as Rayleigh-wing spectroscopy) (8). In its most basic implementation, a reasonably powerful monochromatic laser beam travels through the liquid of interest. Scattered light is collected at a right angle to the laser path and is then dispersed by a monochromator. Thermally excited intermolecular dynamics lead to frequency shifts in the scattered light via the Raman effect, so that the inelastic scattering spectrum yields direct information about the intermolecular modes of the liquid.

Spontaneous scattering is a relatively weak process, and so coherent spectral techniques such as nearly degenerate four-wave mixing (10) and stimulated Raman gain (11) spectroscopies have been developed over the past few decades for studying low-frequency Raman modes. Spontaneous Raman techniques rely on vacuum fluctuations to provide one of the three photons needed to generate the scattered signal; furthermore, because the direction of propagation of the vacuum photon is indeterminate, the spontaneous Raman signal radiates in every direction. In coherent Raman spectroscopies, lasers provide all of the photons needed. These techniques therefore stimulate motion in the intermolecular modes that they probe, rather than relying on thermal fluctuations to excite them. As a result, coherent spectroscopies generate considerably more intense signals than do spontaneous spectroscopies. In addition, the wave vectors of the three incident photons are well defined in coherent spectroscopies, such that the signal propagates in a unique direction. The trade-off is that a minimum of two lasers of different frequencies, at least one of which is tunable, must be employed to accomplish this end.

An alternate approach is to perform coherent Raman spectroscopy in the time domain rather than in the frequency domain. In this case, a single laser that produces short pulses with sufficient bandwidth to excite all of the Raman modes of interest is employed. One pulse or one pair of time-coincident pulses is used to initiate coherent motion of the intermolecular modes. The time dependence of this coherence is then monitored by another laser pulse, whose timing can be varied to map out the Raman free-induction decay (FID). It should be stressed at this point that the information contained in the Raman FID is identical to that in a low-frequency Raman spectrum and that the two types of data can be interconverted by a straightforward Fourier-transform procedure (12–14). Thus, whether a frequency-domain or a time-domain coherent Raman technique should be employed to study a particular system depends only on practical experimental considerations.

While a large number of coherent, time-domain, low-frequency Raman spectroscopies have been developed (15–18), they all fall under the general category of optical Kerr effect (OKE) spectroscopy (19,20). The Kerr effect is a phenomenon that occurs when a polar liquid is subjected to a DC electric field (21). The electric field causes a partial alignment of the molecules in the liquid, thereby breaking the isotropy of the system. As a result, the liquid becomes birefringent. Similarly, the OKE refers to the birefringence created by subjecting a liquid to a fast-oscillating AC electric field (i.e., a pulse of laser light). Liquid molecules are so massive that they cannot respond fast enough for a permanent dipole moment to track an electric field that switches direction every femtosecond or so. Thus, the OKE relies on the polarizability of a liquid molecule, rather than its permanent dipole moment, to create alignment. The electrons in a molecule are light enough that they can respond very quickly to changes in an applied electric field. The induced dipole moment created by the response of the electrons to the electric field can then interact with the electric field. If the molecule has a direction along which the electron cloud is most easily polarized, then a torque is created that drives the molecule towards having this axis aligned with the polarization of the applied field. Thus, a laser pulse can generate net alignment of the molecules in a liquid, resulting in an induced birefringence. In this picture the pump pulse (or pulses) induces coherent motion of the intermolecular modes of the liquid, and the time dependence of this motion is then probed via the induced birefringence. This description of time-domain coherent Raman spectroscopy is complementary to (but completely equivalent to) the Raman FID picture discussed above.

The following section contains a more detailed treatment of the theory behind the nonresonant spectroscopy of liquids. This will be followed by a description of the experimental implementation and data analysis techniques for a common OKE scheme, optical-heterodyne-detected Raman-induced Kerr-effect spectroscopy (22). We will then discuss the application of this technique to the study of the temperature-dependent dynamics of simple liquids composed of symmetric-top molecules.

I. THEORY

Before delving into the Raman spectroscopy of intermolecular modes, it is useful to discuss the IR spectroscopy of these modes briefly. Consider a liquid composed of N molecules, each containing n atoms. The energy of interaction of a liquid with an external electric field $\mathbf{E}$ is given by $-\mathbf{M} \cdot \mathbf{E}$, where $\mathbf{M}$ is the collective dipole moment of the liquid. The dipole moment depends on the coordinates of the molecules $(q_1, q_2, \ldots q_{3nN})$ in a complicated manner, and these coordinates evolve in time in an equally complicated manner. It is customary to describe the time dependence of the dipole moment via a Taylor-series expansion:

$$\mathbf{M}(t) = \mathbf{M}(0) + \sum_{t=1}^{3nN} \left(\frac{\partial \mathbf{M}}{\partial q_i}\right)_{i=0} (q_i(t) - q_i(0))$$

$$+ \frac{1}{2} \sum_{j,k=1}^{3nN} \left(\frac{\partial^2 \mathbf{M}}{\partial q_j \partial q_k}\right)_{t=0} (q_j(t) - q_j(0))(q_k(t) - q_k(0)) + \cdots \quad (1)$$

$$= \mathbf{M}(0) + \sum_{i=1}^{3nN} \mathbf{M}_i^{(1)}(q_i(t) - q_i(0))$$

$$+ \frac{1}{2} \sum_{j,k=1}^{3nN} \mathbf{M}_{j,k}^{(2)}(q_j(t) - q_j(0))(q_k(t) - q_k(0)) + \cdots$$

The FIR spectrum is proportional to the Fourier transform of the dipole correlation function, $\langle \mathbf{M}(t) \cdot \mathbf{M}(0) \rangle$. As discussed above, at short times the correlation function is dominated by pseudo-oscillatory modes. Insofar as the dipole moment of the liquid is relatively weakly dependent on these pseudo-oscillatory coordinates, it is generally safe to truncate Eq. (1) after the second term in describing this short-time behavior. Thus, an IR-active intermolecular mode is considered to be one for which $\mathbf{M}_i^{(1)} \neq 0$.

The total dipole moment of the liquid is the sum of two components, which we will designate permanent and induced:

$$\mathbf{M} = \mathbf{M}_{\text{perm}} + \mathbf{M}_{\text{ind}} \quad (2)$$

The permanent component is the sum of all of the permanent molecular dipole moments $\boldsymbol{\mu}_{\text{perm}}$ within the liquid,

$$\mathbf{M}_{\text{perm}} = \sum_{i=1}^{N} \boldsymbol{\mu}_{\text{perm}}^i \quad (3)$$

The induced component arises from the interaction of the molecular polarizability tensors, $\tilde{\boldsymbol{\pi}}$, with electric fields. The induced dipole moment on molecule i is given by:

$$\boldsymbol{\mu}_{\text{ind}}^i = \mathbf{E}^i \tilde{\boldsymbol{\pi}}^i \quad (4)$$

where $\mathbf{E}^i$ is the electric field at the molecule. In the absence of an external electric field, the electric field at molecule i arises from the dipole moments of the other molecules in the liquid:

$$\mathbf{E}^i = \sum_{j \neq 1} \tilde{\mathbf{T}}_{ij} \boldsymbol{\mu}^j \tag{5}$$

where $\tilde{\mathbf{T}}_{ij}$ is the dipole tensor. This so-called dipole/induced dipole (DID) mechanism can lead to FIR absorption even in fluids composed of molecules that do not have permanent dipole moments, such as noble gases or carbon tetrachloride. Since DID interactions are highly dependent on distance, it is generally only the nearest neighbors of a given molecule that can contribute significantly to its induced dipole moment. Note also that since $\langle \mathbf{M}(t) \cdot \mathbf{M}(0) \rangle$ depends on the total dipole moment at two different times, this correlation function (and therefore its FIR spectrum) will contain a component that arises purely from $\mathbf{M}_{perm}$, one that arises purely from $\mathbf{M}_{ind}$, and one that arises from a cross term between the two.

Raman scattering depends on the time correlation function of the many-body polarizability of the liquid, $\langle \tilde{\mathbf{\Pi}}(t) \cdot \tilde{\mathbf{\Pi}}(0) \rangle$, rather than on that of the collective dipole moment. In the case of Raman scattering, an external electric field (from a laser) generates an induced collective dipole in the liquid:

$$\mathbf{M}_{ind} = \mathbf{E}^{ext} \tilde{\mathbf{\Pi}} \tag{6}$$

The dynamics of this induced dipole can then be probed by another electric field, which has an energy of interaction of $-\mathbf{E}^{ext} \cdot \mathbf{E}^{ext} \tilde{\mathbf{\Pi}}$. Note that these two electric fields may come from the same or two different laser beams.

As in the case of the collective dipole moment, the many-body polarizability of the liquid can be broken up into two distinct contributions, one from single molecules and one from DID effects:

$$\tilde{\mathbf{\Pi}} = \tilde{\mathbf{\Pi}}_{sm} + \tilde{\mathbf{\Pi}}_{DID} \tag{7}$$

The first contribution is merely the sum of the polarizability tensors of all of the individual molecules in the liquid, i.e.,

$$\tilde{\mathbf{\Pi}}_{sm} = \sum_{i=1}^{N} \tilde{\boldsymbol{\pi}}^i \tag{8}$$

The DID contribution arises from the fact that the electric field experienced by a given molecule is the sum of the applied field and the field from the permanent and induced dipoles of the remaining molecules in the liquid, so that

$$E_i^{\text{total}} = E^{\text{ext}} + \sum_{j \neq i} \tilde{T}_{ij} \mu^j \tag{9}$$

The polarizability time correlation function of a liquid will therefore consist of a single-molecule contribution, a DID contribution, and a cross term between the two. The cross term is negative for intermolecular modes, since on the whole the field generated by the molecular dipoles tends to oppose the applied electric field, thus reducing the many-body polarizability of the liquid (23).

The dependence of the many-body polarizability of the liquid on the molecular coordinates can be expressed as a Taylor-series expansion that is entirely analogous to Equation (1):

$$
\tilde{\Pi}(t) = \tilde{\Pi}(0) + \sum_{i=1}^{3nN} \left(\frac{\partial \tilde{\Pi}}{\partial q_i} \right)_{t=0} (q_i(t) - q_i(0))
$$

$$
+ \frac{1}{2} \sum_{j,k=1}^{3nN} \left(\frac{\partial^2 \tilde{\Pi}}{\partial q_j \partial q_k} \right)_{t=0} (q_j(t) - q_j(0))(q_k(t) - q_k(0)) + \cdots \tag{10}
$$

$$
= \tilde{\Pi}(0) + \sum_{i=1}^{3nN} \tilde{\Pi}_i^{(1)} (q_i(t) - q_i(0))
$$

$$
+ \frac{1}{2} \sum_{j,k=1}^{3nN} \tilde{\Pi}_{j,k}^{(2)} (q_j(t) - q_j(0))(q_k(t) - q_k(0)) + \cdots
$$

As in the case of IR spectroscopy, this expansion is often truncated after the second term, in what is known as the Placzek approximation (24). Thus, modes for which $\tilde{\Pi}_i^{(1)} \neq 0$ are described as being Raman active. However, in the case of intermolecular Raman spectroscopy, the Placzek approximation should be applied with some caution. In general, if $M_i^{(1)} = 0$ for a given mode, then the same will be true for all higher-order terms in the Taylor-series expansion of the collective dipole. On the other hand, $\tilde{\Pi}_i^{(1)}$ might be zero simply due to symmetry constraints that will not exist for $\tilde{\Pi}_{ji}^{(2)}$. As an example, consider the CS_2 molecule. The symmetric stretch of this molecule is not IR active because there is no change at all in the dipole moment when moving along this mode. On the other hand, the bending mode is not Raman active because the polarizability of the molecule is identical if the molecule is bent from its equilibrium geometry by angle θ in any direction. The equilibrium geometry must therefore correspond to an extremum of the polarizability with respect to the bending coordinates of the molecule, which means that $\tilde{\Pi}_{ii}^{(2)}$ will not be zero for these modes. It has therefore been argued that $\tilde{\Pi}^{(2)}$ terms may be important in the intermolecular Raman spectroscopy of liquids (25,26), and this question is currently the subject of active study.

It is important to note that the two electric fields that lead to a Raman transition can have different polarizations. Information about how the transition probability is affected by these polarizations is contained within the elements of the many-body polarizability tensor. Since all of the Raman spectroscopies considered here involve two Raman transitions, we must consider the effects of four polarizations overall. In time-domain experiments we are thus interested in the symmetry properties of the third-order response function, $R_{\alpha\beta\gamma\delta}^{(3)}$ (or equivalently in frequency-domain experiments we are interested in the properties of the third-order susceptibility, $\chi_{\alpha\beta\gamma\delta}^{(3)}$), where α, β, γ, and δ denote the polarizations. The third-order response for any set of polarizations can be described by a sum of response functions whose indices are Cartesian axes. Furthermore, in an isotropic medium, the third-order response must be identical if any direction is inverted (27). As a result, the response is only nonzero when any given Cartesian axis appears an even number of times. The response must also be identical under any permutation of Cartesian indices, i.e., $R_{xxyy}^{(3)} = R_{xxzz}^{(3)} = R_{yyzz}^{(3)}$, and so on.

Since the Raman techniques described are sensitive to $\langle \tilde{\Pi}(t) \cdot \tilde{\Pi}(0) \rangle$, we are interested in the properties of products of two elements of Π, subject to the above constraints. Furthermore, it is necessarily true that any tensor elements of Π that are related by reversed indices are identical, e.g., $\Pi_{xy} = \Pi_{yx}$. This leaves us only two independent elements of $R^{(3)}$ to consider. It is conventional to express these independent elements in terms of rotationally invariant features of Π. One of these invariants (usually denoted α) is given by one third of the trace of Π (24). Since this invariant measures the average polarizability of the system, it is known as the isotropic component of Π. The other invariant is generally denoted β, and in the principal axis system of Π it is given by (24)

$$\beta = \sqrt{\tfrac{1}{2}[(\Pi_{xx} - \Pi_{yy})^2 + (\Pi_{xx} - \Pi_{zz})^2 + (\Pi_{yy} - \Pi_{zz})^2]} \tag{11}$$

This invariant can be thought of as measuring the average deviation of the polarizability from its mean value, and is therefore known as the anisotropy of Π.

The different tensor elements of the third-order response can all be described in terms of the isotropic and anisotropic components of Π. The most commonly measured response function tensor elements are $R_{xxxx}^{(3)}(\tau)$ (the "polarized" response) and $R_{xyxy}^{(3)}(\tau)$ (the "depolarized" response), where τ is the experimental delay time. In terms of the invariants of Π, the polarized response is related to the sum of correlation functions (28)

$$\langle \alpha(\tau)\alpha(0) \rangle + \frac{4}{45} \langle \beta(\tau)\beta(0) \rangle \qquad (12)$$

Similarly, the depolarized response (with the same constant of proportionality) is related to the correlation function (28)

$$\frac{1}{15} \langle \beta(\tau)\beta(0) \rangle \qquad (13)$$

Note that the depolarized response is sensitive only to the anisotropic correlation function. To isolate the isotropic correlation function instead, one can measure the "magic angle" response function, $R_{xxmm}^{(3)}(\tau)$, where $m = 54.7°$ (29).

The magnitude and time-dependence of the isotropic and anisotropic portions of the inter-molecular response contain important information about the microscopic properties of a liquid.

To see why this is the case, we first consider the portion of the response that arises from $\tilde{\Pi}_{sm}$. According to Equation (10), we can express $\langle \tilde{\Pi}_{sm}(t) \cdot \tilde{\Pi}_{sm}(0) \rangle$ in terms of derivatives of $\tilde{\Pi}_{sm}$ with respect to the molecular coordinates. Since in the absence of intermolecular interactions the polarizability tensor of an individual molecule is translationally invariant, $\tilde{\Pi}_{sm}$ is sensitive only to orientational motions. Since the trace is a linear function of the elements of $\tilde{\Pi}$, the trace of the derivative of a tensor is equal to the derivative of the trace of a tensor. Note, however, that the trace of a tensor is rotationally invariant. Thus, the trace of any derivative of $\tilde{\Pi}_{sm}$ with respect to an orientational coordinate must be zero. As a result, $\tilde{\Pi}_{sm}$ cannot contribute to isotropic scattering, either on its own or in combination with $\tilde{\Pi}_{DID}$. On the other hand, although the anisotropy is also rotationally invariant, it is not a linear function of the elements of $\tilde{\Pi}$. The anisotropy of the derivative of a tensor therefore need not be zero, and $\tilde{\Pi}_{sm}$ can contribute to anisotropic scattering.

Since DID scattering arises from intermolecular interactions, there is no requirement that the trace of $\tilde{\Pi}_{DID}$ be preserved for any particular molecular motions. Thus, DID interactions can lead to both isotropic and anisotropic scattering. The same orientational fluctuations that are responsible for scattering from single molecules also lead to DID scattering, so there can be a significant cross-term between the two types of scattering for these motions. As mentioned above, because DID effects tend to produce a local electric field around a molecule that is somewhat smaller than the applied field, the cross term is generally negative (23). Furthermore, fluctuations in molecular positions (i.e., density fluctuations) can also generate DID scattering.

DID scattering is particularly sensitive to the degree of microscopic symmetry in a liquid. Consider, for instance, a liquid molecule that oscillates within a cage defined by its surrounding molecules. If this cage is completely symmetric, then the DID polarizability due to this oscillation will have to be at an extremum when the molecule is in the center of the cage. As a result, $\tilde{\Pi}_{\mathrm{DID}}^{(\mathrm{I})}$ will be zero for this motion. Thus, only if the local environment around an average molecule is reasonably asymmetric will there be significant isotropic scattering from the liquid. A useful quantity for assessing the amount of local symmetry is the depolarization ratio ρ, which in frequency-domain spectroscopy is defined as (24)

$$\rho(\omega) = \frac{\chi_{\mathrm{xyxy}}(\omega)}{\chi_{\mathrm{xxxx}}(\omega)} = \frac{3\beta^2(\omega)}{45\alpha^2(\omega) + 4\beta^2(\omega)} \tag{14}$$

In most liquids for which it has been measured, ρ takes on a value near to 0.75, which suggests that DID scattering from density fluctuations is generally not important in liquids and that local environments tend to be reasonably symmetric.

Lastly, we consider the diffusive contribution to the signal. Since this portion of the signal arises from molecular reorientation, it should be completely depolarized unless these diffusive reorientational dynamics also have a significant DID component. The orientational decay will be made up of exponential components, the number of which depends on the molecular symmetry and the relationship between the principal axes of the diffusion and polarizability tensors of the molecules (8). If these tensors share no axes, the orientational decay will be composed of a sum of five exponentials. If the tensors share one axis, the decay will be composed of three exponentials. If the tensors share all three axes, the decay will be composed of two exponentials. If the molecule is further a symmetric top, then reorientation about the axis of symmetry cannot be observed, and the decay will be composed of a single exponential. In principle, considerably more information is available when the principal axes of the diffusion and polarizability tensors are not shared; however, in practice it is virtually impossible to find a unique fit to the sum of five exponentials, some of which may have very small amplitudes. In the remainder of this chapter we will therefore concentrate on symmetric-top liquids.

The orientational diffusion observed by intermolecular Raman techniques is not the diffusion of single molecules, but rather arises from collective effects. For symmetric tops, the collective orientational correlation time τ_{coll} is related to the single-molecule orientational correlation time τ_{sm} via

$$\tau_{coll} = \frac{g_2}{j_2} \tau_{sm} \tag{15}$$

where g_2 is the static orientational correlation parameter and j_2 is the dynamic orientational correlation parameter (30). The static orientational correlation parameter is given by

$$g_2 = 1 + n \frac{\langle P_2(1)P_2(2) \rangle}{\langle P_2(1)P_2(1) \rangle} \tag{16}$$

where $P_2(i)$ is the second-rank Legendre function of the orientation of molecule i, n is the number density of the liquid, θ_{ij} is the angle between the symmetry axes of two molecules, and the averages are over all pairs of molecules in the liquid. The dynamic orientational correlation parameter is usually presumed to be unity for simple liquids (31), and this assumption has generally been borne out in experiments that allow for the independent measurement of g_2 and j_2 (32). As a result, to a good approximation the static orientational correlation parameter is the constant of proportionality that relates the collective correlation time to that for single molecules.

As can be seen from Equation (16), the static orientational correlation parameter is sensitive to the degree of structure in a liquid. In a liquid in which the symmetry axes of molecules tend to be aligned with one another, g_2 can be significantly greater than one. In most simple liquids there is a modest tendency for parallel alignment, so the collective orientational correlation time is generally between one and two times longer than the single molecule orientational correlation time. We should note also that while it might be assumed that any ordering in a simple liquid should be short-ranged, simulation studies have suggested that the value of g_2 can be influenced by molecules that are separated by significant distances (33).

III. EXPERIMENTAL TECHNIQUE

In its simplest implementation, OKE spectroscopy is a form of pump-probe polarization spectroscopy. A powerful pump pulse, linearly polarized at 45° to the vertical, is used to induce a transient birefringence in the liquid sample. A vertically polarized probe pulse travels through the same spot in the sample, after which it is incident on an analyzer polarizer that is set to pass only horizontally polarized light. If the probe pulse arrives at the sample before the pump pulse, no light can pass through the analyzer polarizer. On the other hand, if the probe pulse arrives after the pump pulse, the birefringence of the sample induces depolarization in the probe beam, allowing signal to pass through the analyzer polarizer. By varying the delay time between the two pulses, one measures $|R^{(3)}_{xxxx}(\tau) - R^{(3)}_{xxyy}(\tau)|^2$, which is proportional to $|R^{(3)}_{xyxy}(\tau)|^2$, the square of the depolarized response function (22).

McMorrow and Lotshaw have developed a powerful variant of this polarization-spectroscopy technique, which is often referred to as optically heterodyne-detected Raman-induced Kerr effect spectroscopy (OHD-RIKES) (22). They recognized that the signal could be amplified and linearized (i.e., made proportional to the response function instead of its square) by beating it against a local oscillator. Since the signal arises from a change in the index of refraction in the sample, the signal electric field is 90° out of phase with the electric field of the probe beam. Thus, the ideal local oscillator should be ±90° out of phase with the probe beam and should be of the proper polarization to pass through the analyzer polarizer. This is accomplished by placing a quarter-wave plate after the polarizer that polarizes the probe beam vertically. The quarter-wave plate is oriented such that either its fast or slow axis is vertical. The polarizer is then rotated by a small angle θ. The electric field of the probe pulse after the quarter-wave plate is then given by $\mathbf{E} = \mathbf{E}_{probe} + i\mathbf{E}_{lo}$, where $\mathbf{E}_{probe}$ is vertically polarized and $\mathbf{E}_{lo}$ is horizontally polarized; the factor of i indicates the phase relative to that of the probe beam. Note that since θ is small, $\mathbf{E}_{probe} \gg \mathbf{E}_{lo}$. The electric field that passes through the analyzer polarizer is then given by $i\mathbf{E}_{lo} + i\mathbf{E}_{lo}$. The detected intensity is the magnitude squared of this electric field:

$$I_\theta = |i\mathbf{E}_{lo} + i\mathbf{E}_{sig}|^2 = I_{lo} + 2\mathbf{E}_{lo}\mathbf{E}_{sig} + I_{sig} \tag{17}$$

Note that I_{lo} does not depend on the delay time τ, and so serves only as a constant offset of the detected intensity. Insofar as the electric field of the local oscillator is larger than that of the signal, the heterodyned signal intensity, $2\mathbf{E}_{lo}\mathbf{E}_{sig}$, will be much larger than the homodyned intensity, I_{sig}. However, if one wishes to remove the homodyned contribution to the completely, a second measurement can be made with the polarizer rotated by angle $-\theta$. In this case, the detected intensity is given by

$$I_{-\theta} = |-i\mathbf{E}_{lo} + i\mathbf{E}_{sig}|^2 = I_{lo} - 2\mathbf{E}_{lo}\mathbf{E}_{sig} + I_{sig} \tag{18}$$

Inspection of Equations (17) and (18) shows that the heterodyned contribution can be isolated by subtracting $I_{-\theta}$ from I_{θ}.

We have designed an extremely sensitive experimental apparatus for obtaining OHD-RIKES data (Fig. 1) (34). Our excitation source is an argon-ion-pumped Ti:sapphire laser that produces 40 fs, transform-limited pulses with a center wavelength of 800 nm, a power of several nJ, and a repetition rate of 76 MHz. The laser beam passes through a prism dispersion compensator that is used to adjust for any dispersion introduced by the optics in the apparatus, after which it is spatially filtered and split into a strong pump beam and a weak probe beam. The beams pass through different rings of a chopper wheel that modulates them at a 7:5 frequency ratio. The pump beam then passes through a half-wave plate and a polarizer set at 45° before being focused into the sample by an achromatic lens. The probe beam traverses an optical delay line, which is capable of taking steps as small as 0.1 μm, after which is passes through a vertical polarizer and a quarter-wave plate before being focused into the same spot in

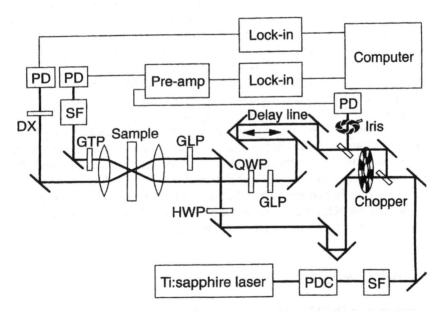

Figure 1 OHD-RIKES experimental setup. DX = Doubling crystal; GLP = Glan laser polarizer; GTP = Glan-Thompson polarizer; HWP = half-wave plate; PD = photodiode; PDC = prism dispersion compensator; QWP = quarter-wave plate; SF = spatial filter. (From Ref. 34.)

the sample by the achromatic lens. After the sample, the probe beam is incident on the analyzer polarizer, and any transmitted light is spatially filtered and then detected by a low-noise amplified photodiode. The first polarizer in the probe path is mounted on a mechanical rotation stage that is computer controlled such that alternate data sets can be obtained at positive and negative heterodyne angles.

Since the pump and probe beams are chopped at different rates, the signal can be detected by a lock-in amplifier referenced to the sum of the two frequencies. This double-modulation scheme can provide a significant increase in the signal-to-noise ratio over more common single-modulation techniques. However, one difficulty associated with this scheme is that I_{lo}, which is the largest portion of the detected intensity, is modulated and therefore cannot be removed simply by employing AC coupling at the lock-in input. To circumvent this problem, we split off a small portion of the probe beam before the delay line and send it to a low-noise amplified detector that is identical to the signal detector. With the pump beam blocked, the gain of this reference detector is set to make the reference intensity identical to that of the local oscillator. Even slight clipping of either the probe or reference beams can lead to imperfect subtraction of these two signals, so we place an adjustable iris in the path of the reference beam to allow us to compensate for any such effects. In addition, because the digital lock-in amplifier used in our apparatus digitizes the signal before implementing the differential detection, we instead use a low-noise, high-gain analog preamplifier to subtract the reference from the signal. In doing so, we greatly increase the dynamic range of the apparatus.

The OHD-RIKES signal is quadratic in the laser intensity, so any laser fluctuations are magnified in the data. To compensate for such fluctuations, the probe beam is sent into a doubling crystal after the sample, and the resultant second-harmonic signal is detected by a separate lock-in amplifier. Second-harmonic generation has the same quadratic intensity dependence as the OHD-RIKES signal, so dividing the RIKES data by the second-harmonic data acts to normalize any intensity fluctuations.

IV. DATA ANALYSIS

Representative OHD-RIKES data obtained in benzene at room temperature are shown in Fig. 2. The spike at zero delay time arises from the electronic hyperpolarizability of the liquid, and its width is indicative of the effective instrument response. Although the pulses used are quite short, appreciable

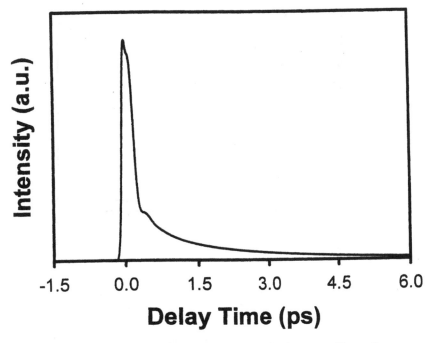

Figure 2 Room-temperature OHD-RIKES data for benzene. The spike at zero delay time arises from the electronic hyperpolarizability of the liquid.

dynamics do occur on the time scale of the pulse duration. In this section we will describe how the deconvolution technique of McMorrow and Lotshaw (22) can be used to compensate for the finite pulse duration.

The experimental signal given pulses of finite duration can be written as

$$S(\tau) = \int_{-\infty}^{\infty} dt' \int_{-\infty}^{\infty} dt \ I_{pump}(t + t')I_{probe}(\tau - t - t')R_{xyxy}^{(3)}(t) \qquad (19)$$

The integral over t' is the convolution of the intensities of the two laser pulses at time separation τ. This function is identical to the noncollinear second-harmonic-generation intensity autocorrelation of the laser pulses,

$$G^{(2)}(\tau) = \int_{-\infty}^{\infty} dt' I_{pump}(t')I_{probe}(\tau - t') \qquad (20)$$

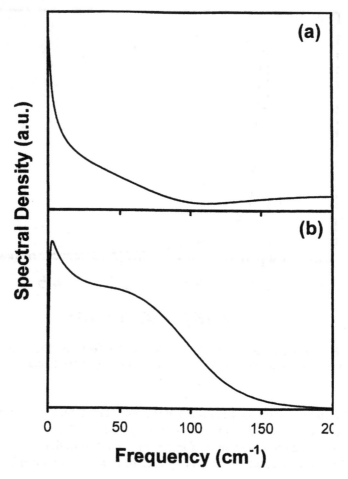

Figure 3   The real (a) and imaginary (b) portions of $D_{xyxy}(\omega)$ for benzene at room temperature.

Thus, the OHD-RIKES signal can be written as a convolution of the auto-correlation function with the response function:

$$S(\tau) = G^{(2)}(\tau) \otimes R_{xyxy}^{(3)}(\tau) \tag{21}$$

According to the convolution theorem (35), the Fourier transform of the convolution of two functions is equal to the product of the Fourier transforms of the functions. Thus, if $S(\tau)$ and $G^{(2)}(\tau)$ have been measured, the effects of the finite pulse duration can be removed completely from the OHD-RIKES data.

The first step in this procedure is to calculate the deconvolved, depolarized spectral density, $D_{xyxy}(\omega)$:

$$D_{xyxy}(\omega) = \frac{\Im[S(\tau)]}{\Im[G^{(2)}(\tau)]} \tag{22}$$

where $\Im$ denotes a Fourier transform. The real and imaginary portions of the $D_{xyxy}(\omega)$ corresponding to the benzene data of Fig. 2 are illustrated in Fig. 3. Note that the electronic response in $S(\tau)$ is completely symmetric in time, and therefore shows up only as a baseline offset in the real portion of $D_{xyxy}(\omega)$. The imaginary portion of $D_{xyxy}(\omega)$, on the other hand, corresponds to the Bose-Einstein-corrected Rayleigh-wing spectrum of benzene (14). We can therefore retrieve the purely nuclear portion of $R_{xyxy}^{(3)}(\tau)$ via

$$R_{xyxy}^{(3)}(\tau) = \Im^{-1}[\text{Im}(D_{xyxy}(\omega))] \tag{23}$$

where $\Im^{-1}$ denotes an inverse Fourier transform. The deconvolved nuclear response function for benzene at room temperature is shown in Fig. 4.

Since benzene is a symmetric top, as discussed above its diffusive reorientational decay would be expected to be exponential in form. Figure 5, which is the natural logarithm of the data in Fig. 2, shows that this is indeed the case. The collective orientational time for this decay is 3.19 ps. To focus on the other intermolecular dynamics, it is useful to remove the diffusive reorientational tail from the nuclear response. Although the diffusive tail can be fit quite accurately, how this response grows from zero at the outset of the decay cannot be determined unambiguously from the data. A common assumption in analyzing such data is that the orientation builds in exponentially after the pump pulse, generally with a time constant of between 100 and 200 fs. This assumption cannot be literally true, since a physically realistic time correlation function must be even in time. However, the intermolecular spectrum is relatively insensitive to the rise time chosen for the diffusive portion of the signal. Changes in the assumed rise time can lead to small differences in the high-frequency portion of the intermolecular spectrum, but these differences are generally within the uncertainty generated by the other steps in the data analysis procedure.

Once the diffusive reorientation contribution has been subtracted from the deconvolved time-domain response, a final Fourier transform yields the intermolecular spectrum (often referred to as the reduced spectral density).

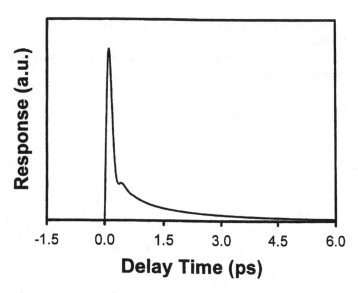

Figure 4 Deconvolved $R_{xyxy}^{(3)}(\tau)$ for benzene at room temperature. Note that, as compared to the undeconvolved data in Figure 2, the deconvolved response function is zero at zero time and has a sharper recurrence near a 400 fs delay time.

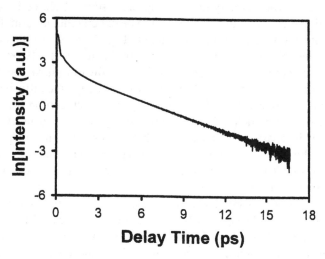

Figure 5 Natural logarithm of the room-temperature OHD-RIKES data for benzene. Note that the tail of the decay is linear, implying that the collective orientational diffusion of benzene is described well by a single-exponential decay, as would be expected for a symmetric-top liquid.

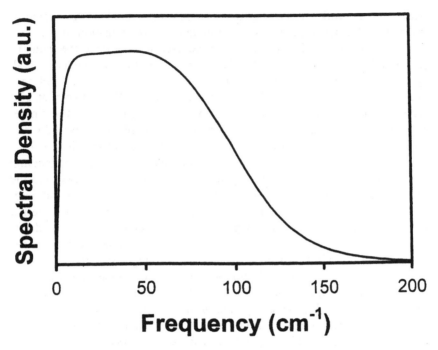

Figure 6 The intermolecular portion (i.e., with the contribution of orientational diffusion excluded) of the Bose-Einstein corrected Rayleigh-wing spectrum of benzene at room temperature.

Figure 6 illustrates the intermolecular spectrum of benzene at room temperature, where a rise time of 150 fs has been assumed in subtracting off the diffusive portion of the nuclear response.

V. SYMMETRIC-TOP LIQUIDS: ORIENTATIONAL DIFFUSION

There is a long history of using Rayleigh-wing and related spectroscopies to study orientational dynamics of symmetric-top liquids (31). Liquids that have been studied include carbon disulfide (11,15,26,36–43), acetonitrile (26,44–47), benzene [(11,32,45,48–52), hexafluorobenzene (32,48,52), 1,3,5-trifluorobenzene (52)], mesitylene (32,44,50,52), chloroform (45,48,53), and methyl iodide (34,45,54,55).

Much of this work has been geared towards understanding the degree of orientational order in pure solvents, through the determination of g_2.

The Rayleigh-wing spectrum can be used to determine the value of g_2 through use of Equation (15). This method requires knowledge of both j_2 and τ_{sm}. As discussed above, it is generally believed that j_2 takes on a value of unity in simple liquids. For the determination of τ_{sm}, two strategies have been employed: direct measurement and dilution studies. The single-molecule orientational correlation time can in principle be derived directly from either Raman scattering or NMR measurements. In practice, making an accurate determination of τ_{sm} using either technique is often challenging; Raman studies can confounded by hot bands, Fermi resonances, and the inherent difficulties of measuring isotropic spectra, while NMR measurements are sensitive to additional rotational degrees of freedom that are not of interest. However, given enough care it is generally possible to determine τ_{sm} quite accurately with either of these techniques. On the other hand, dilution studies only require the use of a single spectroscopic technique. Note that Equation (16) implies that as the number density of the liquid of interest approaches zero, g_2 will approach unity. The same also holds true for j_2 (31). Thus, by measuring τ_{coll} of the liquid of interest at various concentrations in a liquid that gives no reorientational signal (such as carbon tetrachloride), one can in principle determine τ_{sm}. Once again, in practice this determination can be difficult. In particular, nothing in Equation (16) specifies the functional form of the orientational correlation time in going from a pure liquid to an infinitely dilute solution, since both the number density and $\langle P_2(1)P_2(2) \rangle / \langle P_2(1)P_2(1) \rangle$ can change with concentration. Furthermore, corrections must be made for the fact that the single-molecule orientational correlation time may not be the same in the bulk liquid and in dilution, due to changes in viscosity and other parameters. As a result, to achieve accurate measurements with the dilution method it is generally necessary to make measurements on very dilute solutions, which can be experimentally challenging.

The intensity of the reorientational portion of the Rayleigh-wing spectrum of a liquid also depends directly on the static orientational correlation parameter, so g_2 can also be measured without using Equation (15)(31). Since it is much easier to measure a relative intensity than an absolute one, the determination of g_2 through intensity measurements is generally accomplished using dilution studies. Even measuring relative intensities is quite challenging, and these dilution studies are also subject to many of the same difficulties discussed in the preceding paragraph. On the other hand, intensity measurements are not affected at all by j_2, so that a combination of a careful intensity study with an independent determination of τ_{sm} for a liquid can be used to measure the dynamic orientational correlation parameter of a liquid as well (31)]. Indeed, such measurements provide experimental support for the expectation that j_2 takes on a value near unity (32).

Measuring τ_{coll} using frequency-domain techniques such as Rayleigh-wing scattering can at times be challenging, particularly when the relaxation time is long. In such cases, high experimental resolution is needed to separate the narrow central Lorentzian reorientational line from the elastic Rayleigh line. It is somewhat ironic that techniques such as OHD-RIKES, which rely upon ultrafast laser pulses, can offer advantages for studying the slowest components in the relaxation of the system. In general, time-domain techniques run into difficulties at very high frequencies, but given enough dynamic range the time scale of slower processes can be determined with high precision. Thus, OKE spectroscopies have proven to be highly useful in studying collective reorientation in liquids.

To investigate the nature of ordering in liquids, we have studied the temperature dependence of the OHD-RIKES response of a number of symmetric-top liquids, including acetonitrile-d_3, benzene, benzene-d_6, carbon disulfide, chloroform, and methyl iodide (56). These liquids were chosen in particular because data on τ_{sm} were available from other sources, including NMR data for acetonitrile-d_3 (57), Raman data for benzene and benzene-d_3 (45), NMR data for carbon disulfide (58), NMR data for chloroform (59), and Raman data for methyl iodide (45). Since the OHD-RIKES data were not all obtained at the same temperatures as the τ_{sm} data, we used the fact that the single-molecule orientational correlation time generally obeys the Arrhenius equation to interpolate (and, where necessary, extrapolate) values of τ_{sm} at temperatures matching those of the τ_{coll} data.

The Debye-Stokes-Einstein (DSE) equation (60) predicts that the orientational correlation time of a spherical object in a continuum liquid is given by

$$\tau = \frac{4\pi r^3 \eta}{3k_B T} \tag{24}$$

where r is the hydrodynamic radius of the solute, η is the solvent viscosity, and k_B is Boltzmann's constant. Although the DSE equation was not designed to describe the orientational behavior of molecules in a neat liquid, it is generally found that the orientational correlation time in such a system does indeed scale with η/T (31). Figures 7–12 show plots of τ_{coll} and τ_{sm}

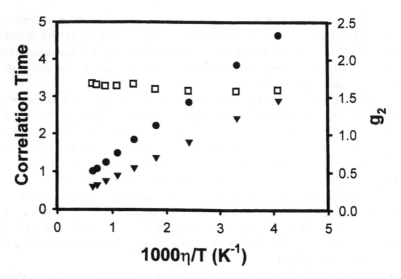

Figure 7 Single-molecule orientational correlation times (circles) from NMR data (57) and collective orientational correlation times (triangles) from OHD-RIKES data (56) as a function of η/T and estimated static orientational correlation parameter (squares) as a function of temperature for acetonitrile-d_3.

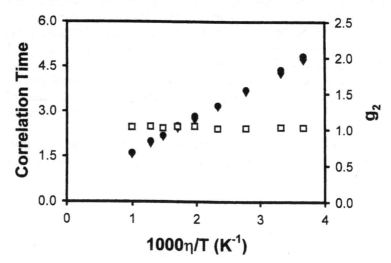

Figure 8 Single-molecule orientational correlation times (circles) from Raman data (45) and collective orientational correlation times (triangles) from OHD-RIKES data (56) as a function of η/T and estimated static orientational correlation parameter (squares) as a function of temperature for benzene.

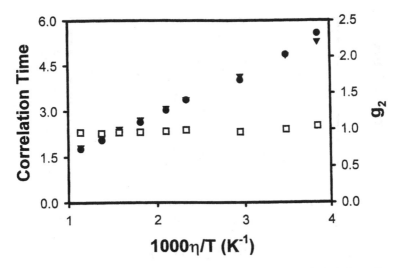

Figure 9 Single-molecule orientational correlation times (circles) from Raman data (45) and collective orientational correlation times (triangles) from OHD-RIKES data (56) as a function of η/T and estimated static orientational correlation parameter (squares) as a function of temperature for benzene-d_6.

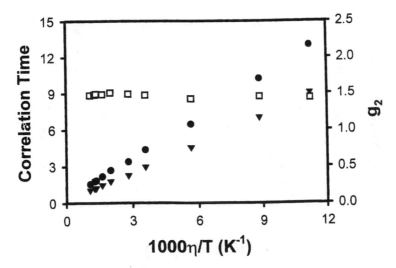

Figure 10 Single-molecule orientational correlation times (circles) from NMR data (58) and collective orientational correlation times (triangles) from OHD-RIKES data (56) as a function of η/T and estimated static orientational correlation parameter (squares) as a function of temperature for carbon disulfide.

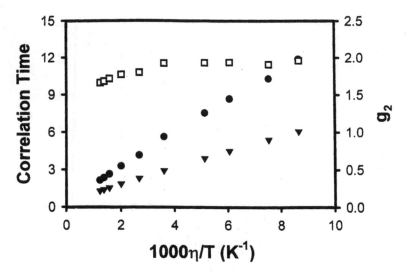

Figure 11   Single-molecule orientational correlation times (circles) from NMR data (59) and collective orientational correlation times (triangles) from OHD-RIKES data (56) as a function of η/T and estimated static orientational correlation parameter (squares) as a function of temperature for chloroform.

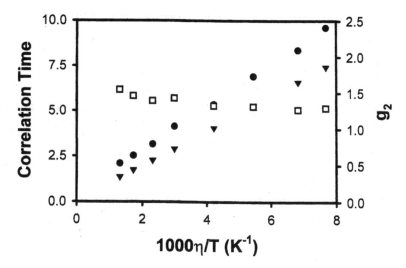

Figure 12   Single-molecule orientational correlation times (circles) from Raman data (45) and collective orientational correlation times (triangles) from OHD-RIKES data (56) as a function of η/T and estimated static orientational correlation parameter (squares) as a function of temperature for methyl iodide.

versus η/T for each of the symmetric-top liquids studied here. It can be seen clearly that both of these orientational correlation times appear to follow DSE behavior in all of these liquids.

The estimated values of g_2 for each liquid are also shown in Figs. 7–12. As might be expected, the static orientational correlation parameter is not highly dependent on temperature for any of these liquids. However, the behavior of g_2 with temperature does appear somewhat surprising in some of the liquids. One would expect that as a liquid is cooled, the microscopic ordering would tend to have a greater resemblance to that of the corresponding crystal. Liquids composed of polar molecules would be expected to have an increased amount of parallel ordering when cooled, as is evidenced by the general trend for dielectric constants to increase with decreasing temperature (61). The data for static orientational correlation parameter of chloroform indeed follow the expected trend, but those for acetonitrile-d_3 and methyl iodide do not. In both of the latter liquids, the data suggest a modest decrease in g_2 with decreasing temperature. Assuming that all of the single-molecule and collective orientational times were measured correctly for these liquids, a likely explanation for the apparent decrease in g_2 with decreasing temperature is that j_2 is not exactly unity at all temperatures for these liquids, but rather increases slightly with decreasing temperature.

The remaining three liquids are nonpolar. In all three liquids, g_2 shows virtually no temperature dependence. The value of g_2 for CS_2 is suggestive of a significant degree of local parallel ordering in this liquid, although it should be stressed that simulations of this liquid suggest that g_2 is influenced by pairs of molecules at relatively large separations (33). Nevertheless, considering the sizable quadrupole moment of CS_2, one might expect that a perpendicular local ordering might be preferred, although a staggered parallel ordering would also have a reasonably favorable quadrupole-quadrupole interaction energy. However, higher multipole moments are believed to have a significant influence on the liquid structure of CS_2. We should also point out that in a dilution study of the collective orientational correlation time of CS_2 in isoheptane at a temperature at which both liquids have the same viscosity, τ_{coll} proved to be insensitive to the mole fraction of CS_2 (39). This result would seem to imply that the static orientational correlation parameter for CS_2 is approximately unity. At least part of this discrepancy probably arises from the difficulties inherent in making an accurate determination of τ_{sm} for the bulk liquid. It is also possible that τ_{sm} for CS_2 changes upon dilution in isoheptane despite the constant viscosity. This subject merits further investigation.

In the case of benzene and benzene-d_6, g_2 takes on a value of approximately unity over the entire temperature range studied. Thus, there must be a significant degree of perpendicular ordering of neighboring molecules in benzene. This result can be explained in terms of the large quadrupole moment of benzene. If quadrupole-quadrupole interactions are the dominant influence in determining local structure, then a benzene dimer should favor a T-shaped perpendicular arrangement by a considerable amount over any possible parallel arrangement (62). Indeed, not only is this structure found in benzene dimers in the gas phase (63), but benzene crystals are also composed of perpendicular nearest neighbors (64).

VI. SYMMETRIC-TOP LIQUIDS: INTERMOLECULAR SPECTRA

Since the development of the Fourier-transform deconvolution procedure for OHD-RIKES data by McMorrow and Lotshaw (22), the intermolecular dynamics of a wide range of liquids have been studied with this technique (26,52,65–85). Figure 13 illustrates representative OKE reduced spectral densities we have recorded in symmetric-top liquids, including acetonitrile, benzene, benzene-d_6, carbon disulfide, chloroform, hexafluorobenzene, mesitylene, and 1,3,5-trifluorobenzene. Although there are conspicuous differences among these spectra, they are all broad and relatively featureless. Indeed, with rare exceptions the reduced spectral densities of simple liquids are devoid of sharp features, which makes it difficult to find an unambiguous interpretation of these spectra.

All of the spectra in Fig. 13 appear to have two "bands": one at low frequency and one at high frequency. The spectra often can be fit reasonably well to the sum of a frequency-weighted exponential that peaks at low frequency,

$$E(\omega) = \omega^\alpha \exp(-\omega/\omega_0) \tag{25}$$

and an antisymmetrized Gaussian that peaks at a higher frequency,

$$G(\omega) = \exp\left[-\frac{(\omega - \omega_0)^2}{(\Delta\omega)^2}\right] - \exp\left[-\frac{(\omega + \omega_0)^2}{(\Delta\omega)^2}\right] \tag{26}$$

There has been a tendency in the OKE literature to ascribe the former feature to DID effects, based in part on a similarity to a prediction for the lineshape of DID scattering in atomic fluids (66). Similarly, the latter feature is often attributed to single-molecule scattering. It is doubtful that any such

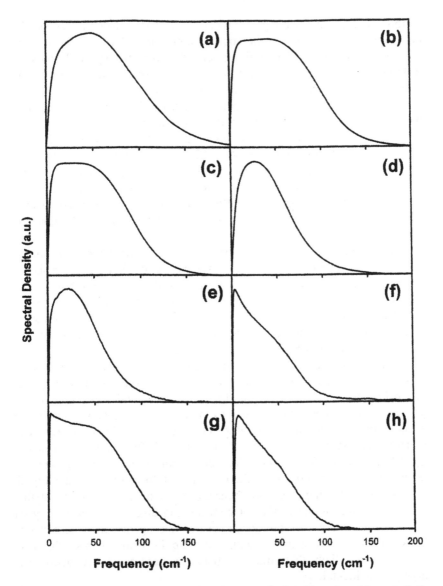

Figure 13 Room-temperature reduced spectral densities for (a) acetonitrile, (b) benzene, (c) benzene-d_6, (d) carbon disulfide, (e) chloroform, (f) hexafluorobenzene, (g) mesitylene, and (h) 1,3,5-trifluorobenzene.

assignments are truly meaningful. First, DID scattering from molecules with anisotropic polarizabilities is far more complex than DID scattering from atoms with isotropic polarizabilities, and therefore the spectral shape of the DID scattering in a molecular liquid is unlikely to resemble that in an atomic fluid. Second, as discussed above there is a strong, negative cross term between DID and single-molecule scattering (23) that is not generally taken into account in bandshape analysis of reduced spectral densities. On the other hand, insofar as this cross term arises from motions that are common to both single-molecule and DID scattering, it may have the same shape as the pure single-molecule contribution to the spectrum (23). If this is the case, and if pure DID scattering is weak [as a recent simulation study of CS_2 has suggested (86)], then the shape of the reduced spectral density should be highly similar to the shape of the pure single-molecule spectrum.

If the reduced spectral densities do indeed mirror the pure single-molecule contribution, then at least for symmetric-top liquids, which should have only one basic type of librational mode, it does not seem that the two observed bands can represent two distinct types of molecular motions. Similarly, the reduced spectral densities of liquids composed of less symmetric molecules also can often be fit to the same two types of bands, despite the existence of multiple possible librational modes.

One possible explanation for the observation of multiple features in reduced spectral densities is that there is considerable microscopic structure in liquids, which leads to multiple librational modes even for highly symmetric molecules. An extreme example of such a viewpoint is the recent suggestion that the reduced spectral density of liquid benzene can be described completely in terms of the collective Raman-active modes of crystalline benzene (87). While this speculation is an intriguing one, we believe it would require an unlikely degree of local order in the liquid. Furthermore, although hexafluorobenzene has a quadrupole moment that is of similar magnitude to that of benzene and also has a similar crystal structure (64), there does not appear to be a strong resemblance between the Raman spectrum of the crystalline and liquid forms of this substance. Thus, it seems likely that similarity between these spectra in benzene is largely coincidental.

To develop a deeper understanding of the contributions to the intermolecular spectrum of symmetric-top liquids, it is useful to study the spectrum as a function of some readily varied parameter. For instance, OHD-RIKES dilution studies of CS_2 in alkanes have proven quite interesting (73,83). The general trend in such studies is that the high-frequency feature moves to lower frequency with increasing dilution, while the low-frequency feature remains virtually unchanged. Furthermore, the spectrum stops changing after a certain degree of dilution has been achieved. These results have been interpreted both in terms of a "softening" of the inter-molecular potential (73) and in terms of changes in the dominant scattering mechanism upon dilution (83).

Studies in which OKE data for pure liquids are obtained as a function of either temperature (26,36,41) or pressure (42) have also proven enlight-ening. Figure 14 shows reduced spectral densities for benzene obtained over a broad range of temperatures. These spectra show two features that are typical for all of the liquids that we have studied: as the tempera-ture is lowered, the low-frequency feature moves to lower frequency and the high-frequency feature moves to higher frequency and broadens. The behavior of the high-frequency feature can be understood readily in terms

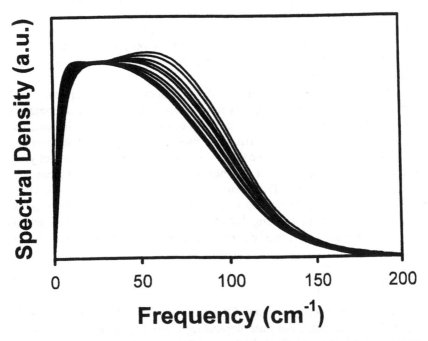

Figure 14 Reduced spectral density for benzene as a function of temperature. The temperatures from top to bottom on the high-frequency side of the spectra are 267, 272, 281, 294, 299, 308, 327, 336, and 341 K. The spectra have been scaled arbitrarily to match in intensity at 30 cm^{-1}.

of the increase in density that accompanies a reduction in temperature. The behavior of the low-frequency feature is rather less intuitive, however.

We have previously suggested that the temperature-dependent behavior of the low-frequency feature in the reduced spectral density is related to damping of the intermolecular modes (26). Assuming that the damping rate is relatively independent of the frequency of a mode, below some critical frequency the spectrum is guaranteed to be overdamped, regardless of the nature of the damping mechanism. Since the damping rate should decrease with decreasing temperature, so will the critical frequency. The end result of this scenario will be increase in intensity at low frequency as the temperature is lowered.

The low-frequency feature in the reduced spectral density corresponds to the long-time tail of the intermolecular response function, which is often denoted the "intermediate" response in the OKE literature (15,51). In most liquids, this portion of the response appears to be exponential over a significant time scale. Why this portion of the response is exponential and what information the time scale of this exponential holds is still poorly understood. For this reason, we have performed detailed temperature-dependent studies of the intermediate relaxation in six symmetric-top liquids: acetonitrile, acetonitrile-d_3, benzene, carbon disulfide, chloroform, and methyl iodide (56).

For each liquid, this portion of the response indeed appears to be exponential over a broad range of temperatures. Given these temperature-dependent data, we can begin to delve into the nature of the intermediate response. Figure 15 shows Arrhenius plots of the intermediate response time, τ_i, for all six liquids. Note that for each liquid the Arrhenius plot yields a straight line, suggesting that the intermediate relaxation arises from some sort of activated process. This result is an important one, because it allows us to eliminate pure dephasing as the cause of the intermediate response. In the diffusive limit, the angular momentum correlation time τ_J and the single-molecule orientational correlation time of a liquid should satisfy the relation

$$\tau_J \tau_{sm} = \frac{I}{6k_B T} \tag{27}$$

where I is the moment of inertia of the molecules (88). Since pure dephasing arises from collisions, the pure dephasing time should be roughly proportional to τ_J. However, since both τ_i and τ_{sm} exhibit Arrhenius behavior, their product clearly cannot be inversely proportional to temperature.

Figure 16 shows that for each liquid, a plot of τ_i versus η/T yields a straight line. This result, which is reminiscent of the DSE equation

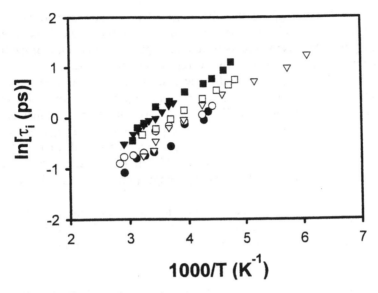

Figure 15 Arrhenius plots of τ_i for acetonitrile (filled circles), acetonitrile-d_3 (open circles), benzene (filled triangles), carbon disulfide (open triangles), chloroform (filled squares), and methyl iodide (open squares). (From Ref. 56.)

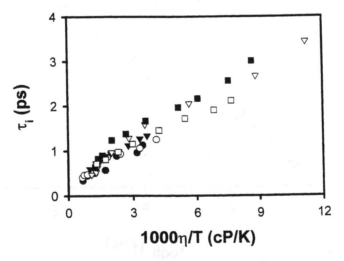

Figure 16 Debye-Stokes-Einstein plots of τ_i for acetonitrile (filled circles), acetonitrile-d_3 (open circles), benzene (filled triangles), carbon disulfide (open triangles), chloroform (filled squares), and methyl iodide (open squares). (From Ref. 56.)

[Equation. (24)], suggests that the intermediate relaxation may be hydrodynamic in origin. If τ_I is indeed related to hydrodynamic effects, it seems unlikely that this decay time is related to population relaxation of the intermolecular modes. Thus we must search for another cause.

Since both τ_{coll} and τ_{sm} for these six liquids follow DSE behavior, for any given liquid a plot of τ_I versus τ_{coll} or τ_{sm} must be linear. However, by comparing such plots for all six liquids, we can determine whether τ_I is more closely related to τ_{coll} or τ_{sm}. The plots of τ_I versus τ_{coll} are shown in Fig. 17 and those versus τ_{sm} in Fig. 18. Strikingly, all of the plots in Fig. 17 fall on the same straight line, which implies that for these liquids the value of τ_i can be predicted at any temperature given knowledge of the value of τ_{coll}.

We have suggested recently that the above data are consistent with motional narrowing being the source of the intermediate OKE relaxation (56). While on short enough time scales the structure of a liquid is relatively

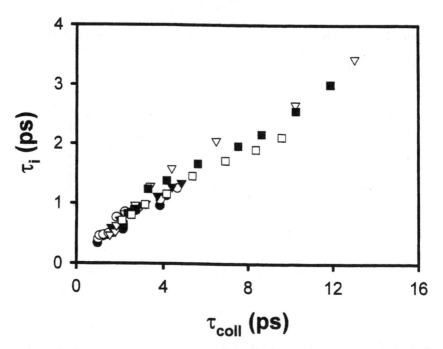

Figure 17 Plots of τ_i versus τ_{coll} for acetonitrile (filled circles), acetonitrile-d_3 (open circles), benzene (filled triangles), carbon disulfide (open triangles), chloroform (filled squares), and methyl iodide (open squares). (From Ref. 56.)

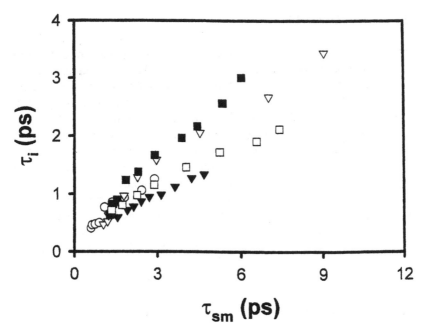

Figure 18   Plots of τ_i versus τ_{sm} for acetonitrile (filled circles), acetonitrile-d_3 (open circles), benzene (filled triangles), carbon disulfide (open triangles), chloroform (filled squares), and methyl iodide (open squares). (From Ref. 56.)

stable, on time scales greater than several hundreds of femtoseconds the structure evolves. Structural evolution in turn leads to "spectral diffusion," i.e., fluctuations in the frequencies of the modes. At low enough frequencies, these fluctuations occur on a time scale that is faster than the characteristic frequencies of the modes. The frequencies of such modes are effectively time-averaged, thus narrowing the spectrum (89).

Since the collective orientational correlation time depends on the structure of a liquid, it is plausible that the rate of structural evolution of the liquid is proportional to this quantity. Thus, at lower temperatures τ_{coll} is longer and therefore the structural fluctuations are slower. As a result, motional narrowing is less effective as the temperature is lowered. While less motional narrowing would normally lead to a slower decay in the time domain, in this case the spectral density goes down to zero frequency. Thus, motional narrowing can reduce the spectral density at low frequencies and thereby decrease the intermediate relaxation time.

VII. CONCLUSIONS

Low-frequency Raman spectroscopies are useful techniques for delving into the microscopic structure and dynamics of liquids. Low-frequency Raman spectroscopy has become all the more powerful a tool with the advent of both reliable femtosecond laser systems and coherent spectroscopic techniques such as OHD-RIKES. However, the relatively broad and featureless nature of low-frequency Raman spectra continues to present challenges for both theorists and experimentalists, although significant progress is being made in developing means of interpreting such spectra. In the meantime, reduced spectral densities from OHD-RIKES data have seen some success in the prediction of solvation dynamics (81,82,90), although such successes have not been universal. The successful implementation of higher-order spectroscopic techniques akin to the photon echo (91,92) should also assist greatly in the resolution of many of the outstanding issues in the low-frequency Raman spectroscopy of liquids.

ACKNOWLEDGMENTS

I wish to acknowledge the members of my group who obtained the data discussed here: Brian Loughnane, Richard Farrer, Alessandra Scodinu, Natalia Balabai, and Tomaso Baldacchini. This work was supported by the National Science Foundation, Grant CHE-9501598. I am also a Research Corporation Cottrell Scholar, an Alfred P. Sloan Research Fellow, and a Dreyfus New Faculty Fellow.

REFERENCES

1. Barbara PF, Jarzeba W. Ultrafast photochemical intramolecular charge and excited state solvation. Adv Photochem 1990; 15:1–68.
2. Fleming GR, Cho M. Chromophore-solvent dynamics. Annu Rev Phys Chem 1996; 47:109–134.
3. Yu J, Fourkas JT, Berg M. Transient hole burning studies of electronic state solvation: phonon and structural contributions. In: Martin J-L, Migus A, Mourou G, Zewail AH, eds. Ultrafast Phenomena VIII. Berlin: Springer-Verlag, 1992.
4. Horng ML, Gardecki JA, Papazyan A, Maroncelli M. Subpicosecond measurements of polar solvation dynamics: coumarin 153 revisited. J Phys Chem 1995; 99:17311–17337.

5. McElroy R, Wynne K. Ultrafast dipole solvation measured in the par-infrared. Phys Rev Lett 1997; 79:3078–3081.
6. Bacon GE, Neutron Diffraction. Oxford: Oxford University Press, 1955.
7. Bratos S, Pick RM, eds. Vibrational Spectroscopy of Molecular Liquids and Solids. New York: Plenum Press, 1980.
8. Berne BJ, Pecora R. Dynamic Light Scattering. New York: Wiley, 1976.
9. Greene BI, Saeta PN, Dykaar DR, Scmitt-Rink S, Chuang SL. Far-infrared light generation at semiconductor surfaces and its spectroscopic applications. IEEE J Quantum Electron 1992; 28:2302.
10. Trebino R, Barker CE, Siegman AE. Tunable-laser-induced gratings for the measurement of ultrafast phenomena. IEEE J Quantum Electron 1986; QE-22:1413–1430.
11. Friedman JS, Lee MC, She CY. Depolarized stimulated gain spectra of liquid CS_2 and benzene at room temperature. Chem Phys Lett 1991; 186:161–169.
12. Yan Y-X, Nelson KA. Impulsive stimulated light scattering. II. Comparison to frequency-domain light-scattering spectroscopy. J Chem Phys 1987; 87:6257–6265.
13. Mukamel S. Principles of Nonlinear Optical Spectroscopy. New York: Oxford University Press, 1995.
14. Kinoshita S, Kai Y, Yamaguchi M, Yagi T. Direct comparison between ultrafast optical Kerr effect and high-resolution light scattering spectroscopy. Phys Rev Lett 1995; 75:148–151.
15. Kalpouzos C, Lotshaw WT, McMorrow D, Kenney-Wallace GA. Femtosecond laser-induced Kerr responses in liquid CS_2. J Phys Chem 1987; 91:2028–2030.
16. Deeg FW, Fayer MD. Analysis of complex molecular dynamics in an organic liquid by polarization selective subpicosecond transient grating experiments. J Chem Phys 1989; 91:2269–2279.
17. Dhar L, Rogers JA, Nelson KA. Time-resolved vibrational spectroscopy in the impulsive limit. Chem Rev 1994; 94:157–193.
18. Chang YJ, Cong P, Simon JD. Isotropic and anisotropic intermolecular dynamics of liquids studied by femtosecond position-sensitive Kerr lens spectroscopy. J Chem Phys 1997; 106:8639–8649.
19. Righini R. Ultrafast optical Kerr effect in liquids and solids. Science 1993; 262:1386–1390.
20. Kinoshita S, Kai Y, Ariyoshi T, Shimada Y. Low frequency modes probed by time-domain optical Kerr effect spectroscopy. Int J Mod Phys B 1996; 10:1229–1272.
21. Hellwarth RW. Third-order optical susceptibilities of liquids and solids. Prog Quantum Electron 1977; 5:1–68.
22. McMorrow D, Lotshaw WT. Intermolecular dynamics in acetonitrile probed with femtosecond Fourier transform Raman spectroscopy. J Phys Chem 1991; 95:10395–10406.

23. Keyes T, Kivelson D, McTague JP. Theory of k-independent depolarized Rayleigh wing scattering in liquids composed of anisotropic molecules. J Chem Phys 1971; 55:4096–4100.

24. Koningstein JA. Introduction to the Theory of the Raman Effect. Dordrecht, Holland: Reidel, 1972.

25. Palese S, Mukamel S, Miller RJD, Lotshaw WT. Interrogation of vibrational structure and line broadening of liquid water by Raman-induced Kerr effect measurements within the multi-mode Brownian oscillator model. J Phys Chem 1996; 100:10380–10388.

26. Farrer RA, Loughnane BJ, Deschenes LA, Fourkas JT. Level-dependent damping in intermolecular vibrations: linear spectroscopy. J Chem Phys 1997; 106:6901–6915.

27. Butcher PN, Cotter D. The Elements of Nonlinear Optics. Cambridge: Cambridge University Press, 1990.

28. Tokmakoff A. Orientational correlation functions and polarization selectivity for non-linear spectroscopy of isotropic media: I. Third order. J Chem Phys 1996; 105:1–12.

29. Murry RL, Fourkas JT. Polarization selectivity of nonresonant spectroscopies in isotropic media. J Chem Phys 1997; 107:9726–9740.

30. Keyes T, Kivelson D. Depolarized light scattering: theory of the sharp and broad Rayleigh lines. J Chem Phys 1972; 56:1057–1065.

31. Kivelson D, Madden PA. Light scattering studies of molecular liquids. Annu Rev Phys Chem 1980; 31:523–558.

32. Bauer DR, Braumann JI, Pecora R. Depolarized Rayleigh scattering and orientational relaxation of molecules in solution. IV. Mixtures of hexafluorobenzene with benzene and mesitylene. J Chem Phys 1975; 63:53–60.

33. Ladanyi B, Keyes T. The role of local fields and interparticle pair correlations in light scattering by dense fluids. I. Depolarized intensities due to orietational fluctuations. Mol Phys 1977; 33:1063–1097.

34. Loughnane BJ, Fourkas JT. Geometric effects in the dynamics of a nonwetting liquid in microconfinement: an optical Kerr effect study of methyl iodide in nanoporous glasses. J Phys Chem B 1998; 102:10288–10294.

35. Gradshteyn IS, Ryzhik IM. Tables of Integrals, Series, and Products. San Diego: Academic Press, 1980.

36. Ruhman S, Kohler B, Joly AG, Nelson KA. Intermolecular vibrational motion in CS_2 liquid at $165 < T < 300$ K observed by pemtosecond time-resolved impulsive stimulated scattering. Chem Phys Lett 1987; 141:16–24.

37. Whittenburg SL, Wang CH. Raman scattering study of CS_2 in CCl_4. J Chem Phys 1979; 70:3141–3142.

38. Geiger LC, Ladanyi BM. Higher order interaction-induced effects on Rayleigh light scattering by molecular liquids. J Chem Phys 1987; 87:191–202.

39. Kalpouzos C, McMorrow D, Lotshaw WT, Kenney-Wallace GA. Femtosecond laser-induced optical Kerr dynamics in CS_2/alkane binary solutions. Chem Phys Lett 1989; 155:240–242.

40. McMorrow D, Lotshaw WT. Dephasing and relaxation in coherently excited ensembles of intermolecular oscillators. Chem Phys Lett 1991; 178:69–74.

41. Ruhman S, Nelson KA. Temperature-dependent molecular dynamics of liquid carbon disulphide: polarization-sensitive impulsive stimulated light-scattering data and Kubo line shape analysis. J Chem Phys 1991; 94:859–867.

42. Kohler B, Nelson KA. Femtosecond impulsive stimulated light scattering from liquid carbon disulfide at high pressure: experiment and computer simulation. J Phys Chem 1992; 96:6532–6538.

43. Rivoire G, Wang D. Dynamics of CS_2 in the large spectral bandwidth stimulated Rayleigh-wing scattering. J Chem Phys 1993; 99:9460–9464.

44. Danninger W, Zundel G. Reorientational motion and orientational correlation functions in weakly associated organic liquids by depolarized Rayleigh scattering. Chem Phys Lett 1982; 90:69–75.

45. Patterson GD, Griffiths JE. Raman and depolarized Rayleigh scattering in the liquid state: reorientational motions and correlations in orientation for symmetric top molecules. J Chem Phys 1975; 63:2406–2413.

46. Whittenburg SL, Wang CH. Light scattering studies of rotational and vibrational relaxations of acetonitrile in carbon tetrachloride. J Chem Phys 1977; 66:4255–4262.

47. Vohringer P, Arnett DC, Westervelt RA, Feldstein MJ, Scherer NF. Optical dephasing on femtosecond time scales: direct measurement and calculation from solvent spectral densities. J Chem Phys 1995; 102:4027–4036.

48. Schmidt RL. Temperature dependence of Rayleigh light scattering and depolarization in pure liquids. J Colloid Interface Sci 1968; 27:516–528.

49. Alms GR, Bauer DR, Brauman JI, Pecora R. Depolarized Rayleigh scattering and orientational relaxation of molecules in solution. I. Benzene, toluene and para-xylene. J Chem Phys 1973; 58:5570–5578.

50. Bauer DR, Alms GR, Braumann JI, Pecor R. Depolarized Rayleigh scattering and ^{13}C NMR studies of anisotropic molecular reorientation of aromatic compounds in solution. J Chem Phys 1974; 61:2255–2261.

51. Lotshaw WT, McMorrow D, Kalpouzos C, Kenney-Wallace GA. Femtosecond dynamics of the optical Kerr effect in liquid nitrobenzene and chlorobenzene. Chem Phys Lett 1987; 136:323–328.

52. Neelakandan M, Pant D, Quitevis EL. Structure and intermolecular dynamics of liquids: femtosecond optical Kerr effect measurements in nonpolar fluorinated benzenes. J Phys Chem A 1997; 101:2936–2945.

53. Rothschild WG, Rosasco GJ, Livingston RC. Dynamics of molecular reorientational motion and vibrational relaxation in liquids. Chloroform. J Chem Phys 1975; 62:1253–1268.

54. Cheung CK, Jones DR, Wang CH. Single particle reorientation and pair correlations of methyl iodide solutions studied by depolarized Rayleigh and Raman scattering. J Chem Phys 1976; 64:3567–3572.

55. Lee YT, Wallen SL, Jonas J. Effect of confinement on the resonant intermolecular vibrational coupling of the ν_2 mode of methyl iodide. J Phys Chem 1992; 96:7161–7164.

56. Loughnane BJ, Scodiunu A, Farrer RA, Fourkas JT, Mohanty U. Exponential intermolecular dynamics in optical Kerr effect spectroscopy of small-molecule liquids. J Chem Phys 1999; 111:2686–2694.

57. Zhang J, Jonas J. NMR Study of the geometric confinement effects of the anisotropic rotational diffusion of acetonitrile-d_3. J Phys Chem 1993; 97:8812–8815.

58. Korb J-P, Xu S, Cros F, Malier L, Jonas J. Quenched molecular reorientation and angular momentum for liquids confined to nanopores of silica glass. J Chem Phys 1997; 107:4044–4050.

59. VanderHart DL. Study of molecular reorientation: pressure and temperature dependence of deuterium relaxation in liquid $CDCl_3$. J Chem Phys 1974; 60:1858–1870.

60. Debye P. Polar Molecules. New York: Dover, 1929.

61. Weast RC, ed. CRC Handbook of Chemistry and Physics. Boca Raton, FL: CRC Press, 1985.

62. Hernandez-Trujillo J, Costas M, Vela A. Quadrupole interactions in pure nondipolar fluorinated or methylated benzenes and their binary mixtures. J Chem Soc Faraday Trans 1993; 89:2441–2443.

63. Venturo VA, Felker PM. Nonlinear Raman spectroscopy of ground-state intermolecular vibrations in benzene complexes. J Phys Chem 1993; 97:4882–4886.

64. Williams JH. The molecular electric quadrupole moment and solid-state architecture. Acc Chem Res 1993; 26:593–598.

65. Castner EW, Chang YJ, Chu YC, Walrafen GE. The intermolecular dynamics of liquid water. J Chem Phys 1995; 102:653–659.

66. Chang YJ, Castner Jr. EW. Femtosecond dynamics of hydrogen-bonding solvents. Formamide and N-methylformamide in acetonitrile, DMF, and water. J Chem Phys 1993; 99:113–125.

67. Chang YJ, Castner Jr. EW. Fast responses from "slowly relaxing" liquids: a comparative study of the femtosecond dynamics of triacetin, ethylene glycol, and water. J Chem Phys 1993; 99:7289–7299.

68. McMorrow D, Lotshaw WT. Evidence for low-frequency ($\sim$15 cm^{-1}) collective modes in benzene and pyridine liquids. Chem Phys Lett 1993; 201:369–376.

69. Deuel HP, Cong P, Simon JD. Probing intermolecular dynamics in liquids by femtosecond optical Kerr effect spectroscopy: effects of molecular symmetry. J Phys Chem 1994; 98:12600–12608.

70. Castner Jr. EW, Chang YJ, Melinger JS, McMorrow D. Femtosecond Fourier-transform spectroscopy of low-frequency intermolecular motions in weakly interacting liquids. J Lumin 1994; 60&61:723–726.

71. Chang YJ, Castner Jr. EW. Deutetrium isotope effects on the ultrafast solvent relaxation of formamide and N,N-dimethylformamide. J Phys Chem 1994; 98:9712–9722.

72. Cong P, Deuel HP, Simon JD. Structure and dynamics of molecular liquids investigated by optical-heterodyne detected Raman-induced Kerr effect spectroscopy (OHD-RIKES). Chem Phys Lett 1995; 240:72–78.

73. McMorrow D, Thantu N, Melinger JS, Kim SK, Lotshaw WT. Probing the microscopic molecular environment in liquids: intermolecular dynamics of CS_2 in alkane solvents. J Phys Chem 1996; 100:10389–10399.

74. Kamada K, Ueda M, Sakaguchi T, Ohta K, Fukumi T. Femtosecond optical Kerr study of heavy atom effects in the third-order optical nonlinearity of thiophene homologues: purely electronic contribution. Chem Phys Lett 1996; 263:215–222.

75. Kamada K, Ueda M, Sakaguchi T, Ohta K, Fukumi T. Femtosecond optical kerr dynamics of thiophene in carbon tetrachloride solution. Chem Phys Lett 1996; 249:329–334.

76. Quitevis EL, Neelakandan M. Femtosecond optical Kerr effect studies of liquid methyl iodide. J Phys Chem 1996; 100:10005–10014.

77. Chang YJ, Castner Jr. EW. The intermolecular dynamics of substituted benzene and cyclohexane liquids, studied by femtosecond nonlinear-optical polarization spectroscopy. J Phys Chem 1996; 100:3330–3343.

78. Kamada K, Ueda M, Ohta K, Wang Y, Ushida K, Tominaga Y. Molecular dynamics of thiophene homologues investigated by femtosecond optical Kerr effect and low frequency Raman scattering spectroscopies. J Chem Phys 1998; 109:10948–10957.

79. Neelakandan M, Pant D, Quitevis EL. Reorientational and intermolecular dynamics in binary liquid mixtures of hexafluorobenzene and benzene: femtosecond optical Kerr effect measurements. Chem Phys Lett 1997; 265:283–292.

80. Steffen T, Duppen K. Time resolved four- and six-wave mixing in liquids part II: experiments. J Chem Phys 1997; 106:3854–3864.

81. Smith NA, Lin S, Meech SR, Yoshihara K. Ultrafast optical Kerr effect and solvation dynamics of liquid aniline. J Phys Chem A 1997; 101:3641–3645.

82. Smith NA, Lin S, Meech SR, Shirota H, Yoshihara K. Ultrafast dynamics of liquid anilines studied by the optical Kerr effect. J Phys Chem A 1997; 101:9578–9586.

83. Steffen T, Meinders NACM, Duppen K. Microscopic origin of the optical Kerr effect response of CS_2-pentane binary mixtures. J Phys Chem A 1998; 102:4213–4221.

84. McMorrow D, Lotshaw WT, Kenney-Wallace GA. Femtosecond optical Kerr studies on the origin of the nonlinear responses in simple liquids. IEEE J Quantum Electron 1998; 24:443–454.

85. Castner Jr. EW, Maroncelli M. Solvent dynamics derived from optical Kerr effect, dielectric dispersion, and time-resolved Stokes shift measurements: an empirical comparison. J Mol Liq 1998; 77:1–36.

86. Murry RL, Fourkas JT, Keyes T. Nonresonant intermolecular spectroscopy beyond the Placzek approximation. I. Third-order spectroscopy. J Chem Phys 1998; 109:2814–2825.

87. Ratajska-Gadomska B, Gadomski W, Wiewior P, Radzewicz C. A femtosecond snap-shot of crystalline order in molecular liquids. J Chem Phys 1998; 108:8489–8498.

88. Hubbard PS. Theory of nuclear magnetic relaxation by spin-rotational interactions in liquids. Phys Rev 1963; 131:1155–1165.

89. Kubo R. A stochastic theory of line-shape and relaxation. In: Haar DT, eds. Fluctutation, Relaxation and Resonance in Magnetic Systems. Edinburgh: Oliver and Boyd, 1961.

90. Yang T-S, Vohringer P, Arnett DC, Scherer NF. The solvent spectral density and vibrational multimode approach to optical dephasing: two-pulse photon echo response. J Chem Phys 1995; 103:8346–8359.

91. Tanimura Y, Mukamel S. Two-dimensional femtosecond vibrational spectroscopy of liquids. J Chem Phys 1993; 99:9496–9511.

92. Steffen T, Fourkas JT, Duppen K. Time resolved four- and six-wave mixing in liquids. I. Theory. J Chem Phys 1996; 105:7364–7382.

12

Lattice Vibrations that Move at the Speed of Light: How to Excite Them, How to Monitor Them, and How to Image Them Before They Get Away

Richard M. Koehl, Timothy F. Crimmins, and Keith A. Nelson
Massachusetts Institute of Technology, Cambridge, Massachusetts

Some polar lattice vibrations can couple strongly to electromagnetic radiation and move at light-like speeds through the host crystal. In this chapter, new methods are reviewed for impulsive excitation of these modes and for monitoring of their spatial and temporal evolution.

I. INTRODUCTION

In most time-resolved spectroscopy measurements, excitation and probe beams arrive at the same region of the sample so that its time-dependent response to the excitation light can be monitored. Accordingly, "impulsive" excitation of coherent lattice or molecular vibrations is usually followed by probing of time-dependent vibrational oscillations and decay within the excitation region (1,2). This is generally the case irrespective of the excitation mechanism, including impulsive stimulated Raman scattering (ISRS), impulsive absorption, or others, through which excitation is sudden compared to vibrational oscillation periods.

There are two important cases in which the impulsively driven vibrational response travels through the host medium, often leaving the region that was irradiated by excitation light. First and best known is the case of acoustic waves generated through impulsive stimulated thermal scattering, i.e., sudden heating, or through impulsive stimulated Brillouin scattering (3–5). Second is the case of mixed lattice vibrational and electromagnetic modes, called optic phonon-polariton modes, which propagate at light-like speeds through ionic crystal lattices (6–10). In both cases, recent advances in experimental methodology have opened up new possibilities for vastly improved excitation and monitoring of the propagating modes.

Here we review three important advances in the study of optic phonon-polaritons. Some of the developments are applicable to acoustic wave generation or detection as well. The first is an improved general method of crossing femtosecond light pulses with a broad range of applications in ultrafast optics and spectroscopy (11). For optic phonon-polaritons, the method permits greatly improved definition of the spatial excitation pattern and therefore of the excitation wave vector. It also permits very rapid changes to be made in the selection of phonon-polariton (or acoustic wave) wavevector magnitude and therefore frequency (12). The second advance is the development of a heterodyne detection method (13,14) through which phonon-polariton (or acoustic) wave amplitudes can be extracted. This offers many advantages over earlier methods, including separation of phonon-polariton signals from other contributions and the capability for full characterization of traveling waves that have left the excitation region. The third advance is a spatiotemporal imaging method (15,16) that yields real-space images of the propagating lattice responses at various times following photoexcitation. This permits the full spatial and temporal characteristics of the response to be monitored in a simple and convenient manner.

The three methods taken together allow time-resolved spectroscopic characterization of propagating modes with far greater ease and completeness than has been possible in the past. They facilitate a number of important possibilities including spatiotemporal control over propagating modes, excitation and characterization of nonlinear lattice responses, tunable terahertz-frequency spectroscopy, and terahertz frequency and bandwidth signal processing.

II. BACKGROUND

A. What are Phonon-Polaritons?

In crystalline solids, optic phonon modes represent vibrations of nearby ions or molecules against each other within a unit cell, with no net change in the shape or volume of the unit cell. The vibrational motions within different unit cells have well-defined phase relationships, so, for example, at the limit of very low wavevector magnitude q (i.e., long wavelength $\Lambda = 2\pi/q$), the vibrations in all nearby unit cells are synchronized with essentially the same phase. In most crystals, there are no long-range interactions between vibrations within unit cells separated by micrometer or longer distances, so the optic phonon frequency is the same whether the wavelength is, say, 1 or 2 μm. However, polar vibrational modes may have long-range interactions because the oscillating dipole associated with each unit cell acts as an antenna that may radiate. Thus, the polar vibrational motion in one unit cell may be assisted by electromagnetic radiation originating from another unit cell many μm away, and so the energy associated with the coupled vibrational/electromagnetic response may be lower than that of the uncoupled vibrational response alone. Radiation may be produced if the vibrational symmetry, frequency, and wavelength all match those of the electromagnetic mode. Thus, transverse optic phonon modes, whose oscillating dipole moments are perpendicular to the wave vector direction, may be coupled to the corresponding electromagnetic modes, and the coupled vibrational/electromagnetic mode energy may be reduced compared to the longitudinal optic phonon mode of the same wavevector. A prototypical phonon-polariton dispersion curve is shown in Fig. 1.

Note that irradiation of a polar crystal with far-infrared (around 0.5–10 THz frequency range) radiation can be used to produce phonon-polaritons. Outside the crystal, purely electromagnetic radiation propagates. Upon entering the crystal, this radiation drives the resonant lattice vibrations at the same frequency. What propagates through the crystal is a mixed polar lattice vibrational/electromagnetic (phonon-polariton) mode. At the other end of the crystal, some phonon-polariton reflection occurs and some amount of pure electromagnetic radiation emerges.

B. Impulsive Phonon-Polariton Excitation

In our experiments, phonon-polaritons are produced not with far-infrared radiation but with visible light pulses through impulsive stimulated Raman scattering (1,2). However, the same type of coupled excitation is produced as long as the excitation wavevector lies within the phonon-polariton range. In most ISRS experiments, crossed excitation pulses are used so that the excitation wavevector is reasonably well defined. The experimental geometry is illustrated in Fig. 2. The crossed pulses produce an optical interference pattern of fringe spacing Λ, and a spatially periodic, temporally

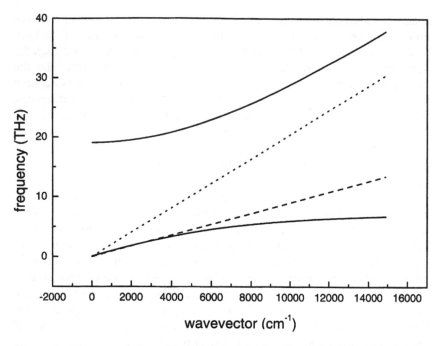

Figure 1 Phonon-polariton dispersion in LiTaO$_3$. The solid lines describe the dispersion of the upper and lower branches of the polariton, the dashed line describes the dispersion of light at frequencies below the phonon resonance, and the dotted line describes the dispersion of light at frequencies above the phonon resonance.

"impulsive" driving force is exerted on the vibrational mode through ISRS. This generates the coherent lattice vibrational response, which consists of phonon-polaritons with wavevectors whose y-components, as shown in Fig. 2, are given by $q_y = \pm 2\pi/\Lambda$. (The z-component will be discussed below.) The two waves form a quasi-standing wave within the excitation region, which persists until they are fully damped or until they leave the region as separated traveling waves.

A variably delayed probe pulse can be used to monitor the time-dependent vibrational oscillations and decay through coherent scattering ("diffraction"), yielding data like that shown in the simulation in Fig. 3a. In this simulation, the excitation and probe regions are overlapped spatially, and the decay of signal is due to damping and dephasing of the phonon-polariton response. From data of this form, the polariton frequency ω and

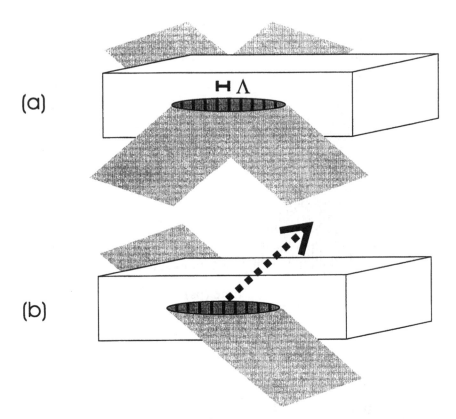

Figure 2 Impulsive stimulated scattering generation of material excitations with crossed beams. (a) Crossed optical beams interfere at the sample and produce a spatially periodic driving force on the material with wavelength Λ. (b) A probe beam incident upon the spatially periodic material response is partially diffracted, with the diffraction efficiency being proportional to the square of the amplitude of the material response.

phase velocity $v = \omega/q$ (with q known from the excitation geometry) and the dephasing rate γ are determined.

Note that the number of optical interference fringes in the excitation light pattern limits the wavevector definition. A small excitation spot size results in a wide range of excitation wavevector components. If the probe beam reaches the same sample region as the excitation beams, the signal will appear to decay rapidly because the phonon-polaritons will leave the region rapidly. This can be thought of equivalently as rapid dephasing

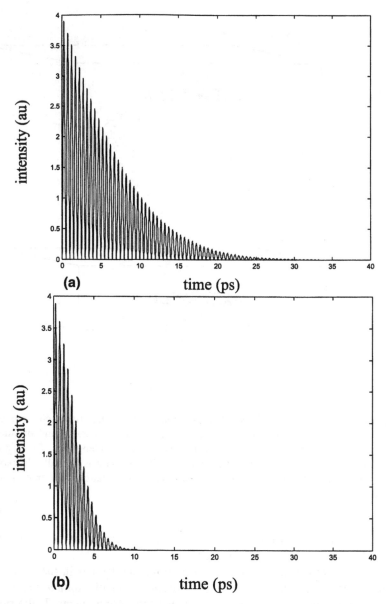

Figure 3 Simulation of ISRS diffraction data from propagating phonon-polaritons. Use of (a) large (1 mm) and (b) small (250 μm) excitation and probe spot sizes gives rise to large differences in apparent damping kinetics for these 1 THz polaritons propagating at a velocity of 50 μm/ps.

of the many different phonon-polariton wave vector components whose frequencies span a wide range. Figure 3b shows simulations in which the parameters are similar to those of Fig. 3a except for the excitation spot size, which was considerably smaller (approximately 1000 μm in Fig. 3a versus 250 μm in Fig. 3b). The more rapid decay of signal is apparent. Similar results obtain in the case of coherent acoustic waves produced through impulsive stimulated scattering with a small excitation region (17,18).

All phonon-polariton modes are infrared active, and in noncentrosymmetrical crystals they also are Raman active. It is in these crystals that ISRS excitation of phonon-polaritons may occur. Far-IR electromagnetic responses may also be introduced in the crystal by difference frequency mixing. Through this mechanism crossed femtosecond pulses generate via electronic nonlinearities far-IR radiation in the crystal with wavevector y-components $q_y = \pm 2\pi/\Lambda$. If the wavevector lies within the phonon-polariton range, this mechanism also results in phonon-polariton excitation. The creation of far-IR frequency components by mixing of optical fields in the presence of ionic and electronic nonlinearities has also been called optical rectification or the inverse electro-optic effect (6,7). It has been shown that the ISRS (i.e., ionic) excitation mechanism is generally more important than the electronic mechanism in LiTaO$_3$ (7,9,19). The equations of motion describing the electrons and the transverse optic phonon and electromagnetic modes (modeled as oscillators with charged masses, whose displacements lie along the x direction and whose wavevectors lie along the y direction), their coupling, and their responses to light appear in the Appendix. Their solutions show that spatially periodic, temporally impulsive excitation results in phonon-polariton wavepackets described by

$$A(\mathbf{r}, t) = A_0(\mathbf{r}, t) \cos\left(\mathbf{k}_r \bullet \mathbf{r} - \frac{ck_r}{\sqrt{\varepsilon(\mathbf{k}_r, \omega)}} t\right) \tag{1}$$

$$A(\mathbf{r}, t) = A_0(\mathbf{r}, t) \cos\left(\mathbf{k}_l \bullet \mathbf{r} - \frac{ck_l}{\sqrt{\varepsilon(\mathbf{k}_l, \omega)}} t\right) \tag{2}$$

The electromagnetic fields of the right- and left-propagating polaritons, respectively, follow the wave equations with the speeds and damping rates of the different frequency components dispersed according to the frequency- and wavevector-dependent complex refractive index $n = \sqrt{\varepsilon(\mathbf{k}, \omega)}$. A typical example of the dispersion of these modes is shown in Fig. 1 for the case of a real permittivity ε. The term $A_0(r,t)$ represents the envelope of the wavepacket on the phonon-polariton coordinate A. Note that this phonon-polariton coordinate is a linear combination of ionic and electromagnetic displacements, which both contribute to the polarization

P and interact with the electric field **E**. The wavevectors $\mathbf{k}_r$ and $\mathbf{k}_l$ [given explicitly in the appendix in Eq. (13)] have equal and opposite y components, so the two waves form a standing wave (fixed peaks and nulls) while they remain in the excitation region. They gradually leave the excitation region in different directions, resulting eventually in two separated traveling waves. The wavevectors include identical z components, so the waves propagate with the same forward component as well as opposite lateral components.

III. RECENT ADVANCES

A. Crossing Femtosecond Pulses

After roughly 20 years of femtosecond laser optics and spectroscopy, the reader might think that there would be little left to learn about the mechanics of spatially overlapping ultrashort pulses. Not true! A familiar effect seen when crossing two ultrashort pulses, illustrated in Fig. 4a, is that even if both spot sizes are large, the overlap region is reduced to a sliver because each pulse is a "pancake" of light whose length in the propagation direction is far exceeded by its transverse dimension. In most pump-probe and transient grating spectroscopy experiments, the only adverse impact of this effect is a reduction in signal level, since there is simply less overlap among the beams. However, in samples whose responses vary with excitation wavevector, the effect is more serious since the reduced number of optical interference fringes also reduces the wavevector definition. This has been discussed above for phonon-polaritons, which may propagate away from the excitation region before they are fully damped. The decay of signal can be viewed as the dephasing of many different phonon-polariton wavevector components with different frequencies. In this case, as noted above, the decay does not reveal the intrinsic phonon-polariton damping or dephasing rate. Not mentioned earlier is the fact that for femtosecond pulses this effect does not disappear even with large excitation spots, if the two excitation beams are produced through partial reflection and crossed in the usual manner. The effect also does not disappear for small excitation angles, at which both the overlap area and the fringe spacing increase correspondingly. The number of interference fringes formed by the two pulses is independent of the intersection angle and is given by (14,19)

$$n_f = \frac{2\tau c}{\lambda \sqrt{\varepsilon_{opt}}} \tag{3}$$

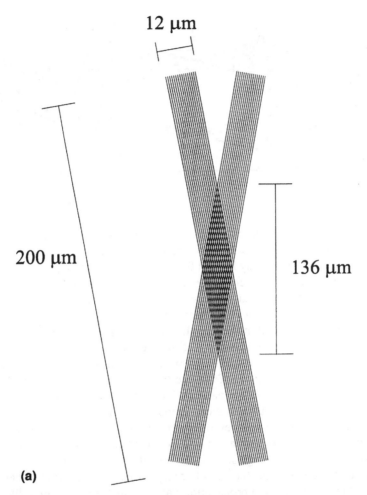

Figure 4 Crossing of ultrashort pulses. (a) The "pancake effect" is illustrated for two crossed 40 fs pulses with 200 μm spot sizes. These two pulses, which travel from left to right, would have been produced from a single pulse with a beamsplitter. The spatial extent of the overlap area is seen to be only a fraction of the incident beam's spatial extent. (b) Pulses split with diffractive optics and recombined with a two-lens telescope overlap over the entire spatial extent of the beam profile. (Adapted from Ref. 14.)

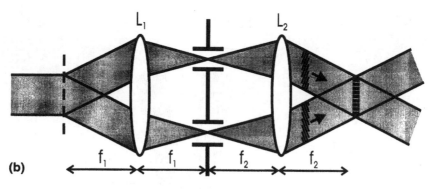

L_1 L_2

(b) f_1 f_1 f_2 f_2

Figure 4   *(continued)*

where τ is the pulse duration, λ is the central wavelength of light, and ε_{opt} is the dielectric permittivity at optical frequencies. Thus, the number of interference fringes is inherently limited for ultrashort excitation pulse durations.

An alternate method for the generation of two excitation pulses is the use of diffractive rather than reflective optics, as shown in Fig. 4b. A phase mask pattern is inserted into the beam path, and the ± 1 orders of diffraction are selected and crossed at the sample with a simple imaging system. This method was first used in impulsive stimulated scattering measurements on picosecond time scales for excitation of acoustic waves (11,12), because it permits convenient changing of the excitation wave vector by simple insertion of different phase grating patterns into the beam path. On femtosecond time scales, an additional and more fundamental advantage is the elimination of the "pancake" effect described above (14). The different wavelength components contained within the bandwidth of the femtosecond pulse are diffracted at different angles, with the result that the phase fronts of the pulses that recombine at the sample are parallel to each other rather than perpendicular to the beam propagation paths. Thus the beams are fully overlapped at the sample, yielding interference fringes whose number is limited only by the spot sizes. In addition, the duration of the pulse incident upon the phase mask is restored at the image plane of the focusing system at which the sample is placed (14). As a result, ISRS excitation is possible with ultrashort pulse durations and highly improved wave vector definition. The simulations shown in Fig. 3 were generated based on a large or small spot size incident on the phase mask (and thus at the sample) in Fig. 3a or 3b, respectively.

The use of diffractive optics for generation of the two excitation pulses results in fixed interference pattern maxima and minima at the sample, in contrast to the frequent fluctuations in the spatial phase of the pattern formed through the use of reflective optics. This has a number of important benefits. First, a substantial source of noise in grating experiments is parasitic scattering of probe light by surface or bulk sample imperfections. Scattered light that goes in the same direction as the signal in the far field (and that therefore reaches the photodetector even after spatial filtering of the diffracted signal) can be viewed as diffraction from the spatial Fourier component of the sample imperfections that happens to match the grating wavevector. This spatial Fourier component of the sample imperfections has a well-defined spatial phase (more imperfections at the peaks, fewer at the nulls), which does not vary with time, and so the optical phase of the scattered light field that is collinear with the diffracted signal does not fluctuate substantially. However, fluctuations in the spatial phase of the excitation grating pattern lead to corresponding fluctuations in the optical phase of the diffracted signal field. The diffracted and parasitically scattered fields are superposed at the photodetector, and shot-to-shot fluctuations in the diffracted field phase lead directly to fluctuations in the signal intensity. This noise contribution cannot be eliminated through subtraction of the parasitically scattered light intensity from the measured signal level. The improvement realized through the use of diffractive optics is often apparent by eye when the measurement is conducted at sufficient repetition rate that the signal appears to be a continuous beam whose spatial quality can be assessed visually. With reflective optics, the diffracted beam that reaches the detector often appears to be milky and undergoing continual fluctuations in its spatial profile. With diffractive optics, the signal beam has the appearance of an ordinary coherent laser beam. The data collected show a corresponding improvement in signal/noise level. This has been illustrated graphically with data from acoustic waves as well as phonon-polaritons.

Fixed spatial phase in the grating pattern also facilitates experiments with multiple excitation pulses (20). A second, delayed pulse incident on the diffractive optic is split in the same manner as the first and results in a second excitation pattern with the same peak and null positions. Thus, multiple excitation gratings, delayed temporally and shifted spatially if desired, can be used for excitation of phonon-polaritons whose coherent superposition is well controlled. A preliminary experiment of this type has been reported (21).

B. Heterodyne Detection

Now we turn to methods for phonon-polariton detection. The signal intensity in a grating measurement like that illustrated in Fig. 2 is given by the square of the grating peak-null difference in refractive index Δn, i.e., by

$$S(t) \propto |E_D(t)|^2 \propto |\Delta n(t)|^2 \propto |A(t) - e^{-\gamma t} \sin \omega t|^2 \qquad (4)$$

where E_D is the diffracted field amplitude and A is the phonon-polariton displacement. In the absence of other signal contributions, the signal oscillates at twice the phonon-polariton frequency ω and decays at twice the damping rate γ, and these parameters may be extracted from the data in a straightforward manner. However, in a wide range of circumstances, including samples with weak phonon-polariton signal levels or with several other contributions to signal, detection of phonon-polariton traveling waves outside the excitation region, or samples with both phase and amplitude grating components (i.e., peak-null variation in both real and imaginary parts of the refractive index), linearization of the signal is desirable. This can be achieved through heterodyne detection of the diffracted signal, which would normally require a precisely aligned and actively phase-stabilized reference beam that is temporally and spatially coincident with the signal. Fortunately, the use of the phase mask approach described above also permits facile heterodyne detection of the grating signal. As shown in Fig. 5, the probe as well as excitation beam must be passed through the phase mask and its ± 1 orders of diffraction also selected and allowed to reach the sample. One of the two beams can be used as the probe and the other, after partial filtering, as the reference that will be properly aligned with the signal by the imaging system. The arrangement does not require

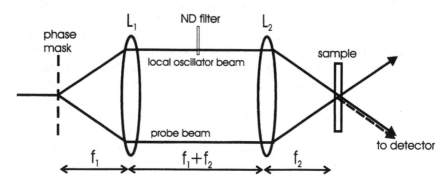

Figure 5 Schematic illustration of optical heterodyne detection of transient gratings. A diffraction grating splits the input pulse into a probe and a reference beam. Diffracted light from the probe beam's interaction with the transient grating propagates collinearly with the reference beam, or local oscillator. (Adapted from Ref. 11.)

active phase stabilization of the reference phase, and static adjustment of the phase can be executed simply through rotation of the partial filter. With the diffracted and reference beams reaching the detector, the total signal is given by

$$S(t) \propto |E_D(t) + E_R \exp i\phi|^2 \approx I_R + 2\sqrt{I_R I_D(t)} \cos \phi \tag{5}$$

where E_R is the reference field amplitude, I_D and I_R are the diffracted and reference intensities, and φ is the phase difference between the two fields. If the phase difference vanishes, then the interference pattern formed by the probe and reference beams has the same spatial phase as that formed by the excitation pulses, i.e., the two sets of interference peaks and nulls coincide. In general, the time-dependent part of the heterodyned signal varies linearly rather than quadratically with the sample response. Typically the reference intensity substantially exceeds the diffracted intensity, often resulting in substantial amplification of the signal and signal/noise levels.

Figure 6 shows examples of heterodyne-detected phonon-polaritons. Figure 6a shows a comparison between ordinary diffraction and heterodyne signal, illustrating the fact that in the latter case the signal oscillations and decay are at the fundamental phonon-polariton frequency and damping rate, not twice these values as in the ordinary diffraction case. Note that the heterodyne signal level is also far higher, a substantial advantage for samples with weak signals. The figure also shows the effects of varying the reference phase by 90 or 180 degrees, yielding, respectively, no interference between reference and diffracted fields or interference with opposite sign of that observed with zero degrees. Note that if there were amplitude grating contributions to signal, these would be heterodyned by the reference field with 90-degree phase shift. In this manner the different contributions could be examined separately (11).

Figure 6b shows heterodyne data recorded with the probe beam displaced from the excitation region. Phonon-polariton oscillations can still be observed because the optical phase of the diffracted field shifts by 180 degrees each time the traveling wave peaks and nulls move by one-half the phonon-polariton wavelength, while the reference phase remains the same and so the interference term between them undergoes time-dependent oscillations. Thus, complete characterization of the phonon-polariton response is possible even with the probe beam displaced from the excitation region. In contrast, homodyne detection permits observation of only the time-dependent envelope showing the traveling wave entry into and exit from the probe region.

Note that there are some advantages in carrying out the measurements with a displaced probe beam. In particular, if the sample shows several responses to the excitation pulses, e.g., electronic, thermal, and other nonpropagating responses in addition to the polariton response, then

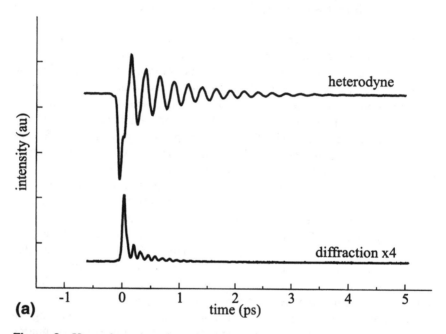

(a)

Figure 6 Heterodyne detection of phonon-polaritons in LiTaO$_3$ (a) Heterodyne data (upper curve) are compared with diffraction data magnified by a factor of four (lower curve). The diffraction data are seen to oscillate and damp at twice the material response rate, while the heterodyne data are linearly related to the material response. The amplitude of the heterodyne signal is also much greater than that of the diffraction signal. (b) The heterodyne signal at a variety of reference beam phases is shown. The polariton is detected in the excitation area. When the reference phase is at 90° with respect to the diffracted field, the two destructively interfere and the heterodyne signal is at a minimum. At 180° or 0° the signal level is at a maximum. (c) The heterodyne signal from a polariton detected with the probe beam substantially but not entirely outside of the excitation area. At early probe delays, standing-wave oscillations of the two polariton responses still within the excitation region are observed. In this case, as illustrated in (b), the amplitude of the signal is a function of reference phase. At later times, a single traveling wave outside the excitation region is observed. In this case the phase of the oscillations is a function of reference phase.

displacement of the probe beam can permit the phonon-polariton response to be examined without any other signal contributions.

The use of heterodyne detection to monitor vibrational oscillations even after traveling wave propagation out of the excitation region has also been demonstrated in the case of acoustic modes in bulk and thin

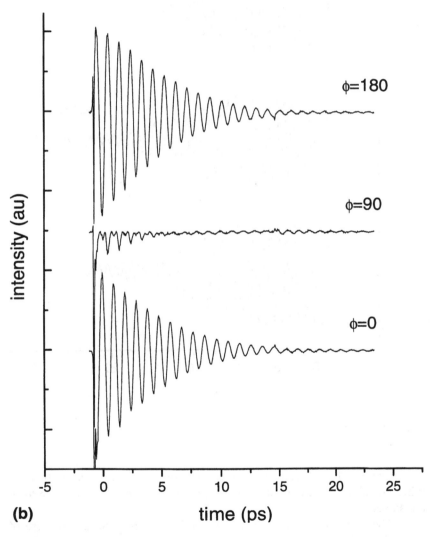

(b)

Figure 6 (*continued*)

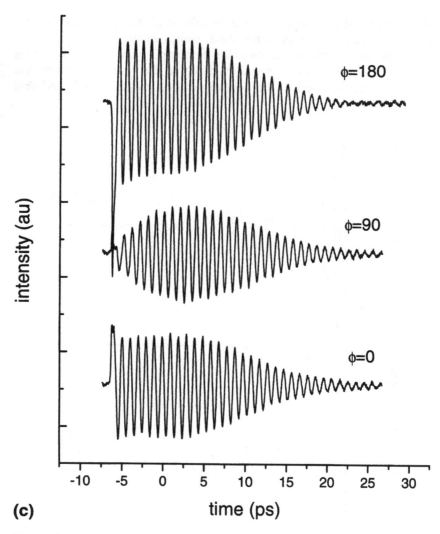

(c)

Figure 6 (*continued*)

film samples (11,22). Here too, separation of the acoustic response from other signal contributions present in the excitation region can be used to advantage.

There are additional practical advantages gained through the use of heterodyne detection as illustrated in Fig. 6. First, the reference beam indicates precisely the direction along which the diffracted signal will go, permitting facile location and spatial filtering of this signal. Second, the probe beam is incident at the phase-matching ("Bragg") angle for diffraction without any alignment, even with different excitation and probe wavelengths (in which case achromatic lenses or mirrors should be used in the imaging system). A third and related advantage is that different wavevectors can be used simply by moving different phase masks into the excitation and probe beam paths, with no further alignment of any beams prior to the sample. Other advantages obtain for transient grating measurements on different types of sample responses. For example, in photorefractive materials the spatial phase of the grating that produces signal may be shifted from that of the excitation pattern due to carrier migration (23); similar effects may occur in photobleached liquids due to molecular diffusion (24). In each case, direct measurement of the shift in spatial phase should be possible.

C. Spatiotemporal Phonon-Polariton Imaging

Monitoring of phonon-polariton propagation with a spatially shifted probe beam is adequate in cases like those treated above, in which the excitation pattern is simple, the spatial characteristics of the response are predictable, and the sample is uniform. More generally, however, phonon-polariton propagation is highly anisotropic and, of course, strongly dependent on the spatial and temporal characteristics of the excitation light pattern. Characterization of the spatial and temporal phonon-polariton response characteristics with many probe beam positions is not convenient and generally not complete.

A more direct approach, spatiotemporal phonon-polariton imaging (15,16,25), is illustrated schematically in Fig. 7. Through this method the full spatial properties of the response are determined at selected times by a variably delayed probe pulse that passes through the sample and the imaging optics and onto a CCD detector. The probe beam size is not comparable to or smaller than that of the excitation beams as usual, but is large enough to reach all of the sample volume into which the phonon-polaritons will ever propagate. The probe beam is not incident on the sample at the phase-matching angle for diffraction from a spatially periodic phonon-polariton response, but is incident approximately normal to the sample irrespective of the phonon-polariton spatial properties. First-order diffraction of the probe beam is not selected for the signal, but rather the zero-order probe light transmitted straight through the sample and many orders of diffraction (i.e., and transmitted light that is scattered over a wide angular range) are all

(a)

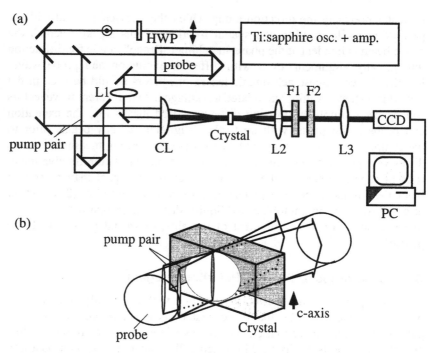

(b)

Figure 7 (a) Experimental layout for spatiotemporal polariton imaging. A 40 fs, 800 nm laser pulse is split into two 20 μJ pump pulses and one 60 μJ probe pulse. The pump pulses are focused by a cylindrical lens CL onto the crystal, where they cross to form a grating pattern. (In some cases only one pump pulse is used.) A spherical lens L1 before the cylindrical lens defocuses the probe pulse, which is incident onto a large region of the crystal. The pump and probe pulses co-propagate through the crystal. A telescope (lenses L2 and L3) images the wide probe spot onto a CCD. Filters F1 and F2 within the telescope remove second harmonic and fundamental pump light, and the data are digitized and stored for later processing. (b) Typical geometry of pump and probe pulses at sample. All pulses are polarized parallel to the optic axis (c-axis) shown here. A pair of pump pulses is focused by CL to form a grating pattern at the sample. The wide probe pulse co-propagates with the pump pulses through the sample. Other geometries, such as a single pump pulse focused to a line or a round spot, are also possible. (From Ref. 16.)

collected. The phonon-polariton response at any time forms a phase pattern (spatial variation in the real part of the refractive index) in the sample, and through the imaging system this is converted to an amplitude pattern at the CCD.

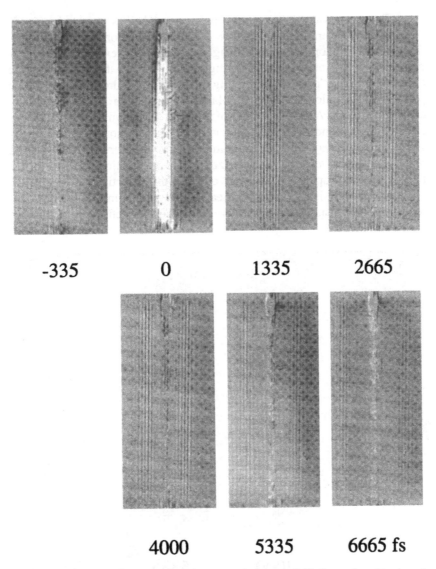

Figure 8 Images of propagating phonon-polaritons in LiTaO₃ produced by impulsive grating excitation with spatial preriod 30 μm. To improve the quality of the images, 11 scans taken well before time t = 0 were averaged and subtracted from the raw data. Image dimensions 2050 μm tall × 895 μm wide. (From Ref. 16.)

Figure 8 shows a series of images recorded as shown in Fig. 7. Crossed pulses were used to produce a grating excitation pattern with few fringes, and phonon-polariton propagation out of the excitation region is monitored at various times following excitation. The figures shown are excerpts from a "movie," consisting of several hundred frames, which permits visualization of phonon-polariton propagation on an essentially continuous basis. A similar series of excerpts is shown in Fig. 9. In this case a single round excitation spot is used, and phonon-polariton propagation radially away from the excitation region is observed. Note that the phonon-polariton response is polarized in the x direction (vertical in the figure), and so propagation does not occur in this direction.

The conversion of phase object to amplitude image required for recording the responses shown can be accomplished through various

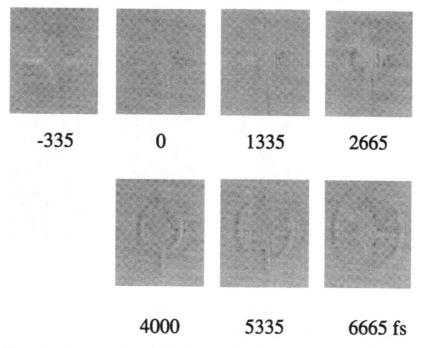

Figure 9 Images from propagating phonon polaritons produced by a single beam focused to a round spot. Background data have been subtracted as in Fig. 8. The splotch in the center that obscures some data near time t = 0 is due to intense scattered pump light, which could not be completely filtered from the imaging setup. Image dimensions 1020 μm tall × 860 μm wide. (From Ref. 16.)

methods (26,27). One method allows all the probe light transmitted through the sample, including the zero-order light and as many diffraction orders (i.e., as wide an angular range) as can be collected by the lenses, to be used. If the detector is in the image plane, the phase pattern is merely reconstructed, and there is no amplitude variation for the detector to record. However, an amplitude pattern is formed away from the image plane by a process called the Talbot effect, and the detector can be moved out of the image plane to record it. Alternatively, in phase-contrast imaging some of the transmitted probe light (e.g., the zero-order light, all the positive diffraction orders, or other choices) can be blocked, in which case the remaining light will form an amplitude pattern in the image plane. Other imaging techniques can be developed based on spectral filtering of the transmitted probe light or, for phonon-polaritons that yield depolarized Raman scattering, based on polarization filtering.

Spatiotemporal imaging permits facile determination of phonon-polariton propagation characteristics in inhomogeneous samples and should be suitable for detection of dielectric heterogeneities including those due to ferroelectric domain reversal or crystalline imperfections, patterning of the materials, static or dynamic electrical signals, and other sources. Imaging of phonon-polariton reflection by and transmission through crystal edges has been demonstrated. Imaging also has been used to facilitate phonon-polariton coherent control through the use of spatially and temporally specified excitation fields.

IV. SUMMARY AND FUTURE PROSPECTS

Recent advances have yielded dramatic fundamental improvements and practical simplifications in impulsive excitation and time-resolved detection of propagating phonon-polariton modes in crystalline solids. These have permitted characterization with dramatic improvements in wavevector definition, separation from other signal features, signal/noise and absolute signal levels, data acquisition time at a single wavevector, and time needed for switching among different wavevectors. They also have permitted far more complete characterization of phonon-polariton spatial characteristics and their time-dependent evolution, with applications in heterogeneous and patterned materials and in coherent optical control.

Finally, they have drastically reduced the complexity of the experimental layout and execution. The current experimental system requires only a few optical elements, almost no beam alignment steps, and little or no searching for signal location.

All of the advances discussed here have evolved quite recently, and their exploitation is far from complete. For example, experiments on phonon-polaritons in other classes of materials such as semiconductors, thin films, and multilayer assemblies will be improved and facilitated by the use of these methods. Coherent optical control and manipulation of phonon-polaritons with spatially and temporally shaped excitation fields has only just begun, with the use of spatiotemporal imaging to direct the excitation field parameters. This may lead to the use of phonon-polaritons as coherent signal carriers for terahertz bandwidth signal processing applications. In addition, ISRS-excited phonon-polaritons provide a source of conveniently tunable terahertz frequency radiation that emerges from the crystal in which it was produced. Such radiation has .been passed through a sample and detected optically after reentry into the crystalline medium, permitting frequency-tunable, narrowband terahertz spectroscopy through the use of visible light, an extremely compact apparatus, and no far-IR optics (28).

Finally, the methods reviewed here have related applications in acoustic wave excitation and monitoring and in transient grating measurements in general. Extremely facile delivery of excitation and probe beams to the sample, detection of signal, and switching among excitation wavevectors have simplified transient grating experimentation to the level of pump-probe transient absorption measurements. At the same time, capabilities for high wavevector definition, linearization of the signal with respect to the material response, assessment of grating spatial phase, and separation of different signal components including those from phase and amplitude grating contributions have improved and added to the information content that can be extracted.

The recent advances are bringing about a broadening of the transient grating experimental user community and the range of areas under study, as well as a deepening of the knowledge that can be gained in each area.

APPENDIX: PHONON-POLARITON EXCITATION: EQUATIONS OF MOTION

The equations of motion describing the transverse optic phonon and electromagnetic modes (modeled as oscillators with charged masses, whose displacements lie along the x direction and whose wavevectors lie mainly along the y direction in the yz plane), their coupling, and their responses to light are as follows (7,19,29–31):

$$(m_i \ddot{x}_i + k_i x_i)\mathbf{x} = k_{ie}(x_i - x_e)^2 \mathbf{x} + q_i \mathbf{E} \tag{6}$$

$$(m_e \ddot{x}_e + k_e x_e)\mathbf{x} = k_{ie}(x_i - x_e)^2 \mathbf{x} + q_e \mathbf{E} \tag{7}$$

$$\nabla \times \nabla \times \mathbf{E} = -\frac{\varepsilon(\mathbf{q}, \omega)}{c^2} \frac{\partial^2 (\mathbf{E} + 4\pi \mathbf{P})}{\partial t^2} \tag{8}$$

$$\mathbf{P}_e = \chi_e \mathbf{E} = N q_e x_e \mathbf{x} \tag{9}$$

$$\mathbf{P}_i = \chi_i \mathbf{E} = N q_i x_i \mathbf{x} \tag{10}$$

In these equations, x_i represents displacements of the polar optic phonon of mass m_i, charge q_i, force constant k_i, and number density N along the crystal axis parallel to the unit vector $\mathbf{x}$; x_e represents electronic displacements with the lowest electronic resonance characterized by mass m_e, charge q_e, and force constant k_e near the Brillouin zone center; $\mathbf{E}$ is the electric field of the terahertz radiation; and $\mathbf{P}_i$ and $\mathbf{P}_e$ are the polarizations induced in the dielectric by the phonon and electronic responses, with $\mathbf{P} = \mathbf{P}_i + \mathbf{P}_e$. The corresponding linear contributions to the susceptibility tensor χ are χ_i and χ_e. Without any nonlinear coupling terms, the solutions to the phonon-polaritons are propagating electromagnetic modes with the dispersion shown, for example, in Fig. 1. These contributions to the susceptibility may be combined with other contributions to yield a perturbative expansion of the complex dielectric permittivity $\varepsilon(\mathbf{q}, \omega)$. In many cases, it is possible to simplify the system into sets of optical and far-infrared modes of the form of Eq. (8) with couplings between the modes based on the coupling term in Eqs. (6) and (7).

This coupling term $k_{ie}(x_i - x_e)^2$ is a phenomenological approximation of the lowest-order (cubic) potential energy terms that gives rise to nonlinear effects. These effects can be described by perturbative expansions of χ (7,30,32,33):

$$\omega_i^2 P_i(\omega_1) + \omega_i^2 P_i(\omega_1) = \omega_i^2 \varepsilon_0 \{\chi_i E(\omega_1) + (-\chi_{iii}(\omega_1, \omega_2, \omega_3)$$
$$- \chi_{iee}(\omega_1, \omega_2, \omega_3) + 2\chi_{iie}(\omega_1, \omega_2, \omega_3))E(\omega_2)E(\omega_3)\} \tag{11}$$

$$\omega_i^2 P_e(\omega_1) + \omega_e^2 P_e(\omega_1) = \omega_e^2 \varepsilon_0 \{\chi_e E(\omega_1) + (\chi_{eii}(\omega_1, \omega_2, \omega_3)$$
$$+ \chi_{eee}(\omega_1, \omega_2, \omega_3) - 2\chi_{eie}(\omega_1, \omega_2, \omega_3))E(\omega_2)E(\omega_3)\} \tag{12}$$

The effect of the $k_{ie}x_e^2$ term on the motion of x_e is written as χ_{eee}, for example, and physically represents optical second harmonic generation (SHG), to which the polar phonons do not contribute because they are not able to drive or follow high-frequency responses. This nonlinearity should have roughly the same magnitude at terahertz frequencies, as can be seen through Miller's rule (30,34), because the electrons can still follow the terahertz fields. Two other perturbative terms out of the four possible symmetry-reduced terms are interesting. (Unfortunately, one limitation of this phenomenological analysis is that the fourth term χ_{eii} is not particularly intuitive in the Born-Oppenheimer approximation.) Garrett first noted (30) that χ_{iee} represents Raman scattering, in which two optical fields drive a lower-frequency resonance by an electronic deformation of the ionic potential; this term has also been described as an inverse electro-optic effect (6,7). Finally, the intriguing term χ_{iii} represents the all-ionic contribution to terahertz-frequency SHG, which occurs at frequencies low enough for both phonons and electrons to contribute to and be affected by the nonlinearity. It is also possible to estimate contributions to this process through a quantum-mechanical perturbation theory and wavevector overtone spectroscopy (35,36). The clearest measurements of χ_{iii} so far (with some contamination by χ_{iie}) have been made at sub-terahertz frequencies on a number of materials, including $LiTaO_3$, $BaTiO_3$, and GaAs (32,33). The corresponding terahertz SHG coefficients are enormous compared to SHG coefficients at optical frequencies due to the substantial ionic susceptibilities in linear response that contribute to both "driving" and "following."

The dielectric is often assumed to be isotropic in order to simplify Eq. (8) by assuming transverse phonon-polaritons; the extension to anisotropic media is straightforward (31). In the limit of very short pulse duration compared to the phonon-polariton oscillation period, the time-dependence of the excitation field can be treated as a delta function, and the phonon-polariton response is given by the impulse response function for the spatial excitation pattern used. If crossed excitation pulses are used, then it is simplest to describe the excitation and response in terms of the excitation wavevector or wavevector range.

It is straightforward to see that the phonon-polaritons generated by crossed excitation pulses will have a wavevector component q_z in the forward (z) direction in addition to the expected wavevector component in the optical grating (y) direction. Note that the excitation pulses arrive first at the front of the sample, then the middle, then the back, so the vibrational phase varies linearly as a function of depth into the crystal (the z direction). In the limit of a very small scattering angle, i.e., a single excitation beam, there is no wavevector component at all in the y direction, and a traveling wave with wavevector entirely in the z direction is produced through forward ISRS. In general, and especially for small scattering angles, the phonon-polariton phase fronts may be tilted substantially into the yz plane. The full wave vector is given by (19,36).

$$\mathbf{k}_{ph} = \pm \frac{\omega_{opt}\sqrt{\varepsilon_{opt}}}{c}(\mathbf{k}_1 - \mathbf{k}_2) + \frac{\omega_{ph}\sqrt{\varepsilon_{opt}}}{2c}(\mathbf{k}_1 + \mathbf{k}_2) \tag{13}$$

where $\mathbf{k}_1 + \mathbf{k}_2$ points in the z direction and $\mathbf{k}_1 - \mathbf{k}_2$ points in the y direction. ω_{opt} is the central frequency of the excitation pulses, $\mathbf{k}_1$ and $\mathbf{k}_2$ are the central wavevectors of the pulses, ω_{ph} is the frequency of the excited phonon-polaritons, and ε_{opt} is the dielectric permittivity at optical frequencies. Equation (13) follows from the twin requirements of energy and momentum conservation (19,36). Although $\omega_{opt} \gg \omega_{ph}$, for small angles the second term cannot be neglected. Figure 10 shows a simulation of the phonon-polariton response to crossed excitation pulses, illustrating the y and z components of the wave vector.

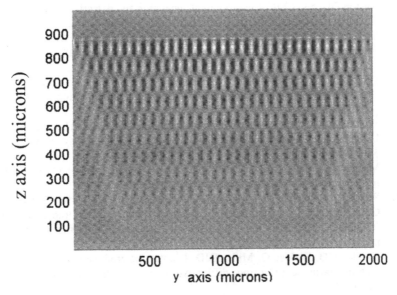

Figure 10 Gray-scale plot of coherent phonon displacements A(y, z, t). The excitation pulses form an interference pattern, which is moving in the +z direction (vertical). The displacement direction is along the x-axis (out of the page). The y component of the grating wavevector, $k_g = 1000 \text{ cm}^{-1}$, is given by the excitation interference pattern, fringe spacing of 63 μm. The z component arises due to the finite propagation time of the excitation pulses through the sample. The total polariton wavevector magnitude is $k_p = 1153 \text{ cm}^{-1}$, corresponding to a wavelength of 54 μm. (Adapted from Ref. 19.)

REFERENCES

1. Dhar L, Rogers JA, Nelson KA. Time-resolved vibrational spectroscopy in the impulsive limit. Chem Rev 1994; 94:157.
2. Yan Y-X, Cheng L-T, Nelson KA. Impulsive stimulated light scattering. In: RJH Clark, RE Hester, eds. Advances in Nonlinear Spectroscopy, Vol. 16. Richester: Wiley, 1988:299–355.
3. Nelson KA, Miller RJD, Lutz DR, Fayer MD. Optical generation of tunable ultrasonic waves. J Appl Phys 1982; 52(3):1144–1149.
4. Fayer MD. Dynamics of molecules in condensed phases: picosecond holographic grating experiments. Ann Rev Phys Chem 1982; 33:63–87.
5. Thomsen C, Grahn HT, Maris HJ, Tauc J. Surface generation and detection of phonons by picosecond light pulses. Phys Rev B 1986; 34(6):4129–4138.
6. Auston DH, Cheung KP, Valdmanis JA, Kleinman DA. Cherenkov radiation from femtosecond optical pulses in electro-optic media. Phys Rev Lett 1984; 53(16):1555–1558.
7. Auston DH, Nuss MC. Electrooptical generation and detection of femtosecond electrical transients. IEEE J Quantum Electron 1988; QE-24(2):184–197.
8. Etchepare J, Grillon G, Antonetti A, Loulergue JC, Fontana MD, Kuge GE. Third-order nonlinear susceptibilities and polariton modes in PbTiO3 obtained by temporal measurements. Phys Rev B 1990; 41(17):12362.
9. Dougherty TP, Wiederrecht GP, Nelson KA. Impulsive stimulated Raman scattering experiments in the polariton regime. J Opti Soc Am B 1992; 9(12): p. 2179.
10. Wiederrecht GP, Dougherty TP, Dhar L, Nelson KA, Leaird DE, Weiner AM. Explanation of anomalous polariton dynamics in LiTaO$_3$. Phys Rev B 1995; 51(2):916.
11. Maznev AA, Nelson KA, Rogers JA. Optical heterodyne detection of laser-induced gratings. Opt Lett 1998; 23(16):1319.
12. Rogers JA, Fuchs M, Banet MJ, Hanselman JB, Logan R, Nelson KA. Optical system for rapid materials characterization with the transient grating technique: application to nondestructive evaluation of thin films used in microelectronics. Appl Phys Lett 1997; 71(2):225–227.
13. Goodno GD, Dadusc G, Miller RJD. Ultrafast heterodyne-detected transient-grating spectroscopy using diffractive optics. J Opt Soc Am B 1998; 15(6):1791.
14. Maznev AA, Crimmins TF, Nelson KA. How to make femtosecond pulses overlap. Opt Lett 1998; 23:1378.
15. Koehl RM, Adachi S, Nelson KA. Real-space polariton wave packet imaging. Chem J Phys 1999; 110(3):1317.
16. Koehl RM, Adachi S, Nelson KA. Direct visualization of collective wavepacket dynamics. J Phys Chem 1999; 111:3559–3571.
17. Yan Y-X, Nelson KA. Impulsive stimulated light scattering. I. General theory. J Chem Phys 1987; 87(11):6240–6256.

18. Yan Y-X, Nelson KA. Impulsive stimulated light scattering. II. Comparison to frequency-domain light-scattering spectroscopy. J Chem Phys 1987; 87(11):6257–6265.

19. Brennan CJ. Femtosecond wavevector overtone spectroscopy of anharmonic lattice dynamics in ferroelectric crystals. Cambridge, MA: Massachusetts Institute of Technology, 1997.

20. Weiner AM, Leaird DE, Wiederrecht GP, Nelson KA. Femtosecond pulse sequences used for optical manipulation of molecular motion. Science 1990; 247(4948):1317.

21. Koehl RM, Adachi S, Nelson KA. Multiple-pulse control and bispectral 2D Raman analysis of nonlinear lattice dynamics. In: Ultrafast Phenomena XI. T Elsaesser et al., eds. Berlin: Springer-Verlag, 1998:136–137.

22. Rogers JA, Maznev AA, Banet MJ, Nelson KA. Optical generation and characterization of acoustic waves in thin films: fundamentals and applications. Ann Rev Mater Sci 2000; 30:115–157.

23. Gunter P, Huignard J-P. Photorefractive Materials and Their Applications I. Berlin: Springer-Verlag, 1988.

24. Johnson J. Structural and dynamic origins of intensity in holographic relaxation spectroscopy. J Opt Soc Am B 1985; 2:317–321.

25. Adachi S, Koehl RM, Nelson KA. Real-space and real-time imaging of propagating polariton-wavepackets. Butsuri 1999; 54(5):357.

26. Born M, Wolf E. Principles of Optics. Oxford: Pergamon Press, 1980.

27. Hecht E. Optics. Reading, MA: Addison-Wesley, 1987.

28. Crimmins TF, Nelson KA. Phys Rev B, submitted.

29. Born M, Huang K. Dynamical Theory of Crystal Lattices. Oxford: Clarendon Press, 1954.

30. Garrett CGB. Nonlinear optics, anharmonic oscillators, and pyroelectricity. IEEE J Quantum Electron 1968; QE-4(3):70.

31. Barker AS, Loudon R. Response functions in the theory of Raman scattering by vibrational and polariton modes in dielectric crystals. Rev Modern Phys 1972; 44:18.

32. Boyd GD, Bridges TJ, Pollack MA, Turner EH. Microwave nonlinear susceptibilities due to electronic and ionic anharmonicities in acentric crystals. Phys Rev Lett 1971; 26(7):387.

33. Pollack MA, Turner EH. Determination of absolute signs of microwave nonlinear susceptibilities. Phys Rev B 1971; 4(12):4578.

34. Boyd RW. Nonlinear Optics. San Diego, CA: Academic Press, Inc., 1992.

35. Brennan C, Nelson KA. Direct time-resolved measurement of anharmonic lattice vibrations in ferroelectric crystals. J Chem Phys 1997; 107:9691–9694.

36. Romero-Rochin V, Koehl RM, Brennani CJ, Nelson KA. Theory of anharmonic phonon-polariton excitation in $LiTaO_3$ by ISRS and detection by wavevector overtone spectroscopy. J Chem Phys 1999; 111(8):3559–3571.

13

Vibrational Energy Redistribution in Polyatomic Liquids: Ultrafast IR-Raman Spectroscopy

Lawrence K. Iwaki*, John C. Deàk†, Stuart T. Rhea‡, and Dana D. Dlott
University of Illinois at Urbana–Champaign, Urbana, Illinois

I. INTRODUCTION

In this chapter we discuss recent measurements of vibrational energy redistribution in polyatomic liquids using the ultrafast IR-Raman technique (1,2). In the IR-Raman technique, a tunable midinfrared (mid-IR) pulse, is used to pump a selected vibrational transition of a polyatomic liquid, and a time-delayed visible pulse is used to monitor the instantaneous vibrational population via incoherent anti-Stokes Raman scattering. The energy redistribution processes that will be considered are vibrational energy relaxation (VER) and vibrational cooling (VC) (3). VER refers to the elementary process of energy loss from a specific vibrational mode (the "system") to some or all of the other mechanical degrees of freedom (the "bath"). VC refers to the process where a vibrationally hot molecule loses its excess vibrational energy to the surroundings. VC is not an elementary process like VER. It is a complex process that generally involves many VER steps (4,5).

The VER and VC processes considered in this chapter will all involve polyatomic liquids at ambient temperature. It is useful to contrast this problem

---

* *Current affiliation*: National Institute of Standards and Technology, Gaithersburg, Maryland
† *Current affiliation*: Procter & Gamble Company, Ross, Ohio
‡ *Current affiliation*: CMI, Inc., Owensboro, Kentucky

with two fundamentally different problems that have been studied to a greater extent: VER of isolated molecules or high-pressure gases, for which a vast literature exists, and VER of diatomic molecules in condensed phases.

Isolated polyatomic molecules can undergo only *intramolecular* vibrational energy redistribution (IVR). Isolated molecules cannot lose their vibrational energy except by (slow) radiative processes. IVR is a type of *vibrational dephasing* process, where the initially prepared vibrational excitation time-evolves into other isoenergetic states. Neglecting radiative processes, gas-phase molecules can lose vibrational energy only through collision, so VER in gases is usually associated with high-pressure gases (6). In high-pressure gases, a target molecule heated with a laser is observed to lose its energy by many binary collisions with a buffer gas. Many authors have tried to draw a close analogy between VER in high-pressure gases and VER in condensed phases (7). For example, at STP a hot gas-phase molecule will experience collisions with a buffer gas at an average frequency of $\sim 10^{10}$ s^{-1}, whereas at liquid densities a hot molecule will experience collisions at an average frequency of $\sim 10^{13}$ s^{-1}. Theories that try to model condensed phase VER as simply faster gas-phase VER are termed "binary collision theories" (7). VER measurements have been made that bridge the gap between gases and fluids, including the remarkable works of Troe (8–10), Fayer (11–13), and Harris (14–17). After many years of studying binary collision models, a consensus opinion is emerging that binary collision models fail to capture the complexity of condensed phase VER, because many-body interactions are intrinsically significant and cannot be neglected (18,19).

Diatomic molecules are the simplest condensed phase VER systems, for example, a dilute solution of a diatomic such as I_2 or XeF in an atomic (e.g., Ar or Xe) liquid or crystal. Other simple systems include neat diatomic liquids or crystals, or a diatomic molecule bound to a surface. VER of a diatomic molecule can occur only by energy transfer to the collective vibrations of the bath, i.e., the phonons. Ordinarily VER is a high-order multiphonon process. Consequently there is an enormous variability in VER lifetimes, which may range from 56 s [liquid N_2 (20)] to 1 ps [e.g., XeF in Ar (21)], and a high level of sensitivity to environment. *Diatomic molecules have simple structures but complex VER mechanisms.*

Polyatomic molecules represent a major step up in complexity because every molecule has *several vibrational modes*. This feature guarantees enormous qualitative differences between diatomic and polyatomic VER and casts doubt on the possibility of extrapolating insights gained from studying diatomics to polyatomic systems. However, in what is at first glance a paradox, *polyatomic molecules have a complex structure but ordinarily much simpler VER mechanisms than diatomic molecules.* That is because the "ladder" VER process discussed in Section II, which often dominates in polyatomics, is lower-order and thus much simpler than higher-order multi-phonon processes involved in diatomic VER. A ladder process is one where an excited vibration loses its energy to a lower energy vibration plus a small number of phonons, via lower-order anharmonic couplings. The "rungs" of the vibrational ladder are the vibrations of the polyatomic molecule.

VER occurs as a result of fluctuating forces exerted by the bath on the system at the system's oscillation frequency (22). We will use the upper-case Ω to denote the system's vibrational frequency and lower-case ω to denote other vibrations. It may also be useful to look at fluctuating forces exerted on a particular chemical bond (23). Fluctuating forces are characterized by a force-force correlation function. The Fourier transform of this force correlation function at Ω, denoted $\eta(\Omega)$, characterizes the quantum mechanical frequency-dependent friction exerted on the system by the bath (19,22). This friction, especially at higher (i.e., vibrational) frequencies, plays an essential role in condensed phase chemical reaction dynamics (24,25).

The multiple roles of VER (friction) and VC (dissipation) in essentially all condensed-phase chemical processes have been extensively discussed (18,19). In chemical reactions (Fig. 1), the "system" is the specific mode of the reactant [or a coupled set of reactant and solvent modes (19)] associated with the reaction coordinate. Chemical reactions are "catalyzed" by vibrational energy. The system becomes activated by vibrational energy from the bath. Then the barrier is crossed. Then that vibrational energy plus the enthalpy of reaction is returned to the bath (Fig. 1). Much has been written about the dynamical effects of VER on barrier crossings (19,25,26). If VER is too slow, there is too little friction and chemical reactions are slower due to multiple barrier recrossings. If VER is too fast, there is too much friction and chemical reactions are again slower because barrier crossing becomes slow. If the VER rate is just right, the rate is the maximum possible. The maximum attainable rate, which is exactly that given by transition-state theory (19,26), occurs at the Kramers turnover (Fig. 1). The VER rate is systematically varied in practice by pressure-tuning the solvent density, as in classic studies of photoisomerization of stilbene (8,27–29) and boat-chair isomerization of cyclohexane (30,31).

A major breakthrough in the measurement of VER occurred in 1972. Laubereau et al. (32) used picosecond laser pulses to pump molecular vibrations via stimulated Raman scattering (SRS) and time-delayed inco-herent anti-Stokes probing to study VER of C–H groups in ethanol and methanol ($\sim$3000 cm^{-1}). Alfano and Shapiro (33) used the same technique to monitor both the decay of the initially excited (parent) C–H stretch excitation and the appearance and subsequent decay of a daughter vibration,

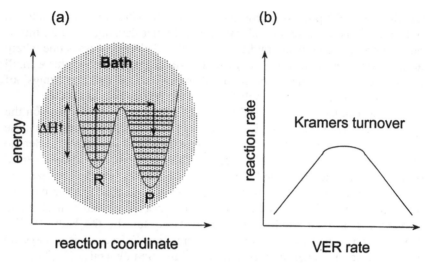

Figure 1 (a) Vibrational energy catalyzes chemical reactions. The reactant R is activated by multiphonon up-pumping, when R takes up the enthalpy of activation $\Delta H^\dagger$ from the bath. That energy plus the heat of reaction is returned to the bath after barrier crossing. (b) VER influences chemical reaction rates. When VER is just the right rate, the reaction rate is a maximum at the Kramers turnover. When VER is too slow or too fast, the reaction rate decreases. (From Ref. 50.)

a C–H bending vibration ($\sim$1460 cm^{-1}). Several reviews have described these early studies of liquids (2,7,22,34,35).

Another important breakthrough occurred with the 1974 development by Laubereau et al. (36) of intense tunable ultrashort mid-IR pulses. IR excitation is more selective and reliable than SRS, so SRS pumping is hardly ever used any more. At present the most powerful methods for studying VER in condensed phases use IR pump pulses. The most common (and complementary) techniques to probe nonequilibrium vibrational dynamics induced with mid-IR pump pulses are anti-Stokes Raman probing (the IR-Raman method) or IR probing (IR pump-probe experiments).

In the early days of IR-Raman measurements (2,34) with older laser technology [e.g., Nd:glass (2)], the technique was mainly limited to studying energy leaving higher frequency parent vibrations pumped by mid-IR laser pulses, such as C–H stretching transitions ($\sim$3000 cm^{-1}), OH stretching transitions ($\sim$3600 cm^{-1}), or metal carbonyl C$\equiv$O stretching transitions ($\sim$2000 cm^{-1}) (34). In a few cases it was possible to observe the first generation of high-frequency (hv $\gg$ kT) daughter vibrations, e.g., energy transfer from parent C–H stretch ($\sim$3000 cm^{-1}) to daughter C$\equiv$C stretch (1968 cm^{-1}) of acetylene (34,37), energy transfer from N–H stretch of pyrrole (38) ($\sim$3400 cm^{-1}) in benzene solution to ring stretching modes of pyrrole ($\sim$1400 cm^{-1}) or to the benzene solvent ($\sim$1000 cm^{-1}), energy transfer from C–H stretch to C–C stretching ($\sim$1400 cm^{-1}) of naphthalene (39), and energy redistribution among nearly degenerate C$=$O stretching vibrations ($\sim$2000 cm^{-1}) of W(CO)$_6$ (40).

Recent advances in laser instrumentation have, for the first time, permitted researchers to monitor the redistribution of vibrational energy throughout all Raman-active vibrations of a polyatomic molecule. In noncentrosymmetrical molecules, that would be essentially all the molecule's vibrations. Dlott and coworkers (41), using a 300 Hz Nd:YLF laser system, were the first to observe vibrational energy flow in a polyatomic liquid from parent C–H stretching vibrations all the way down to the lowest energy doorway vibrations in their studies of VER in nitromethane (NM). However the time resolution of $\sim$30 ps was not good enough to resolve all relevant VER processes. The same authors also studied *intermolecular* vibrational energy transfer between alcohols and nitromethane (42). Graener et al. used a 50 Hz Nd:YLF system with 1.5 ps resolution to study VER in dichloromethane (43) (CH$_2$Cl$_2$) and chloroform (44) (CHCl$_3$) after C–H stretch excitation. The latter work is especially notable for being the first where *every one* of the vibrations of a particular polyatomic molecule was monitored during a VER process. Recently our group at the University of Illinois has developed an IR-Raman instrument based on a Ti:sapphire laser, which provides 1 ps time resolution and unprecedented sensitivity (45). This system has been used to investigate VER in a variety of polar, nonpolar, and associated liquids and mixtures (46–50). In particular, we have extensively investigated acetonitrile (ACN) (46,47), which is widely viewed as a model for polyatomic liquids (18,51) and throughout this chapter we will illustrate general concepts with specific examples from our ACN studies.

In this chapter, we first discuss the theoretical framework needed to understand VER measurements, including force-force correlation function methods and perturbative techniques. We discuss experimental aspects of the IR-Raman technique, paying attention to the laser instrumentation, the experimental setup, the nature of the pumping and probing processes, detection sensitivity and optical background, and the interpretation of results including spectroscopic artifacts. Then we provide examples from recent research by our group, focusing on timely problems such as the vibrational cascade, the dynamics of doorway vibrations, methods for probing the build-up of bath excitation during VC, dynamics of overtones with Fermi resonance, multiple vibrational excitations via combination band pumping, and spectral evolution in associated liquids.

II. THEORETICAL SECTION

A. Hamiltonian

Consider an excited condensed-phase quantum oscillator Ω, with reduced mass μ and normal coordinate q_Ω. The bath exerts friction on the oscillator, which causes it to lose vibrational energy to its surroundings (Fig. 2).

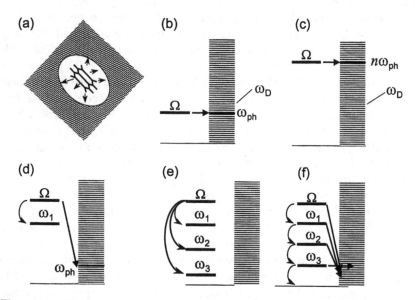

Figure 2 Vibrational energy relaxation (VER) mechanisms in polyatomic molecules. (a) A polyatomic molecule loses energy to the bath (phonons). The bath has a characteristic maximum fundamental frequency ω_D. (b) An excited vibration $\Omega < \omega_D$ decays by exciting a phonon of frequency $\omega_{ph} = \Omega$. (c) An excited vibration $\Omega \gg \omega_D$ decays via simultaneous emission of several phonons (multiphonon emission). (d) An excited vibration Ω decays via a ladder process, exciting lower energy vibration ω_1 and a small number of phonons. (e) Intramolecular vibrational relaxation (IVR) where Ω simultaneously excites many lower energy vibrations. (f) A vibrational cascade consisting of many steps down the vibrational ladder. The lowest energy doorway vibration decays directly by exciting phonons. (From Ref. 96.)

The friction is described as a fluctuating force exerted by the bath on the oscillator. The quantum mechanical Hamiltonian is [52, 53]

$$\hat{H} = \hat{H}_\Omega(q_\Omega) + \hat{H}_B(Q) + \hat{V}(q_\Omega, Q) \tag{1}$$

where $\hat{H}_\Omega(q_\Omega)$ is the Hamiltonian for the oscillator Ω, $\hat{H}_B(Q)$ is the Hamiltonian for the bath, where Q represents collective bath coordinates, and $\hat{V}$ represents the Hamiltonian for interaction between Ω and the bath. In solids these collective bath states are phonons, which extend from zero frequency to a cut-off frequency, the Debye frequency, ω_D. In liquids, the collective states are not so easily defined. Several authors use the term "instantaneous normal modes" (54–56). Since these play the same role as phonons in VER (54,57–59), we will call them liquid phonons or simply phonons. In the liquid state ω_D is used to indicate the *characteristic maximum fundamental frequency for collective motion in the liquid.*

B. Force Correlation Function Approach

The fluctuating forces F(t) on the rigid oscillator Ω are characterized by a time-dependent force-force correlation function (52,53),

$$\langle \hat{F}(t)\hat{F}(0) \rangle_Q = \frac{Tr[e^{-\hat{H}_B/kT}\hat{F}(t)\hat{F}(0)]}{Tr[e^{-\hat{H}_B/kT}]} \tag{2}$$

where

$$\hat{F}(t) = e^{i\hat{H}_B t}\hat{f}e^{-i\hat{H}_B t} \tag{3}$$

is the Heisenberg operator for the fluctuating forces and Tr denotes trace.

Equation (2) is a function of bath coordinates only. The VER rate constant is proportional to the Fourier transform, at the oscillator frequency Ω, of the bath force-correlation function. This Fourier transform is proportional as well to the frequency-dependent friction $\eta(\Omega)$ mentioned previously. For example, the rate constant for VER of the fundamental ($v = 1$) to the ground ($v = 0$) state of an oscillator with frequency Ω is (52)

$$k_{1\rightarrow 0} = \frac{1}{2\mu\hbar\Omega} \int_{-\infty}^{+\infty} dt\, e^{i\Omega t} \langle \hat{F}(t)\hat{F}(0) \rangle_Q \tag{4}$$

Equation (4) provides a prescription for computing the VER rate, which can be used for diatomic and polyatomic molecules alike: the force-force correlation function can be determined from a molecular simulation, and its Fourier transform at the desired frequency can be numerically computed. But there are two significant problems. First, it is presently impossible to do the fully quantum mechanical simulation on a system of any reasonable size. Classical mechanics may be used instead, but then a "quantum correction" must be introduced (52,60,61) to Equation (4). The quantum correction may be quite large (orders of magnitude) (52), and computing it accurately for anharmonic baths is problematic. Second, we often need to know the VER rate constant for vibrations with frequencies $\Omega \gg \omega_D$. At higher frequencies far above the fundamental characteristic bath frequency, the Fourier transform is very small. For example, in liquid O_2, the vibrational frequency $\Omega = 1552$ cm^{-1}, whereas $\omega_D \sim 50$ cm^{-1}, so the Fourier transform is required at a frequency about 30 times greater than the highest fundamental bath frequency (52). In such cases, numerical errors in the calculation of the correlation function lead to large inaccuracies in the Fourier transform.

C. Perturbation Approach

An alternative approach widely used in polyatomic molecule studies is based on the Golden Rule and a perturbative treatment of the anharmonic coupling (57,62). This approach is not much used for diatomic molecules. In the liquid O_2 example cited above, the Hamiltonian must be expanded to 30th order or so to calculate the multiphonon emission rate. But for vibrations of polyatomic molecules, which can always find relatively low-order VER pathways for each VER step, perturbation theory is very useful. In the perturbation approach, the molecule's entire ladder of vibrational excitations is the "system" and the phonons are the "bath." Only lower-order processes are ordinarily needed (57) because polyatomic molecules have many vibrations ranging from higher to lower frequencies and only a small number of phonons, usually one or two, are excited in each VER step. The usual practice is to expand the interaction Hamiltonian $\hat{V}(q_\Omega, Q)$ in Equation (2) in powers of normal coordinates (57,62):

$$\hat{V} = \sum_\alpha \frac{\partial \hat{V}}{\partial Q_\alpha}\bigg|_{\{Q=0\}} Q_\alpha + \frac{1}{2!} \sum_{\alpha,\gamma} \frac{\partial^2 \hat{V}}{\partial Q_\alpha \partial Q_\gamma}\bigg|_{\{Q=0\}} Q_\alpha Q_\gamma$$

$$+ \frac{1}{3!} \sum_{\alpha,\gamma,\beta} \frac{\partial^3 \hat{V}}{\partial Q_\alpha \partial Q_\gamma \partial Q_\partial}\bigg|_{\{Q=0\}} Q_\alpha Q_\gamma Q_\partial$$

$$+ \frac{1}{4!} \sum_{\alpha,\gamma,\partial,\varepsilon} \frac{\partial^4 \hat{V}}{\partial Q_\alpha \partial Q_\gamma \partial Q_\partial \partial Q_\varepsilon}\bigg|_{\{Q=0\}} Q_\alpha Q_\gamma Q_\partial Q_\varepsilon + \cdots \quad (5)$$

For polyatomics, usually only the last two terms in Equation (5), the cubic and quartic anharmonic terms, need be considered (57). The lowest-order process involving cubic anharmonic coupling involves excited vibration Ω relaxing by interacting with two other states, say another vibration ω and one phonon ω_{ph}, or alternatively a pair of phonons. For example, the total rate constant for energy loss from Ω for cubic coupling was given by Fayer and coworkers as (57)

$$K_\Omega = \sum_\omega [n_\omega (1 + n_{\Omega+\omega}) \rho_{\Omega+\omega} C_{\Omega+\omega}$$
$$+ (1 + n_\omega)(\alpha + n_{|\Omega-\omega|}) \rho_{\Omega-\omega} C_{|\Omega-\omega|}] \tag{6}$$

where n_ω is the thermal occupation number,

$$n_\omega = (e^{\hbar\omega/kT} - 1)^{-1} \tag{7}$$

ρ_ω is the density of phonon states at ω, C_ω is a product of coupling constants that contains factors such as $\hbar/2\mu\omega$ and the derivatives of V in Equation (5), and $\alpha = 1$ if $\Omega > \omega$ or $\alpha = 0$ if $\Omega < \omega$. When T $\rightarrow$ 0, all the thermal occupation factors in Equation (6) vanish, but the VER rate does not vanish. VER is then said to occur via spontaneous emission of phonons. As T is increased, two new thermally activated processes turn on. One involves stimulated phonon emission and the other phonon absorption. Spontaneous and stimulated emission processes convert Ω to a lower energy vibration ω (down-conversion). Phonon absorption converts Ω to a higher energy vibration ω (up-conversion).

Some representative examples of common zero-temperature VER mechanisms are shown in Fig. 2b–f. Figures 2b,c describe the decay of the lone vibration of a diatomic molecule or the lowest energy vibrations in a polyatomic molecule, termed the "doorway vibration" (63), since it is the doorway from the intramolecular vibrational ladder to the phonon bath. In Fig. 2b, the excited doorway vibration Ω lies below ω_D, which can be the case for large molecules or macromolecules. In the language of Equation (4), fluctuating forces of fundamental excitations of the bath at frequency Ω are exerted on the molecule, inducing a spontaneous transition to the vibrational ground state plus excitation of a phonon at $\omega_{ph} = \Omega$. The rate of this transition is proportional to the Fourier transform of the force-force correlation function at frequency Ω, denoted C(Ω).

In Fig. 2c, the vibration Ω lies well above the phonon cut-off ω_D, as for example the 379 cm^{-1} doorway vibration in ACN (46), where ω_D is in the 100–150 cm^{-1} range. Fluctuating forces exerted by the bath at frequency Ω cause the doorway vibration to decay. In the language of

Equation (5), VER involves a higher-order anharmonic coupling matrix element, which gives rise to decay via simultaneous emission of several phonons $n\omega_{ph}$ (multiphonon emission). In the ACN case, three phonons must be emitted simultaneously via quartic anharmonic coupling (or four phonons via fifth-order coupling, etc.).

In Fig. 2d, Ω is one vibrational fundamental of a polyatomic molecule, whose relaxation involves exciting ω_1, a lower energy vibration of the same molecule. In the language of Equation (5), Ω, ω_1, and other intramolecular vibrations are part of the system and the phonons are the bath. Fluctuating forces exerted by the bath at frequency $\omega_{ph} = \Omega - \omega_1$ induce a transition from Ω to ω_1 via cubic anharmonic coupling. This mechanism, which ought to be most important in polyatomic molecules where the spacings between adjacent vibrations are generally less than ω_D (4), is the "ladder relaxation" (57) process mentioned previously. The name derives from the motion of the vibrational excitation, which hops downward from one vibration to another, which are the "rungs" of the "ladder" of vibrational fundamentals of the polyatomic molecule.

In Fig. 2e, Ω decays by an intramolecular vibrational redistribution (IVR) process, involving lower energy vibrations $\omega_1, \omega_2, \omega_3, \dots$, via a higher-order anharmonic coupling which causes Ω to decay by spontaneous emission of several lower energy vibrations $\omega_1, \omega_2, \omega_3, \dots$. In condensed phases, phonons may play a role in IVR as well, which is analogous to the role of rotations in gas-phase IVR, by dynamically modulating and broadening the vibrational energy levels, making it more likely for a resonance to occur.

D. Vibrational Cascade

The "vibrational cascade" (64) illustrated in Fig. 2f is widely believed to be a prominent mode of vibrational cooling (VC). The general properties of vibrational cascades in large molecules at finite temperature were studied theoretically by Hill and Dlott (4,5). The vibrational cascade occurs when the lowest-order ladder processes dominate. In a large molecule, the rungs of the vibrational ladder are on average closely spaced. Here closely spaced means the average energy difference is less than ω_D, so that a step from one rung to another can occur with just one phonon. In this case, energy loss from a polyatomic molecule at zero temperature, with vibration Ω initially excited, occurs by a sequence of VER processes, each involving the emission of just one phonon via cubic anharmonic coupling. A vibrational cascade in an intermediate-size molecule such as ACN or NM might involve some steps where two or three phonons are emitted via quartic or fifth-order coupling. In a vibrational cascade at zero temperature, a vibrational excitation descends the ladder, losing a small amount of energy in each step. At the bottom rung of the ladder, the doorway vibration, the final VER step cannot occur by a ladder process since there are no more lower energy rungs (4). The final VER step occurs by a single or multiphonon emission process (e.g., Fig. 1b or 1c). At finite temperature, each step along the ladder might go up or down, depending on whether phonons are absorbed or emitted. However the net motion of excess vibrational energy is always downward, to states of lower energy (4,5).

Hill and Dlott (5) illustrated the properties of vibrational cascades in model calculations of VC in crystalline naphthalene. Naphthalene ($C_{10}H_8$) has 48 normal modes. Forty of these vibrations (all except the eight C–H stretching vibrations) lie in the frequency range 1627–180 cm^{-1}. In the calculation, one unit of excitation is input to the highest vibration in this range, 1627 cm^{-1}. The ensemble-averaged population of the ith mode is determined by a master equation:

$$\frac{d\mathbf{P}(t)}{dt} = \mathbf{K} \cdot \mathbf{P}(t), \tag{8}$$

where $\mathbf{P}(t)$ is a vector of vibrational populations and $\mathbf{K}$ is a matrix of transition rate constants. The elements of the rate matrix were computed using Equation (6), which assumes that VER occurs solely by cubic anharmonic coupling. In naphthalene, the phonon density of states ρ_ω^{ph} is accurately known from neutron scattering measurements (65,66). The coupling factors C_Ω in Equation (6) were determined using the density of states and VER lifetimes determined by low-temperature vibrational lineshape measurements (5,67). Where VER measurements were not available, C_Ω was taken to be the average of the known values.

The average energy jump ΔE for every step up or down the ladder in a large molecule is approximately equal to the average phonon frequency,

$$\Delta E_{avg} = \frac{\int \hbar\omega\rho_\omega^{ph}d\omega}{\int \rho_\omega^{ph}d\omega} \tag{9}$$

For naphthalene (5), $\Delta E_{avg} = 95$ cm^{-1}.

Figure 3 shows the results of a calculation (5) at T = 0 assuming an initial condition of unit vibrational excitation at 1627 cm^{-1}. As time progresses, the center of the vibrational population distribution moves toward lower energy. The vibrational population distribution spreads out because the size of each step is distributed over the range 0–$\hbar\omega_D$, with the average step being 95 cm^{-1}. Vibrational cooling is essentially complete

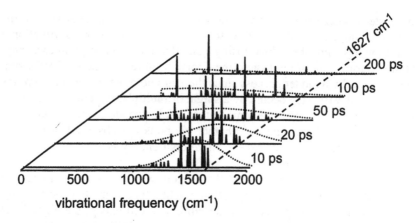

Figure 3 Calculated vibrational cascade in crystalline naphthalene at $T = 0$, for initial excitation at 1627 cm^{-1}. The calculation uses Equation (6), which assumes that cubic anharmonic coupling dominates. From Ref. 5.

by about 200 ps. When the vibrational density of states is a constant, the population distribution should eventually approach a Gaussian distribution (4,5). The peak of the distribution will move toward lower energy at a constant velocity (the "vibrational velocity," with units of energy dissipated per unit time) and the width will increase as the square root of time. The dashed curves in Fig. 3 are the best fits to this Gaussian distribution.

The temperature dependence of VC discovered by Hill and Dlott (4,5), as a consequence of the temperature dependence predicted by Equation (6), is very interesting. Equation (6) shows that the lifetime, that is the rate of leaving a particular state, decreases with increasing T. Equation (6) has three parts: *temperature-independent* spontaneous emission (downward) and *temperature-dependent* stimulated emission (down) and absorption (up). It is the increase in the rates of the latter two temperature-dependent processes that causes the lifetime to decrease with increasing T. The two temperature-dependent processes by themselves do not cause vibrationally hot molecules to cool, since they are as likely to drive an excited vibration to higher energy states as to lower energy states. Increasing the temperature only increases the rate of vibrational energy jumping up and down. Any actual cooling, which is caused by the net motion of vibrational excitation to lower energy, is driven by the temperature-independent spontaneous emission processes. Thus the VC process is largely independent of temperature when only lower order anharmonic coupling processes dominate. This point is illustrated in Fig. 4, which plots

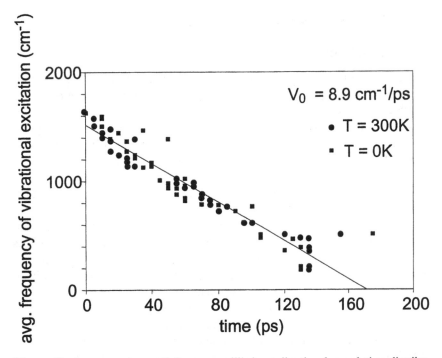

Figure 4 Average energy of the nonequilibrium vibrational population distribution computed for the vibrational cascade in crystalline naphthalene in Fig. 3. At T = 0, the peak moves toward lower energy at a roughly constant rate, the "vibrational velocity" of 8.9 cm^{-1} ps. The initial 1627 cm^{-1} of vibrational energy is dissipated in ~180 ps. The vibrational velocity is the same at 300 K. In the limit that cubic anharmonic coupling dominates [Equation (6)], increasing the temperature increases the rates of up- and down-conversion processes, but has no effect on the net downward motion of the population distribution. Although the lifetimes of individual vibrational levels will decrease with increasing temperature, VC is not very dependent on temperature in this limit. (From Ref. 5.)

the time dependence of the average energy of the population distribution after 1627 cm^{-1} excitation at T = 0 and T = 300 K. At both temperatures the population distribution moves down at an approximately constant rate, with a vibrational velocity $V_0 = 9$ cm^{-1}/ps. That is to say, the average rate of energy lost from the molecule is 9 cm^{-1} per picosecond; losing 1627 cm^{-1} takes about 180 ps. This calculation shows that although the rates of VER processes may increase dramatically with T, the overall rate of VC ought not to be much affected by temperature.

III. THE IR-RAMAN TECHNIQUE

A. The Method

A tunable mid-IR pulse at frequency ω_{IR} pumps vibrational excitations in a polyatomic liquid (all work discussed here is at ambient temperature $\sim$295 K). A time-delayed visible probe pulse at frequency ω_L generates incoherent anti-Stokes Raman scattering. For an instantaneous pump pulse arriving at time t = 0, the change in the anti-Stokes intensity of transition i, with frequency ω_i, the "anti-Stokes transient," is (44)

$$\Delta I_i^{AS}(t) = \text{const}\Delta n_i(t) g_i \sigma_{Ri} (\omega_L + \omega_i)^4 \tag{10}$$

where the constant depends on the experimental set up, $\Delta n_i(t)$ is the instantaneous change in vibrational population, g_i is the degeneracy, and σ_{Ri} is the Raman cross section. Equation (10) shows that the intensity of an anti-Stokes transient is proportional to the population change in the vibrational transition during the VER process induced by the pump pulse (2).

The IR-Raman experiment is difficult. An ultrashort tunable mid-IR pulse is needed that produces a substantial number of vibrational excitations. An intense visible pulse is also needed that generates enough anti-Stokes signal photons despite the small magnitude of the Raman cross section. Detecting small numbers of anti-Stokes photons is no problem with today's 90% quantum efficiency CCD detectors. The real problem arises from optical background, as discussed below. The laser pulses must be short enough to time-resolve VER processes of interest, but if they are too short undesirable effects occur (45): (1) the spectral bandwidth of the pulses becomes too broad to resolve individual vibrational transitions and (2) the short pulses at fluence levels needed to generate and detect transient vibrational populations will generate optical background in the sample via nonlinear optical interactions. We designed our apparatus to produce an $\sim$0.8 ps pulse, which is short enough to time-resolve most VER processes of interest. The transform limited spectral bandwidth of a 0.8 ps pulse is $\sim$20 cm^{-1}, but practical bandwidths in our system are 25–35 cm^{-1} (45). Figure 5 shows IR and Raman spectra of neat liquid ACN, obtained using conventional spectrometers with resolution better than the natural linewidths (46). In Fig. 5 we also show a Stokes Raman spectrum obtained with the ultrafast laser system. Most of the Raman transitions can be resolved, although the C–H bending modes in particular tend to run together.

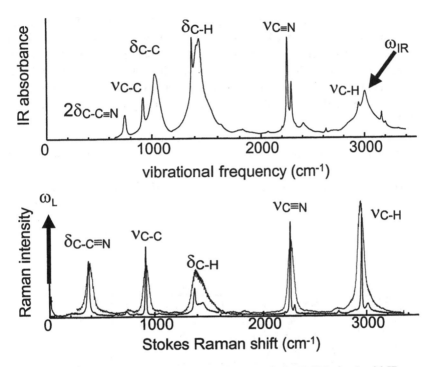

Figure 5 Vibrational spectra of neat liquid acetonitrile (ACN): (top) mid-IR spectrum; (bottom) Stokes Raman spectrum using a conventional spectrometer (solid line). The dashed spectrum obtained with the ultrafast laser system has somewhat lower resolution (From Ref. [46].)

B. The Laser

Since the early IR-Raman experiments with Nd:glass lasers, Ti:sapphire lasers with chirped-pulse amplification (CPA) (68,69) have revolutionized ultrafast spectroscopy. Ti:sapphire lasers ordinarily run in femtosecond mode (pulse duration ~100 fs) where the spectral bandwidth (>140 cm^{-1}) is too large for Raman spectroscopy, so methods have to be found to lengthen the Ti:sapphire pulses and reduce the spectral bandwidth. That is a bit ironic, since so much work has been devoted to producing pulses with ever-shorter duration. Efforts are being made today to produce longer-duration, spectrally narrower bandwidth pulses from Ti:sapphire (45) or to convert femtosecond pulses into spectrally narrower picosecond pulses in a more efficient manner than simply removing energy with a narrow bandpass filter (70).

The experimental setup used at the University of Illinois is diagrammed in Fig. 6. A CPA laser from Clark-MXR Corp. was modified by the manufacturer to run in picosecond mode by substituting more dispersive gratings (2000 lines/mm for 1400 lines/mm) and by adding birefringent filters (71) in the regenerative amplifier cavity. The system outputs ~0.8 ps duration pulses with 1.0 mJ energy at a repetition rate of 1 kHz.

The most common technology used today to produce intense pulses at wavelengths other than 800 nm and its harmonics is optical parametric amplification (OPA) (72). In an OPA, a "signal" laser pulse at frequency ω_S propagates through a nonlinear crystal along with an intense pump pulse at

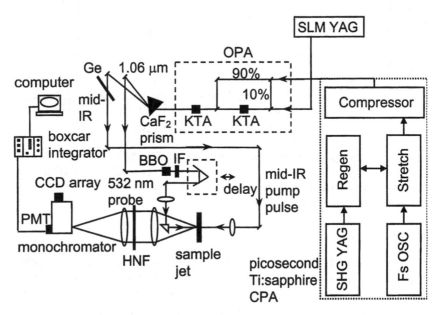

Figure 6 Block diagram of the two-color optical parametric amplifier (OPA) and IR-Raman apparatus. CPA = Chirped pulse amplification system; Fs OSC = femtosecond Ti:sapphire oscillator; Stretch = pulse stretcher; Regen = regenerative pulse amplifier; SHGYAG = intracavity frequency-doubled Q-switched Nd:YAG laser; YAG = diode-pumped, single longitudinal mode, Q-switched Nd:YAG laser; KTA = potassium titanyl arsenate crystals; BBO = β-barium borate crystal; PMT = photomultiplier tube; HNF = holographic notch filter; IF = narrow-band interference filter; CCD = charge-coupled device optical array detector. (From Ref. 96.)

ω_P. The pump pulse amplifies the signal while simultaneously producing an "idler" pulse at frequency $\omega_I = \omega_P - \omega_S$. For 800 nm Ti:sapphire pump pulses, the idler will be tunable throughout the mid-IR (broadly 400–4000 cm^{-1}) for signal pulses in the 820–1176 nm range, provided the nonlinear crystal is transparent in all these spectral regions. We needed a design that would produce tunable mid-IR pulses, but we wanted the probe pulses to have a fixed wavelength so that readily available interference filters and holographic Raman notch filters could be used in the detection setup. In our development phase we took anti-Stokes spectra from various liquids using ~0.8 ps pulses at 800 and 400 nm. The longer wavelength 800 nm pulses had poor scattering efficiency due to the ω^4 dependence in Equation (10). The 400 nm second harmonic pulses, in principle 16 times better than the fundamental, produced too much multiphoton fluorescence and ionization. Believing that a probe pulse in the ~500 nm region was optimal, we designed at two-color OPA system (45) as in Fig. 6.

Our OPA is based on potassium titanyl arsenate (KTA) crystals (73). KTA is quite similar to KTP which is widely used in OPAs, but KTA has better mid-IR transparency than KTP. The two-color OPA is seeded (the "seed" is the signal pulse) in a quasi-CW fashion (45,74) by a diode-pumped Q-switched Nd:YAG laser running in a single longitudinal mode at 1.064 μm, which generates ~10 μJ in a 50 ns pulse (45). Only a ~1 ps duration slice of this seed pulse is amplified in the OPA. The seed power is about 1 kW. Laser seeding produces narrower bandwidths than the more usual white-light seeding, and it avoids the problems of generating a seed supercontinuum with the picosecond pulses (75), which is more difficult than with femtosecond pulses. With the Nd:YAG seed, the center frequency of the amplified signal output will remain pinned at 1.064 μm. The signal pulses are frequency doubled in a BBO crystal to produce the fixed-frequency Raman probe pulses at 532 nm, where off-the-shelf optics and filters are readily available. The mid-IR idler pulses can be tuned over an ~100 cm^{-1} range by simply tilting the crystals and over an ~1000 cm^{-1} range by tuning the CPA laser. When the CPA pump pulses are tuned in the 770–820 range, the mid-IR output ranges from 2800–3600 cm^{-1}. This mid-IR range allows us to pump almost any molecule containing at least one hydrogen atom, including common functional groups such as C–H, O–H, S–H, N–H, etc. (76).

The performance of the two-color OPA (45) is illustrated in Fig. 7. Mid-IR pulse energies of 40–50 μJ are typically obtained in a nominal 0.8 ps pulse with a 35 cm^{-1} spectral bandwidth. These are very large mid-IR energies for a kilohertz laser (45). It is easier to obtain high conversion

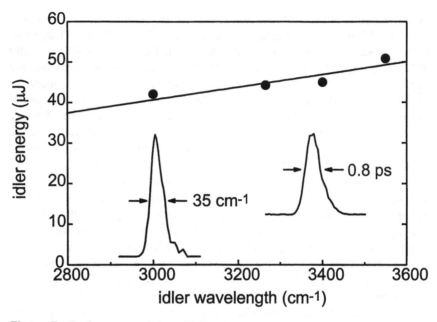

Figure 7 Performance of the mid-IR OPA. The insets show a cross-correlation between mid-IR pulses and 532 nm pulses from the OPA, with FWHM of 0.8 ps, and the 35 cm^{-1} FWHM spectrum of mid-IR pulses at 3000 cm^{-1}. (From Ref. 45.)

efficiency with our 0.8 ps pulses than with 100 fs pulses. The higher pump power of femtosecond pulses is not really an advantage because the power at the nonlinear crystal must be kept below a critical level, typically a few hundred GW/cm^2, which is determined mainly by the onset of super-continuum generation. For a 100 fs pulse with ~150 cm^{-1} bandwidth, our calculations have shown that the KTA crystal length is limited to <2 mm. The crystal length is ordinarily limited either by group-velocity mismatch or by the acceptance bandwidth (72). For a 100 fs pulse, the acceptance bandwidth becomes limiting (i.e., it is ~150 cm^{-1}) at ~2 mm thickness, and the group-velocity mismatch becomes significantly limiting at a slightly greater thickness. For 0.7–1.0 ps duration pulses, the band-width and the group-velocity mismatch becomes limiting in the 7–10 mm thickness range. Because the OPA gain in the small-depletion limit is an exponential function of the usable crystal length (72), the ability to use longer (7–10 mm) crystals in our picosecond OPA provides us with enor-mous small-signal gain. With the gain so great, the OPA can be run deeply into saturation, which depletes the pump pulses and produces large pulse energies. When the signal pulses at 1.064 µm are frequency doubled in a BBO crystal, ~25 µJ pulses are produced at 532 nm with a 50 cm^{-1} bandwidth. These pulses are usually frequency narrowed with optical interference filters (Fig. 6), typically with 25 cm^{-1} bandpass filters, at the expense of lost energy.

C. Experimental Setup

The experiments described in this chapter are performed with the liquid sample in a thin (50–150 µm) flowing jet stream. The high velocity of the jet minimizes the effects of heating or multiple-pulse photochemical decomposition of the liquid sample under the intense irradiation of the pump and probe pulses. The jet stream works very well with higher viscosity liquids, and it is manageable with lower viscosity liquids. We use the jet because the optical windows on sealed cells (see next section) are problematic. However, the jet limits us to liquids that are inexpensive enough to fill the jet pumping system (~100 mL are needed) and that have low toxicity. Using a jet precludes studying very expensive or very volatile liquids, solid samples, or cryogenic solids. Ultimately it will be necessary to extend our work to these systems, which will be held in sealed flow cells with thin near-IR windows or in a cryostat mounted on a motorized mover. As discussed below, with windows we must use less laser pulse energy and subtract off the anti-Stokes spectrum of the window, which might be more than 10 times thicker than the sample.

The optical geometry at the sample is shown in Fig. 8. In the counterpropagating setup used here, the time resolution can be degraded by the sample thickness. The sample should be <50 µm thick to introduce negligible time broadening. For a 50 µm sample with refractive index n = 1.5, the broadening is only ~0.25 ps. Ordinarily we make the jet thickness slightly greater than the characteristic absorption length for the pumped transition, so almost all the pump photons are absorbed, as shown in Fig. 8. In this limit it is not the physical thickness of the sample but the absorption length that determines the time-broadening factor. Since most samples have absorption lengths of <50 µm, the time broadening is not a problem if the jet is a bit too thick. The pump pulse is focused to a diameter of ~200 µm (the diameter refers to the $1/e^2$ Gaussian beam diameter). The probe pulse has a smaller diameter. Backscattered Raman light is collected with f/1.2 optics, transmitted through a pair of holographic Raman notch filters (Kaiser Optical) and sharp-cut short pass filters (Omega Optical) to remove Raleigh scattering, expanded to f/4 and directed into

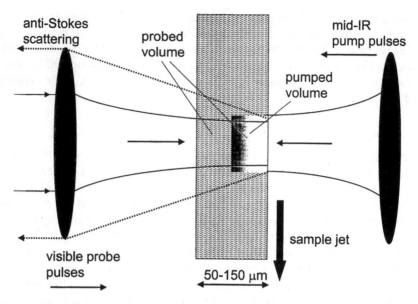

Figure 8 Expanded view of the pump-probe geometry and the sample jet. The pumped volume is 200 μm in diameter. The depth of the pumped volume is equal to the absorption depth of the liquid sample, typically 20–50 μm. The probed volume includes some sample that is not pumped, which is minimized by minimizing the jet thickness. (From Ref. 96.)

a monochromator/spectrograph equipped with both a fast photomultiplier (PMT) and a charge-coupled array detector (CCD). Transient Raman spectra at a particular delay time are obtained with the CCD (Fig. 9). Anti-Stokes Raman transients are obtained by scanning the optical delay line while a computer acquires the signal from a PMT and a boxcar integrator (Fig. 10).

D. Optical Background

Assuming that all fluorescent impurities are removed from the sample, there are three main sources of optical background: (1) Rayleigh scattered probe light, (2) the thermal background of excited vibrational states, and (3) visible light created via nonlinear interactions by the pump and probe pulses. Rayleigh scattering can be efficiently removed with holographic Raman notch filters. Short of cooling the sample, the only way to minimize the thermal background is to minimize the probed sample volume. Figure 8 shows that is accomplished by using a jet that is only slightly thicker than the mid-IR absorption length.

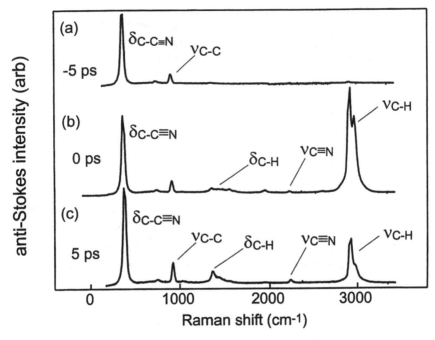

Figure 9 Transient anti-Stokes Raman spectra of acetonitrile (ACN) with 3000 cm^{-1}C–H stretch pumping, (a) At t = −5 ps when probe precedes pump, only the thermal background is seen. (b) At t = 0, there is a large jump in C–H stretch population and a small increase in some other vibrations. (c) At t = 5 ps, the C–H stretch has partially decayed and the other vibrational populations have increased. (From Ref. 96.)

Figures 9 and 10 illustrate the thermal background problem. Figure 9 shows transient anti-Stokes spectra from ACN pumped at 3000 cm^{-1}. The Raman signal in Fig. 9a, where the probe pulses precede the pump pulses by 5 ps, is due solely to the thermal background, which is substantial below ~1000 cm^{-1}. The thermal background is negligible for higher frequency vibrations (>2000 cm^{-1}), but it is a serious problem for lower frequency vibrations. Compounding the problem is that generally only a fraction of the excitations generated by the pump pulse end up being transferred to any specific lower energy vibration. These points are illustrated in Figs. 9 and 10, where we compare anti-Stokes transients in ACN from the (pumped) highest energy fundamental, a C–H stretch at ~3000 cm^{-1}, and from the lowest energy fundamental, a doubly degenerate C–C≡N

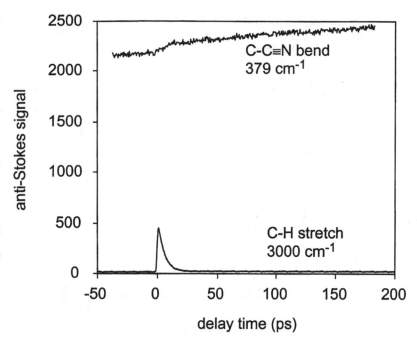

Figure 10 Anti-Stokes Raman transients in acetonitrile (ACN) after C–H stretch pumping at ~3000 cm⁻¹. There is almost no thermal background for the C–H stretch. There is a large background for the C–C≡N bend, and the signal (the jump in anti-Stokes intensity induced by the pump pulse at t = 0) is smaller. The C–H stretch data is obtained with ~30 minutes of averaging. The C–C≡N data required 10 hours to overcome intensity fluctuations in the thermal background. (Adapted from Ref. 46.)

bend at 379 cm⁻¹. The laser-induced nonequilibrium population increase of the C–H stretch is about 0.02 (see Section III.F.). The background thermal population is $\sim 4 \times 10^{-7}$, so there is essentially no background, and the signal-to-noise ratio obtained in a 10-minute scan is about 100:1. By contrast, for the C–C≡N bend, the laser-induced nonequilibrium population around t = 0 (generated via C–H stretch decay) is about one-fifth the C–H stretch population, or about 0.004. The thermal background population is 0.38, so in Fig. 9 hardly any increase in the C–C≡N bend population can be seen near t = 0. In fact, the fluctuations in the thermal background are about the same size as the signal. The anti-Stokes transient for the C–C≡N bend in Fig. 10 was obtained by averaging the results of several hours of scanning, and still the signal-to-noise ratio is only 5:1.

The most serious obstacle to detecting anti-Stokes transients is visible light generated by the pump or probe pulses. There are a myriad of sources. Mid-IR generation in the OPA is accompanied by the production of small amounts of visible light at practically every imaginable sum and difference combination of the three interacting waves and their harmonics. This light is removed from the mid-IR idler pulse by a Ge Brewster window (Fig. 6). Frequency doubling of the 1.064 μm signal pulses in BBO also generates small amounts of light at other visible wavelengths, so before the 532 nm pulses are directed to the sample, they are sent (Fig. 6) through either a color filter or optical interference filters (Omega Optical) with either 10 or 25 cm^{-1} bandpass.

The largest signals are obtained when the pump and probe pulses are tightly focused in the sample, but then the optical intensities become huge. A typical pump pulse intensity (45 μJ, 200 μm diameter, 1 ps) is 160 GW/cm^2. The pump and probe pulse intensities must be kept below the threshold for supercontinuum generation (77) in the sample; otherwise the continuum overwhelms the anti-Stokes signal. To maximize our signal-to-noise ratio, we try to keep the intensities only a little bit below the continuum threshold. The worst case occurs at t = 0, when the two pulses overlap in time and the instantaneous intensity in the sample is a maximum. The supercontinuum threshold increases for decreasing sample thickness, so the highest pulse intensities can be used with the thinnest samples — hence our preference for a thin sample jet with no windows. A final problem that occurs at lower intensity than supercontinuum generation is the onset of self–phase modulation (77) of the 532 nm probe pulses, which causes the probe pulse to spectrally broaden. When the probe pulse spectrum begins to broaden in the sample, the Rayleigh scattering begins to leak past the Raman notch filters into the detector. Given the amount of pulse energy we can generate, we use trial and error to determine the minimum laser beam diameter that reduces all these troublesome factors to acceptable levels.

E. Pumping Vibrations

In an ACN sample jet ~100 μm thick, we determined that the pump pulses can be focused no more tightly than ~200 μm diameter. Table 1 lists typical parameters for the laser pulses used in this experiment, given that the probe pulse is focused to a slightly smaller diameter of ~150 μm. Table 2 lists some properties of ACN relevant to experiments with C–H stretch pumping at ~3000 cm^{-1}. The fluence of the mid-IR pulses on the jet is ~0.13 J/cm^2,

Table 1 Typical Laser Parameters for IR-Raman Experiments

	Pulse energy E_p	Gaussian $(1/e^2)$ beam diameter	Area	Fluence	Intensity
Mid-IR pump pulses	40 µJ $\sim 7 \times 10^{14}$ photons	200 µm	3×10^{-4} cm^2	0.13 J/cm^2	160 GW/cm^2
Visible (532 nm) probe pulses	5 µJ[a] $\sim 1.3 \times 10^{13}$ photons	150 µm	1.8×10^{-4} cm^2	0.03 J/cm^2	35 GW/cm^2

[a]With 25 cm^{-1} bandpass interference filter. Probe pulse energy may be 20 µJ with full 50 cm^{-1} bandwidth.

Table 2 Acetonitrile (CH$_3$–C≡N) Properties

Property	Symbol	Value or typical value
Molecular weight	MW	41
Density	ρ	0.79 g/cm^3
Number density	n	1.2×10^{22} cm^{-3}
Heat capacity	ρC	2 J cm^{3}K^{-1}
Decadic molar extinction coefficient for mid-IR C–H stretch transition ($\lambda \sim 3.3$ µm)	ε	4 liters mol^{-1}cm^{-1}
Absorption cross section for C–H stretch	σ	1.5×10^{-20} cm^2
Unsaturated absorption coefficient for C–H stretch	$\alpha = n\sigma$	180 cm^{-1}
Unsaturated absorption depth for C–H stretch	α^{-1}	55 µm
Saturation fluence (for pulse durations shorter than the VER lifetime)	$J_{sat} = h\nu/\sigma$	4 J/cm^2
Pumped volume (~ 200 µm diameter $\times 100$ µm deep)	V	3.2×10^{-6} cm^3
Density of pumped C–H stretching excitations	ρ_e	2×10^{20} cm^{-3}
Raman cross section	σ_R	2×10^{-28} cm^{2}sr^{-1}

whereas the saturation fluence is $J_{sat} \sim 4$ J/cm^2. Optical pumping is almost totally unsaturated. Since the unsaturated absorption depth is ~ 55 µm, most of the mid-IR pump photons are absorbed in the first 50–100 µm of the sample jet (see Fig. 8). The density of C–H stretching excitations in the pumped volume (V $\sim 3.2 \times 10^{-6}$ cm^3) is $n_e \sim 2 \times 10^{20}$ excitations/cm^3, so about 2% of the ACN molecules are vibrationally excited by the pump pulse (46). After VC is complete (~ 200 ps) but before thermal conduction cools the pumped volume (~ 100 µs), all the pump pulse energy is converted into a bulk temperature increase ΔT in the irradiated volume. The temperature jump averaged over the pumped volume is computed using the relation

$$\Delta T = J\alpha/\rho C \tag{11}$$

where C is the heat capacity. Using the parameters in Tables 1 and 2, $\Delta T \sim 12$ K (46).

F. Probing Vibrations

The number of detected anti-Stokes photons per shot induced by the pump pulse is given by

$$\text{number of detected photons} = Jn_e\sigma_R\Theta\eta\phi$$

where J is the probe fluence, n_e is the number of vibrational excited states, σ_R is the Raman cross section, Θ is the solid angle of light collected, η is the overall transmission of the optical setup including spectrograph, and ϕ is the detector quantum efficiency. From Table 1, the probe pulse fluence is $J \sim 7.5 \times 10^{16}$ photons/cm^2. The number of vibrational excitations is $n_e = \rho_e V \sim 7 \times 10^{14}$. The f/1.2 lens in our apparatus (Fig. 6) collects light over a solid angle of $\Theta \sim 0.05$ sr. The overall transmittance of the interference filters (we use two holographic notch filters and two short-wavelength pass filters) and spectrograph (with a holographic diffraction grating with $\sim 70\%$ efficiency) is $\eta \sim 10\%$. The quantum efficiency of the CCD detector is $\phi = 0.9$; for the photomultiplier it is $\phi = 0.2$. The number of detected anti-Stokes photons from the parent C–H stretch excitations detected by the CCD is therefore about 100 per shot. With a 1 kHz repetition rate, the signal from the pumped C–H stretch transition is $\sim 10^5$ photons/s. There is a minimal optical background equal to a few photons/s. By stepping the optical delay line once each second, a signal-to-noise ratio of $\sim 100:1$ is obtained for pumped C–H stretching vibrations, as shown in Fig. 10. However, the signal-to-noise ratio worsens when daughter vibrations are probed, since the signal is smaller and the thermal background is larger, as discussed in Section III.D.

G. Coherent Artifacts

Because the effects of VER are contained in the *incoherent* anti-Stokes signal (2), any coherent emission resulting from coupling between the pump and probe pulses may be regarded as an artifact. Coherent coupling artifacts are well known in pump-probe measurements of population dynamics (78). In the IR-Raman experiment, the dominant artifact in the anti-Stokes region is produced by coherent sum-frequency generation (79) (SFG) at frequency $\omega_L + \omega_{IR}$, where ω_L is the frequency of the Raman probe laser. The incoherent anti-Stokes signal from the vibration being pumped is also at frequency $\omega_L + \omega_{IR}$. Thus SFG coherent artifacts interfere with the observation of the population dynamics of the pumped vibrational transition, but not with the observation of daughter vibrational transitions.

The SFG process, where the mid-IR pulse is resonant with a vibrational transition, is a well-known process termed "vibrational sum-frequency generation" (79). In the dipole approximation, SFG does not occur in centrosymmetric media such as liquids, but it is allowed at surfaces or interfaces of centrosymmetric media (80), and in recent years vibrational SFG has emerged as a powerful interface-selective spectroscopic tool. Our experimental geometry minimizes surface SFG (Fig. 8), because both pump and probe electric field vectors lie in the plane of the sample jet surface. In fact in the dipole approximation, the surface SFG signal vanishes in this geometry. But when quadrupolar and higher-order terms in the multipole expansion for the susceptibility are considered, both surface and bulk SFG processes become allowed in our experimental geometry. In certain cases the SFG signal due to these higher-order terms can compete with or even overwhelm the desired incoherent anti-Stokes emission.

Examples of anti-Stokes data contaminated by an SFG artifact are shown in Fig. 11a and c, where the higher energy C–H stretching transition of neat methanol is pumped at $\omega_{IR} = 3020$ cm^{-1}. The artifact will be centered at $\omega_L + \omega_{IR}$, which appears at the same frequency as the incoherent anti-Stokes emission from methanol vibrational transitions at 3020 cm^{-1} (actually the higher energy tail of the C–H stretch transition at 2940 cm^{-1}). The spectral and temporal properties of the artifact can be independently characterized by purposely generating SFG in a thin ($\sim$50 μm) slab of KTP placed at the location of the sample. However, the amplitude of the SFG artifact in the spectrum is unknown.

In order to subtract off the SFG artifact, we used a model to fit the anti-Stokes spectrum. The methanol Stokes Raman spectrum can be fit with two Lorentzian lineshapes to represent the C–H stretching transitions centered at 2830 and 2940 cm^{-1}, and a Gaussian lineshape to represent the broad O–H stretching transition in the 3200–3700 cm^{-1} region (L. K. Iwaki et al., unpublished). The anti-Stokes data in Fig. 11a and c were fit by a computer program (Microcal Origin™) to the lineshapes that fit the Stokes spectrum plus a Gaussian function centered at 3020 cm^{-1}, to represent the SFG artifact (e.g., see the spectrum in Fig. 7). This fitting procedure gives the amplitude of the Gaussian component needed to fit

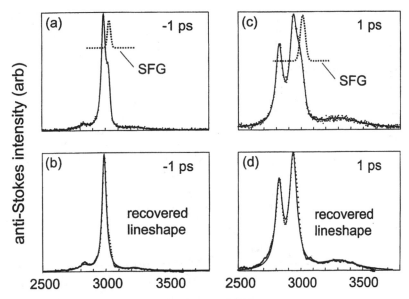

Figure 11 Examples of transient anti-Stokes spectra with coherent SFG artifacts, obtained from methanol at 300 K with C–H stretch (3020 cm^{-1}) pumping. (a) and (c) Experimental results at two times. The dashed curves represent the (nearly Gaussian) spectrum of the SFG artifact. (b) and (d) Recovered lineshapes with SFG contribution subtracted away. At early time (−1 ps) mainly the higher frequency C–H stretch is seen. As times passes, energy is redistributed among the two C–H stretch and the O–H stretch transitions. (From L. K. Iwaki, unpublished.)

the SFG artifact, which can be subtracted from the data to recover the lineshapes shown in Fig. 11b and d.

A second example is from our study of benzene with C–H stretch excitation (49). In earlier work, Fendt et al. (81) studied benzene by pumping with 3030 cm^{-1} mid-IR and probing with 532 nm pulses. They reported a C–H stretch lifetime of 1 ps. Our benzene experiments used the same pump and probe wavelengths. Our data are shown in Fig. 12. A fast build-up and decay on the order of 1 ps are observed, consistent with the earlier report. However, with the higher signal-to-noise ratio from our Ti:sapphire system, we could also see a long tail of the decay stretching over >10 ps. That led us to investigate the intensity dependence of the benzene anti-Stokes signal. We found that the fast part near t = 0 was a coherent artifact, which could be eliminated by attenuating the probe pulses. This artifact is the likely cause of the earlier (81) report of a 1 ps

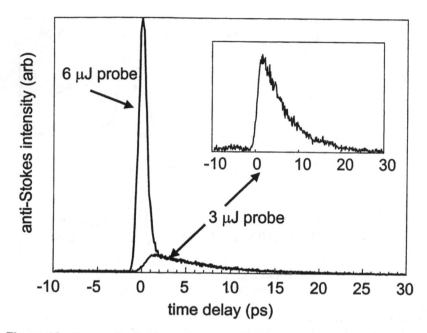

Figure 12 Decay of a C–H stretching transition in liquid benzene, pumped by a mid-IR pulse. When a 6 μJ visible probe is used, an intense artifact is seen near t = 0, which might lead to the erroneous interpretation of a fast (~1 ps) VER lifetime. By lowering the probe pulse energy to 3 μJ, the artifact is greatly reduced, and an 8 ps decay is seen clearly (inset). The strong dependence on the intensity of the visible 532 nm probe pulse is attributed to an SFG process enhanced by benzene's two-photon absorption at 532 nm. (Adapted from Ref. 49.)

lifetime. The anti-Stokes signal with lower intensity probe pulses (inset to Fig. 12) shows an 8 ps decay, which is the true lifetime of benzene C–H stretch. The relatively intense artifact in benzene and its strong dependence on the intensity of the visible (532 nm) probe pulse is attributed to an SFG process enhanced by the well-known two-photon absorption of benzene at this wavelength.

IV. EXAMPLES FROM CURRENT RESEARCH

A. Vibrational Energy Redistribution in Acetonitrile

Figure 13 shows ACN anti-Stokes transients after C–H stretch excitation (3000 cm^{-1}) obtained by tuning the spectrometer to a particular transition

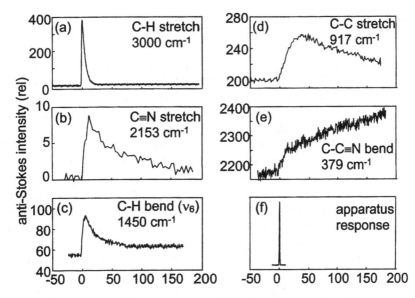

Figure 13 IR-Raman measurements of vibrational energy flow in acetonitrile (ACN). The pumped C–H stretch ($\sim$3000 cm^{-1}) decays in 5 ps. Only 2% of the energy is transferred to the C≡N stretch, which has an 80 ps lifetime. Most of the energy is transferred to the C–H bend plus about 4 quanta of C–C≡N bend. The daughter C–H bend vibration relaxes by exciting the C–C stretch. The build-up of energy in the C–C≡N stretch mirrors the build-up of energy in the bath, which continues for about 250 ps after C–H stretch pumping. The apparatus response is determined by placing a noncentrosymmetric solid at the sample location and monitoring the SFG signal. (From Ref. 46.)

and scanning the delay (46). C–H stretch excitation rises instantaneously and decays exponentially with a time constant of 5 ps. C–H stretch pumping is also associated with excitation of the first overtone (v = 2) of a C–H bending transition at about one-half the energy (see Section IV.H). The transient for the C≡N stretch at 2173 cm^{-1} in Fig. 13b has a 5 ps rise, which mirrors the C–H stretch decay, and an 80 ps decay. The C≡N stretch transient has a small amplitude, indicating that only $\sim$2% of the C–H stretch energy is transferred to the C≡N stretch. The transient for a C–H bending vibration at 1440 cm^{-1} shows a 5 ps rise and an $\sim$30 ps decay to a plateau. The plateau results from the temperature jump (46). The transient for the C–C stretch at 917 cm^{-1} in Fig. 13d rises with the 30 ps decay of the C–H bend and decays with an $\sim$50 ps lifetime. The transient for the C–C≡N bend at 379 cm^{-1} in Fig. 13e has a 5 ps rise. After the

5 ps rise, the C–C≡N population continues to rise as the other vibrations decay. The behavior of the ACN doorway vibration is discussed in more detail in Section IV.C.

VER in ACN after C–H stretch excitation is summarized as follows (46). The mid-IR pump pulse (3000 cm^{-1}) excites a combination of C–H stretching fundamentals and C–H antisymmetrical bending overtones, which are coupled by Fermi resonance. Vibrational excitation decays from the coupled C–H stretch and bend overtones with a 5 ps time constant. Almost none of the C–H stretch energy is transferred to the C≡N stretch (2253 cm^{-1}). The C≡N stretch behaves as a VER "blocking group", which blocks vibrational energy from leaving the CH$_3$–C moiety. We might speculate that molecules designed with specific C≡N groups might be able to channel vibrational energy to a particular location. The little energy that does get into the C≡N moiety is trapped for a relatively long time, since the C≡N decays with an ∼80 ps time constant.

About one-half of the C–H stretch energy is transferred to the C–H bending fundamentals (∼1500 cm^{-1}). The other half of the energy is taken up by the doorway vibration, which equilibrates rapidly with the bath. After the first ∼5 ps, most of the remaining nonequilibrium vibrational excitation is in the daughter C–H bending vibrations. The C–H bending vibrations go on to decay with a 30 ps lifetime, by exciting C–C stretching vibrations (918 cm^{-1}), and possibly the CH$_3$ rock (1040 cm^{-1}), which was not observed due to its very small Raman cross section. The C–C stretch decays with a 45 ps time constant. The entire VC process in acetonitrile takes about 250 ps.

B. Pseudo-vibrational Cascade in Nitromethane

The vibrational cascade described in Section II is a common paradigm for polyatomic molecule VC. Several years ago we studied VER and VC in nitromethane (NM) using a Nd:YLF laser system with 30 ps time resolution (41). VER and VC are particularly interesting in NM because it is a model system for understanding high explosives, and both VER and VC have been shown to play an important role in shock-wave initiation to detonation (41,63). In Ref. 41, the C–H stretch (∼3000 cm^{-1}) pumped by the laser decayed instantaneously (compared to 30 ps). An instantaneous rise was also seen in the daughter C–H bending and NO$_2$ stretching vibrations (∼1500 cm^{-1}). After several tens of ps, the only remaining transient response was seen at the even lower energy C–N stretch (918 cm^{-1}), which decayed on the 50 ps time scale. With these data, which show the average vibrational excitation continously moving to lower energy, it appeared as if a vibrational cascade occurred after C–H stretch excitation.

With the Ti:sapphire system, it became possible to see all relevant VER processes in NM 48. Some representative data are shown in Fig. 14, where the pumped 3000 cm^{-1} C–H stretch decays with a 2.6 ps time constant. C–H stretch excitation is also associated with excitation in the first overtone of the antisymmetrical C–H bend and the antisymmetrical NO_2 stretch. As the C–H stretch decays, energy builds up with a 2.6 ps time constant in the symmetrical C–H bending and NO_2 stretching vibrations at ~1400 cm^{-1}, as shown in Fig. 14a. These daughter vibrations decay with a 15 ps time constant. Looking to lower energy vibrations such as the C–H

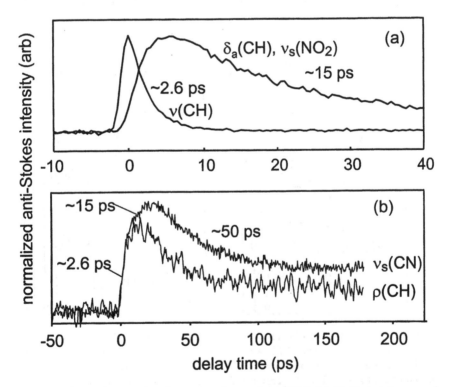

Figure 14 IR-Raman data from neat nitromethane (NM), (a) C–H stretch pumped by the laser decays with 2.6 ps time constant. C–H bending and NO_2 stretching vibrations at ~1500 cm^{-1} build up in 2.6 ps and decay in 15 ps. (b) All the lower frequency vibrations (≤918 cm^{-1}) have a two-part build-up (2.6 and 15 ps) and decay time constants in the 30–50 ps time range. (From Ref. 96.)

rock (1100 cm^{-1}), the C–N stretch (918 cm^{-1}), and NO$_2$ symmetrical bend and rock (657 and 480 cm^{-1}; omitted for clarity), as seen in Fig. 14b, there is a two-stage build-up. The first stage occurs with the 2.6 ps decay of the C–H stretch and the second with the 15 ps decay of the C–H bend and NO$_2$ stretch. These lower energy vibrations subsequently decay with lifetimes in the 30–50 ps range. A 2.6 ps rise is seen in every vibration of NM other than the pumped C–H stretch (48).

A level diagram for NM (48) in Fig. 15 provides an overview of VER and VC in this system. VC occurs in three stages. First the C–H stretch decay, a fast IVR process, populates every other vibration. Then the intermediate vibrations decay in 15 ps by populating all lower energy vibrations. Finally, the longest-lived lower energy vibrations decay into

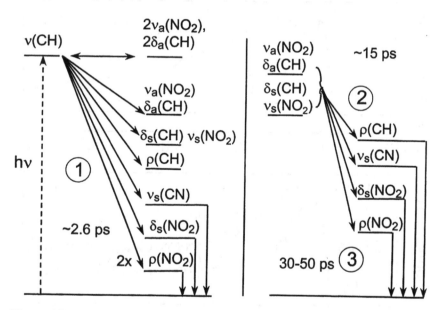

Figure 15 Energy level diagram showing the three stages of vibrational energy relaxation of nitromethane (NM) after C–H stretch excitation. The C–H stretch fundamental and the first overtones of antisymmetrical C–H bend and NO$_2$ stretch are pumped. In stage **1**, this energy is redistributed among all other vibrations with a 2.6 ps time constant. The population increase in the lowest vibration $\rho(NO_2)$ is twice as great as in the others. In stage **2**, the intermediate energy vibrations decay by exciting the lower energy vibrations with a 15 ps time constant. In stage **3**, the lower energy vibrations, which build up in the first two stages, decay in ~100 ps by exciting the bath. (From Ref. 48.)

phonons. It was easy to mistake this complicated VC process for a vibrational cascade because in NM, the vibrational lifetimes increase as the vibrational frequency decreases. Thus, after C–H stretch excitation, the highest energy vibrations vanish first, the intermediate vibrations vanish next, and the lowest energy vibrations vanish last. But neither of the first two processes, which cause the two-stage build-up in the lower vibrations, would occur in a true vibrational cascade.

C. Dynamics of Doorway Vibrations

Doorway vibration decay is particularly interesting because it is the one situation where polyatomic molecule VER looks just like diatomic molecule VER. The doorway vibrations of polyatomic molecules decay by exactly the same multiphonon mechanism as the VER of a diatomic molecule. Diatomic molecules have been extensively studied (7). One prediction for diatomic molecules is an exponential energy-gap law (2). As the vibrational frequency is increased, with everything else held constant, the number of emitted phonons increases (the order of the multiphonon process increases) and the VER rate should decrease exponentially with increasing vibrational frequency.

Representative data for the doorway vibrations of ACN (C–C$\equiv$N bend; 379 cm^{-1}), NM (NO$_2$ rock; 480 cm^{-1}), and benzene (ring deformation; 606 cm^{-1}) are shown in Fig. 16. It would be preferable to pump the doorway vibration directly, but suitably powerful ultrashort pulse sources are not yet available in the needed range (here 16–26 μm). We have been able to understand the behavior of doorway vibrations by watching energy run in and out of these vibrations after C–H stretch pumping; however, this indirect method of excitation complicates the problem somewhat.

The doorway vibration data in Fig. 16 at a glance shows how fast VC occurs in each molecule. The end of the doorway vibration population build-up denotes the end of the VC process. Fig. 16 shows that in ACN, VC takes ~250 ps. In NM VC takes ~100 ps, and in benzene VC takes ~150 ps.

The VER lifetime of the doorway vibration in ACN is quite short. It was estimated to be <5 ps (46). This estimate was obtained by molecular thermometry. As discussed above, the occupation number of C–H stretching excitations produced by the laser is 0.02. Since about one half of the C–H stretch excitation energy (~1500 cm^{-1}) is transferred to the doorway vibration at 379 cm^{-1}, about 4 quanta of doorway excitations will be produced in the first ~5 ps due to C–H stretch decay. If the doorway vibration were long-lived (i.e., if $T_1 > 5$ ps), then the doorway

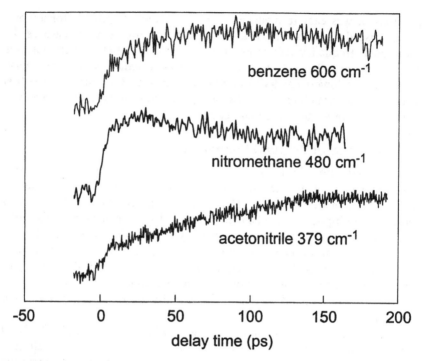

Figure 16   IR-Raman data for the lowest frequency doorway vibrations of three liquids after C–H stretch pumping at ∼3000 cm⁻¹. The build-up reflects the complicated vibrational cooling (VC) processes of each liquid. The higher frequency doorway vibrations have longer lifetimes. (From Ref. 96.)

vibration occupation number would jump from its thermal equilibrium value of n = 0.38 to a value n = 0.38 × 0.02 × 4 = 0.46. That would represent a jump in doorway vibration occupation number of ∼20% in the first few picoseconds. What is actually observed in Fig. 10 is a jump of only a few percent, which can be used to show the doorway vibration lifetime is considerably less than 5 ps. After the ∼5 ps jump, doorway vibration excitation builds up with a complicated functional form, which reflects heat build-up in the bath due to subsequent processes of C–H bend and C–C stretch relaxation.

In the NM data (48) in Fig. 16, energy builds up in the 480 cm⁻¹NO₂ rock in two stages, as described in Section IV.B. The subsequent decay of NO₂ rock excitation seen in Fig. 16 indicates this doorway vibration has a much longer lifetime than in ACN. The lifetime is about 50 ps.

In the benzene data (49) in Fig. 16, the 606 cm^{-1} ring deformation mode has a two-part rise. The first part mirrors the $\sim$10 ps decay of the C–H stretch. The second part mirrors the $\sim$40 ps decay of intermediate energy vibrations such as the ring breathing mode at 992 cm^{-1}. The decay lifetime is difficult to determine accurately because it is so slow, but is estimated at 100 ($\pm$20) ps.

As these examples show, the rather complicated behavior of the doorway vibrations in each liquid opens a window directly into the complicated VC process following C–H stretch pumping. The doorway vibration data by itself is not sufficient to unravel the entirety of the VC process, but when combined with IR-Raman data from a few other vibrations, most of the puzzle can be solved (49). It is interesting to note that by comparing ACN, NM, and benzene, the doorway vibration lifetime increases with increasing frequency, which is qualitatively what would be expected for the predicted (82) energy-gap law. Of course things are not really that simple, since the comparison involves changing both the vibrational frequency and the bath. Ideally one should calculate the fluctuating forces on each doorway vibration in its own bath and in comparing different liquids consider both the change in vibrational frequency and the change in Debye frequency ω_D.

D. Monitoring the Bath

In these studies we attempt to develop a consistent picture of VER in a given system by watching energy move among a polyatomic molecule's intramolecular vibrations. It would help a great deal to know at all times how much of the system's energy had been dissipated to the bath. A technique we have developed to monitor the build-up of bath excitation involves spiking the liquid with carbon tetrachloride (CCl$_4$). CCl$_4$ is a nonpolar, noncomplexing liquid, which is miscible with most other liquids. CCl$_4$ has three lower frequency vibrations with large Raman cross sections, as shown in Fig. 17, a Raman spectrum of a mixture of benzene and CCl$_4$. CCl$_4$ has little effect on the vibrations of benzene (49) or most other liquids. The primary effects of CCl$_4$ are to slightly modify the phonon density of states (54), usually by shifting the density of states a bit to lower energy.

Intermolecular energy transfer from a vibrationally excited molecule such as benzene to CCl$_4$ could occur by two mechanisms termed "direct" and "indirect," as diagrammed in Fig. 18. In a liquid mixture, the phonons are collective excitations of the mixture. But the coupling between intramolecular vibrations on adjacent molecules is ordinarily quite weak. Direct intermolecular transfer is important primarily in the long-lived vibrations of cryogenic liquids (7,84). In polyatomic liquid mixtures, VER

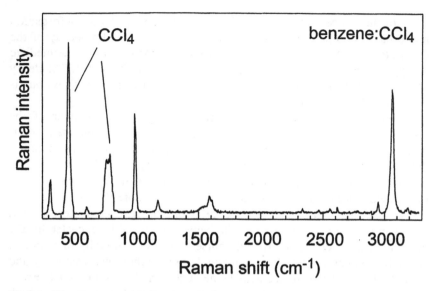

Figure 17 Raman spectrum of liquid benzene with CCl. The CCl$_4$ has little effect on the benzene vibrational spectrum or the benzene VER rates. Monitoring the CCl$_4$ vibrational transitions while pumping benzene vibrations provides an indication of the energy build up in the bath. (From Ref. 49.)

occurs on the picosecond time scale, so there is rarely enough time for direct transfer to occur. Direct transfer in polyatomic liquid mixtures has been observed in a few cases, most notably pyrrole to benzene (38) and alcohol to nitromethane (42). In both systems, there is a relatively strong noncovalent interaction between the two components, which would not be the case with CCl$_4$ mixtures.

Indirect transfer occurs by a two-part mechanism, as shown in Fig. 18. First a vibrational excitation decays by generating phonons. The phonons then produce vibrational excitation on other molecules by multiphonon up-pumping. Indirect transfer will not occur unless the density of vibrational excitations is large enough to produce a real increase in the bath temperature.

Our experiments with mixtures of CCl$_4$ and other liquids indicate it does not matter much which of the three CCl$_4$ vibrations shown in Fig. 17 are monitored, because all three pump up at about the same rate. Some CCl$_4$ data are shown in Fig. 19, where the transient CCl$_4$ population build-up results from C–H pumping of methanol, benzene, or NM. In methanol and benzene, all three CCl$_4$ transitions could be observed, and they all

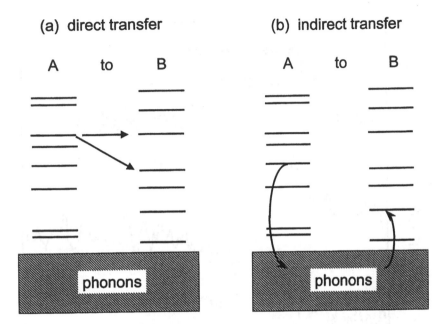

Figure 18 Direct and indirect intermolecular vibrational energy transfer in a poly-atomic liquid mixture consisting of molecules A and B. In polyatomics, direct transfer does not often compete efficiently with VER. Indirect transfer from A to B occurs when A undergoes VER, which produces phonons, which pump vibrations on B. Indirect transfer is efficient only when the density of excited vibrations is large enough to significantly increase the phonon population.

behaved similarly. In NM, the 479 cm^{-1} CCl$_4$ transition is obscured by the 480 cm^{-1} NM doorway vibration, so the 315 cm^{-1} CCl$_4$ transition was monitored. This CCl$_4$ technique proved problematic with ACN since all three useful CCl$_4$ transitions were obscured by ACN transitions similar frequencies.

The data in Fig. 19 show that the build-up of bath excitation, as measured by CCl$_4$, is complete in about 20 ps in methanol, in about 60 ps in benzene, and in about 150 ps in NM. By comparing Fig. 19 to Fig. 16, we can compare the behavior of the doorway vibrations in benzene and NM to the behavior of CCl$_4$ vibrations at about the same frequency in solution with benzene or NM. The doorway vibration and the CCl$_4$ data in a particular solution both attain a stable plateau value at about the same time, which indicates the time when VC has ceased and all vibrational levels have come to equilibrium at the new temperature. The doorway vibration

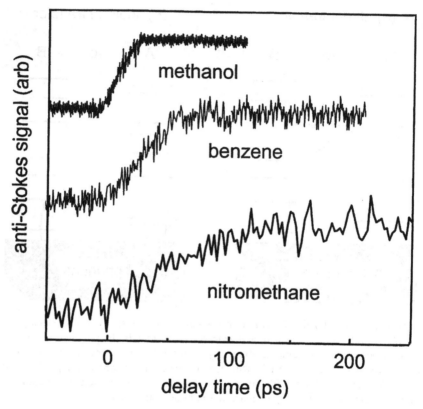

Figure 19 Build-up of bath excitation, monitored via CCl₄ vibrations, in methanol, benzene, and nitromethane after C–H stretch excitation.

and CCl₄ transients differ substantially at short times. At short times, the fast decay of the C–H stretch in benzene and NM generates vibrational population in the doorway vibrations but does not generate much bath excitation. The CCl₄ data for these systems shows that the initial C–H stretch decay is primarily an intramolecular energy redistribution process and that it is the subsequent decay of daughter processes that generates most of the bath excitation.

E. Fermi Resonance and Overtones

A sizable literature exists describing how C–H stretch decay in polyatomic liquids usually takes a few ps (34) and occurs by exciting daughter C–H bending vibrations (22,33,34). Since C–H bends have about one-half the energy of the stretch, two quite different C–H stretch decay processes are possible, depending on whether the daughter excitation is the nearly isoenergetic $v = 2$ first overtone of the bend or the $v = 1$ bending fundamental. A common occurrence in molecules with methyl groups is degeneracy between the fundamental C–H stretch and the first overtone of the C–H bend, that is, Fermi resonance.

Broadly speaking, there are two ways of picturing C–H stretch decay with Fermi resonance (Fig. 20a,b). One possibility (Fig. 20a) is that the

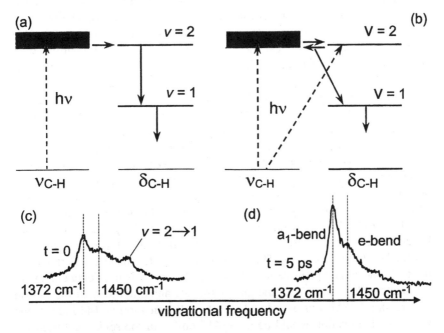

Figure 20 Fermi resonance and overtone pumping in acetonitrile (ACN). (a) The laser might pump a pure C–H stretch, which decays into the degenerate first overtone of the C–H bend. (b) The laser might pump a coupled bend-stretch state, which decays into lower energy levels. (c) $t = 0$ spectrum in the C–H stretching region of ACN after 3000 cm^{-1} pumping. The two C–H bending fundamentals are at 1372 and 1440 cm^{-1}. The new peak at ~1500 cm^{-1} is due to the C–H bending overtone transition $v = 2 \rightarrow 1$, which is shifted out of the region between the two fundamentals by pumping the higher energy tail of the bend overtone. (d) The bend overtone decays along with the C–H stretch, indicating that scheme (b) is correct. (From Ref. 96.)

laser prepares a nearly pure C–H stretch, whose decay populates the nearly isoenergetic C–H bend overtone. Overtone decay would subsequently populate lower energy vibrations such as the C–H bend fundamentals. Another is that the pulse pumps a coupled C–H stretch and C–H bend overtone state (Fig. 20b), which goes on to decay by populating lower energy vibrations. In the former case, the bend overtone would build up as the C–H stretch decayed. In the latter case, C–H stretch and bend excitations would decay together.

The problem in investigating these two mechanisms is spectroscopically distinguishing the C–H bend overtone ($\nu = 2$) from the fundamental ($\nu = 1$). The Raman cross section of the two-quantum $\nu = 2 \to 0$ transition is orders of magnitude smaller than one-quantum transitions. The overtone is more easily detected via its $\nu = 2 \to 1$ transition. Unfortunately that transition is likely to be near the fundamental $\nu = 1 \to 0$ transition. With the relatively low spectral resolution of ultrafast laser systems, it is often impossible to distinguish the two transitions. In that case, overtone excitation appears as excitation of the fundamental with twice the amplitude (44), since the Raman cross section for the $\nu = 2 \to 1$ transition is twice as large as the cross section for the $\nu = 1 \to 0$ transition.

In the specific example of ACN (46) (point group C_{3v}), there is one C–H stretch and one C–H bend of a_1-symmetry and a pair of doubly degenerate stretches and bends of e-symmetry (84). In Fig. 5, the ACN Raman spectrum in the C–H bending region can be seen. The spectrum consists of a sharper a_1-symmetry bend 1372 cm^{-1} and a broader e-symmetry bend at 1440 cm^{-1}. The e-symmetry bend is broadened by Fermi resonance because the e-bend overtones ($2 \times \sim 1440$ cm^{-1}) are degenerate with the e-stretches (~ 3000 cm^{-1}). The a_1-bend and stretch have no Fermi resonance because the bend overtone ($2 \times \sim 1372$ cm^{-1}) is not degenerate with the stretch (2943 cm^{-1}). In the gas phase, the anharmonicity of the e-bend is ~ 25 cm^{-1} [85]. For an e-bend $\nu = 1 \to 0$ transition at 1440 cm^{-1}, the $\nu = 2 \to 1$ transition would be at ~ 1415 cm^{-1}, in the region between the a_1- and e-bend fundamentals, and it would be very difficult to see.

It would help if the e-bend overtone could be frequency-shifted away from the fundamentals. That would be possible if the e-bend were inhomogeneously broadened (see Section IV.G.). Then we could pump the C–H stretch around 3000 cm^{-1} and excite only bend overtones in the ~ 1500 cm^{-1} region, away from the region between the two fundamentals. Figure 20c,d shows the results of pumping ACN at 3000 cm^{-1}. At $t = 0$, a new peak is observed in the C–H bend spectrum near 1500 cm^{-1}, whose spectral width is about the same as the apparatus-limited bandwidth. This peak cannot be an SFG artifact. It is attributed to bend overtone transitions of a subset of ACN molecules with bending vibrations at the higher end of the spectral range. The overtone transition (Fig. 20c) is observed only while C–H stretch excitations are present (near $t = 0$), and the overtone decays along with the C–H stretch. These results are consistent with the picture of a mixed C–H stretch and bend overtone state, which decays into bend fundamentals, as indicated in Fig. 20b.

F. Multiple Vibrational Excitations

It is interesting to study molecules where two or more vibrational excitations are simultaneously present. Since excited vibrations interact via cubic, quartic, or even higher order anharmonic coupling, a complicated nonlinear interaction can be expected, which cannot be understood solely by studying molecules with only a single vibrational excitation. This nonlinear interaction will play a role in the exothermic reaction dynamics of polyatomic molecules, since unless the reaction proceeds solely along a simple one-dimensional reaction coordinate, nascent product molecules will be created with several vibrational excitations. By using the pump pulse to excite a combination band (47), a pair of interacting vibrational excitations can be produced on the same molecule.

In the IR absorption spectrum of ACN (Fig. 5) there is a small relatively sharp transition at ~ 3150 cm^{-1}. This transition has previously been assigned as a combination of C–C stretch (918 cm^{-1}) and C≡N(2253 cm^{-1}) stretch (84). Since the combination transition overlaps the higher energy tail of the C–H stretch fundamental, pumping at this frequency is expected to produce vibrationally excited molecules where all three vibrations interact. The excitations may be viewed as weakly interacting independent vibrations (47). The weak interaction is assured since the anharmonicity is relatively small (<0.3% of the vibrational frequency). Any coherent vibrational states produced by the pump pulse will lose coherence very rapidly, since the dephasing time constant ($T_2 \approx 0.5$ ps) (51) is much faster than the VER lifetime.

Some results are shown (47) in Fig. 21. The left-hand column of Fig. 21 is a reminder of how VER occurs in ACN with only the C–H stretch fundamental (~ 3000 cm^{-1}) pumped. The C–H stretch decays in ~ 5 ps, and only $\sim 2\%$ of this energy populates the 2153 cm^{-1} C≡N stretch, which has an 80 ps lifetime. The 918 cm^{-1} C–C stretch rises in ~ 30 ps because it is populated by the pathway C–H stretch $\to$ C–H bend $\to$ C–C stretch. Subsequent C–C stretch decay occurs with an ~ 45 ps time constant. With combination band pumping (right-hand column of Fig. 21), the C–H stretch

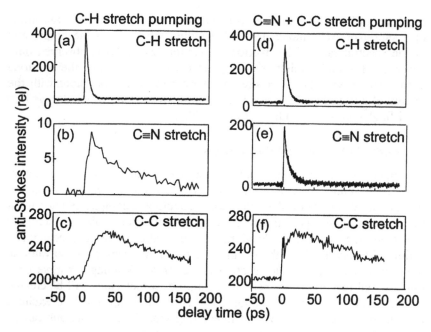

Figure 21 Anti-Stokes transients for acetonitrile (ACN) with C–H stretch pumping (left) and with combination band (C≡N + C–C stretch) pumping (right). With combination band pumping, population builds up instantaneously in the C≡N stretch and C–C stretch, the decay of the C≡N stretch is about 10 times faster, and a population oscillation is seen in the C–C stretch. (From Ref. 47.)

decay is unchanged. A much larger build-up is seen in the C≡N stretch, whose lifetime dramatically drops from 80 to 5 ps. The C–C stretch shows a population oscillation. Its population rises instantaneously, decays a bit over 5 ps, rises again over ~30 ps, and then decays with an ~45 ps time constant.

Since the combination band anharmonicity (<10 cm^{-1}) (84) is less than our spectral resolution (~60 cm^{-1} in this measurement), excitation of the combination C–C stretch and C≡N stretch is seen as excitation of both fundamentals (44). Pumping the combination band causes an instantaneous jump in the populations of the C–H stretch, C–C stretch, and C≡N stretch. The directly pumped C≡N stretch excitation rises to a level about 20 times greater than when it is indirectly populated with C–H stretch pumping. A C≡N stretch and a C–C stretch on the same molecule can annihilate each other via cubic anharmonic coupling to create C–H stretch excitations (47).

This is an up-conversion process (57). The C≡N stretch and C–C stretch excitations are in fast equilibrium with C–H stretch excitations, but the C–H stretch decays with a 5 ps time constant. That explains the C≡N stretch decay in 5 ps. The population oscillation in the C–C stretch data occurs as follows. First we see C–C stretch excitation pumped by the laser. This is actually a sign of combination band excitation (C≡N + C–C stretch). Then the C–C stretch starts to be annihilated by the ~5 ps up-conversion and C–H stretch decay process. C–H stretch decay produces C–H bending excitations, which subsequently repopulate the C–C stretch, causing the second rise in the C–C stretch transient (47).

These results illustrate some of the very interesting phenomena that can be observed with combination band pumping. The pump pulses create a large population at 2253 and 917 cm^{-1}, which would ordinarily require intense ultrashort pulses at 4.43 and 10.9 μm — hard to do with today's technology. With combination band pumping, the lifetime of the C≡N stretch was reduced from its normal value of 80 ps to about 5 ps because C≡N stretches were annihilated by C–C stretches on the same molecule. The C–C stretch population was caused to undergo an oscillation. These new phenomena suggest some of the complicated behavior that might result when a chemical reaction releases enough chemical energy to excite several different vibrations on the same molecule.

G. Spectral Evolution in Associated Liquids

When a vibrational transition is inhomogeneously broadened, optical line-shape studies provide no dynamical information (78). A variety of nonlinear and coherent techniques, collectively called line-narrowing techniques, have been developed to study the dynamics underneath an inhomogeneously broadened transition (78). These include hole burning, fluorescence line-narrowing, photon and vibrational echoes, etc. The IR-Raman experiment sees the excited vibrational population during the probe pulse, so it is closely analogous to the fluorescence line-narrowing (FLN) technique with time resolution (85). In time-resolved FLN, a narrow-band laser is used to excite a selected isochromatic packet of an inhomogeneously broadened transition. Fluorescent emission from the excited isochromat is time resolved and spectrally dispersed. The emission is at first narrower than the inhomogeneous width, but it may broaden or otherwise evolve with time due to the system's dynamics (85).

Neat liquid methanol is an associated liquid characterized by an extended hydrogen bonding network (86,87). Much of the width of the broad O–H stretching transition is due to a distribution of hydrogen-bonded environments. There is a direct inverse correlation between the strength of the hydrogen bond and the O–H stretching frequency: the lowest frequency O–H stretch is associated with the strongest hydrogen bond and vice versa (86,87). Pumping the O–H stretch of associated alcohol oligomers is known to result in a vibrational predissociation process which breaks the hydrogen bond (88–91), so the transition from $\nu = 1 \to 0$ is accompanied by hydrogen bond cleavage. Vibrational predissociation in associated alcohols reduces the vibrational lifetime from tens or more picoseconds to ~1 ps (88–92).

Figure 22 shows some data on methanol pumped in the O–H stretching region by a 35 cm^{-1} wide pulse at 3400 cm^{-1}. When the pump and probe pulses are time coincident, a coherent artifact is observed at an anti-Stokes shift of 3400 cm^{-1}. By about 2 ps this artifact has vanished

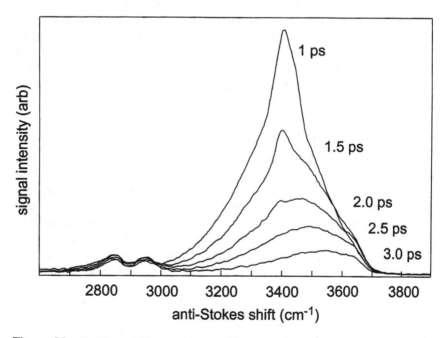

Figure 22 O–H stretching region of neat methanol with 3400 cm^{-1} pumping. The peak at 3400 cm^{-1} is a coherent artifact. After the artifact has disappeared, the spectrum of O–H stretching excitations decays and the peak of the OH stretching spectrum shifts to higher energy. Methanol molecules with higher frequency O–H stretching transitions and weaker hydrogen bonding decay more slowly. (From L. K. Iwaki et al., unpublished.)

and the spectrum of the remaining methanol excitations can easily be seen. The O–H stretching population decays in just a few picoseconds. During this short lifetime, the O–H spectrum undergoes a time-dependent change in shape, with the spectral peak continuously moving toward the region of higher energy. A similar motion toward higher O–H stretching frequency is seen in Fig. 11. There is no noticeable spectral evolution in the daughter C–H stretch excitations, between 2750 and 3000 cm^{-1}, which are excited by O–H stretch decay. The shape change in the O–H spectrum is attributed to frequency-dependent vibrational relaxation rates. Decay of vibrational excitations at the higher frequency end of the spectrum appears to be slower than at the lower frequency end. Therefore, methanol molecules with the weakest hydrogen bonds have the longest VER lifetimes.

This effect is understood by considering the dispersal of the $\sim$3400 cm^{-1} of energy liberated in the $v = 1 \rightarrow 0$ transition. Figure 22 shows that very little of the O–H stretching excitation is transferred to the nearby C–H stretching region. Instead, the energy is used to populate fundamental excitations of vibrations in the 1000–1500 cm^{-1} range, such as O–H bending, C–H bending, and C–O stretching excitations (L. K. Iwaki et al., unpublished). Exciting these vibrations requires the simultaneous emission of a large number of phonons equal to 1000–2000 cm^{-1}. This would be a high-order inefficient multiphonon process. Populating these vibrations while simultaneously breaking a hydrogen bond is a much faster and more efficient process, because breaking a hydrogen bond in methanol releases 1000–2000 cm^{-1} of energy (87).

V. SUMMARY AND CONCLUSIONS

The IR-Raman technique has been used for over 20 years, but only in the last few years has suitable laser instrumentation been developed to monitor the flow of vibrational energy throughout polyatomic molecules in liquids. The new results are remarkable. For example, although it had been known since the earliest transient spontaneous Raman experiments that C–H and O–H stretch decay takes a few picoseconds, now we can see that the VC process initiated by pumping higher frequency vibrations can last for more than 150 ps (e.g., ACN in Fig. 16). New results also show us the need to move beyond the vibrational cascade picture of VC. A quite new development is the ability to study the interactions between vibrations on the same molecule via combination band pumping. Although experiments so far have been mainly limited to polyatomic liquids consisting of relatively small molecules, the CCl_4 results suggest a new way of studying VC in larger molecules — or even macromolecules, molecular aggregates, or nanoparticles — by monitoring the flow of energy from the large molecules to CCl_4 or another molecular probe in the surroundings.

The IR-Raman technique is one of several new ultrafast two-dimensional vibrational spectroscopies (78,93–95), which have begun to revolutionize our understanding of condensed phase vibrational dynamics. Although IR-Raman methods are complementary to IR pump-probe measurements, especially two-color pump-probe, a stellar advantage of the IR-Raman technique is the ability to simultaneously obtaining the entire vibrational spectrum with an optical array, as illustrated in Figs. 9, 11, 20, and 22. The big disadvantage of the IR-Raman technique is the weak Raman signal, which is easily overwhelmed by even tiny amounts of optical background. However, the IR-Raman technique allows us to study the time evolution of vibrational populations even at lower frequencies inaccessible to today's ultrashort IR lasers and to study the time evolution of the optical lineshape with unprecedented accuracy.

ACKNOWLEDGMENTS

This work was supported by Air Force Office of Scientific Research contract F49620-97-1-0056, U.S. Army Research Office contract DAAH04-96-1-0038, and National Science Foundation grant DMR-9714843. L.K.I. and S.T.R. acknowledge support from an AASERT fellowship, DAAG55-98-1-0191, from the Army Research Office.

REFERENCES

1. Spanner K, Laubereau A, Kaiser W. Vibrational energy redistribution of polyatomic molecules in liquids after ultrashort infrared excitation. Chem Phys Lett 1976; 44:88–92.
2. Laubereau A, Kaiser W. Vibrational dynamics of liquids and solids investigated by picosecond light pulses. Rev Mod Phys 1978; 50:607–665.
3. Dlott DD. Dynamics of molecular crystal vibrations. In: Yen W, ed. Laser Spectroscopy of Solids II. Berlin: Springer-Verlag, 1988:167–200.
4. Hill JR, Dlott DD. A model for ultrafast vibrational cooling in molecular crystals. J Chem Phys 1988; 89:830–841.
5. Hill JR, Dlott DD. Theory of vibrational cooling in molecular crystals: application to crystalline naphthalene. J Chem Phys 1988; 89:842–858.
6. Flynn GW, Parmenter CS, Wodtke AM. Vibrational energy transfer. J Phys Chem 1996; 100:12817–12838.

7. Chesnoy J, Gale GM. Vibrational energy relaxation in liquids. Ann Phys Fr 1984; 9:893–949.

8. Schroeder J, Troe J, Vöhringer P. Photoisomerization of trans-stilbene in compressed solvents: Kramers-turnover and solvent induced barrier shift. Z Phys Chem 1995; 188:287–306.

9. Schwarzer D, Troe J, Votsmeier M, Zerezke M. Collisional deactivation of vibrationally highly excited azulene in compressed liquids and supercritical fluids. J Chem Phys 1996; 105:3121–3131.

10. Schwarzer D, Troe J, Zerezke M. The role of local density in the collisional deactivation of vibrationally highly excited azulene in supercritical fluids. J Chem Phys 1997; 107:8380–8390.

11. Cherayil BJ, Fayer MD. Vibrational relaxation in supercritical fluids near the critical point. J Chem Phys 1997; 107:7642–7650.

12. Myers DJ, Urdahl RS, Cherayil BJ, Fayer MD. Temperature dependence of vibrational lifetimes at the critical density in supercritical mixtures. J Phys Chem 1997; 107:9741–9748.

13. Myers DJ, Chen S, Shigeiwa M, Cherayil BJ, Fayer MD. Temperature dependent vibrational lifetimes in supercritical fluids near the critical point. J Chem Phys 1998; 109:5971–5979.

14. Harris CB, Smith DE, Russell DJ. Vibrational relaxation of diatomic molecules in liquids. Chem Rev 1990; 90:481–488.

15. Paige ME, Harris CB. A generic test of gas phase isolated binary collision theories for vibrational relaxation at liquid state densities based on the rescaling properties of collision frequencies. J Chem Phys 1990; 93:3712–3713.

16. Paige ME, Harris CB. Ultrafast studies of chemical reactions in liquids: validity of gas phase vibrational relaxation models and density dependence of bound electronic state lifetimes. Chem Phys 1990; 149:37–62.

17. Russell DJ, Harris CB. Vibrational relaxation in simple fluids: a comparison of experimental results to the predictions of isolated binary collision theory. Chem Phys 1994; 183:325–333.

18. Stratt RM, Maroncelli M. Nonreactive dynamics in solution: the emerging molecular view of solvation dynamics and vibrational relaxation. J Phys Chem 1996; 100:12981–12996.

19. Voth GA, Hochstrasser RM. Transition state dynamics and relaxation processes in solutions: a frontier of physical chemistry. J Phys Chem 1996; 100:13034–13049.

20. Brueck SRJ, Osgood Jr. RM. Vibrational energy relaxation in liquid $N_2 - CO$ mixtures. Chem Phys Lett 1976; 39:568–572.

21. Hoffman GJ, Imre DG, Zadoyan R, Schwentner N, Apkarian VA. Relaxation dynamics in the B(1/2) and C(3/2) charge transfer states of XeF in solid Ar. J Chem Phys 1993; 98:9233–9240.

22. Oxtoby DW. Vibrational population relaxation in liquids. In: Jortner J, Levine RD, Rice SA, eds. Photoselective Chemistry Part 2. Vol. 47. New York: Wiley, 1981:487–519.

23. Vergeles M, Szamel G. A theory for dynamic friction on a molecular bond. J Chem Phys 1999; 110:6827–6835.
24. Grote RF, Hynes JT. The stable states picture of chemical reactions. II. Rate constants for condensed and gas phase reaction models. J Chem Phys 1980; 73:2715–2732.
25. Grote RF, Hynes JT. Reactive modes in condensed phase reactions. J Chem Phys 1981; 74:4465–4475.
26. Kramers HA. Brownian motion in a field of force and the diffusion model of chemical reactions. Physica 1940; VII:284–304.
27. Lee M, Holtom GR, Hochstrasser RM. Observation of the Kramers turnover region in the isomerism of trans-stilbene in fluid ethane. Chem Phys Lett 1985; 118:359–363.
28. Fleming G, Hänggi P. Activated Barrier Crossing. River Edge, NJ: World Scientific, 1993.
29. Nikowa L, Schwarzer D, Troe J. Transient hot UV spectra in the collisional deactivation of highly excited trans-stilbene in liquid solvents. Chem Phys Lett 1995; 233:303–308.
30. Hasha DL, Eguchi T, Jonas J. Dynamical effects on conformational isomerization of cyclohexane. J Chem Phys 1981; 75:1571–1573.
31. Hasha DL, Eguchi T, Jonas J. High-pressure NMR study of dynamical effects on conformational isomerization of cyclohexane. J Am Chem Soc 1982; 104:2290–2296.
32. Laubereau A, von der Linde D, Kaiser W. Direct measurement of the vibrational lifetimes of molecules in liquids. Phys Rev Lett 1972; 28:1162–1165.
33. Alfano RR, Shapiro SL. Establishment of a molecular-vibration decay route in a liquid. Phys Rev Lett 1972; 29:1655–1658.
34. Seilmeier A, Kaiser W. Ultrashort intramolecular and intermolecular vibrational energy transfer of polyatomic molecules in liquids. In: Kaiser W, ed. Ultrashort Laser Pulses and Applications. Vol. 60. Berlin: Springer-Verlag, 1988: 279–315.
35. Oxtoby DW. Vibrational relaxation in liquids. Ann Rev Phys Chem 1981; 32:77–101.
36. Laubereau A, Greiter L, Kaiser W. Intense tunable picosecond pulses in the infrared. Appl Phys Lett 1974; 25:87–89.
37. Zinth W, Kolmeder C, Benna B, Irgens-Defregger A, Fischer SF, Kaiser W. Fast and exceptionally slow vibrational energy transfer in acetylene and phenylacetylene in solution. J Chem Phys 1983; 78:3916–3921.
38. Ambroseo JR, Hochstrasser RM. Pathways of relaxation of the N-H stretching vibration of pyrrole in liquids. J Chem Phys 1988; 89:5956–5957.
39. Gottfried NH, Kaiser W. Redistribution of vibrational energy in naphthalene and anthracene studied in liquid solution. Chem Phys Lett 1983; 101:331–336.
40. Tokmakoff A, Sauter B, Kwok AS, Fayer MD. Phonon-induced scattering between vibrations and multiphoton vibrational up-pumping in liquid solution. Chem Phys Lett 1994; 221:412–418.

41. Chen S, Hong X, Hill JR, Dlott DD. Ultrafast energy transfer in high explosives: vibrational cooling. J Phys Chem 1995; 99:4525–4530.

42. Hong X, Chen S, Dlott DD. Ultrafast mode-specific intermolecular vibrational energy transfer to liquid nitromethane. J Phys Chem 1995; 99:9102–9109.

43. Hofmann M, Zürl R, Graener H. Polarization effects in time resolved incoherent anti-Stokes Raman spectroscopy. J Chem Phys 1996; 105:6141–6146.

44. Graener H, Zürl R, Hofmann M. Vibrational relaxation of liquid chloroform. J Phys Chem 1997; 101:1745–1749.

45. Deàk JC, Iwaki LK, Dlott DD. High power picosecond mid-infrared optical parametric amplifier for infrared-Raman spectroscopy. Opt Lett 1997; 22:1796–1798.

46. Deàk JC, Iwaki LK, Dlott DD. Vibrational energy relaxation of polyatomic molecules in liquids: acetonitrile. J Phys Chem 1998; 102:8193–8201.

47. Deàk JC, Iwaki LK, Dlott DD. When vibrations interact: ultrafast energy relaxation of vibrational pairs in polyatomic liquids. Chem Phys Lett 1998; 293:405–411.

48. Deàk JC, Iwaki LK, Dlott DD. Vibrational energy redistribution in polyatomic liquids: ultrafast IR-Raman spectroscopy of nitromethane. J Phys Chem A 103:971–979.

49. Iwaki LK, Deàk JC, Rhea ST, Dlott DD. Vibrational energy redistribution in liquid benzene. Chem Phys Lett 1999; 303:176–182.

50. Iwaki L, Dlott DD. Vibrational energy transfer in condensed phases. In: Moore JH, Spencer ND, eds. Encyclopedia of Chemical Physics and Physical Chemistry. Philadelphia: Institute of Physics, 2000.

51. Berg M, Vanden Bout DA. Ultrafast Raman echo measurements of vibrational dephasing and the nature of solvent-solute interactions. Acc Chem Res 1997; 30:65–71.

52. Everitt KF, Egorov SA, Skinner JL. Vibrational energy relaxation in liquid oxygen. Chem Phys 1998; 235:115–122.

53. Velsko S, Oxtoby DW. Vibrational energy relaxation in liquids. J Chem Phys 1980; 72:2260–2263.

54. Moore P, Tokmakoff A, Keyes T, Fayer MD. The low frequency density of states and vibrational population dynamics of polyatomic molecules in liquids. J Chem Phys 1995; 103:3325–3334.

55. Seeley G, Keyes T. Normal-mode analysis of liquid-state dynamics. J Chem Phys 1989; 91:5581–5586.

56. Xu B-C, Stratt RM. Liquid theory for band structure in a liquid. II. p Orbitals and phonons. J Chem Phys 1990; 92:1923–1935.

57. Kenkre VM, Tokmakoff A, Fayer MD. Theory of vibrational relaxation of polyatomic molecules in liquids. J Chem Phys 1994; 101:10618–10629.

58. Goodyear G, Stratt RM. The short-time intramolecular dynamics of solutes in liquids. I. An instantaneous-normal-mode theory for friction. J Chem Phys 1996; 105:10050–10071.

59. Goodyear G, Larsen RE, Stratt RM. Molecular origin of friction in liquids. Phys Rev Lett 1996; 76:243–246.

60. Bader JS, Berne BJ. Quantum and classical rates for classical simulations. J Chem Phys 1994; 100:8359–8366.

61. Egelstaff PA. Neutron scattering studies of liquid diffusion. Adv Phys 1962; 11:203–232.

62. Califano S, Schettino V, Neto N. Lattice Dynamics of Molecular Crystals. Berlin: Springer-Verlag, 1981.

63. Dlott DD, Fayer MD. Shocked molecular solids: vibrational up pumping, defect hot spot formation, and the onset of chemistry. J Chem Phys 1990; 92:3798–3812.

64. Nitzan A, Jortner J. Vibrational relaxation of a molecule in a dense medium. Molec Phys 1973; 25:713–734.

65. Bokhenkov ÉL, Rodina EM, Sheka EF, Natkaniec I. Inelastic incoherent neutron scattering spectra at different temperatures and computer experiment for external phonon modes of naphthalene crystals. Phys Status Solidi B 1978; 85:331–342.

66. Belushkin AV, Bokhenkov ÉL, Kolesnkiov AI, Natkaniec I, Righini R, Sheka EF. Spectrum of external phonons of a naphthalene crystal at 5K. Sov. Phys Solid State 1991; 23:1529–1533.

67. Hill JR, Chronister EL, Chang T-C, Kim H, Postlewaite JC, Dlott DD. Vibrational relaxation and vibrational cooling in low temperature molecular crystals. J Chem Phys 1988; 88:949–967.

68. Backus S, Durfee III CG, Murnane MM, Kapteyn HC. High power ultrafast lasers. Rev Sci Instrum 1998; 69:1207–1223.

69. Rudd JV, Korn G, Kane S, Squier J, Mourou G, Bado P. Chirped-pulse amplification of 55 fs pulses at a 1 kHz repetition rate in a $Ti:Al_2O_3$ regenerative amplifier. Opt Lett 1993; 18:2044–2046.

70. Raoult F, Boscheron ACL, Husson D, Sauteret C, Modena A, Malka V, Dorchies F, Migus A. Efficient generation of narrow-bandwidth picosecond pulses by frequency doubling of femtosecond chirped pulses. 1998; Opt Lett 23:1117–1119.

71. Gabl EF, Walker DR, Pang Y. Regenerative amplifier incorporating a spectral filter within the resonant cavity. USA Patent 5,572,358 (1996).

72. Zhang J-Y, Huang JY, Shen YR. Optical Parametric Generation and Amplification. Harwood Academic Publishers, 1995.

73. Dierlein JD, Vanherzeele H, Ballman AA. Linear and nonlinear optical properties of flux-grown $KTiOAsO_4$. Appl Phys Lett 1989; 54:783–785.

74. Petrov V, Noack F. Mid-infrared femtosecond optical parametric amplification in potassium niobate. Opt Lett 1996; 19:1576–1578.

75. Gragson DE, Alavi DS, Richmond GL. Tunable picosecond infrared laser system based on parametric amplification in KTP with a Ti:sapphire amplifier. Opt Lett 1995; 20:1991–1993.

76. Schrader B. Raman/Infrared Atlas of Organic Compounds. Weinheim: VCH, 1989.
77. Alfano RR. The Supercontinuum Laser Source. New York: Springer-Verlag, 1989.
78. Mukamel S. Principles of Nonlinear Optical Spectroscopy. New York: Oxford University Press, 1995.
79. Shen YR. Surfaces probed by nonlinear optics. Surface Sci 1994; 299/300:551–562.
80. Shen YR. Surface properties probed by second-harmonic and sum-frequency generation. Nature 1989; 337:519–525.
81. Fendt A, Fischer SF, Kaiser W. Vibrational lifetime and Fermi resonance in polyatomic molecules. Chem Phys 1981; 57:55–64.
82. Nitzan A, Mukamel S, Jortner J. Energy gap law for vibrational relaxation of a molecule in a dense medium. J Chem Phys 1975; 63:200–207.
83. Chandler DW, Ewing GE. Transfer and storage of vibrational energy in liquids: liquid nitrogen and its solutions with carbon monoxide. J Chem Phys 1980; 73:4904–4913.
84. Herzberg G. Molecular Spectra and Molecular Structure II. Infrared and Raman Spectra of Polyatomic Molecules. New York: Van Nostrand Reinhold, 1945.
85. Bai YS, Fayer MD. Time scales and optical dephasing measurements: Investigation of dynamics in complex systems. Phys Rev B 1989; 39:11066–11084.
86. Matsumoto M, Gubbins KE. Hydrogen bonding in liquid methanol. J Chem Phys 1990; 93:1981–1994.
87. Liddel U, Becker ED. Infra-red spectroscopic studies of hydrogen bonding in methanol, ethanol, and t-butanol. Spectrochem Acta 1957; 10:70–84.
88. Graener H, Ye TQ, Laubereau A. Ultrafast vibrational predissociation of hydrogen bonds: mode selective infrared photochemistry in liquids. J Chem Phys 1989; 91:1043–1046.
89. Graener H, Ye TQ, Laubereau A. Ultrafast dynamics of hydrogen bonds directly observed by time-resolved infrared spectroscopy. J Chem Phys 1989; 90:3413–3416.
90. Laenen R, Rauscher C. Time-resolved infrared spectroscopy of ethanol monomers in liquid solution: molecular reorientation and energy relaxation times. Chem Phys Lett 1997; 274:63–70.
91. Laenen R, Rauscher C, Laubereau A. Vibrational energy redistribution of ethanol oligomers and dissociation of hydrogen bonds after ultrafast infrared excitation. Chem Phys Lett 1998; 283:7–14.
92. Woutersen S, Emmerichs U, Bakker HJ. A femtosecond midinfrared pump-probe study of hydrogen-bonding in ethanol. J Chem Phys 1997; 107:1483–1490.
93. Zhao W, Wright JC. Doubly vibrationally enhanced four-wave mixing-the optical analogue to 2D NMR. Science, Phys Rev Lett 2000; 84:1411–1414.

94. Tokmakoff A, Fleming GR. Two-dimensional Raman spectroscopy of the intermolecular modes of liquid CS_2. J Chem Phys 1997; 106:2569–2582.
95. Rector KD, Fayer MD, Engholm JR, Crosson E, Smith TI, Schwettman HA. T_2 selective scanning vibrational echo spectroscopy. Chem Phys Lett 1999; 305:51–56.
96. Deak JC, Iwaki LK, Rhea ST, Dlott DD. Ultrafast infrared-Raman studies of vibrational relaxation in polyatomic liquids. J Raman Spectrosc 2000; 31:263–274.

14

Coulomb Force and Intramolecular Energy Flow Effects for Vibrational Energy Transfer for Small Molecules in Polar Solvents

James T. Hynes
University of Colorado, Boulder, Colorado, and Ecole Normale Supérieure, Paris, France

Rossend Rey
Universitat Politècnica de Catalunya, Barcelona, Spain

I. INTRODUCTION

Molecular vibrational energy transfer (VET) in solution is a phenomenon of long standing and — as the present volume attests — continuing interest and importance (1–11). In this chapter, we discuss two aspects of solution phase VET: (1) the role of *electrostatic* Coulomb forces and (2) *polyatomic* VET, i.e., the involvement of noninitially excited vibrational modes in a polyatomic in the vibrational relaxation. While perhaps not lying currently in the mainstream of theoretical effort in solution phase VET — which is often focused on diatomics immersed in simple fluids — these topics represent rivulets that we anticipate will soon emerge in full flood, as more complex molecular solutes and solutions come under increasing modern experimental and theoretical scrutiny.

Both of these topics are readily motivated, since obviously most vibrating bonds or modes are polar to some degree and most common molecular solvents are polar, and most molecules are polyatomic. (But we should hasten to add that in this chapter, "polyatomic" will usually turn out to mean triatomic, a first foray into the area which opens a window on the possibilities but which remains within the reach of current theoretical and (detailed) experimental probes.) However, there are further aspects to the motivation, here illustrated by just two examples. First, to the degree that the Coulomb force issue is important, the isolated binary collision ideas so fruitful in simple systems (1,3,4,6) will need to be replaced, since long-ranged solute-solvent interactions will certainly differ in character from short-range binary collisions. Second, the polyatomic aspect brings in the question — pervasive in chemical dynamics — of the competition of intramolecular versus intermolecular energy transfer. Indeed, an alternate title for this topic could be intramolecular vibrational redistribution (IVR) in solution, although we will mainly focus on this topic at low excitation energies, rather than the higher energies more commonly associated with IVR. And while chemical reactions in solution are beyond the scope of this chapter, contemplation of the microscopic level course of a paradigm solution reaction such as the S_N2 reaction between chloride ion and the methyl chloride molecule in water solvent (12) quickly shows that these two VET topics must be comprehended if we are ever to have a detailed molecular level description of the pathway and dynamics of solution reactions. As a further connection to reactions in solution, it is evident that the issues of the present chapter lie at the very heart of the ability to induce, and to induce in any selective way, such reactions (13–15). Finally — but assuredly not the least of motivations — assorted advances in experimental techniques are following the probing of a variety of molecular systems in which the issues of this chapter come to the fore.

It should also be frankly acknowledged here that there are a variety of theoretical challenges associated with these problems that are not highlighted at all in this chapter. These range from formulation questions involving quantum versus classical issues in calculating rates (see, for example, Chapter 16) to the quantum chemical electronic structure issues of solute intramolecular force fields. These and other difficulties certainly impede the theoretical ability to confidently predict VET rates and mechanisms, but not the desire to try.

The outline of this chapter is as follows. In Section II we treat the Coulomb force issue, progressing from a polar molecule in water to the case of the cyanide ion in water, concluding with a brief discussion of the special effects arising when the solute charge distribution is not fixed. In Section III we deal with the polyatomic issue, for which much less information is currently available. Section IV attempts to identify a few important open questions and avenues for further research.

Finally, this short chapter makes no pretense to completeness and is restricted to a focus on a few systems where detailed results are available, largely drawn from our own work, with some brief commentary on related work.

II. COULOMBIC FORCE EFFECTS ON VET

In most standard discussions of VET for high-frequency diatomics, attention is focused — quite appropriately — on the short-range repulsive forces between the diatomic and the collision partner, since those rapidly spatially varying forces are the most effective in the VET. In this perspective, longer ranged, slowly varying, electrostatic Coulomb forces — such as those between dipolar molecules — would not be effective. Accordingly, with the important exception of long-range resonant or near-resonant VV transfer between polar molecules in the gas phase (1), the role of Coulombic forces has evidently not attracted much attention in neutral VET systems. (A discussion of VET for ions in the gas phase is given in Ref. 16). On the other hand, when a polar molecule is immersed in a polar solvent, the multitude of significant Coulomb interactions present might be expected to have some role to play in VET.

A rather less opaque indication that Coulomb force effects could be significant is the short population relaxation time (T_1) found in pioneering experimental measurements of the CN^- ion in water (17): an extrapolation to infinite dilution indicated a T_1 value of about 25 ps for a vibration of about 2100 cm^{-1}, a time much shorter than for comparable frequency diatomics in, for example, rare gas solvents (8). That this was not a singularity was subsequently shown for several triatomic ions (8,18).

At all events, the role of Coulombic forces for VET in solution was first examined in a molecular dynamics simulation for the ~ 680 cm^{-1} C–Cl vibration of the CH_3Cl molecule (modeled as a diatomic) in water solvent by Whitnell et al. (19). The (classical) relaxation time T_1 was determined both by nonequilibrium simulations and by use of the classical Landau-Teller (LT) formula (1,3,19,20).

$$T_1^{-1} = \frac{1}{\mu k_B T} \int_{-\infty}^{\infty} dt \; e^{i\omega_0 t} \langle \delta F \delta F(t) \rangle \tag{1}$$

in terms of the Fourier component at the oscillator frequency (ω_0) of the equilibrium time correlation function (tcf) of the fluctuating force on the bond, taken fixed in length. We will often just call this the force (power) spectrum. (One could also term it the vibrational frequency–dependent friction, a term we like less, since it is associated with a generalized Langevin equation and Equation (1) has a validity far beyond that of the latter.) As a side note, the demonstration that these two routes were in good agreement was the first explicit simulation demonstration that the classical LT formula gives correct results (for the classical relaxation).

By the artificial simulation device of alternately removing and doubling the assigned point charges on the C and Cl moieties, it was shown that Coulomb effects are decisive: T_1 shrinks from $\sim$100 to $\sim$5 ps as the charges are turned on to their assumed standard value, and this time declines to 1.4 ps as those charges are doubled. This latter behavior is in fact just the behavior expected from a simple scaling of the charges for the amplitude of the fluctuating force in the LT formula. The importance of the Coulomb forces was further established in extensive studies of both the diatomic vibrational phase distributions as well as the surrounding water molecule spatial distributions sampled in the nonequilibrium dynamics. For example, it was shown that the range of the Coulomb forces was critical in allowing the transfer, that water molecules in and beyond the first solvent shell were involved, and that there was significant participation of water molecules in locations *perpendicular* to the C–Cl bond, a feat that would be fairly Herculean for short-ranged non-Coulomb forces. It is also worth pointing out that the shape of the force spectrum in the neighborhood of the CCl frequency is rather different in the absence versus presence of the Coulomb forces.

While a short-range isolated binary collision (IBC) picture, in its most extreme form — with the emphasis on "collision" — is on its face ludicrous for long-range force interactions, the corresponding isolated binary interaction (IBI) idea (5) could be examined. The IBI concept is that it is only the direct interaction with the solute of a single solvent molecule at a time that determines the VET; this is of course also a key ingredient in the IBC picture, but in the IBI there is no *a priori* specification of exactly what the interaction force is. If the Fourier component of the fluctuating force tcf is decomposed as

$$\langle \delta \hat{F} \delta \hat{F}(\omega) \rangle = \sum_i \langle \delta \hat{F}_i \delta \hat{F}_i(\omega) \rangle + \sum_{i \neq j} \langle \delta \hat{F}_i \delta \hat{F}_j(\omega) \rangle \tag{2}$$

where the sums are over the solvent molecules interacting with the solute, the first contribution is a self-contribution — which would be the only term retained in the IBI approximation — while the second cross-correlation term indicates the importance of nonbinary interaction forces for the VET. With the Coulomb forces turned off, the IBI approximation was found to be nearly perfect, while its quality wanes as the charges are increased: for example, the IBI approximation gives a T_1 about two thirds of the full value for the standard CH_3Cl charge case. While it would be hyperbolic to claim this as a "breakdown" of an IBI picture, it seems fair to say that an IBI approach, while valuable, is clearly missing a significant part of the picture. (See also Ref. 21 and Chapter 4.)

Finally, it was shown in various ways that it is the water librational motions that are important in the VET and that these involve coupled water molecular motions, since there is a significant contribution from non-IBI terms here. In view of the remarks above about the shape of the force spectrum itself differing in the absence and presence of the solute charges, and the validity of the IBI perspective in the absence of charges, the implication is that for the hypothetical no charge CCl vibration at the same frequency, the librations would still be important for the VET, but they would involve only pair effects for the VET and would perforce interact significantly more feebly with the mode.

Among the signatures, alluded to above, of the importance of the solvent librations was the interpretation of the deuterium isotope effect by which the simulated VET is slowed, via a shift of the solvent D_2O librational band away from the CH_3Cl frequency, while the H_2O librational band is well overlapped with that frequency. While one might think that this *direction* of isotope effect arising from the librations is obvious for any solute vibration frequency in this frequency range, we need to add a cautionary note. In an LT formula study of the ClO^- ion, of frequency 713 cm^{-1}, in aqueous solution, Lim et al. (22) found a negligible solvent H_2O/D_2O isotope effect, in contrast to the CH_3Cl findings, but in agreement with these authors' experimental results (22). Even further, it seems from Fig. 9 of Lim et al. that the direction of the isotope effect would invert for hypothetical ClO^- molecular ions of lower and higher frequency than the 713 cm^{-1} value! It seems that such behavior arises from the fact that the red shift upon deuteration of the solvent librational band in the force spectrum will only slow the VET if the spectrum is monotonic in frequency in the relevant frequency region — as it is in Ref. 19 — in contrast with Ref. 22, where it is not. In this connection, it should be recalled that it is the dynamical solvent librations occurring in *the presence of the solute* that are germane, and these can certainly be different for different solutes. It should also be borne in mind that not even the equilibrium static structures of neat H_2O and D_2O liquids are the same (23).

Before returning to the main Coulomb force theme, it is important to stress that in LT formula simulations focused on the force on the solute vibrational mode, any energy transfer to implicated solvent modes, e.g., the water librations above, is actually inferred, rather than directly demonstrated. That is to say, such transfer — which is of course just vibration-vibration (VV) transfer — is not directly probed. We return to this issue at the conclusion of this chapter.

Two subsequent simulation studies for low-frequency vibration systems clearly show that the strong and striking dominance of Coulomb force effects found for CH_3Cl in water is by no means so clear-cut (or even true) in low-frequency diatomic systems. In the first of these, related to experiments by the Barbara group (24,25), Benjamin and Whitnell (26) found that for the diatomic I_2^- of frequency ~ 115 cm^{-1}, with a vibrational relaxation time of about 1 ps, the presence of Coulomb forces accelerated the VET in water by about a factor of 4, a noticeable but somewhat muted effect considering that one is comparing an ion to a neutral (with the same frequency). The authors noted the importance of the fact that the short-range non-Coulomb forces themselves are quite efficient at the low I_2^- frequency.

The Coulomb forces are even more subdued in the LT formula simulations by Gnanakaran and Hochstrasser (27) for HgI, of frequency 125 cm^{-1}, in ethanol. At the same frequency, the charges only reduced the relaxation time from 3 to 2 ps. Further, by examining the force spectrum as a function of frequency, it was shown that the Coulomb force contribution at the HgI frequency is negligible compared to the short-range non-Coulomb force contribution. In extended calculations where the spectrum of forces was examined as a function of frequency, the Coulomb force contribution only becomes the uniformly dominant one above about 1200 cm^{-1}, although it becomes transiently dominant in a frequency range near 700 cm^{-1}. Returning to the actual HgI frequency case, the authors noted an indirect Coulomb force effect, in which the Coulomb force brings the polar solvent molecule closer into the diatomic, thereby increasing the repulsive non-Coulomb forces. It is of interest to note that a strong qualitative similarity was observed between the OKE spectrum of pure ethanol and the force spectrum on the HgI bond fixed in the solvent. More such comparisons would help clarify to what degree pure solvent aspects are directly correlated to the motions of the solvent molecules in the presence of the solute that are key for VET.

The indirect Coulomb force effect noted above had also been pointed out in a LT formula simulation study by Bruehl and Hynes (28) for a model AH $\cdots$ B hydrogen-bonded complex in a model polar CH_3Cl solvent for a low-frequency AB vibration of the complex in the range $100-300$ cm^{-1} (since the AH $\cdots$ B complex was not intended to be any particular molecular system, a representative range of frequencies for this and other vibrations was considered). As in both studies above, the non-Coulomb forces dominate the VET in this low-frequency region, but there is nonetheless an important indirect Coulomb effect, tightening the local solvent cage around the solute and thereby inducing a more effective non-Coulomb force for the VET.

From a general point of view, it is mildly ironic that in both the just-described AB vibration and the HgI examples, the *direct* slowly varying Coulomb force impact on the VET is unimportant in this low-frequency region; the reason would appear to be that in this region, a standard perspective is evidently correct, even the $100-300$ cm^{-1} frequency range being sufficiently high that only the repulsive non-Coulomb forces vary rapidly enough to be effective.

The story changed when the higher frequency vibrations of the AH $\cdots$ B complex were examined (28). For the H-bending vibration in the $1000-1700$ cm^{-1} range, the direct Coulomb force was essentially entirely responsible for the VET, a phenomenon in part due to the fact that the light H motion is a dominant component in the bend coordinate, and in the model the H had exclusively Coulomb interaction with the CH$_3$Cl molecules. Finally, for the proton stretch vibration in the $2500-3500$ cm^{-1} range, the situation proved to be rather complex, involving a combination of non-Coulomb, Coulomb, and quite important cross-correlation effects, all shifting as a function of the stretch frequency. This complexity defies a simple summary here, but a few points can be noted. At the highest frequencies ($\sim 3000-3500$ cm^{-1}) the direct Coulomb forces are not important, probably due to the lack of any real possibility of involvement of strongly coupled librational overtones in the model CH$_3$Cl solvent (which might not be the case in, e.g., H$_2$O). Nonetheless, the Coulomb/non-Coulomb force cross correlation — whose physical origin was described (28) — remains quite significant and is in competition with the dominant direct non-Coulomb force effects. In the lower frequency range (~ 2500 cm^{-1}), all the force correlation spectra — non-Coulomb, Coulomb, and cross — are comparable in magnitude.

This model study — which included neither the charge shifting or intramolecular coupling effects of the sort described elsewhere in this chapter — suggests a quite rich variety to be sorted out in more realistic molecular studies. An important general lesson of the hydrogen-bonded complex study is that the impact of Coulombic forces is in general likely to be decidedly *mode-specific*.

Since experimental results were available for the high-frequency ($\sim$2080 cm^{-1}) diatomic CN$^-$ in water (as opposed to CH$_3$Cl) (17), with an estimated T_1 value of some 25 ps, an MD study was undertaken by Rey and Hynes (29) to clarify the role of Coulomb forces for VET in this accesible case. The charge distribution of CN$^-$ in the solvent was modeled by a negative charge on N and a finite dipole located on the C site (30). The equilibrium solvent structure about this ion involved greater solvation number on the N end compared to the C end, a result consistent with some small cluster calculations (31). Since the frequency shift from the vacuum and the anharmonicity in the CN$^-$ bond are both relatively small (29), the static vibrational aspects of the ion are evidently fairly "clean."

Analysis of the bond force power spectrum revealed the rather remarkable feature that the Coulomb force effects dominated the VET, with a small contribution from the Coulomb/non-Coulomb cross correlation. This central result is in stark contrast to the common thought that the slowly varying Coulomb forces could not fail to be ineffective at high solute vibrational frequencies. In this context, and in this system, it is not the case that Coulomb force effects are absolutely large at high frequencies — they are not; rather they simply decline less rapidly than do the non-Coulomb forces [a situation totally reversed in the artificial case of lower (200–300 cm^{-1}) solute frequencies, all other things remaining the same]. The only possible accepting modes in the rigid water molecule solvent of the simulation would be combinations of the water librations, a feature supported in a subsequent and entirely different analysis (32) not addressing the Coulomb force. In a paper essentially confirming the original experimental estimation of T_1 for this system, Hochstrasser and coworkers (33) suggested that a water libration-bend combination band — a band apparent, though weak, in experimental water spectra (34) and absent in most water simulation models, and certainly absent in the rigid model used in Ref. 29 — could play some role in the relaxation, a suggestion supported by some model calculations (32).

The CN$^-$ problem deserves reinvestigation, not only to establish dynamically the involvement of the named solvent molecule combination bands and to confirm in this context the Coulomb force dominance for the VET, but also to explore the possible role in the VET of any counter ions in the CN$^-$ first solvation shell at higher concentrations of the solute (17). The latter issue, which is in principle ubiquitous in ionic solute VET studies, has yet to receive any theoretical attention. Some improvement is also in order on the quantitative side: no available calculation is closer to the experimental result than to within a factor of 2 (29,32).

All the above discussion has focused on the situation in which the electrical charges in the vibrationally relaxing solute are fixed. Interesting and important additional effects for VET can arise when instead those charges vary strongly through a variation of the molecular bond length(s) and/or the surrounding polar solvent molecule configurations, an aspect suggested in Ref. 28. An early indication of the potential importance of such effects arose in the $Cl^- + CH_3Cl \rightarrow ClCH_3 + Cl^-$ S_N2 reaction in water (12,35): the solvent generalized frictional force exerted on the (unstable) antisymmetric stretch reaction coordinate was completely dominated by the "polarization" or "charge shift" (or "flow") force associated with the rapid shift of the negative charge from one chlorine to the other *as a function of the antisymmetric stretch.*

One illustration of the impact of such a charge shifting force in the vibrational relaxation context arises in the I_2^- system. On the experimental side, Lineberger and coworkers (36) found strikingly rapid relaxation of highly vibrationally excited I_2^- in clusters. (The high vibrational energy molecular ions were produced via initial electronic excitation of bound ground state I_2^- and subsequent transition to high vibrational energy portions of the ground electronic state surface.) This rapid relaxation feature was also subsequently seen in corresponding experimental solution studies (25,37). Since the I_2^- molecular anion provides a relatively simple vehicle for describing a charge shifting force and its consequences, we focus our initial discussion on it.

Briefly, I_2^- can be described in a two valence bond (VB) state scenario corresponding to the two charge localized structures $I \cdots I^-$ and $I^- \cdots I$ (25,38). At any given I-I nuclear separation (r) and arrangement of the surrounding solvent molecules, the ground electronic state will be some mixture of these structures. In particular, the resonance electronic coupling mixing these two localized states is a strong, approximately exponential, function of r. At large r, this coupling is small, and I_2^- closely resembles a (weak coupling, electronically nonadiabatic) outer sphere electron transfer reaction system. The two charge localized structures are separately stable; they are separated by an activation barrier in a solvent polarization coordinate associated with the cost to rearrange the solvent molecules between the two asymmetrical arrangements, appropriate to equilibrium solvation of the two electronic structures $I \cdots I^-$ and $I^- \cdots I$. On the other hand, at smaller separations r, the electronic coupling is large, and overcomes the preference of the solvent for the more favorable (from the solvent's point of view) interaction with localized charge distributions. Now the electronic adiabatic ground state is completely delocalized, schematically represented by $I^{-1/2}I^{-1/2}$.

The above considerations indicate that at some intermediate value of r, the I_2^- system on its way to form the completely equilibrated ground state will experience a significant charge flow, as charge localized I_2^- converts to charge-delocalized I_2^-. Associated with this shift is a corresponding force that potentially can be quite effective in the transfer of vibrational energy.

The analytic theory (38) of this force for I_2^-, based on the above ideas, was followed by an MD simulation (39) involving the molecular level transcription and showing that the charge shifting or polarization force could indeed lead to quite rapid vibrational energy transfer from I_2^- to the surrounding solvent, in reasonable agreement with the companion experiments (37) on I_2^- in water. It should be stressed that this VET is of an unusual character little resembling the more stately progress familiar in VET at low excitation energy. In particular, significant energy is transferred on a time scale that is short compared to a vibrational period. Nonetheless, this type of behavior might be prevalent in a number of high-energy systems, (22,40,41),* although the phenomenon might be convoluted with dynamics related to the conversion from the excited electronic state to the ground electronic state (39,42).

In the I_2^- system, the charge shifting force is only important for VET at higher internal energies, for the reasons discussed above. A recent study (43) of the azide ion N_3^- clearly shows the charge shift force at work in low-energy relaxation. In particular, it was concluded that in this system the charge shifting force associated with the antisymmetric stretch ($\sim 2050 \text{ cm}^{-1}$) is dominant in the $v = 1 \rightarrow v = 0$ vibrational relaxation of this ion in water. This effect, together with intramolecular effects to be discussed in Section III, suffices to reduce the calculated relaxation time by about an order of magnitude below a previous simulation, not including those effects (44). The final result of 0.87 ps is in good agreement with experiment (1.2 ps) (18). While the authors did not explicitly invoke any picture involving VB resonance structures for the charge shift force, the simple three VB state picture of the azide ion described by Pauling (45) might give useful insight here.

It seems clear that charge shifting force effects on VET should be quite common in a variety of systems, although whether this effect is

---

* This same large energy transfer effect has been found to occur in simulations of I_2^- in some CO_2 clusters (R. Parson, personal communication).

important at low energies will depend on molecular details of the system. Certainly there is a good chance for such effects in some energy range for any molecule in which there should be important contributions of several VB structures to the electronic structure. There are many significant candidates here, as indicated in our concluding section.

III. SOLUTE INTRAMOLECULAR EFFECTS ON VET

As indicated in Section I, our second major topic is the involvement of intramolecular solute effects in VET, i.e., the involvement of internal energy transfer pathways within the vibrationally excited molecule. This statement, however, is not very precise and could in principle include a wide range of behavior including for example, IVR not even involving interaction with the surrounding solvent molecules (46), or solvent-induced transfer of energy between solute modes that are not themselves intramolecularly coupled (47–51). Recent studies of azulene in several solvents by Heidelbach et al. (51) can be consulted for some important insight on the cornucopia of possibilities. The limit of interest in the present chapter is the one in which those pathways are not open to the excited molecule in the absence of the solvent interactions. Even more specifically, we have in mind the case where the excited vibrational mode in the solute would in the isolated molecule be anharmonically coupled (more properly, "nonlinearly") to another mode within the same molecule, but the energy gap between the two modes is so far off resonance that transfer between the modes would be effectively feeble. On the other hand, in the presence of the interaction with the solvent, transfer to the surroundings can occur to take care of the deficit gap.

A number of examples in the gas phase are known, some of which are summarized in Refs. 1, 52, and 53. In solution, there is much early experimental work on VET for the CH stretch fundamental in assorted molecules, which invoked involvement of a Fermi resonance between the CH stretch and the overtone of the CH bend (54); reviews of this work can be found in Refs. 1, 55, and 56.

Here we focus on the case of the excited OH vibration in the HOD molecule in liquid D_2O solvent, the first solution example, of which we are aware, studied in detail via MD by the present authors (57). This VET was first studied experimentally by Vodopyanov (58) and by Graener et al. (59) in 1991, an effort that has continued to the present (60–62), with, as we will see later, evolving results. For the moment, we quote the initial experimental result of 8 ± 2 ps (59) as a guideline for our discussion. Beyond the obvious general interest in the water system, no simple isolated binary collision picture for VET would be expected to apply to this strongly hydrogen-bonded liquid system, even whether the VET pathway in solution will be the same as in the gas phase.

The basic formulation of this solution phase problem is in the Landau-Teller type language originally generally formulated by Oxtoby (53). As applied to the HOD in D_2O system, the system Hamiltonian can be written in the form $H = H_{HOD} + H_{coupling} + H_{Bath}$. In the semiclassical approach adopted, a quantum representation of the vibrational HOD modes is retained, while the remaining coordinates are described classically. H_{HOD} is the quantum mechanical (anharmonic) Hamiltonian for the vibrational motions of the HOD molecule; it includes the static effect of the solvent and, as a consequence, can be expressed in terms of the normal modes characteristic of the molecule in solution. Such an approach is directly related for example to that of Berkowitz and Gerber (63) in their theoretical study of vibrational relaxation in matrices, where the molecule effective Hamiltonian included the static perturbation from the lattice. Although not stressed inordinately hereafter, it is clear that such renormalization effects will always have to be taken into account, or at least examined, when the vibrationally relaxing solute is in fairly strong interaction with the surrounding environment.

H_{Bath} represents the classical Hamiltonian for the D_2O solvent together with the translational and rotational contribution of the HOD molecule (see below). $H_{coupling}$, the nonequilibrium coupling of the molecular vibrational coordinates to the rest of degrees of freedom, is split into several contributions: $H_{coupling} = H_{V-B} + H_{Cor} + H_{Cen}$. Here H_{V-B} represents the coupling of the molecular vibrational coordinates to the surrounding solvent molecules and turns out to be, not surprisingly, one of the main factors in the VET. In addition, the influence of the anharmonicity of the intramolecular potential of HOD will prove to be critical. The remaining contributions to $H_{coupling}$ involve vibration-rotation couplings, which, while not the focus in this chapter, are briefly included for completeness. H_{Cor} represents the Coriolis coupling between normal modes, which — as known from simple classical arguments (64) — can be especially strong between bending and stretching for the water molecule. Finally, H_{Cen} represents centrifugal coupling, i.e., the effect that the variation of moment of inertia with the vibrational motion may have on the relaxation.

Since the vibrational relaxation rate is evidently some 2 orders of magnitude smaller than the OH vibrational frequency, it was reasonable to assume that time-dependent perturbation theory is appropriate to handle the coupling. In addition, $H_{coupling}$ was expressed as a Taylor expansion in the harmonic normal modes, so that the coupling for a fixed (equilibrium) configuration of the solute can be computed. It is this last aspect that allows the performance of fully classical simulations in which the solute is frozen in its equilibrium configuration.

In addition, terms higher than first order in the expansion of the interaction potential in the normal modes of HOD were neglected. In view of the claim above that anharmonic effects are paramount in the problem, this linear approximation for the coupling certainly requires justification. A first argument comes from the fact that (as follows from simple considerations based on the Fermi Golden Rule) the second order contribution is zero [for the transition found to be dominant, i.e., $(001) \rightarrow (020)$] if harmonic states are taken for the solute (so that its contribution would be second order using the anharmonic states). A second reason to truncate the coupling at first order is that the HOD vibrational contribution to the Hamiltonian depends on the effective solution normal modes. Therefore, while there are important distortions from the gas phase geometry, so that the coupling would probably not be well represented (to first order) in terms of gas phase normal modes, one would not expect important deviations in terms of solution normal modes. It is important to emphasize that this truncation does not ignore the importance of the intramolecular HOD force field anharmonicities.

The V-B coupling Hamiltonian to first order in the three HOD dimensionless normal coordinates is $H_{V-B} = -\Sigma_{i=1}^3 q_i F_i$, where F_i is the intermolecular force due to the solvent exerted on the harmonic normal coordinate, evaluated at the equilibrium position of the latter. This force obviously depends on the relative separations of all molecules, and on their relative orientations. In the most rigorous quantum description of rotations, this term would depend on the excited molecule rotational eigenstates and of the solvent molecules. Instead rotation was treated classically, a reasonable approximation for water at room temperature. With this form for the coupling, the formal conversion of the Golden Rule formula into a rate expression follows along the lines developed by Oxtoby (2,53), with a slight variation to maintain the explicit time dependence of the vibrational coordinates (57),

$$k_{if} = \gamma_{if} \, \text{Re} \sum_{s} \sum_{s'} \int_{-\infty}^{\infty} dt \, \langle i|\delta q_s|f\rangle \langle f|\delta q_{s'}(t)|i\rangle \times \langle \delta F_s \delta F_{s'}(t)\rangle \qquad (3)$$

Here δF_s signifies the fluctuating solvent force on the coordinate q_s, while $\delta q_s(t)$ is the Heisenberg time-dependent operator, with dynamics governed by the full internal anharmonic molecular Hamiltonian, associated with the fluctuation $\delta q_s = q_s - \langle i|q_s|f \rangle$. Finally, the prefactor γ_{if} is (2)

$$\gamma_{if} = \frac{2\hbar^{-2}}{1 + \exp(-\beta\hbar\omega_{if})}$$

Each molecular vibration factor in Equation (3) is a type of molecular time correlation function for the internal vibrational dynamics. In the harmonic approximation, $|i\rangle$ and $|f\rangle$ would reduce to the harmonic vibrational eigenstates and the $\{q_s\}$ would be the actual molecular normal modes. Then one has the simplification

$$\text{Re}\langle i|\delta q_s|f \rangle \langle f|\delta q_{s'}(t)|i \rangle \propto \cos(\omega_{if}^0 t) \tag{4}$$

where $\{\omega_{if}^0\}$ are the harmonic frequencies and, in the case of the harmonic O–H stretch state, only transitions $1 \rightarrow 0$ occur in this mode, since the V-B Hamiltonian induces no intramode mixing of the harmonic normal modes. But in the presence of anharmonicity, this is no longer true. The H_{V-B} derived vibrational coupling factors $\{\delta q_s\}$ couple the anharmonic vibrational eigenstates $|i\rangle$ and $|f\rangle$. There are two consequences of this. First, transitions formerly forbidden are now allowed, e.g., as seen below, the transition to the bend overtone $[(001) \rightarrow (020)]$ is now possible: a Fourier factor $\exp(i\omega_{if}t)$ will appear at the corresponding frequency gap ω_{if}, and the Fourier component of the corresponding force tcf at this much lower frequency can be effective in the relaxation. Second, even formerly allowed transitions will have their rates altered due both to the anharmonic shift in the O–H mode frequency and to the coupling to the remaining internal modes. In short, the key point is that the vibrational eigenstates and modes in the absence of H_{V-B} are *not* the harmonic ones, due to the internal molecular anharmonicities. Thus, the states $|i\rangle$ and $|f\rangle$ refer to the anharmonic vibrational eigenstates of the molecule, and not to their harmonic approximation.

Finally, if we expand Equation (3) we can rewrite it such that the Fourier transforms appear explicitly, together with the quantum matrix elements that have to be computed (2,53),

$$k_{if} = \gamma_{if} \int_{-\infty}^{\infty} dt \, e^{i\omega_{if}t} \sum_{s,s'=1}^{3} \langle i|q_s|f \rangle \langle f|q_{s'}|i \rangle \langle F_s F_{s'}(t) \rangle$$

$$= \gamma_{if} \sum_{s,s'=1}^{3} \langle i|q_s|f \rangle \langle f|q_{s'}|i \rangle \langle F_s \hat{F}_{s'}(\omega) \rangle \tag{5}$$

where the total rate constant for depletion of state is the sum over all possible final states $k_i = \Sigma_f k_{if}$. As discussed above, the intramolecular potential of the HOD solute is required in order to compute the anharmonic states and the relevant matrix elements. In Ref. 57, the intramolecular potential for water proposed by Sceats and Rice (65), based on a previous gas phase model due to Smith and Overend (66), was employed; partial support for the model includes the fact that, e.g., the overtone spectra of various H_2O and D_2O mixtures can be satisfactorily assigned with this force field (65). The energies and eigenstates corresponding to this potential can be computed by applying standard perturbation theory, producing a strong mixing of harmonic states: each anharmonic eigenstate is a superposition of more than 40 harmonic eigenstates. This mixing is critical in the VET, but mercifully the pattern in terms of the anharmonic eigenstates is fairly simple.

Figure 1 shows the resulting calculated pathway for the VET. It is not the direct conversion of the OH stretch down to its ground state with a VET of some 3430 cm^{-1} into the solvent. Rather, the transition is to the first overtone of the HOD bend some 530 cm^{-1} below the OH stretch, followed by relaxation of the bend down to its ground state. Here we summarize some of the important features of the VET pathway, emphasizing the very first step, which turns out to be rate limiting.

We devote a few paragraphs to how all this works out in some detail, to give an impression of the important issues that can arise in such problems. To begin, Fig. 2 presents the Fourier transforms of the correlation functions of the forces on the solute vibrational modes for two different frequency scales. The main common characteristic is the almost exponential decay exhibited (note the semilog scale), with the small spikes reflecting the bending and stretching modes of the D_2O solvent. This rapid frequency decay is to be sure highly relevant since, as indicated in Equation (5), the relaxation rate is proportional to the Fourier component at the given transition frequency. This of course embodies the hardly revolutionary idea that transitions to states close in energy to that of the $v = 1$ OH stretch state are strongly favored, and it so happens that it is mainly for this reason that the relaxation proceeds to the $v = 2$ state of the HOD bend.

But we will later see a violation of a simple energy gap behavior such as the above. It is critical to appreciate that this frequency effect is modulated by the factors that depend on the intramolecular potential; it suffices to recall that in an harmonic approximation only the transition $(001) \rightarrow (000)$ would be allowed. To place this issue in perspective, we write the rate constant [Equation (5)] for transitions out of the anharmonic

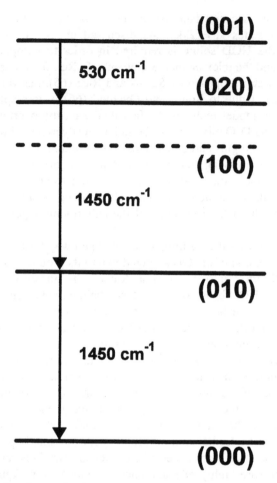

Figure 1 Relaxation path, determined from MD simulation, for the vibrational relaxation of the OH stretch for HOD dissolved in D_2O. The levels are labeled according to the standard ordering: (OD stretch, bend, OH stretch).

OH stretch state (001) in some detail, ignoring for the purposes of discussion the off-diagonal terms and focusing on just two final states: (000), the ground anharmonic OH stretch state, and (020), the overtone of the anharmonic HOD bend. Then we have

$$k_{001} = \gamma_{001,000} \langle 001|q_{OH}|000\rangle^2 \langle F_{OH}\hat{F}_{OH}(\omega_{001-000})\rangle$$
$$+ \gamma_{001,020} \langle 001|q_{OH}|020\rangle^2 \langle F_{OH}\hat{F}_{OH}(\omega_{001-020})\rangle + \cdots$$

$$+ \gamma_{001,000} \langle 001|q_{HOD}|000 \rangle^2 \langle F_{HOD} \hat{F}_{HOD}(\omega_{001-000}) \rangle$$
$$+ \gamma_{001,020} \langle 001|q_{HOD}|020 \rangle^2 \langle F_{HOD} \hat{F}_{HOD}(\omega_{001-020}) \rangle$$
$$+ \cdots \tag{6}$$

Here we have used the labels OH for the harmonic OH stretch mode coordinate ($s = 3$) and HOD for the harmonic HOD bend mode coordinate ($s = 2$). We have not written the corresponding two terms involving the matrix elements of the harmonic OD stretch normal mode, which were found to have a minor contribution. Figure 3 displays the prefactors of the Fourier transforms for the two illustrative cases. Figure 3a shows the

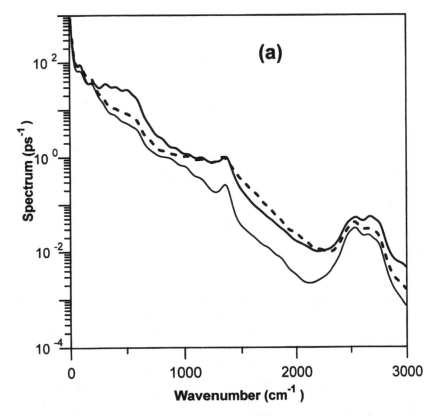

Figure 2 Power spectra corresponding to the forces on the three HOD vibrational modes. Thick solid line: bend (principal contribution to the relaxation); dashed line: OH stretch; thin solid line: OD stretch. (a) Semilog plot; (b) not including intramolecular frequencies.

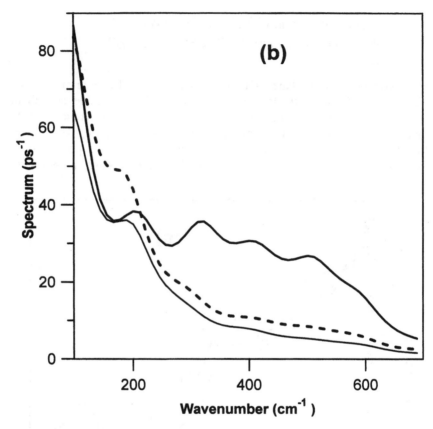

Figure 2   (*continued*)

prefactors that modulate the Fourier transform of the O–H stretch force autocorrelation function. As could be expected, there is a large value for the direct transition to the ground state [in Equation (6)]. The rest of the bars represent the new pathways of relaxation that have been opened up by anharmonicity: to the $v = 1$ and $v = 2$ bend states (at ~ 1980 and ~ 530 cm^{-1}, respectively; the latter is the term in Equation (6), which reflects the rather weak, but nonetheless vital, *Fermi resonance* interaction between the OH stretch fundamental and the HOD bend overtone), and to the $v = 1$ state of the O–D stretch (at ~ 900 cm^{-1}). Even though the prefactors corresponding to transition to the latter two states [(020) and (100)] are just barely noticeable on the plot, their net contributions are actually larger because of the several orders of magnitude difference in the respective Fourier components (see Fig. 2). Figure 3b is a similar plot for the bending mode. Here, given that all the prefactors have very similar values, the dominant role of the Fourier components is even clearer. It turns out that the contribution of the first transition of this plot [at ~530 cm^{-1}; the $\gamma_{001,020}\langle001|q_{HOD}|020\rangle^2$ term in Equation (6)] is the one that almost completely determines the relaxation time for the transition $(001) \rightarrow (020)$, giving a time of 9.4 ps. Accordingly, Equation (6) simplifies to the formula

$$k_{001} \approx k_{001,020} \approx \gamma_{001,020}\langle001|q_{HOD}|020\rangle^2\langle F_{HOD}\hat{F}_{HOD}(\omega_{001-020})\rangle \quad (7)$$

Taking into account the contributions of the q_{OH} and q_{OD} modes, a time of 9.7 ps results for this transition (the contribution to the relaxation

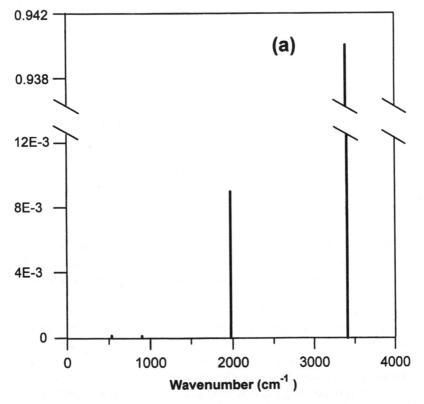

Figure 3 Prefactors of the power spectra [see Equation (6)]: (a) OH stretch; (b) bend.

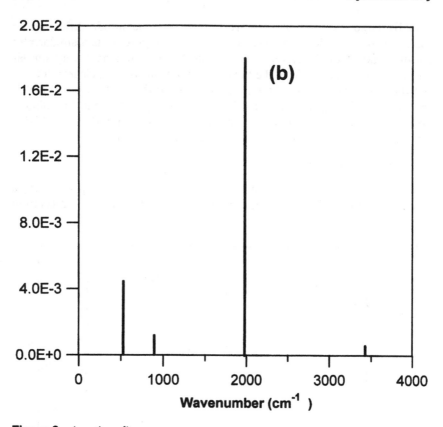

Figure 3 (*continued*)

rate arising from the prefactors of the cross-correlation functions can be negative). It was determined that the rest of other possible transitions have much longer relaxation times and only reduce the total relaxation time to 9.3 ps.

   The vital importance of anharmonicity can also be revealed by computation of the relaxation time resulting when only the harmonic part of the intramolecular potential is considered. This gives a relaxation time of ~3 ns for the direct transition to the ground state, the only allowed transition: anharmonicity reduces the relaxation time by *three orders of magnitude*.

   It is worth pointing out that there is a small but nonnegligible acceleration of the OH fundamental stretch VET by Coriolis coupling — this being analyzed via a basic formulation due to Velsko and Oxtoby (67) — but a negligible effect of centrifugal coupling, the latter consistent with its predicted general unimportance (68). The calculated overall time for flow out of the OH stretch was found to be 7.5 ps, compared to the experimental result in Ref. 59 of 8 ± 2 ps; more on this below!

The remaining VET flow process is found to be a faster and thus non–rate limiting, sequential VET in the HOD bend manifold: $v = 2 \rightarrow v = 1 \rightarrow v = 0$. An important general point here is that this illustrates that prediction of VET pathways requires careful attention not only to frequency gaps, but *also* to the coupling factors. In particular, the OD stretch is largely bypassed in the calculated decay route: while the energy gap from the bend overtone is about a quarter of that to the $v = 1$ bend state, the coupling is sufficiently weak to mute its participation.*

Which solvent motions represented in the bend force spectrum [see Equation (6)] are important in inducing the rate limiting transition from the OH stretch into the HOD bend? For the same D_2O model, Marti et al. (69) have found that the D_2O librational spectrum is peaked at ~ 400 cm^{-1} with a FWHM of ~ 300 cm^{-1}. Thus, the dominant motions in the bend force spectrum at 530 cm^{-1} responsible for the calculated relaxation are the solvent librations.

Briefly returning to the Coulomb force theme of Section II, although not presented in Ref. 57, subsequent (unpublished) results have shown that Coulombic forces are dominant in the OH stretch relaxation. Figure 4 displays the results for the bend power spectrum for two different frequency ranges, including the contribution of Coulomb and non-Coulomb forces. From Fig. 4a, Coulomb forces clearly dominate over the full range of interest; a similar behavior has also been found for the stretchings spectra (not shown).†

Expanding our horizons for a moment, although there are some theoretical studies of the VET for Si–H bonds at solid surfaces (70), the only

---

* In recent experiments (D. Dlott, personal communication), some OD stretching vibration is detected (although not unambiguously assignable to the HOD molecule and/or to a D_2O solvent molecule). In the computed decay pathways (57), a minor energy transfer from the OH stretch to the OD stretch in the same HOD molecule was found, both directly and via the second overtone of the bend. This in turn could decay by near-resonant transfer to an OD stretch in the solvent.

† In these calculations a slightly different intramolecular force field for D_2O has been used (69), designed to exactly match the experimental liquid frequencies, the effect on the computed relaxation time is rather small, yielding an almost equal time, within numerical accuracy.

other solution polyatomic VET simulation of which we are aware is that of
Morita and Kato (43), who used electronic structure calculations and MD
simulation to analyze the relaxation path and rate for the initial relaxation
of the azide ion N_3^- antisymmetric stretch, of frequency $\sim$2050 cm^{-1}, in
water, a study already mentioned in Section II in connection with charge
shifting electrostatic acceleration of VET. The N_3^- relaxation appears to
be more subtle than that of HOD. The picture resulting from this work is
that there are two pathways out of the initially excited state, of comparable
contribution to the rate: the direct route to the ground state and an IVR
route. The former is of special interest in that a peak in the force spectrum
shows up in the 2000 cm^{-1} region; evidently the flexible polarizable SPC

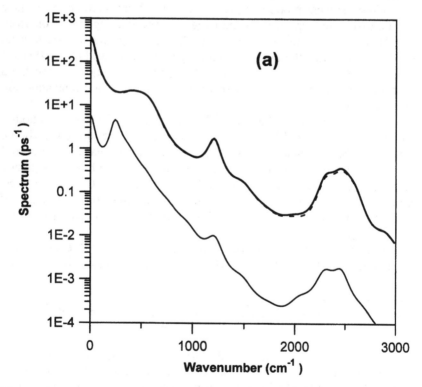

Figure 4 Power spectrum corresponding to the bend. Thick solid line: total;
dashed line: Coulomb contribution; thin solid line: non-Coulomb contribution;
dash-dot line: Coulomb–non-Coulomb cross-correlation contribution. (a) Semilog
plot [does not include cross correlation — see (b)]; (b) not including intramolecular
frequencies.

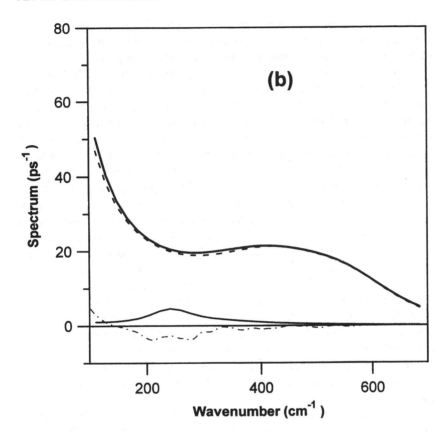

Figure 4   (*continued*)

water model used by these authors incorporates the water bend-water libration activity referred to in connection with CN^- relaxation in Section II. The latter, IVR, route depends first on the fact that the excited vibrational eigenstate is not exactly a pure antisymmetric stretch, but also has a small component, due to internal anharmonicity, involving a single excitation of the antisymmetric stretch *and* the symmetric stretch. Due to Fermi resonance coupling, this component allows transitions, each of comparable rate, to two states, roughly 750 cm^{-1} lower in energy. These two states are approximately plus and minus combinations of the pure symmetric stretch and the first overtone of the bend. As with the cases of CH_3Cl in H_2O and HOD in D_2O, the water librations are implicated as the solvent accepting modes in this route. The lesson stressed above concerning the importance of simultaneous attention to energy gaps and coupling factors again appears here: transitions from the excited state to the two states that are the closest in energy are in fact unimportant due to the feeble coupling.

What is the relationship of the OH (in HOD) stretch VET pathway in D_2O solution to that in the gas phase or in other solvents? This query cannot be precisely answered, since evidently *only* VET for H_2O or D_2O has been studied. Nonetheless, there are a number of intriguing indications. One important difference from HOD is that in both the H_2O and D_2O molecules, the IR active antisymmetric stretch is very close in energy to the symmetric stretch. In both gas phase self-relaxation of H_2O (71) and H_2O and D_2O in assorted organic solvents (72,73), rapid equilibration between these two stretches is inferred (itself an interesting theoretical problem, not yet examined). In organic solvents, a relaxation pathway involving the HOH or DOD bend overtone is also suggested (72), although the evidence is indirect (see also Ref. 74). But for H_2O in the gas phase it seems clearly established that the subsequent VET is exactly (71) the same as that of the liquid simulation above — through the bend overtone which then deactivates sequentially to its ground state. This striking commonality — which extends to similar step rate ratios (57) — would seem to be a tribute to the robust importance of the intramolecular force fields. In a different context, a related conclusion was reached in experimental studies (75) of the amide I modes vibrational relaxation in peptides: the results are interpreted as dominated by IVR, an intrinsic aspect of the peptide group independent of the environment.

It should be noted that in a model calculation (76) for the gas phase H_2O self-relaxation, the accepting mode for the transition to the bend overtone is the O–O stretch in the modeled hydrogen-bonded pair, rather than the D_2O librations invoked in the solution case. This might represent a real difference between relaxation in the two phases, but the O–O frequency adopted in the model seems to be (57,77) about a factor of 2 too high, which would rather overemphasize the ease of transfer into this mode.

It is now time to return to the close agreement of the calculated (57) and one (59) of the two first experimental relaxation times for the OH stretch in D_2O. There are several points to make. The first is that subsequent experiments (60–62) have largely confirmed the initial measure of Vodopyanov (58) (0.3 ps < T_1 < 0.6 ps), so that now it is estimated to be about 1 ps or so. The second is that, as was stressed in Ref. 57, the (former) agreement could be fortuitous, owing for example to the uncertain character of the force fields employed. The third is that, as recounted in Refs. 57 and 78, one original motivation of the calculation of the VET pathway and time was to discover if instead the mechanism of loss of OD population might in fact be a quite *different* route of vibrational predissociation (79), in which energy would flow from the OH stretch into an OH · · · O low-frequency hydrogen bond with a neighboring water molecule, breaking that bond; several experiments on other hydrogen-bonded systems have been interpreted as involving such a predissociation pathway (80). In view of this, it is perhaps ironic that the current interpretation of the ~1 ps OH stretch lifetime by Woutersen et al. (62) is that it is indeed due to this predissociation mechanism, based on the striking and unusual temperature independence of T_1 over a wide temperature range in the solid phase and a lengthening of T_1 in the liquid phase, both adequately accounted for via the T_1-OH stretch frequency red shift correlation that follows from a simple model of predissociation (79). The low-frequency hydrogen bond OH · · · O was argued to be the accepting mode, rather than the HOD bend, via arguments related to the observation of blue shifts (to be expected from weakened hydrogen bonds). If this is indeed the route for energy transfer in this system, it must be counted as remarkable that the predissociation time scale could be so rapid as 1 ps (and a similar remark could be made about an alternate, nondissociative, intramolecular vibrational energy flow pathway). In any event, it is certain that this system will continue to attract intense effort to provide increasing detail on the answer to the seemingly simple question of how energy exits the OH stretch in this system.

IV. SOME PERSPECTIVES

In closing, we comment on a few directions for the future. This listing is decidedly not even remotely exhaustive, but rather is just a sampling, with a personal taste, from an extensive menu.

To begin with a fairly general issue, it is to be noticed that many of the vibrational relaxation times discussed within are quite short, a few ps or even less. In the usual discussion of vibrational dynamics, one assumes (typically justifiably) that the population relaxation time T_1 far exceeds the vibrational dephasing time T_2. But it is clear that this assumption is no longer secure in such rapidly relaxing systems, and one needs to look more carefully at the overall formulation of vibrational dynamics to account for the fact that the relevant solute frequency is fluctuating on the same time scale as the population relaxation (81). (Of course, vibrational dephasing for systems involving, e.g., strong Coulomb interactions (29,33,82) is itself of considerable independent interest for issues of lineshapes and coherences).

Concerning the influence of Coulomb forces on VET of a fixed charge solute in solution, the survey in Section II would appear to provide ample indication that this impact can often be significant or even dominant but is not catholic. Its influence can be a fairly sensitive function of the solute vibration, the solute itself, and both the overall spatial and orientational structure of the solvent as well as the intramolecular degrees of freedom of the solvent molecules. Perhaps it is fair to say that Coulomb force effects should always be considered as a possibility in plausible solutes and solvents, but that much of the overall tapestry (which may be rather baroque) remains to be filled in before we have a complete and reliable picture. Certainly one useful avenue here would be to more clearly and directly address the relevant solvent degrees of freedom serving as plausible energy acceptors, a point taken up again below.

As for charge shifting force effects on VET, it seems clear that these should be quite common in a variety of systems, although whether this effect is important at low or high energies will depend on molecular details of the system. Certainly, there is a good chance for such effects in some energy range for any molecule in which there should be important contributions of several VB structures to the electronic structure. Thus such examples as the carboxylate ion (45) — indeed a perusal of Pauling's *Nature of the Chemical Bond* (45) is probably a good source of possibilities — and sufficiently strongly hydrogen-bonded systems (18) are significant candidates here.

As for the general area of intramolecular effects in solution phase VET, there is without doubt a true embarrassment of riches for future efforts. But perhaps due to its offer of the challenging mix of the effects of Coulomb forces, possible charge shifts and intramolecular degrees of freedom, solute molecular VET in aqueous solvents arouses special interest, holding some promise to shed light on the microscopic pathways of chemical reactions in aqueous solution. This interest is heightened by the further possibility of vibrational predissociation. The precise role of infrared-induced hydrogen bond breaking in, e.g., HOD in D_2O and other (83–88) hydrogen-bonded systems will no doubt attract considerable effort to unravel the dynamics of this important and ubiquitous bond type.

An intriguing question awaiting further scrutiny is whether VET mechanisms and pathways could differ in different solvents or between solute and the gas phase (or various size cluster systems). Certainly diatomic solute VET can depend on VV pathways differently available in different solvents (6,89). But there is little information available for the molecules and solvents of the type highlighted in this chapter (for some examples, see Chapter 13 and Refs. 1 and 90). While the water VET problem may provide an example, establishing this will require resolution of the issues noted above. It has been argued that the decay routes of the first CH overtone in CHI_3 differ in the strongly polar solvent acetone and the less polar solvent chloroform (91), surely there must be more examples to be found.

Throughout much of the work summarized in this chapter, frequent use was made of theoretical arguments identifying relevant solvent modes or motions based on overlap of a solvent peak or band with the *solute* vibrational frequency, e.g., the involvement of water vibrations for CH_3Cl VET in Section II, or with some intramolecular frequency difference in the polyatomic case, e.g., in the HOD VET problem of Section III. Clearly, a more direct demonstration would be desirable that a certain mode, or modes — for example, the librational modes of the solvent water molecules — are the energy-receiving modes in VET from the solute molecule. One possible avenue here is to do a simulation using the solution phase generalization of the VV energy transfer rate formula familiar from gas phase work (1,51,89,92,93).* Another, more challenging, route would be direct simulation monitoring of those *quantum* modes; this kind of thing has been done for isolated molecules (46,98) and seems within reach for solution problems. Indeed, it might be a fruitful perspective for some strongly coupled systems to view the solute and, e.g., some portion of the first solvation shell as a "supermolecule" and exploit isolated molecules IVR ideas. The urgency for theoretical advances here is clearly enhanced by the advent of ultrafast experiments that directly probe the solvent receiving modes (see, e.g., Chapter 13 in this volume and Ref. 99).

ACKNOWLEDGMENTS

JTH acknowledges support of this work by NSF grants CHE88-07852, CHE93-12267, and CHE-9700149. A portion of the work for this chapter was done while JTH was an invited Condorcet Professor at ENS. RR acknowledges DGICYT grants PB93-0971-C03 and PB96-0170-C03-02.

---

* It would of course be of interest to apply such formulas to experiments that explicitly deal with intermolecular VV transfer in solution; beyond those experiments already mentioned in this chapter (1,56,73,91), a few further representative examples can be cited (94–97).

REFERENCES

1. Yardley JT. Introduction to Molecular Energy Transfer. New York: (Academic Press, 1980).
2. Oxtoby DW. Adv Chem Phys 47:487, 1981.
3. Oxtoby DW. Ann Rev Phys Chem 32:77, 1981.
4. Harris AL, Brown JK, Harris CB. Annu Rev Phys Chem 39:341, 1988.
5. Chesnoy J, Gale GM. Adv Chem Phys 70:297, 1988.
6. Harris CB, Smith DE, Russell DJ. Chem Rev 90:481, 1990.
7. Elsaesser T, Kaiser W. Ann Rev Phys Chem 42:83, 1991.
8. Owrutsky JC, Raftery D, Hochstrasser RM. Annu Rev Phys Chem 45:519 1994.
9. Voth GA, Hochstrasser RM. J Phys Chem 100:13034, 1996.
10. Stratt RM, Maroncelli M. J Phys Chem 100:12981, 1996.
11. Bagchi B, Biswas R. Adv Chem Phys 109:207, 1999.
12. Gertner BJ, Whitnell RM, Wilson KR, Hynes JT. J Am Chem Soc 113:74, 1991.
13. Kim HJ, Staib A, Hynes JT. In: V Sundstrom, ed. Ultrafast Reaction Dynamics at Atomic-Scale Resolution Femtochemistry and Femtobiology. London: Imperial College Press, 1998, pp 510–527.
14. Ando K, Staib A, Hynes JT. In: A Tramer, ed. Fast Elementary Processes in Chemical and Biological Systems. New York: AIP, 1996, pp 326–332.
15. (a) Hammes-Schiffer S, Tully JC. J Chem Phys 101:4657, 1994; (b) J Phys Chem 99:5793, 1995.
16. Ferguson EE. J Phys Chem 90:731, 1986.
17. Heilweil EJ, Doany FE, Moore R, Hochstrasser RM. J Chem Phys 76:5632, 1982.
18. Li M, Owrutsky J, Sarisky M, Culver JP, Yodh A, Hochstrasser RM. J Chem Phys 98:5499, 1993.
19. (a) Whitnell RM, Wilson KR, Hynes JT. J Phys Chem 94:8625, 1990; (b) J Chem Phys 96:5354, 1992.
20. Landau L, Teller E. Z Sowjetunion 10:34, 1936.
21. Ladanyi BM, Stratt RM. J Phys Chem A 102:1068, 1998.
22. Lim M, Gananakaran S, Hochstrasser RM. J Chem Phys 106:3485, 1997.
23. Kuharski RA, Rossky PJ. J Chem Phys 82:5164, 5289, 1985.
24. Johnson AE, Levinger NE, Barbara PF. J Phys Chem 96:7841, 1992.
25. Kliner DAV, Alfano JC, Barbara PF. J Chem Phys 98:5375, 1993.
26. Benjamin I, Whitnell RM. Chem Phys Lett 204:45, 1993.
27. Gnanakaran S, Hochstrasser RM. J Chem Phys 105:3486, 1996.
28. Bruehl M, Hynes JT. Chem Phys 175:205, 1993.
29. Rey R, Hynes JT. J Chem Phys 108:142, 1998.
30. (a) Ferrario M, McDonald IR, Symons MCR. Mol Phys 77:617, 1992; (b) Ferrario M, McDonald IR, Klein ML. J Chem Phys 84:3975, 1986.
31. Ikeda T, Nishimoto K, Asada T. Chem Phys Lett 248:329, 1996.

32. Shiga M, Okazaki S. Chem Phys Lett 292:431, 1998.
33. Hamm P, Lim M, Hochstrasser RM. J Chem Phys 107:10523 1997.
34. (a) Bertie JE, Ahmed MK, Eysel HH. J Phys Chem 93:2210, 1989; (b) Maréchal Y. J Chem Phys 95:5565, 1991.
35. (a) Bergsma JP, Gertner BJ, Wilson KR, Hynes JT. J Chem Phys 86:1356, 1987; (b) Gertner BJ, Bergsma JP, Wilson KR, Lee S, Hynes JT. J Chem Phys 86:1377, 1987.
36. Papanikolas JM, Vorsa V, Nadal ME, Campagnola PJ, Gord JR, Lineberger WC. J Chem Phys 97:7002, 1992.
37. Walhout PK, Alfano JC, Thakur KAM, Barbara PF. J Phys Chem 99:7568, 1995.
38. Gertner BJ, Ando K, Bianco R, Hynes JT. Chem Phys 183:309, 1994.
39. Benjamin I, Barbara PF, Gertner BJ, Hynes JT. J Phys Chem 99:7557, 1995.
40. Walhout PK, Silva C, Barbara PF. J Phys Chem 100:5188, 1996.
41. (a) Faeder J, Parson R. J Chem Phys 108:3903, 1998; (b) Ashkenazi G, Banin U, Bartana A, Kosloff R, Ruhman S. Adv Chem Phys 100:317, 1997.
42. (a) Faeder J, Delaney N, Maslen PE, Parson R. Chem Phys 239:525, 1998; (b) Delaney N, Faeder J, Parson R. J Chem Phys 111:651, 1999.
43. Morita A, Kato S. J Chem Phys 108:6809, 1998.
44. Ferrario M, Klein ML, McDonald IR. Chem Phys Lett 213:537, 1993.
45. Pauling L. The Nature of the Chemical Bond and the Structure of Molecules and Crystals: An Introduction to Modern Structural Chemistry. Ithaca, NY: Cornell University Press, 1960.
46. Nesbitt DJ, Field RW. J Phys Chem 100:12735, 1996.
47. Bakker HJ, Planken PCM, Kluipers L, Lagendijk A. J Chem Phys 94:1730, 1991.
48. Kenkre VM, Tokmakoff A, Fayer M. J Chem Phys 101:10618 1994.
49. Graener H, Zürl R, Hoffmann M. J Phys Chem B 101:1745, 1997.
50. Poulsen JA, Thomsen CL, Keiding SR, Thogersen J. J Chem Phys 108:8461, 1998.
51. (a) Vikhrenko VS, Heidelbach C, Schwarzer D, Nemtsov VB, Schroeder J. J Chem Phys 110:5273, 1999; (b) Heidelbach C, Vikhrenko VS, Schwarzer D, Schroeder J. J Chem Phys 110:5286, 1999.
52. Schatz GC, Mulloney T. J Chem Phys 71:5257, 1979.
53. Velsko S, Oxtoby DW. J Chem Phys 72:2260, 1980.
54. (a) Laubereau A, von der Linde D, Kaiser W. Phys Rev Lett 28:1162, 1972; (b) Laubereau A, Fisher SF, Spanner K, Kaiser W. Chem Phys 31:335, 1978.
55. Fendt A, Fischer SF, Kaiser W. Chem Phys 57:55, 1981.
56. Laubereau A, Kaiser W. Rev Mod Phys 50:607, 1978.
57. Rey R, Hynes JT. J Chem Phys 104:2356, 1996.
58. Vodopyanov KL. J Chem Phys 94:5389, 1991.
59. Graener H, Seifert G, Laubereau A. Phys Rev Lett 66:2092, 1991.
60. Seifert G, Weidlich K, Graener H. Phys Rev B 56:R14231, 1997.
61. (a) Laenen R, Rauscher C, Laubereau A. Phys Rev Lett 80:2622, 1998; (b) J Chem Phys B 102:9304, 1998.

62. (a) Woutersen S, Emmerichs U, Nienhuys HK, Bakker HJ. Phys Rev Lett 81:1106, 1998; (b) Nienhuys HK, Woutersen S, Vansanten RA, Bakker HJ. J Chem Phys 111:1494, 1999.

63. Berkowitz M, Gerber RB. Chem Phys 37:369, 1979.

64. Herzberg G. Infrared and Raman Spectra of Polyatomic Molecules. Princeton Van Nostrand, NJ: 1968.

65. Sceats MG, Rice SA. J Chem Phys 71:973, 1979.

66. Smith DF, Overend J. Spectrochim Acta 28A:471, 1972.

67. Velsko S, Oxtoby DW. Chem Phys Lett 69:462, 1980.

68. Berne BJ, Gordon RG, Sears VF. J Chem Phys 49:475, 1968.

69. Martí J, Padró JA, Guàrdia E. Mol Sim 11:321, 1993.

70. (a) Gai H, Voth GA. J Chem Phys 99:740, 1993; (b) Sun Y-C, Gai H, Voth GA. Chem Phys 205:11 1996; (c) Ermoshin VA, Kazansky AK, Smirnov KS, Bougeard D. J Chem Phys 105:9371, 1996; (d) Lu H-F, Ho K-S, Hong S-Ch, Liu A-H, Wu P-F and Sun Y-Ch, J Chem Phys 109:6898, 1998.

71. Zittel PF, Masturzo DE. J Chem Phys 90:977, 1989.

72. Graener H, Seifert G, Laubereau A. Chem Phys 175:193, 1993.

73. Graener H, Seifert G. J Chem Phys 98:36, 1993.

74. Engholm JR, Rella CW, Schwettman HA, Happek U. J Non-Crist Solids 203:182, 1996.

75. Hamm P, Lim M, Hochstrasser RM. J Phys Chem B 102:6123, 1998.

76. Shin HK. J Chem Phys 98:1964, 1993.

77. (a) Dyke TR. J Chem Phys 66:492, 1977; (b) Reimers JR, Watts RO. Chem Phys 85:83, 1984.

78. Staib A, Rey R, Hynes JT. In: Gauduel Y, Rossky PJ, eds. Ultrafast Reaction Dynamics and Solvent Effects. New York: AIP, 1994, pp 173–190.

79. Staib A, Hynes JT. Chem Phys Lett 204: 197, 1993.

80. (a) Laubereau A, Kehl G, Kaiser W, Optics Commun 11: 74, 1974; (b) Fendt A, Fischer SF, Kaiser W. Chem Phys 57: 55, 1981; (c) Graener H, Ye TQ, Laubereau A. J Chem Phys 90: 3413, 1989; (d) J Chem Phys 91: 1034, 1989.

81. Staib A, Borgis D. J Chem Phys 103: 2642, 1995.

82. Cho M. J Chem Phys 105: 10755, 1996.

83. (a) Heilweil EJ, Casassa MP, Cavanagh RR, Stephenson JC. J Chem Phys 81: 2856, 1984; (b) Heilweil EJ. Chem Phys Lett 129: 48, 1986.

84. (a) Arrivo SM, Heilweil EJ. J Phys Chem 100: 11975, 1996; (b) Grubbs WT, Dougherty TP, Heilweil EJ. J Am Chem Soc 117: 11989, 1995.

85. Brugmans MJP, Kleyn AW, Lagendijk A, Jacobs W, van Santen RA. Chem Phys Lett 217: 117, 1994.

86. Laenen R, Rauscher C. Chem Phys Lett 274: 63, 1997.

87. Staib A. J Chem Phys 108: 4554, 1998.

88. Heilweil EJ. Science 283: 1467, 1999.

89. Nesbitt DJ, Hynes JT. J Chem Phys 77: 2130, 1982.

90. Hong X, Chen S, Dlott DD. J Phys Chem 99: 9102, 1995.
91. Bonn M, Brugmans MJP, Bakker HJ. Chem Phys Lett 249: 81, 1996.
92. Zawadzki AG, Hynes JT. J Mol Liq 48: 197, 1991.
93. (a) Herman MF. Int J Quant Chem 70: 897, 1998; (b) Velev P, Herman MF. Chem Phys 240: 241, 1999.
94. Laubereau A, Kirschner L, Kaiser W. Opt Comm 9: 182, 1973.
95. Ambroseo JR, Hochstrasser RM. J Chem Phys 89: 5956, 1988.
96. Turnidge ML, Reid JP, Barnes PW, Simpson CJSM. J Chem Phys 108: 485, 1998.
97. Iwata K, Hamaguchi HO. Laser Chem 19: 367, 1999.
98. Sibert EL, III, Reinhardt WP, Hynes JT. J Chem Phys 77: 3583, 1982.
99. (a) Chudoba C, Nibbering ETJ, Elsaesser T. Phys Rev Lett 81: 3010, 1998; (b) J Phys Chem A 103: 5625, 1999.

15

Vibrational Relaxation of Polyatomic Molecules in Supercritical Fluids and the Gas Phase

D. J. Myers, Motoyuki Shigeiwa, and M. D. Fayer
Stanford University, Stanford, California

Binny J. Cherayil
Indian Institute of Science, Bangalore, India

I. INTRODUCTION

Vibrational energy relaxation (VER) is central to many chemical processes and reactions. A final event in a reaction pathway is often the release of excess vibrational energy. VER can be incredibly fast (<1 ps) or amazingly slow (>100 ms), depending on the molecules involved and the physical nature of the system (1–5). Since VER has been the subject of intense study over the years (2,5), many of the broad outlines are now generally regarded as well understood, at least qualitatively. Consider a classical oscillator, a mass connected to a spring. The classical oscillator has a natural frequency determined by the spring constant and the mass. If a time-dependent force is applied to the oscillator on-resonance and in phase with the oscillation, an increase in amplitude will result. If the force is applied out of phase with the oscillation, the oscillator's amplitude will decrease. Thus, the oscillator can increase in energy or decrease in energy when an external driving force is applied. A classical oscillator can be driven by an off-resonance force. However, even a classical oscillator has a sharply peaked response at its resonance frequency.

A quantum mechanical oscillator will only respond to a driving force at its resonance frequency. The response to an external driving force can cause a quantum oscillator to change its quantum state. For molecules in a solvent, the external driving force is provided by time-dependent intermolecular interactions of the oscillator with the solvent. The solvent undergoes constant thermal motion. These motions exert a broad frequency distribution of fluctuating forces on the oscillator of interest (vibrational mode of a solute molecule). The time dependence of the fluctuating forces is described by the force-force correlation function (6–10). The Fourier transform of the force-force correlation function is the spectrum of fluctuating forces. The Fourier component of the force-force correlation function that is on resonance with the molecular oscillator can produce transitions between quantized vibrational states. If a vibrational mode of a molecule is placed in an excited vibrational state by excitation with a short resonant infrared pulse of light, the system will be driven back to thermal equilibrium by vibrational relaxation induced by the resonant Fourier component of the spectrum of forces exerted by the bath.

If the solute molecule is a diatomic, all of the energy of the excited mode must be transferred to the bath (the solvent). If the diatomic is very massive, e.g., I_2, the vibrational frequency may be within the spectrum of fluctuating forces, and rapid vibrational relaxation can occur. If the diatomic has a high frequency mode, e.g., N_2, vibrational relaxation is very slow. Because the bath is composed of anharmonic oscillators, it will have high-frequency Fourier components that go beyond the spectrum of the fundamental modes of the bath. While the low-frequency oscillator can relax through a single quantum excitation of the bath, the high-frequency oscillator can only relax through multi-quanta excitations of the bath. Processes that require the excitation of many quanta of the bath result in slow VER.

Polyatomic solutes differ from diatomics because a mode of a polyatomic can be coupled to other internal modes of the solute as well as to the external bath modes. A low-frequency mode of a polyatomic can relax directly into the bath by exciting one or a small number of bath modes. Unlike a high-frequency diatomic, a high-frequency mode of a polyatomic molecule may relax rapidly. The high-frequency mode relaxes by excitation of a number of internal lower frequency modes of the molecule along with excitation of one or more modes of the bath. The bath participates by enhancing internal intramolecular pathways. Because molecular vibrational potentials are anharmonic, the internal modes of molecules are coupled. However, for vibrational relaxation to take place, energy must be conserved. The initially excited mode may be strongly coupled to a set of other internal modes, but these modes may have energies that do not sum to the initial energy. The bath provides a continuum of modes that makes energy conservation possible. In this case, it is the Fourier component of the force-force correlation function at the frequency required to conserve energy (the frequency associated with the energy that is deposited directly into the bath) that drives the VER.

Many questions remain concerning the details of vibrational relaxation, particularly for polyatomic solutes in polyatomic solvents. What are the solute and solvent parameters that dictate the relaxation rate? What are the dominant relaxation mechanisms in a large molecule? Do internal solvent modes participate? What are the influences of temperature and density on VER? What is the nature of VER of a large polyatomic in the limit of zero solvent density (the gas phase)?

The lifetime (T_1) of a vibrational mode in a polyatomic molecule dissolved in a polyatomic solvent is, at least in part, determined by the interactions of the internal degrees of freedom of the solute with the solvent. Therefore, the physical state of the solvent plays a large role in the mechanism and rate of VER. Relaxation in the gas phase, which tends to be slow and dominated by isolated binary collisions, has been studied extensively (11). More recently, with the advent of ultrafast lasers, vibrational lifetimes have been measured for liquid systems (1,4). In liquids, a solute molecule is constantly interacting with a large number of solvent molecules. Nonetheless, some systems have been adequately described by isolated binary collision models (5,12,13), while others deviate strongly from this type of behavior (14–18). The temperature dependence of VER of polyatomic molecules in liquid solvents can show complex behavior (16–18). It has been pointed out that a change in temperature of a liquid solute-solvent system also results in a change in the solvent's density. Therefore, it is difficult to separate the influences of density and temperature from an observed "temperature" dependence.

However, if the solvent is a supercritical fluid (SCF), it is possible to examine the effects of temperature and density on VER independently. The role of other solvent properties, such as viscosity, dielectric constant, and correlation length, can also be studied. A supercritical fluid is a substance that has been heated above its critical temperature (T_c) and, therefore, no longer undergoes the liquid/gas phase transformation. A typical phase diagram for an SCF is shown in Fig. 1. In an SCF, it is possible to fix the temperature and vary the density continuously (by varying the pressure) from gas-like densities to liquid-like densities. It is also possible to vary the temperature at fixed density.

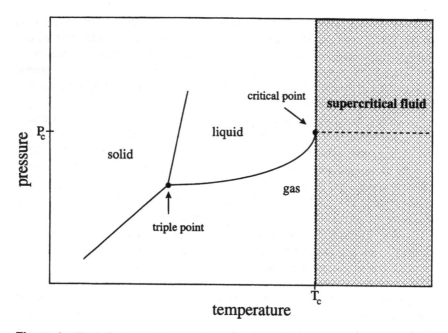

Figure 1 Typical phase diagram for compounds like ethane, fluoroform, and carbon dioxide. The fluid is technically supercritical when $T > T_c$ and $P > P_c$, but many authors refer to the fluid as supercritical when only the first condition is met. Once the temperature is raised above the T_c, the density can be varied continuously and there is no phase transformation.

The physical properties of supercritical fluids tend to lie between those of gases and liquids. The increased density relative to a gas, and the decreased viscosity relative to a liquid, allow supercritical fluids to be used as excellent solvents in many laboratory and industrial applications (19–25). Also, some notable solvation peculiarities of supercritical fluids have been discovered. For example, supercritical water can dissolve nonpolar oils because the dielectric constant of supercritical water decreases drastically near the critical point (26).

There is considerable interest in the nature of solute-solvent interactions in SCFs, and a variety of experimental and theoretical techniques have been brought to bear on the subject. Experimental work near the critical point often shows interesting anomalies. One of the groundbreaking studies found that partial molal volumes of infinitely dilute solutes become large and negative in the compressible region around the critical point

(27,28). Theoretical work shows that these partial molal volumes should diverge to negative infinity at the critical point (29,30). Many different types of spectroscopic experiments have been performed on supercritical fluids. Absorption and/or fluorescence wavelengths often show a characteristic "three-regime" density dependence, which will be discussed below (31–34). In addition, some studies have observed greatly enhanced excimer emission near the critical point (35–37). Orientational motion has been studied near the critical point (38–40). Some research indicates that orientational motion slows down substantially (39), though these results are debated by others (40). Electron paramagnetic resonance experiments show that spin exchange and hyperfine coupling deviate from normal behavior near the critical point in some fluids (41,42). These experimental results are detailed in several excellent recent reviews (34,43). Finally, x-ray (44) and neutron scattering (45) experiments performed on supercritical fluids demonstrate the existence of long density correlation lengths in the near-critical fluid. The observation of long correlation lengths is consistent with theory that shows the density correlation length diverges to infinity at the critical point (46).

Absorption and fluorescence experiments observe the solvatochromic shift of molecular (usually electronic) transitions. Several authors have reported similar density dependences for the shift of fluorescence or absorbance spectral lines in different solute/SCF solvent systems (31–33). This density dependence is characterized by three regions. At low densities, the absorption/emission line red shifts strongly with density. Around $\rho_c/2$, the slope changes and the line position becomes approximately independent of density. At higher densities, somewhat above ρ_c, the slope changes again, and the spectral line shifts further to the red with increasing density.

Very few experiments have been performed on vibrational dynamics in supercritical fluids (47). A few spectral line experiments, both Raman and infrared, have been conducted (48–58). While some studies show nothing unique occurring near the critical point (48,51,53), other work finds anomalous behavior, such as significant line broadening in the vicinity of the critical point (52,54–60). Troe and coworkers examined the excited electronic state vibrational relaxation of azulene in supercritical ethane and propane (61–64). Relaxation rates of azulene in propane along a near-critical isotherm show the three-region dependence on density, as does the shift in the electronic absorption frequency. Their relaxation experiments in supercritical carbon dioxide, xenon, and ethane were done farther from the critical point, and the three-region behavior was not observed. The measured density dependence of vibrational relaxation in these fluids was modeled well by a modified isolated binary collision approach. They treated the azulene/supercritical solvent as a system of attractive hard spheres. Monte Carlo simulations obtained the radial distribution function, which was factored in with the hard sphere collision rate in order to obtain a more realistic collision frequency in the high-density fluid. This model gave excellent agreement for vibrational relaxation far from the critical point.

One of the fundamental questions involves the role of attractive solute-solvent interactions in SCFs. Strong solute-solvent attraction may give rise to enhanced solvent density around a solute, particularly near the critical point (34,43,65,66). This enhancement is referred to as local density augmentation or solvent clusters. These clusters are usually suggested as the source of any anomalous experimental observations of solute properties in SCFs (34,43). For example, it has been postulated that the three-region behavior described above for solvatochromic shift data results from local density enhancement (formation of solute-solvent clusters). At low densities, the fluid is gas-like and binary collisions predominate. It has been suggested that at intermediate densities somewhat below ρ_c, attractive interactions produce an environment around the solute in which the solvent density is higher than in the bulk fluid. This local environment dominates the solvent's influence on the frequency shift. Increasing the solvent density changes the bulk fluid but does not affect the environment surrounding the solute. At sufficiently high densities above ρ_c, the nature of the clusters changes because of packing effects, and the solute properties are once again dependent on the bulk solvent density. Local density augmentation is accepted by many as the explanation for a variety of experimental observables near the critical point. Whether solvent clustering is responsible for experimental observations is an open question (67).

In this chapter, we describe the density- and temperature-dependent behavior of the vibrational lifetime (T_1) of the asymmetric CO stretching mode of $W(CO)_6$ (~ 2000 cm^{-1}) in supercritical ethane, fluoroform, and carbon dioxide (CO_2). The studies are performed from low density (well below the critical density) to high density (well above the critical density) at two temperatures: one close to the critical temperature and one significantly above the critical temperature (68–70). In addition, experimental results on the temperature dependence of T_1 at fixed density are presented. T_1 is measured using infrared (IR) pump-probe experiments. The vibrational absorption line positions as a function of density are also reported in the three solvents (68,70) at the two temperatures.

As discussed briefly above, T_1 in a polyatomic molecule can arise from both intramolecular and intermolecular processes. To separate the solute-solvent intermolecular contributions to T_1 from intramolecular vibrational relaxation, measurements of T_1 at zero density are reported. While the IR pump-probe measurements of the vibrational relaxation in SCFs are all single exponential decays, the decay at zero density (pure gas phase $W(CO)_6$) is a tri-exponential (71). The nature of the tri-exponential decay is discussed, and a theoretical analysis provides insight into the nature of the fast decay component. In addition, measurements in Ar as a function of density show that T_1 is density independent, in contrast to the density dependence observed in the polyatomic SCFs. Possible differences in solute-solvent interactions responsible for distinct behavior of T_1 in polyatomic SCF solvents versus Ar are discussed.

The results of the density-dependent vibrational relaxation measurements are compared to a density functional hydrodynamic/thermodynamic theory (72). The theory involves solvent thermodynamic parameters that are obtained from the SCF equations of state. Thus, detailed properties of solvent, including behavior near the critical point, are included in the theory. However, the solute-solvent spatial distribution is obtained from a hard sphere calculation. No specific solute-solvent interactions or clustering are included in the description of the distribution of the solvent about the solute. Evaluation of the resulting theoretical expression shows good agreement with the T_1 data over a wide range of densities at two temperatures. In addition, important features of the temperature dependence at fixed density are reproduced. An important point is that the theory is able to do a good job of reproducing the data without invoking solute-solvent local density augmentation.

II. EXPERIMENTAL PROCEDURES

The apparatus used to perform vibrational relaxation experiments in supercritical fluids consists of a picosecond mid-infrared laser system and a variable-temperature, high-pressure optical cell (68,73). Because the vibrational absorption lines under study are quite narrow (<10 cm^{-1}), a source of IR pulses is required that produces narrow bandwidths. To this end, an output-coupled, acousto-optically Q-switched and mode-locked Nd:YAG laser is used to synchronously pump a Rhodamine 610 dye laser. The Nd:YAG laser is also cavity-dumped, and the resulting 1.06 µm pulse is doubled to give an $\sim$600 µJ pulse at 532 nm with a pulse duration of $\sim$75 ps. The output pulse from the amplified dye laser ($\sim$35 µJ at 595 nm, 40 ps FWHM) and the cavity-dumped, frequency-doubled pulse at 532 nm serve as the signal and pump inputs, respectively, in a $LiIO_3$ optical parametric amplifier (OPA) used to generate the idler output near 5 µm. The IR wavelength is determined to within ± 0.10 cm^{-1} using a FTIR spectrometer. The bandwidth of the IR pulses is narrow (<1 cm^{-1}). The energy of the pulses is typically ~ 1.5 µJ, which, for the vibrational mode of the solute under study, is enough to create a significant transfer of population from the $v = 0$ to $v = 1$ vibrational levels but not enough to populate higher vibrational levels of the solute.

The IR pulse is split into a weak probe beam, which passes down a computer-controlled variable delay line with up to 12 ns of delay and a strong pump beam. The pump and probe pulses are counterpropagating and focused into the center of the SCF cell. Typical spot sizes (1/e radius of E-field) were $\omega_0 \sim 120$ µm for the pump beam and $\omega_0 \sim 60$ µm for the probe beam. A few percent of the transmitted probe beam is split off and directed into an InSb detector. A reference beam is sent through a different portion of the sample. The reference beam is used to perform shot-to-shot normalization. The pump beam is chopped at half the laser repetition rate (900 Hz). The shot-to-shot normalized signal is measured with a lock-in amplifier and recorded by computer.

The high-pressure optical cell is comprised of a Monel 400 body, gold or teflon O-rings, and CaF_2 or sapphire windows. The two windows are secured using an opposed force–type seal incorporating Belleville spring washers. The cell is essentially leak-free at pressures up to 3000 psia (~ 200 atm).

Stable uniform temperatures are produced using an air-circulated oven controlled by a fuzzy-logic controller/power supply. Two 100Ω platinum resistance temperature detectors are inserted in the body of the cell to permit careful measurement and control of the temperature to within $\pm 0.1°$C. Great care was taken to ensure thermal homogeneity of the sample, an important criterion for working near the critical point. A syringe pump was used to generate the variable pressures required to compress the fluid to the desired density. The pressure was accurately monitored with a precision strain-gauge transducer.

The experiments were conducted on the asymmetric CO stretching mode of $W(CO)_6$ near 1990 cm^{-1} (5.03 µm). The concentrations were $\sim 10^{-5}$ mol/L. The laser frequency was tuned to the absorption peak at each density and temperature. A typical sample was prepared by placing a few micrograms of solid $W(CO)_6$ on the tip of a syringe needle, inserting the powder into the center of the SCF cell through a side port, sealing the cell, and then compressing the system to the final operating pressure with high-purity CO_2 [99.995%; $T_c = 304.1$ K, $P_c = 1070$ psia, $\rho_c = 10.63$ mol/L (74)], ethane [99.99%; $T_c = 305.4$ K, $P_c = 706$ psia, $\rho_c = 6.88$ mol/L (74,75)], or fluoroform [99.9%; $T_c = 299.1$ K, $P_c = 699$ psia, $\rho_c = 7.56$ mol/L (76)]. The SCF cell was flushed repeatedly with gas before final pressurization to eliminate any air introduced into the system during insertion of the solid solute.

The optical density (OD) of the sample was varied by repeatedly diluting the mixture with fresh gas until a value of roughly 0.8–1.2 was obtained; in some cases, scans were acquired with ODs as low as $\sim$0.1. Absorbance measurements were made directly in the cell using a Mattson Research Series FT-IR spectrometer (0.25 cm^{-1} resolution) configured for external beam operation. The optical layout makes it possible to easily switch from making ps pump-probe measurements to recording an IR spectrum.

Vibrational lifetimes in supercritical fluids were obtained by fitting the data to a convolution of the instrument response and an exponential using a grid-search fit method. Vibrational peak positions were obtained by subtracting a background spectrum of the pure SCF, taken at the experimental pressure and temperature, from the solute-solvent sample spectrum. This technique removes small solvent peaks that can distort the spectrum.

Gas phase experiments utilized the same laser system but employed a different sample cell, a 1.5 cm long stainless steel cell with CaF_2 windows. A turbomolecular pump was used to evacuate the cell down to the vapor pressure of the $W(CO)_6$. The cell was heated slightly to 326 K to increase the vapor pressure and produce an optical density of $\sim$1.0. The gas phase decays are not single exponentials, but rather tri-exponentials. This tri-exponential character will be discussed in the results section.

In all of the experiments, careful power-dependent studies were performed to ensure that the data are free of artifacts. Studies at various optical densities were performed that confirmed that the experimental results are independent of the concentration of $W(CO)_6$ used.

III. RESULTS

A. Density Dependence

Using supercritical fluids as solvents in the study of vibrational relaxation permits the independent control of temperature and density. Figure 2 shows a typical pump-probe scan of the asymmetric CO stretching mode of $W(CO)_6$ in a supercritical fluid. This decay was taken in supercritical ethane

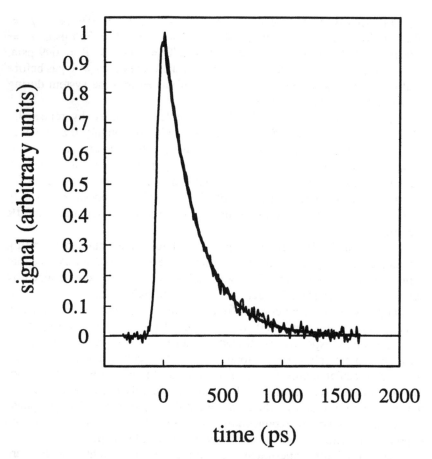

Figure 2 Pump-probe data for the asymmetric stretching CO mode of $W(CO)_6$ ($\sim$1990 cm^{-1}) in supercritical ethane at the critical density (6.88 mol/L) and 343 K. The heavy line is a fit to a single exponential. The lifetime, T_1, equals 278 ps. Data scans in other solvents, temperatures, and densities were of similar quality.

at a temperature of 70°C and a density of 6.88 mol/L. A single exponential fit with a T_1 value of 278 ps is also shown in the figure. Single exponentials yield excellent fits for decays in supercritical ethane, carbon dioxide, and fluoroform at all temperatures and densities studied (densities $\geq$1 mol/L).

Figure 3 shows the lifetime data as a function of density in the three SCF solvents at $\sim$2 K above the critical temperature, T_c (upper panels), and at $\sim$20 K above T_c (lower panels): $T_c = 32.2, 25.9,$ and

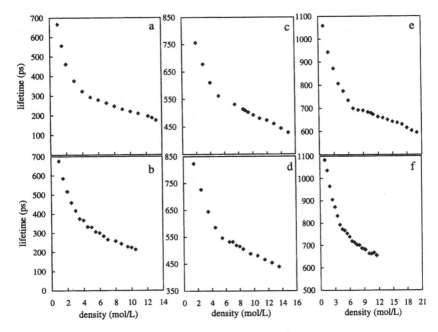

Figure 3 Vibrational lifetimes for the asymmetric CO stretching mode of $W(CO)_6$ vs. density along two isotherms of three polyatomic supercritical fluids: ethane (34°C panel a and 50°C panel b), fluoroform (28°C panel c and 44°C panel d), and carbon dioxide (33°C panel e and 50°C panel f). The upper panel for each solvent is an isotherm at 2°C above the critical temperature. In all six data sets, error bars (representing one standard deviation) are approximately the size of the points.

31.0°C, for ethane, fluoroform, and CO_2, respectively. The near-critical isotherm temperatures for the three solvents were chosen so that the reduced temperatures (T/T_c) are essentially the same. For the near-critical isotherm, the lifetime decreases rapidly as the density is increased from low density. As the critical density, ρ_c, is approached, the rate of change in the lifetime with density decreases ($\rho_c = 6.88$, 7.56, and 10.6 mol/L, for ethane, fluoroform, and CO_2, respectively). The change in slope is more pronounced for CO_2. This is even more evident when the gas phase (zero density) contribution to the lifetime is removed from the data (see below). The difference between the CO_2 data and the other data will be discussed in Section V. The higher temperature isotherm data show a somewhat more uniform density dependence. Although the slope decreases with increasing density, there is a somewhat smaller change in slope than in the near-critical isotherm data.

A number of studies of electronic transitions (31–33) have displayed a three-regime density dependence. At low densities, the observables change rapidly with density. Over an intermediate density range around ρ_c, a plateau region exists where the observables are essentially density independent. At still higher densities, the observables once again change with increasing density. The vibrational lifetime data do not display a distinct plateau region on the near-critical isotherm. This will be discussed further in Section V.

In addition to measuring T_1, the CO asymmetric stretch absorption line position was measured as a function of density in the three solvents at the same two temperatures. The results are presented in Fig. 4 as the line

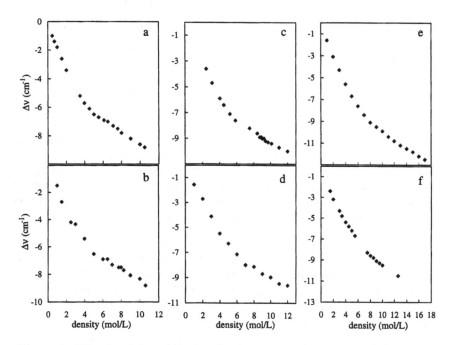

Figure 4 Vibrational line shift data for the asymmetric CO stretching mode of W(CO)$_6$ vs. density along two isotherms of three polyatomic supercritical fluids, ethane (34°C panel a and 50°C panel b), fluoroform (28°C panel c and 44°C panel d), and carbon dioxide (33°C panel e and 50°C panel f). The data are presented as the difference between the absorption peak frequency and the gas phase (zero density) peak in cm^{-1}. The negative sign means that the shift is to lower energy. The upper panel for each solvent is an isotherm at 2°C above the critical temperature. Error bars on each point are approximately 0.1 cm^{-1}.

position shift relative to the peak position at zero density. The absorption line position at zero density (gas phase spectrum) is 1997.3 cm^{-1}. The absorption line red shifts as the density is increased.

B. Gas Phase Vibrational Dynamics

To analyze the density-dependent vibrational lifetime data displayed in Fig. 3, it is necessary to separate the contributions to T_1 from intramolecular and intermolecular vibrational relaxation. The intermolecular component of the lifetime arises from the influence of the fluctuating forces produced by the solvent on the CO stretching mode. This contribution is density dependent and is determined by the details of the solute-solvent interactions. The intramolecular relaxation is density independent and occurs even at zero density through the interaction of the state initially prepared by the IR excitation pulse and the other internal modes of the molecule. Figure 5 shows the extrapolation of six density-dependent curves (Fig. 3; three solvents, each at two temperatures) to zero density. The spread in the extrapolations comes from making a linear extrapolation using only the lowest density data, which have the largest error bars. From the extrapolations, the zero density lifetime is $\sim$1.1 ns. To improve on this value, measurements were made of the vibrational relaxation at zero solvent density.

A gas phase pump-probe decay on the Q-branch of the CO asymmetric stretching mode of $W(CO)_6$ at 326 K is displayed in Fig. 6. The infrared frequency was tuned so that it was at the peak of the Q-branch at $\sim$1997.3 cm^{-1}. The pressure is so low that the average collision time is many orders of magnitude longer than the time scale of the measurements. The data are clearly nonexponential. The data can be fit very well with a triexponential. The average values obtained from a number of measurements are 140 ± 25 ps, 1.28 ± 0.1 ns, and >100 ns. These values are substantially different, permitting reliable separation of the decay components. The error bar corresponds to one standard deviation. The slowest component is too slow to obtain an accurate measurement of the decay time, but it is clearly very long. Because the extrapolation of the density-dependent curves to zero density (Fig. 5) gave $\sim$1.1 ns for T_1, we identify the 1.28 ± 0.1 ns decay component as the zero density T_1. This value will be used in the subsequent analysis of the density dependent data.

To explain the fastest and slowest components, a number of experiments were conducted to rule out various possibilities. Careful intensity studies showed no power effects. When the sample is at room temperature, $W(CO)_6$ has very little vapor pressure. No signal could be detected,

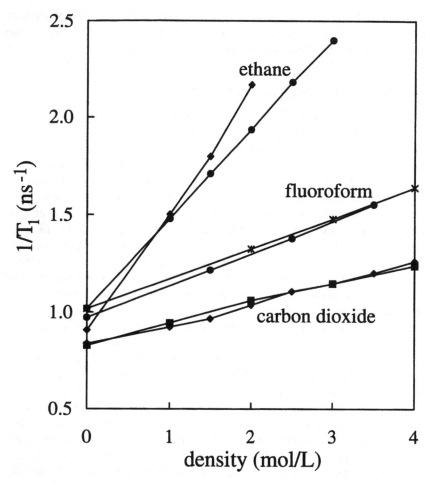

Figure 5 Vibrational relaxation rate data for the CO asymmetric stretch of $W(CO)_6$ in various supercritical fluids at low densities. Extrapolation of the data to zero density gives an estimate of the gas phase (collisionless) lifetime of $\sim$1.1 ns.

showing that the pump-probe signal is due solely to gaseous $W(CO)_6$ and not $W(CO)_6$ crystallized on the window surfaces. Experiments with the probe at the magic angle and a variety of other angles showed no change either in the decay times or the relative amplitudes of the components. Therefore, orientational dynamics are not contributing to the decays. (Orientational dynamics should occur on a time scale much faster than that of the

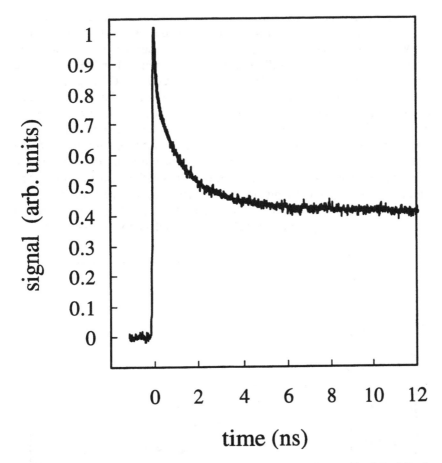

Figure 6 Typical pump-probe data for the CO asymmetric stretch of $W(CO)_6$ in the gas phase at 326 K. The fit (barely discernible) is a tri-exponential with components 113 ps, 1.26 ns, and 157 ns. The values reported in the text are averages of many different data sets taken on many samples.

experiments.) The only significant gas in the cell is $W(CO)_6$ itself. $W(CO)_6$ is at a very low concentration of 10^{-5} mol/L, which dictates a hard sphere time between collisions of $\sim 10^{-6}$ s. The possibility of very long range interactions was tested. By varying the temperature a few degrees, the $W(CO)_6$ concentration (vapor pressure) can be varied by more than an order of magnitude. The change in concentration did not affect the tri-exponential decay.

The explanation for the slowest component in the tri-exponential decay is relatively straightforward. The frequencies of all of the modes of W(CO)$_6$ are known (77–79). Using these frequencies, the average total internal vibrational energy (all modes) was determined. For T = 326 K, the average energy is 2900 cm^{-1}. When the high-frequency CO stretch relaxes, 2000 cm^{-1} of additional energy are deposited into the low-frequency modes. Because the molecules are collision-free on the time scale of the experiments, the extra 2000 cm^{-1} cannot leave the molecule. The deposition of energy raises the vibrational temperature for the average molecule to ~450 K. The data in Fig. 7 show the measured temperature dependence of the Q-branch peak position as a function of temperature. The figure shows that as the temperature is increased, the spectrum red shifts slightly. By 450 K, we estimate that the peak of the Q-branch has shifted ~1.1 cm^{-1} from its position at 326 K, the initial sample temperature.

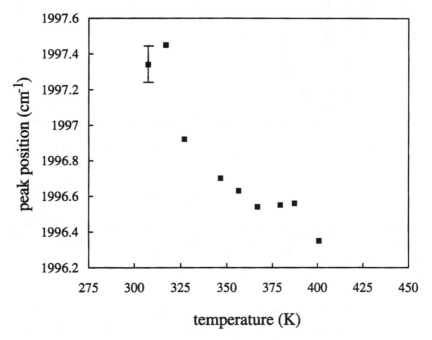

Figure 7 Absorption frequency (Q-branch) vs. temperature for the CO asymmetric stretch of W(CO)$_6$ in the gas phase. A representative error bar is shown. Extrapolation to 450 K (internal vibrational temperature following relaxation of the 2000 cm^{-1} CO stretch) yields a temperature-dependent shift of ~1.1 cm^{-1} from the peak position at 326 K, the initial sample temperature.

Following a pump pulse that produces vibrational excitations, the probe pulse transmission through the sample is increased. A pump-probe signal has two contributions, stimulated emission and ground state depletion. These two contributions are equal in magnitude; each contribution is 50% of the signal. As the excited state population relaxes into a combination of low-frequency modes, the stimulated emission is reduced. Normally, it is expected that relaxation of the excited state population fills in the ground state depletion. Thus, the stimulated emission and the ground state depletion decay together.

However, in the system under study here, relaxation of the 2000 cm^{-1} CO stretch into a combination of low-frequency modes heats the molecule substantially and shifts the absorption spectrum to the red. First, consider the case in which the absorption shift is very large. Although the CO stretch is no longer excited, the ground state absorption *at the laser frequency* does not recover because the molecules that have undergone excitation and relaxation do not absorb at their initial frequency. The stimulated emission contribution to the signal is lost, but the ground state depletion at the probe frequency still exists. Thus, the signal will decay to 50%, and, in the absence of other processes, the signal will remain at 50%. Examining Fig. 6, it is seen that the slow component of the decay is somewhat below 50%. The decay to a value of less than 50% occurs because the heating-induced spectral shift is relatively small. The shift of ~ 1.1 cm^{-1} does not completely shift the relaxed population off of the probe wavelength because the probe bandwidth is ~ 0.8 cm^{-1}. Thus the ground state depletion within the probe bandwidth recovers to some extent, and the signal decays to less than 50%. The signal will eventually recover as infrequent collisions cool the molecules and cold molecules from other parts of the cell move into the probe volume. We estimate that the time scale for these processes is approximately the same, i.e., ~ 1 μs. Thus, over the 12 ns of delay used in the experiments, the long component appears essentially flat.

To more fully understand the nature of this long component, we performed experiments with various argon pressures in the cell. Figure 8a shows a decay curve with a moderate pressure (18 psia, ~ 0.05 mol/L) of Ar. The slow decay component is now noticeably faster, with a decay time of 8.1 ns. Collisions with Ar cool the low-frequency modes, allowing the spectrum to return to its initial wavelength, which results in ground state recovery. Figure 8b shows data with an even higher Ar pressure (1100 psia, ~ 3 mol/L). The decay is now a single exponential. The collision-induced cooling is fast compared to T_1, and the stimulated emission and ground state depletion decay together, eliminating the long-lived component. It

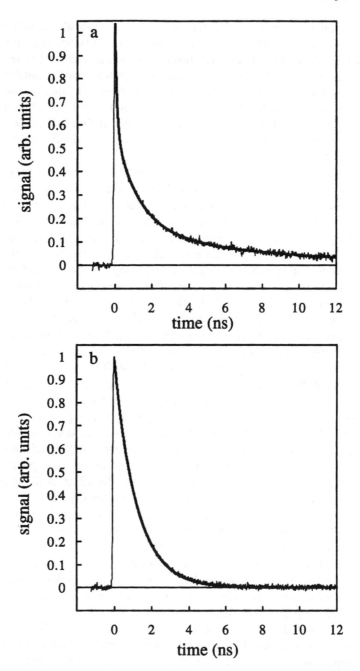

is interesting to note that while Ar is effective in causing the energy in the low-frequency modes to relax into the bath, the value of T_1 remains unchanged. This is in contrast to the experiments in polyatomic supercritical fluids, where T_1 becomes shorter as the solvent density is increased. The difference between Ar and polyatomic solvents will be discussed further below.

We propose that the fast component of the tri-exponential decay (140 ps) can be explained by the influence of the low-frequency modes of the molecule on the high-frequency CO stretch. In the gas phase vibrational experiment, prior to application of the pump pulse, the initial state of the molecule is prepared by its last collision with the wall of the cell or another molecule. The initial state is a complex superposition of eigenstates. Each mode, λ, will have some occupation number, n_λ. Under collision-free conditions, for a given molecule the n_λ are fixed. At the 326 K sample temperature, the average total internal vibrational energy of a molecule is 2900 cm^{-1}, and the density of states at this energy (calculated with the harmonic approximation) is 5×10^5 states/cm^{-1} (78,80). Thus, there are a vast number of initial states of the molecules that comprise the experimental ensemble. Absorbing a photon, which takes a molecule from the 0 to 1 state of the high-frequency CO stretch, changes the potential of the low-frequency modes. The change in potential produces a time evolution that influences the CO stretch optical transition probability measured by the probe pulse.

As a first approach to describing the fast time evolution of the system quantitatively (71), a description using a harmonic basis set with anharmonic coupling of the high-frequency CO stretch and the other modes can

---

Figure 8 Pump-probe scans on the CO asymmetric stretch of $W(CO)_6$ in argon at two different pressures at 333 K (60°C). (a) Pressure = 18 psia (~1 atm). Argon density of ~0.05 mol/L and inverse collision frequency of ~0.2 ns. The decay is fit with tri-exponential components 107 ps, 1.29 ns, and 8.1 ns. The long component has become substantially faster compared to collision-free conditions. Collisions with argon cause vibrational cooling of the low-frequency modes of $W(CO)_6$. The cooling reduces the spectral shift, which is responsible for the long decay component, and the decay becomes faster. (b) Pressure = 1100 psia. Argon density of ~3 mol/L. The decay is now a single exponential with lifetime 1.20 ns. The fastest and slowest tri-exponential components observed in the gas phase (Fig. 6) are eliminated by rapid collisions with argon at this high pressure (inverse collision frequency of ~3 ps). The lifetime is unchanged from the gas phase value, within experimental error.

be employed. The Hamiltonian for the system is

$$H = \sum_\lambda \left(\frac{1}{2} P_\lambda^2 + \frac{1}{2} \omega_\lambda^2 q_\lambda^2 \right) + \left(\frac{1}{2} P^2 + \frac{1}{2} \Omega^2 Q^2 \right) + V(Q, \{q_\lambda\}) \qquad (1)$$

where the sum is over the harmonic part of all modes, q_λ, other than the high-frequency asymmetric CO stretching mode, the second term in brackets is for the CO stretching mode, Q, and the last term is the potential for the anharmonic interactions among modes. Since the experiments only involve the ground and first excited vibrational states of the CO stretch, the Hamiltonian can be restricted to that space. Then,

$$H = \sum_\lambda \left(\frac{1}{2} P_\lambda^2 + \frac{1}{2} \omega_\lambda^2 q_\lambda^2 \right)$$
$$+ E_0 |0\rangle\langle 0| + (E_0 + \Omega) |1\rangle\langle 1| + V_0 |0\rangle\langle 0| + V_1 |1\rangle\langle 1| \qquad (2)$$

where

$$V_0 = \langle 0|V(Q, \{q_\lambda\})|0\rangle \equiv V_0(\{q_\lambda\}) \qquad (3)$$
$$V_1 = \langle 1|V(Q, \{q_\lambda\})|1\rangle \equiv V_1(\{q_\lambda\}) \qquad (4)$$

V_0 is the anharmonic interaction when the CO stretch is in the ground vibrational state, and V_1 is the anharmonic interaction when the CO stretch is in its first excited vibrational state. To simplify the analysis, take $E_0 = 0$, and transform into the interaction representation with respect to all of the $\{q_\lambda\}$. Furthermore, take $V_0 = 0$, which restricts the analysis to the excited vibrational state but does not fundamentally change the nature of the results. Then, the effective Hamiltonian for the CO stretch is

$$H_{\text{eff}}(t) = \Omega |1\rangle\langle 1| + V_1(\{q_\lambda(t)\}) |1\rangle\langle 1| \qquad (5)$$

Initially the CO stretch is in the ground state, so $\Psi = |0\rangle$. The laser prepares the state $|1\rangle$, and the system evolves as

$$|\Psi(t)\rangle = U(t)|1\rangle = e^{-i\Omega t} e_0^{-i \int_0^t V_1(q_\lambda(\tau))d\tau} |1\rangle \qquad (6)$$

where $U(t)$ is the time propagator and e_0 is the time-ordered exponential. For a pump-probe experiment, we need the probability, $P(t)$, of making a transition back to $|0\rangle$. $P(t)$ depends on the dynamics of the low-frequency modes and is obtained through an average over the initial states of $\{q_\lambda\}$

$$P(t) \propto \overline{|\langle 0|\mu|\Psi(t)\rangle|^2} \propto \overline{|\langle 1|\Psi(t)\rangle|^2} \qquad (7)$$

where μ is the transition dipole operator and the bar over the expression indicates the average over initial states. Substituting Equation (6) into Equation (7) gives

$$P(t) = A\left|\overline{\langle 1|e_0^{-i\int_0^t V_1(q_\lambda(\tau))d\tau}|1\rangle}\right|^2 \tag{8}$$

with A a constant. As an initial approximation, the average in Equation (8) was replaced by a second cumulant approximation, and a simple model was used in which the potential V_1 depends on the coupling between modes as

$$V_1(\{q_\lambda\}) = \sum_{\lambda,\mu} c_{\lambda\mu} q_\lambda q_\mu \tag{9}$$

where $c_{\lambda\mu}$ is the coupling constant and $\lambda \neq \mu$. Evaluation of the resulting expressions yields (71)

$$
\begin{aligned}
P(t) = A\exp\Bigg[&-2\sum_{\lambda,\mu} |c'_{\lambda,\mu}|^2 \Big\{ [n_\lambda n_\mu + (n_\lambda + 1)(n_\mu + 1)] \\
&\times \left(\frac{1 - \cos[(\omega_\lambda + \omega_\mu)t]}{(\omega_\lambda + \omega_\mu)^2}\right) + [n_\lambda(n_\mu + 1) + n_\mu(n_\lambda + 1)] \\
&\times \left(\frac{1 - \cos[(\omega_\lambda - \omega_\mu)t]}{(\omega_\lambda - \omega_\mu)^2}\right) \Big\} \Bigg]
\end{aligned}
\tag{10}
$$

Even for the simple model of cubic anharmonic coupling ($Qq_\lambda q_\mu$) of the high-frequency mode Q and the low-frequency modes $\{q_\lambda\}$, Equation (10) demonstrates that the pump-probe signal will have a time dependence in addition to the vibrational lifetime, T_1. (Note that for $W(CO)_6$, $\omega_\lambda = \omega_\mu$ will not appear in the sum since the coupling coefficient for such terms will vanish. Q is an antisymmetric mode. For $\omega_\lambda = \omega_\mu$, the modes q_λ and q_μ are members of a degenerate mode. Therefore, the direct product of their irreducible representations is symmetric, and the matrix element vanishes by symmetry.) Equation (10) describes the time dependence of a single molecule. The time dependence depends on the frequencies of the low-frequency harmonic modes, the coupling strengths of the modes, and the occupation numbers of the modes. While the harmonic mode frequencies and the coupling strengths will be identical for all molecules, the occupation numbers will differ. Each molecule will display a complex time dependence. $W(CO)_6$ has 27 low-frequency modes ranging in frequency from 60 to 590 cm^{-1}. Thus, the time dependence will be complicated and recurrences will occur. An ensemble average of Equation (10) was performed numerically by properly selecting a large number of sets of occupation numbers and evaluating Equation (10) for each set and summing the results. An initial partial decay of P(t) was observed followed by oscillations. The oscillations occur because of the harmonic nature of the theoretical model. The occupation number of a given mode varies from one molecule to another, but because the modes are harmonic, the frequency does not change with the occupation number. It is anticipated that a more detailed theory that includes anharmonicity will eliminate the recurrences. Nonetheless, the theoretical treatment illustrates that even in the gas phase in the absence of collisions, a fast component can appear in the pump-probe signal.

The decay of the CO stretch is a single exponential when $W(CO)_6$ has substantial interactions with a solvent. A single exponential (aside from orientational relaxation in liquids) is observed even when very fast pulses are used in the experiments (81). In the gas phase, the transition frequency of the CO stretch evolves over a range of frequencies because of its time-dependent interaction with the low-frequency modes. When a buffer gas or solvent is added, collisions cause the coherent evolution of the slow modes to be interrupted frequently, possibly averaging away the perturbation responsible for the observed fast time dependence. Thus, the fastest and slowest components of the tri-exponential decay are inherently low-pressure, gas phase phenomena.

IV. THEORY OF T_1 IN SUPERCRITICAL FLUIDS

Using a Fermi's Golden Rule approach, if the coupling between the oscillator and the bath modes is weak, then, to first order, the transition rate from the first excited vibrational level to the ground state is given by (3)

$$k(\rho, T) = T_1^{-1} = \frac{1}{\hbar^2} \int_{-\infty}^{\infty} dt \ e^{i\omega t} \langle V_{10}(t) V_{01}(0) \rangle \tag{11}$$

where $\hbar$ is Planck's constant divided by 2π. For a diatomic, ω is the oscillator frequency. However, for a polyatomic, in which some fraction of the energy of the initially excited mode can be transferred to other internal modes of the molecule, ω is the frequency corresponding to the amount of energy that is transferred to the bath (solvent). V_{ij} is the ij matrix element of the perturbation coupling the bath and oscillator. The brackets denote the thermally weighted average over all the quantum states of the bath. The perturbation V is usually assumed to be linear in the solute coordinate, i.e.,

$$V = -qF \tag{12}$$

so that F is the solvent force on the oscillator. If one makes the additional assumption that the oscillator is harmonic, q can be rewritten in terms of raising and lowering operators and Equation (11) can be reduced to

$$T_1^{-1} = \frac{1}{2m\hbar\omega} \int_{-\infty}^{\infty} dt \ e^{i\omega t} \langle F(t)F(0) \rangle_{qm} \tag{13}$$

where m is the oscillator mass. The "qm" subscript is a reminder that this is a quantum mechanical trace over the solvent. This force-force correlation function is difficult to calculate quantum mechanically, so that the relaxation rate is often calculated by

$$T_1^{-1} = \frac{1}{2m\hbar\omega} \int_{-\infty}^{\infty} dt \ e^{i\omega t} \langle F(t)F(0) \rangle_{cl} \tag{14}$$

where cl indicates that the force-force correlation function is classical.

However, as several authors have pointed out (5,7,82), it is incorrect to directly replace the quantum mechanical correlation function with its classical analog because the detailed balance condition will not be met. Therefore, the correct expression is

$$T_1^{-1} = \frac{Q}{2m\hbar\omega} \int_{-\infty}^{\infty} dt \ e^{i\omega t} \langle F(t)F(0) \rangle_{cl} \tag{15}$$

where Q is called the quantum correction factor and accounts for the limitations of the semi-classical approximation. The nature of the correction factor is a topic of considerable recent interest (82). Several forms of Q have been suggested, and they are generally functions of temperature and oscillator frequency. The choice of an appropriate correction factor will be discussed in greater detail below.

The calculation of the lifetime is thus reduced to the problem of calculating $\langle F(t)F(0) \rangle$. This is a problem that has had a fairly long association with studies of solvation dynamics, where it usually appears in the context of efforts to model friction coefficients. A great deal of activity in this field has been directed at using the methods of density functional theory (83) to derive expressions for the correlation function that involve the thermodynamic parameters of the system (72,84), which themselves are often amenable to further analytical treatment or else may be determined experimentally or through simulations. In the treatment of vibrational relaxation in supercritical fluids, the density functional approach permits the use of known density- and temperature-dependent properties of the solvent in the calculation of T_1.

Accordingly, we express $\langle F(t)F(0) \rangle$ in terms of an effective thermodynamic potential A, which itself is calculated from the following many-component density functional expansion (84):

$$\beta A[\rho_i(\mathbf{r})] = \sum_i \int d\mathbf{r} \rho_i(\mathbf{r})[\ln \rho_i(\mathbf{r})/\rho_i^0 - 1]$$

$$- \frac{1}{2} \sum_{i,j} \int d\mathbf{r} \int d\mathbf{r}' C_{ij}(\mathbf{r} - \mathbf{r}')\delta\rho_i(\mathbf{r})\delta\rho_j(\mathbf{r}') + \cdots \quad (16)$$

where $\beta = 1/k_B T$ with k_B as Boltzmann's constant and T as the temperature, $\rho_i(\mathbf{r})$ is the local density of the ith component at the point $\mathbf{r}$, ρ_i^0 is the mean density of that component, $\delta\rho_i(\mathbf{r})$ is the density fluctuation at $\mathbf{r}$, i.e., $\delta\rho_i(\mathbf{r}) = \rho_i(\mathbf{r}) - \rho_i^0$, and $C_{ij}(\mathbf{r} - \mathbf{r}')$ is the two-particle direct correlation function between components i and j located at positions $\mathbf{r}$ and $\mathbf{r}'$. The right-hand side of Equation (16) represents the first two terms of an infinite expansion in powers of the density fluctuation, the coefficients of which are multiparticle direct correlation functions. This expression for the free energy is most often used in studies of dense fluids (85), where density fluctuations are typically small, and the series may be safely truncated at second order. If density fluctuations are large, as is the case near a critical point, it is not entirely clear whether cubic and quartic contributions to Equation (16) are important. In the absence of any definite information that would resolve this issue, we have employed Equation (16) as it stands.

The effective potential $V^{(i)}(\mathbf{r})$ that determines the equilibrium density distribution $\rho_i(\mathbf{r})$ of the ith component is obtained from the requirement of the vanishing of the local chemical potential, which is the functional derivative of the free energy with respect to $\rho_i(\mathbf{r})$. This yields

$$\rho_i(\mathbf{r}) = \rho_i^0 \exp(-\beta V^{(i)}(\mathbf{r})) \quad (17)$$

where

$$\beta V^{(i)}(\mathbf{r}) \equiv -\sum_j \int d\mathbf{r}' C_{ij}(\mathbf{r} - \mathbf{r}')\delta\rho_j(\mathbf{r}') \quad (18)$$

If $i = 1$ denotes the solvent and $i = 2$ the solute, and if, further, the solution is assumed to be infinitely dilute (as is the case in the experiments with

$W(CO)_6$), then the potential acting on the fixed solute at $\mathbf{r}$ is

$$\beta V^{(2)}(\mathbf{r}) = -\int d\mathbf{r}' C_{21}(\mathbf{r} - \mathbf{r}')\delta\rho_1(\mathbf{r}') \tag{19}$$

and the corresponding force (which we shall denote $\delta F^{(2)}(\mathbf{r})$) is $-\partial V^{(2)}(\mathbf{r})/\partial \mathbf{r}$. The generalization of these results to time t involves the replacement of $\delta\rho_1(\mathbf{r})$ by $\delta\rho_1(\mathbf{r}, t)$.

We now make the identification

$$\langle F(t)F(0)\rangle \equiv \frac{1}{3V}\int d\mathbf{r}\langle \delta F^{(2)}(\mathbf{r}, t) \cdot \delta F^{(2)}(\mathbf{r}, 0)\rangle \tag{20}$$

where V is the volume of the system, and the integral over $\mathbf{r}$ represents an average over all spatial locations of the fixed solute. Rewriting this expression in Fourier space and substituting the result into Equation (15), we can show that

$$T_1^{-1} \propto Q \int_0^\infty dt \cos(\omega t) \int dk k^2 |\hat{C}_{21}(\mathbf{k})|^2 \hat{S}_1(\mathbf{k}, t) \tag{21}$$

where constants independent of density and temperature have been omitted. $\hat{C}_{21}(\mathbf{k})$ is the Fourier transform of the solute-solvent direct correlation function and $\hat{S}_1(\mathbf{k}, t)$ is the dynamic structure factor.

The direct correlation function is k dependent and is calculated explicitly through the binary hard sphere expression (86)

$$\hat{C}_{21}(\mathbf{k}) = -2\pi(I_1 + I_2) \tag{22}$$

The variables I_1 and I_2 involve functions of R_1 and R_2 (the solvent and solute hard sphere diameters, respectively) and ρ_1 and ρ_2 (the solvent and solute number densities, respectively). I_1 and I_2 are defined in the Appendix.

For small k, $\hat{S}_1(\mathbf{k}, t)$ can be determined from hydrodynamics as the solution to the coupled equations for the conservation of mass, momentum and energy, which leads to (87)

$$\hat{S}_1(\mathbf{k}, t) = \hat{S}_1(\mathbf{k})\left[\left(1 - \frac{1}{\gamma}\right)e^{-D_T k^2 t} + \frac{1}{\gamma}\cos(c_s kt)e^{-\Gamma k^2 t}\right] \tag{23}$$

Here $\hat{S}_1(\mathbf{k})$ is the equilibrium structure factor of the solvent, which we approximate by the Ornstein–Zernike expression, given by (87)

$$\hat{S}_1(\mathbf{k}) = \frac{\rho_1 \kappa_T/\kappa_T^0}{1 + k^2\xi^2} \tag{24}$$

where ρ_1 is the number density of the solvent, κ_T is its isothermal compressibility, κ_T^0 is the isothermal compressibility of the ideal gas, and ξ is the correlation length of density fluctuations. Further, $\gamma \equiv C_p/C_V$ is the ratio of specific heats, D_T is the thermal diffusivity, c_s is the adiabatic speed of sound, and Γ is the sound attenuation constant. In light scattering experiments, the first term inside the brackets in Equation (23) corresponds to the zero-frequency Rayleigh peak, which is associated with thermal diffusion, while the second term corresponds to twin Brillouin peaks at the frequencies $\pm c_s k$ from the Rayleigh peak, which are associated with sound propagation.

Numerical analysis of $\hat{S}_1(\mathbf{k}, t)$ shows that contributions to the k integral in Equation (23) are overwhelmingly dominated by the function at large k, i.e., $k \approx 1 \text{ Å}^{-1}$. The first exponential decay time constant in Equation (23), $1/\tau_1(k)$, [$D_T k^2$ in Equation (23)] can be generalized to include large k using an expression by Kawasaki (88):

$$\frac{1}{\tau_1(k)} = \frac{k_B T}{8\pi\eta\xi^3} \left[1 + k^2\xi^2 + \left(k^3\xi^3 - \frac{1}{k\xi} \right) \tan^{-1}(k\xi) \right] \qquad (25)$$

where η is the viscosity. For small k, this expression reduces to $D_T k^2$.

Since we need to evaluate Equation (21) only at large k, we have determined that $1/\tau_2(k)$, [Γk^2 in Equation (23)] becomes sufficiently large that the second term in brackets in Equation (23) can be neglected in comparison to the first. Neglecting $1/\tau_2(k)$ at large k is supported by both experiment and theory. For small k, acoustic waves can propagate in the fluid as in a continuum, because the wavelength is much longer than the average intermolecular spacing. In a crystal, the shortest wavelength for a propagating wave is twice the lattice spacing. In a fluid, as the wavelength approaches the average solvent intermolecular spacing, a propagating wave will experience the disorder of the fluid, and such a wave is expected to damp rapidly. For even shorter wavelengths, wave propagation is not possible. These qualitative ideas are borne out by neutron scattering experiments (89–92) and molecular dynamics simulations (93–95). Neutron scattering experiments show that the Brillouin lines either are absent or are extremely broad at large k. The broad linewidths are attributed to very rapid damping of large k acoustic waves, i.e., Γ in Equation (23) becomes very large. Hard sphere molecular dynamics simulations also show that Γ becomes large at large k and that at sufficiently large k (k $\approx$ 1.5/solvent diameter Å^{-1}), acoustic waves do not exist (94). These considerations demonstrate that the propagating wave term in Equation (23) becomes negligible for sufficiently large wavevectors.

Therefore, for the large values of k that play a role in the calculation, the dynamic structure factor is given by

$$\hat{S}_1(\mathbf{k}, t) = \hat{S}_1(\mathbf{k}) \left[\left(1 - \frac{1}{\gamma}\right) e^{-t/\tau_1(k)} \right] \qquad (26)$$

As mentioned above, careful numerical inspection of the k integral in Equation (21) shows that all of the contribution to the integral occurs for k sufficiently large that Equation (26) is the appropriate form for the dynamic structure factor for evaluating the integral. Equation (23) is never used in the calculations, but it was presented to motivate the form of Equation (26).

Q in Equation (21) is a quantum correction factor. A number of quantum corrections have been suggested depending on the nature of the system. Most involve a function of the temperature and the oscillator frequency and apply to simple model theoretical systems. One form that seems to be more generally applicable is due to Egelstaff (96). Egorov and Skinner recently tested several correction factors on exactly solvable systems, as well as on vibrational relaxation in liquid oxygen (82,97). The Egelstaff correction performed the best overall. The Egelstaff correction is given by a prefactor

$$Q = e^{\hbar\omega/2k_B T} \qquad (27a)$$

and the time variable, t, is replaced as

$$t \rightarrow \sqrt{t^2 + (\hbar/2k_B T)^2} \qquad (27b)$$

in the classical force-force correlation function. In Equation (21), only the dynamic structure factor component of the force-force correlation function, $\hat{S}_1(\mathbf{k}, t)$, contains t, and the substitution, Equation (27b), is made in this term. Equation (21) is evaluated using Equations (22), (26), and (27). The time integral in Equation (21), including the Egelstaff correction factor, can be performed analytically. The k integral is performed numerically.

Before preceding, it is useful to consider the form of the force-force correlation function, which is given in Equation (21), with Equations (22), (24), (25), (26) and (27). The form of the force-force correlation function, derived using density functional formalism, is employed because it permits the use of very accurate equations of state for solvents like ethane and CO_2 to describe the density dependence and temperature dependence of the solvent properties. These equations of state hold near the critical point as well as away from it. Using the formalism presented above, we are able to build the known density and temperature-dependent properties of the solvent into the theory of vibrational relaxation. There is a choice inherent in the approach we have employed. We build in a very accurate description of the solvent properties and how they change with density and temperature. This is not doable using other approaches for the development of the force-force correlation function (9,10). In exchange for the detailed description of the solvent, we have made more approximate the description of the force-force correlation function. All theories of vibrational relaxation are approximate. The real question is what is necessary to provide a good description of the system under study.

A proper force-force correlation function is Gaussian at short time (the inertial dynamics of the solvent) (10,98). The Gaussian behavior is usually modeled as $1-at^2$, where a is a constant. Following this initial rapid drop, the decay becomes slower, reflecting structural dynamics of the solvent. The longer time behavior can have a complicated functional form. If the Fourier transform in Equation (21) is taken at very high frequency, the very short time inertial component of the correlation function will make the major contribution. Thus, to calculate the vibrational relaxation of a high-frequency diatomic, properly describing the inertial part of the correlation function is essential. However, for a polyatomic molecule, in which most of the energy flows into a combination of intramolecular modes, the Fourier transform frequency, ω, will be substantially lower than the initially excited oscillator frequency, and the details of the very short time behavior of the correlation function will be less important.

The force-force correlation function used here has a complicated form that can be determined by numerical evaluation. We examined the correlation function for ethane at 50°C as well as the critical density. With the Egelstaff quantum correction, the correlation function initially decays as $1-at^2$ for a very short time ($\sim$15 fs). It then slows and becomes progressively slower at longer times. As mentioned above and as will be discussed in detail in connection with the experiments, the Fourier transform is taken at a relatively low frequency (150 cm^{-1}), not the 2000 cm^{-1} oscillator frequency. For low frequencies, the very short time details of the correlation function are not of prime importance. Without the quantum correction, the strictly classical correlation function does not begin with zero slope at zero time, but, rather, it initially falls steeply. However, the quantum corrected function and the classical function have virtually identical shapes after $\sim$15 fs. As will be demonstrated below, the force-force correlation function contained in Equation (21) with Equations (22), (24), (25), (26), and (27) does a remarkable job of reproducing the density dependence observed experimentally. The treatment also works very well for describing the temperature dependence of the vibrational relaxation as discussed below. While the correlation function does not rigorously have the correct short time functional form, by incorporating a detailed description of the solvent properties it is able to provide a very good description of the vibrational relaxation as a function of density and temperature.

A wide variety of density- and temperature-dependent input parameters are required. These include ρ_1, the number density of the solvent, κ_T, the isothermal compressibility, ξ, the correlation length of density fluctuations, $\gamma \equiv C_p/C_v$, the ratio of specific heats, and η, the viscosity. Very accurate equations of state for ethane (74,75,99) and CO_2 (74) are available that provide the necessary input information. The necessary input parameters for fluoroform were obtained by combining information from a variety of sources (76,100,101). There is somewhat greater uncertainty in the fluoroform parameters.

The thermodynamic and hydrodynamic parameters that enter the theory build in a detailed description of the SCF solvent. All of these input parameters vary substantially with density and temperature. In the near-critical region, the variations of the parameters are enormous. In comparing the theory with experiment in the next section, the solvent thermodynamic/hydrodynamic parameters feed into the calculations. As discussed below, the zero density T_1 is removed from the data, and the resulting density- and temperature-dependent lifetimes, $T_1(\rho, T)$, are compared to the theory. The theoretical curves are scaled to match the data at one particular temperature and density point. This accounts for constants not included in the calculation, some of which are related to the magnitude of the solute-solvent coupling. In addition, the solute and solvent hard sphere diameters are input parameters, though they should be approximately equal to estimates of the hard sphere diameters from crystallography data and other means. The results are not highly sensitive to the solute diameter but are more sensitive to the solvent diameter. Small reductions in the solvents' hard sphere diameters improve the agreement between theory and experiment. The adjustable parameter in Equation (21) that has a substantial effect on the shape of the curves is the frequency, ω. In a diatomic, ω is the frequency of the vibrational transition since all of the energy must be deposited into the solvent for relaxation to occur. For a high-frequency mode of a polyatomic with many internal lower frequency modes, a substantial amount of energy can go into internal modes of the molecule (102). This is clear from the gas phase decays, which show that vibrational relaxation of the ~ 2000 cm^{-1} CO asymmetric stretch can occur even in the absence of solvent. For a polyatomic, $\hbar\omega$ corresponds to the energy that is deposited into the solvent when vibrational relaxation occurs. Since the details of the pathway or pathways for relaxation are not known, the frequency ω is used as an adjustable parameter to fit the data.

Figure 9 shows calculations of $T_1(\rho, T)$ versus ρ for several ω. The input parameters are those for the solvent ethane at 34°C (see below for more details). The calculations are shown as reduced lifetimes, i.e., each curve is normalized at 1 mol/L. The frequency for each curve is given in the figure caption. The results show that both the shape of the curves and the magnitude of the density dependence change with ω. As the frequency is increased, the lifetime changes less with increasing density. The highest frequency curves may have substantial error because they will depend on

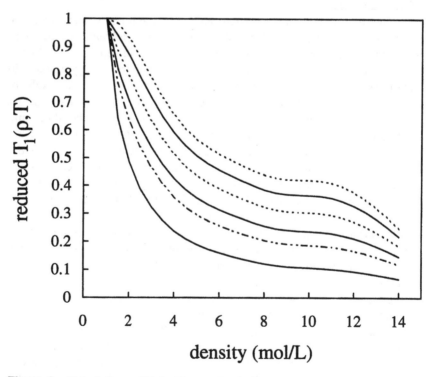

Figure 9 Calculations of $T_1(\rho, T)$ vs. ρ for various choices of ω. The input parameters are those for the solvent ethane at 34°C. The calculations are shown as reduced lifetimes, i.e., each curve is normalized at 1 mol/L. The curves, from top to bottom, correspond to ω values 2000, 1000, 500, 250, 150, and 50 cm^{-1}.

the details of the very short time portion of the force-force correlation function.

V. COMPARISON OF THEORY AND EXPERIMENT

A. Density Dependence

Figure 10a shows the density dependence of the solvent-induced lifetime $T_1(\rho, T)$ in ethane along an isotherm at 34°C, very close to the critical temperature (T_c). The values in the figure are obtained from the measured lifetimes and the intrinsic (zero density) lifetime using

$$\frac{1}{T_1(\rho, T)} = \frac{1}{T_1^m} - \frac{1}{T_1^g} \tag{28}$$

T_1^g is the gas phase lifetime, 1.28 ± 0.1 ns (see above), and T_1^m are the measured lifetimes given in Fig. 3. Also given in Fig. 10a is the best theoretically calculated curve. The theory is scaled to match the data at the critical density. The solute hard sphere diameter is taken to be 6.70 Å, the hard sphere value obtained from recent x-ray data (103). The solvent hard sphere diameter is 3.94 Å, which is a small ($\sim 7\%$) reduction from the literature value (104). To obtain the good agreement displayed in the figure, the frequency ω was chosen to be 150 cm^{-1}. Figure 10b shows the density-dependent data along a higher temperature isotherm, 50°C, in ethane. The equation of state for ethane at 50°C is used to obtain the fluid parameters at this temperature. The other parameters are the same as those used in the 34°C calculation. Thus, the calculated curve for the 50°C data is obtained without recourse to adjustable parameters. As can be seen from the figure, the theory does a very good job of reproducing the data.

The $T_1(\rho, T)$ data in ethane show only a mild change in going from 34 to 50°C. However, the theory's ability to reproduce the data at both temperatures is not because the system is little changed in going from one temperature to another. Figure 11 shows plots of the solvent parameters that enter the theory versus density at the two temperatures. As can be seen, most of the curves change substantially with temperature. However, these changes balance out in the theory to give curves that are in accord with the small changes in the data.

Figure 12 shows $T_1(\rho, T)$ measured in fluoroform. Figure 12a shows data on the near critical isotherm of 28°C, i.e., 2 K above T_c. The calculated curve is scaled to match the data at the critical density, 7.5 M. (In

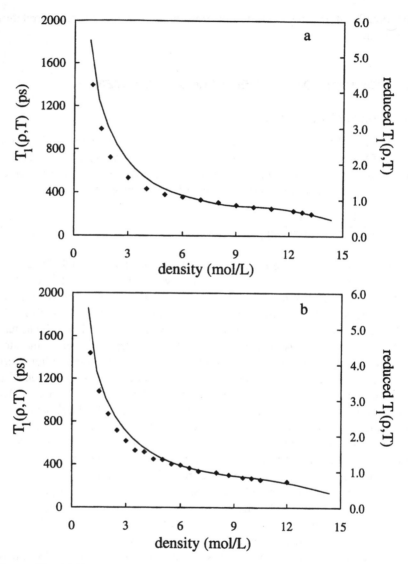

Figure 10 (a) $T_1(\rho, T)$ vs. density for the solvent ethane at 34°C and the best fit theoretically calculated curve. The theory was scaled to match the data at the critical density, 6.88 mol/L. The best agreement was found for $\omega = 150$ cm^{-1}. (b) $T_1(\rho, T)$ vs. density for the solvent ethane at 50°C and the theoretically calculated curve. The scaling factor, frequency ω, and the hard sphere diameters are the same as those used in the fit of the 34°C data. Given that there are no free parameters, the agreement is very good.

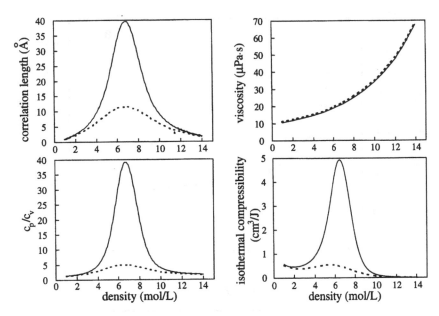

Figure 11 Plots of the ethane solvent parameters vs. density for the thermodynamic properties that enter the theory at the two temperature, 34°C (solid lines) and 50°C (dashed lines), the temperatures for the data and theory curves in Fig. 10. Most of the parameters change dramatically with temperature.

fluoroform, 2 mol/L is the lowest density for which $W(CO)_6$ has sufficient solubility to perform the experiments at 28°C.) ω is not adjusted, but is set equal to 150 cm^{-1}, the value obtained in the fit of the ethane data. An accurate value of the effective hard sphere diameter is not available for fluoroform. Hard sphere diameters tend to be slightly temperature dependent, becoming smaller as the temperature is increased (104). X-ray data are available at 70 K (105), well below the experimental temperature. The 70 K x-ray data yield a hard sphere diameter of 4.60 Å. The best agreement with the $T_1(\rho, T)$ at 28°C is obtained with 3.28 Å. The hard sphere diameter used is ~30% smaller than the 70 K x-ray determined value; it is not known how this value would compare to a good value of the effective hard sphere diameter at 301 K. The theory does a remarkably good job of reproducing the shape of the data with only the adjustment in the solvent size as a fitting parameter that affects the shape of the calculated curve. Figure 12b shows data taken at 44°C, which is the equivalent increase in temperature above T_c as the higher temperature data taken in ethane (Fig. 10). The calculated

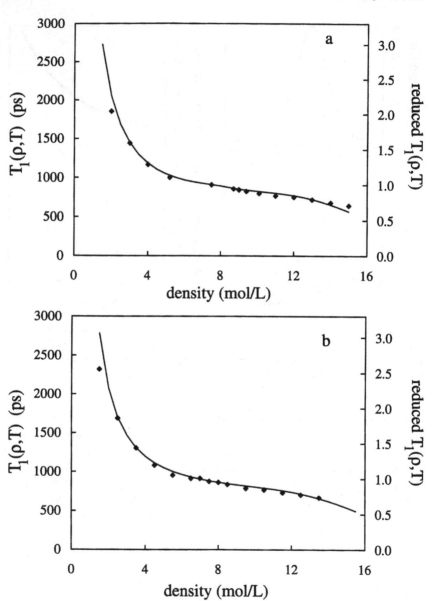

curve (solid line) is obtained using the thermodynamic/hydrodynamic input parameters for the higher temperature and using the same scaling factor as at 28°C. ω is 150 cm^{-1} and the fluoroform diameter is 3.28 Å. Therefore, there are no free parameters in the calculation of the lifetimes along this isotherm. As with the data taken in ethane, the theory does a very good job of reproducing the density-dependent data.

As shown in Fig. 9 and discussed in the theory section, the calculations are quite sensitive to the choice of ω. The fact that the same ω gives good theoretical agreement for the $T_1(\rho, T)$ data in both ethane and fluoroform suggests that this is not an arbitrary value, but, rather, it reflects the energy deposited to the solvent upon vibrational relaxation. A number of calculations were performed to examine the influence of ω on the calculation of $T_1(\rho, T)$. The curves in Fig. 9 are all for single values of ω. Calculations were also performed in which distributions of ω were used. It was found that ω need not be a single frequency. For example, an ensemble average over relaxation occurring with a Gaussian distribution of ω centered about 150 cm^{-1} with standard deviation of, e.g., 20 cm^{-1}, gave the identical results. We also tried to reproduce the data with various bimodal distributions. Such distributions were not able to reproduce the data. $\omega = 150$ cm^{-1} may be a single frequency or the average of a spread of frequencies about 150 cm^{-1}. This frequency is most likely located within the single "phonon" density of states (DOS) of the continuum of low-frequency modes of the solvents. Instantaneous normal mode calculations in CCl_4, $CHCl_3$, and CS_2 show cut-offs in the DOS at $\sim$150, $\sim$180, and $\sim$200 cm^{-1}, respectively (106,107). The higher frequency portions of the DOS are dominated by orientational modes. Since ethane, fluoroform, and CO_2 are much lighter

---

Figure 12   (a) $T_1(\rho, T)$ data measured in fluoroform on the near critical isotherm (28°C), 2 K above T_c, and the calculated curve, which is scaled to match the data at the critical density, 7.56 mol/L. The fluoroform hard sphere diameter was adjusted since a good value at experimental temperatures is not available. A diameter of 3.28 Å yielded the optimal fit. ω is not adjustable. It is set equal to 150 cm^{-1}, the value obtained in the fit of the ethane data. The theory does a very good job of reproducing the shape of the data with only the adjustment in the solvent size as a fitting parameter that affects the shape of the calculated curve. (b) $T_1(\rho, T)$ data taken at 44°C, which is the equivalent increase in temperature above T_c as the higher temperature data taken in ethane (Fig. 10b). The theory curve is calculated using the same scaling factor, frequency, and solvent hard sphere diameter as at the lower temperature. Considering that there are no free parameters, the theory does an excellent job of reproducing the higher temperature data.

than the molecules that comprise the liquids mentioned above, it is expected that their DOS will extend to higher frequencies.

In the gas phase, the asymmetric CO stretch lifetime is 1.28 ± 0.1 ns. The solvent can provide an alternative relaxation pathway that requires single phonon excitation (or phonon annihilation) (102) at 150 cm^{-1}. Some support for this picture is provided by the results shown in Fig. 8. When Ar is the solvent at 3 mol/L, a single exponential decay is observed with a lifetime that is the same as the zero density lifetime, within experimental error. While Ar is effective at relaxing the low-frequency modes of W(CO)$_6$, as discussed in conjunction with Fig. 8, it has no affect on the asymmetric CO stretch lifetime. The DOS of Ar cuts off at $\sim$60 cm^{-1} (108). If the role of the solvent is to open a relaxation pathway involving intermolecular interactions that require the deposition of 150 cm^{-1} into the solvent, then in Ar the process would require the excitation of three phonons. A three-phonon process would be much less probable than single phonon processes that may occur in the polyatomic solvents. In this picture, the differences in the actual lifetimes measured in ethane, fluoroform, and CO$_2$ (see Fig. 3) are attributed to differences in the phonon DOS at $\sim$150 cm^{-1} or to the magnitude of the coupling matrix elements.

Figure 13 shows a comparison of $T_1(\rho, T)$ data with calculated curves for the CO$_2$ solvent at two temperatures, 33°C (2 K above T_c) and 50°C. The calculated curves are scaled to the data at 33°C and 2 mol/L. Again, ω is set to 150 cm^{-1}. The solvent hard sphere diameter is the literature value, 3.60 Å (104). The agreement between theory and experiment, while not poor, is clearly not as good as that displayed for ethane and fluoroform. Adjusting ω does not improve the agreement between theory and data, nor does a further variation of the solvent diameter. The $T_1(\rho, T)$ data appear to differ from the data in the other two solvents. After the initial rapid decrease in the lifetime at low densities, the data curve with CO$_2$ as the solvent is much flatter than the data obtained in the other two solvents. The theory does well up to $\sim$6 mol/L but then overestimates the decrease in lifetime at higher densities.

An interesting question arises as to why the theory does not do as good a job of reproducing the data with CO$_2$ as the solvent. The CO$_2$ data becomes much flatter above $\sim$6 mol/L than the data taken in the other two solvents. The theoretical curves for the CO$_2$ solvent have shapes that are similar to the theoretical and data curves for the other two solvents. This might suggest that the CO$_2$ solvent data is modified by some special chemical interaction. The source of the discrepancy between theory and data does

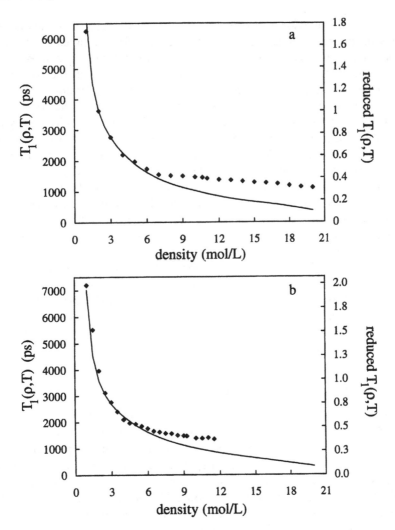

Figure 13 (a) $T_1(\rho, T)$ data with calculated curves for the CO_2 solvent at 33°C (2 K above T_c). The calculated curve is scaled to the data at 2.0 mol/L. $\omega = 150$ cm^{-1}, and the solvent hard sphere diameter is 3.60 Å. The agreement between theory and experiment, while not terrible, is clearly not as good as that displayed for ethane (Fig. 10) and fluoroform (Fig. 12). The data in CO_2 appear to be different than in ethane and fluoroform, as discussed in the text. (b) $T_1(\rho, T)$ data with calculated curves for the CO_2 solvent at 50°C. The scaling factor, frequency ω, and the hard sphere diameters are the same as those used in the fit of the 33°C data. Again, the theory drastically overestimates the slope at higher densities.

not appear to be associated with near-critical phenomena since the discrepancy for the near-critical isotherm is not reduced for the higher temperature isotherm. There may be a specific interaction between CO_2 and $W(CO)_6$. Supercritical CO_2 tends to form T-shaped structures with the oxygen of one CO_2 coordinated with the carbon of another CO_2, forming a T (109). It is possible that CO_2 has a tendency to coordinate with the oxygens of $W(CO)_6$ in the same manner. Such a specific interaction could modify the density dependence of the vibrational lifetime. Substantial association of CO_2 with the COs of $W(CO)_6$ would not be reflected in the hard sphere direct correlation function used in the calculations, nor would it occur in a Lennard-Jones description of the interactions. The specific $W(CO)_6/CO_2$ chemical interaction is an appealing explanation given the good agreement between theory and experiment for the other solvents. On the other hand, the reduced quality of the agreement for the CO_2 data may reflect inadequacies in the theory.

Figure 4 shows the density dependence of the vibrational absorption line shift in the three solvents at the same temperatures used in the T_1 measurements. The line shift behavior is seen to be similar to the vibrational lifetime trends. Notice that there are no substantial density-independent plateaus like those frequently observed in solvatochromic shifts of electronic transitions.

B. Temperature Dependence

The isothermal data in the three supercritical solvents all show similar trends. For the near-critical isotherm, the lifetime initially decreases rapidly with density. At an intermediate density less than ρ_c, the slope changes and develops a much weaker density dependence. Along the higher temperature isotherm, the density dependence is somewhat smoother and does not exhibit as substantial a change of slope. The isopycnic (constant density) data, however, display an interesting solvent dependence.

Figure 14 shows the lifetime dependence on temperature in ethane at a constant density equal to ρ_c. The data have a region, extending well above the critical temperature, in which the lifetime becomes longer as the temperature is increased. This behavior in ethane is unexpected and is in contrast to theoretically proposed trends of vibrational lifetimes with temperature at constant density (102,110). We will refer to this temperature range as inverted (16,111).

In general, an increase in temperature (at constant density) should decrease lifetimes. Vibrational relaxation is caused by fluctuating forces in

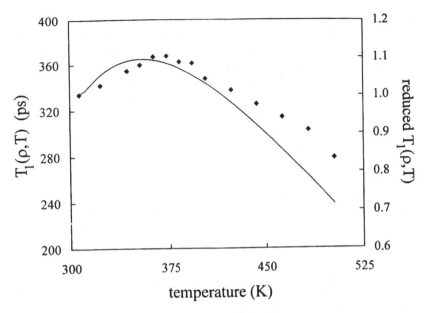

Figure 14 $T_1(\rho, T)$ vs. temperature for the solvent ethane at fixed density (the critical density) and the theoretically calculated curve. The scaling factor, frequency ω, and the hard sphere diameters are the same as those used in the fit of the 34°C density-dependent data, i.e., there are no free parameters in this calculation. Notice the presence of an inverted regime, i.e., a range of temperatures for which the lifetime increases with temperature, contrary to expected behavior. The lifetime peaks at ~375 K before decreasing with temperature. Remarkably, the theory captures this phenomenon, though it overestimates the drop in the lifetime with temperature after 375 K.

the bath acting on the vibrational oscillator, i.e., by the Fourier component of the force-force correlation function at ω, the frequency associated with energy deposition into the bath. Since the bath is incoherent, there are no well-defined phase relationships between modes that could lead to destructive interference at some frequencies. Thus, when the temperature is increased, all Fourier components of the force-force correlation function might be expected to increase in amplitude because all modes increase in population. The Fourier component of the force at ω would then increase with temperature, and vibrational relaxation, at constant density, would be expected to become faster.

It is important to emphasize again that the data displayed in Fig. 14 are at fixed density. At each temperature, the pressure was adjusted to bring the density back to 6.88 mol/L, the required pressure being determined from an accurate equation of state for ethane (74,75). Had the density not been held constant, the increase of the lifetime with temperature would not have been as remarkable. For instance, the lifetime of the CO asymmetric stretching mode of $W(CO)_6$ in liquid $CHCl_3$ has been observed to lengthen as the temperature is increased from the melting point to the boiling point of the liquid (16). But in this system, as the temperature increases, the liquid density decreases significantly. For the experiments in $CHCl_3$ liquid, the increase in lifetime with temperature was ascribed to effects arising from the decrease in density (16,106).

The solid line in Fig. 14 is the theoretically calculated temperature dependence of the lifetime using Equation (21). The frequency, the hard sphere diameter, and the scaling factor used in the calculation are those obtained from the fitting of the density dependent data in ethane at 34°C. Therefore, there are no adjustable parameters in the calculation of the theoretical curve in Fig. 14. The theory does a good job of predicting the overall temperature dependence of the lifetime. The inverted region between T_c and $\sim$375 K is reproduced, though the predicted lifetimes drop too rapidly with temperature past the inverted region.

The calculations include many detailed temperature-dependent properties of ethane (see Fig. 11). Many of these properties are highly temperature dependent in the range from the critical temperature to $\sim$70 K above T_c, and they come into the calculation in a complex manner. The temperature also enters explicitly in the Egelstaff quantum correction [Equation (27)] and in the Kawasaki expression for $1/\tau_1(k)$. It is the complex interplay of these changing parameters, which are determined by the equation of state of ethane, that is responsible for the initial increase of T_1 with temperature. By $\sim$70 K above the critical point, however, the various physical properties of ethane are no longer changing as rapidly. It is at this point that the lifetime decreases with increasing temperature. Once the last vestiges of the influence of the critical point are gone, the expected constant density temperature dependence is manifested, i.e., the lifetime decreases with increasing temperature. From a microscopic perspective, when the temperature is increased, the occupation numbers of all modes of the solvent increase. The amplitude of all solvent modes will increase including the Fourier component of the bath spectral density at ω. However, vibrational relaxation depends on the spectrum of forces, not on the spectral density of the solvent fluctuations. As the temperature is initially increased above the critical temperature, both the data and the theory show that the relevant Fourier component of the force-force correlation function initially decreases, leading to an increase in the lifetime. In ethane, once the temperature is well above the critical temperature, the physical properties of the fluid are not changing rapidly, and the normal temperature dependence (decreasing lifetime with increasing temperature) occurs.

Figure 15 shows the lifetime as a function of temperature at the critical density of carbon dioxide. With CO_2 as the solvent there is no inverted region in which the lifetime becomes longer as the temperature is increased. Instead, the lifetime decreases approximately linearly. Thus the inverted behavior is not universal but is specific to the properties of the particular solvent. The fact that the nature of the temperature dependence changes fundamentally when the solvent is changed from ethane to CO_2 demonstrates the sensitivity of the vibrational relaxation to the details of the solvent properties. The solid line is the theoretically calculated curve. The calculation of the temperature dependence is done with no adjustable

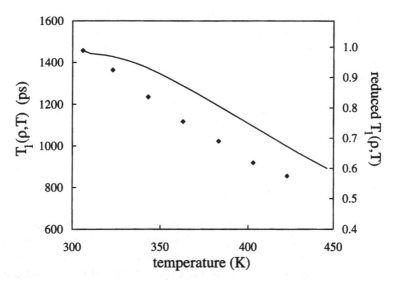

Figure 15 $T_1(\rho, T)$ vs. temperature for the solvent carbon dioxide at the critical density and the theoretically calculated curve. The frequency ω and the hard sphere diameters are the same as those used in the fit of the 33°C data. The theory is scaled to match the data at 33°C and the critical density, 10.6 mol/L. Unlike ethane at the critical density, there is no inverted region, and the vibrational lifetime decreases nearly linearly with temperature. The theory does not quantitatively fit the data, but it does show the correct general behavior. Most importantly, the hydrodynamic/thermodynamic theory shows the existence of the inverted region in ethane and the lack of one in carbon dioxide.

parameters other than the scaling factor used to match the theory with the data at 33°C. The input parameters are taken from the physical properties of the solvent and the parameters obtained from fitting the isothermal density dependent data at 33°C. The theory does a reasonable job of reproducing the data. The most important feature of the theoretical calculation is the absence of an inverted region. Aside from the hard sphere diameters, the only differences between the ethane calculations and the CO_2 calculations are the physical properties of the solvents obtained from the equations of state of the two fluids. The interplay of parameters that give rise to an inverted region in ethane and no inverted region in CO_2 are complex. The fact that the theory can reproduce this qualitative difference in the vibrational relaxation temperature dependence in the two solvents shows that it is capturing the essential features of vibrational relaxation as a function of temperature and density in supercritical fluids.

The theory reflects the solvent properties through the thermodynamic/hydrodynamic input parameters obtained from the accurate equations of state for the two solvents. However, the theory employs a hard sphere solute-solvent direct correlation function (C_{12}), which is a measure of the spatial distribution of the particles. Therefore, the agreement between theory and experiment does not depend on a solute-solvent spatial distribution determined by attractive solute-solvent interactions. In particular, it is not necessary to invoke local density augmentation (solute-solvent clustering) (31,112,113) in the vicinity of the critical point arising from significant attractive solute-solvent interactions to theoretically replicate the data.

VI.   CONCLUDING REMARKS

We have presented experimental and theoretical results for vibrational relaxation of a solute, $W(CO)_6$, in several different polyatomic supercritical solvents (ethane, carbon dioxide, and fluoroform), in argon, and in the collisionless gas phase. The gas phase dynamics reveal an intramolecular vibrational relaxation/redistribution lifetime of 1.28 ± 0.1 ns, as well as the presence of faster (140 ps) and slower (>100 ns) components. The slower component is attributed to a heating-induced spectral shift of the CO stretch. The fast component results from the time evolution of the superposition state created by thermally populated low-frequency vibrational modes. The slow and fast components are strictly gas phase phenomena, and both disappear upon addition of sufficiently high pressures of argon. The vibrational lifetime component does not change with the argon density, however, which is distinctly different from the behavior of the lifetime in the polyatomic supercritical solvents.

In all three SCF polyatomic solvents studied, the vibrational lifetime decreases with increasing density at constant temperature. The lifetimes show a significant dependence on the solvent, with the lifetimes shortest in ethane and longest in carbon dioxide. The spectral peak positions of the $W(CO)_6$ asymmetric CO stretching vibration in the three solvents shift with density in a manner similar to the trends observed for the lifetimes. The vibrational lifetime versus temperature data at fixed solvent density display a remarkable solvent-dependent behavior. Vibrational relaxation with ethane as the solvent actually slows down with increasing temperature for a significant range of temperatures above T_c. At sufficiently high temperature, the trend changes direction, and the lifetime decreases with further increases in temperature. The inverted temperature dependence does not occur when the solvent is CO_2. The lifetime decreases as the temperature is increased above T_c.

The theory that we have developed to explain the vibrational relaxation data in supercritical fluids is based on relations between the microscopic events and bulk solvent properties given by accurately known equations of state. The theory has a minimum of free parameters to fit: a scaling factor to account for various constants, the Fourier transform frequency, ω, which corresponds to the energy deposited into the solvent in the vibrational relaxation process, and the hard sphere diameter of the solvent, which is heavily constrained. All three variables were optimized once (the solvent diameter once for each solvent) and then fixed. The Egelstaff quantum correction is used to account for the calculation of a classical force-force correlation function. Overall, the agreement between theory and experiment is remarkably good. The theory can do a very respectable job of reproducing the density dependence at fixed temperature. It is also able to reproduce the temperature dependence at fixed density. Of particular importance is the theoretical reproduction of the inverted temperature dependence in ethane and the lack of an inverted region in CO_2.

Studying vibrational dynamics in supercritical fluids can aid in understanding the vibrational relaxation process as well as providing insights into the nature of solute-solvent interactions in supercritical fluids. One important result that emerged from the studies is the success of the theory to describe the data over a range of solvents, densities, and temperatures. The theory contains details of the solvent through the input of thermodynamic and hydrodynamic parameters obtained from the solvent's equation of state. However, the solute-solvent spatial distribution, which comes into the theory through the solute-solvent direct correlation function, is obtained with a hard sphere description. Therefore, it is possible to describe the data without invoking attractive solute-solvent interactions, which can give rise to local density augmentation (solute-solvent clustering). Other theoretical treatments can describe some of these same data using attractive solute-solvent interactions (114).

Local density augmentation is a common theme in the description of many experiments. It is a topic of a great deal of current research on super-critical fluids. The appearance of mild changes in experimental observables with large changes in density around the critical density is often ascribed to local density augmentation. The experiments presented here exhibit the type of behavior that might be explained by solute-solvent clustering. The theoretical analysis demonstrates that the vibrational relaxation experiments can be described with a theory that does not include attractive solute-solvent interactions and local density augmentation. This is an area of rapid growth in understanding. It is clear that the study of vibrational dynamics in super-critical fluids, with both experiment and theory, will continue to be an active area of research for some time to come.

ACKNOWLEDGMENTS

We would like to thank Professor J.L. Skinner, Department of Chemistry, University of Wisconsin at Madison, for many informative conversations pertaining to vibrational relaxation. We would also like to thank Professor R. Silbey, Department of Chemistry, Massachusetts Institute of Technology, for helping us with insights into the gas phase dynamics. We thank the Air Force Office of Scientific Research, which made this work possible through grant #F49620-94-1-0141. DJM acknowledges the NSF for a graduate fellowship. MS would like to thank the Mitsubishi Chemical Corporation for supporting his participation in this research.

APPENDIX

The following equations are used in the calculation of the direct correlation function, Equation (22) (86).

$$I_1 = \frac{2a_1}{k} \left[-\frac{\lambda}{k} \cos(k\lambda) + \frac{1}{k^2} \sin(k\lambda) \right] \tag{A1}$$

$$I_2 = a_1 I_{21} + \lambda^2 (b - 3d\lambda^2) I_{22} - 2\lambda(b - 4d\lambda^2) I_{23}$$
$$+ (b - 6d\lambda^2) I_{24} + d I_{25} \tag{A2}$$

$$a_1 = A \left[B_2 + \frac{\pi R_1^3}{6} (\rho_1 + \rho_2)(1 + 2\xi) \right.$$

$$\left. - 2\pi\lambda^2 \rho_2 \left\{ R_1 + R_2 + \frac{\pi R_1 R_2}{6} (2\rho_1 R_1^2 + \rho_2 R_2^2) \right\} \right]$$

$$+ (B - C) \frac{\pi}{2} R_1^3 (1 - \xi)^{-4} \tag{A3}$$

$$\lambda = \tfrac{1}{2}(R_2 - R_1) \tag{A4}$$

$$A = (1 - \xi)^{-3} \tag{A5}$$

$$\xi = \eta_1 R_1^3 + \eta_2 R_2^3 \tag{A6}$$

$$\eta = \frac{\pi \rho_i}{6} \qquad i = 1, 2 \tag{A7}$$

$$B_2 = 1 + \xi + \xi^2 \tag{A8}$$

$$B = B_2(\rho_1 + \rho_2) \tag{A9}$$

$$C = \frac{72}{\pi} \eta_1 \eta_2 \lambda^2 [R_1 + R_2 + R_1 R_2(\eta_1 R_1^2 + \eta_2 R_2^2)] \tag{A10}$$

$$I_{21} = I_{23} = -\frac{2R_{21}}{k^2} \cos(kR_{21}) + \frac{2\lambda}{k^2} \cos(k\lambda)$$

$$+ \frac{2}{k^3} \sin(kR_{21}) - \frac{2}{k^3} \sin(k\lambda) \tag{A11}$$

$$I_{22} = -\frac{2}{k^2} [\cos(kR_{21}) - \cos(k\lambda)] \tag{A12}$$

$$I_{24} = -\frac{2R_{21}^2}{k^2} \cos(kR_{21}) + \frac{2\lambda^2}{k^2} \cos(k\lambda) + \frac{4R_{21}}{k^3} \sin(kR_{21}) \tag{A13}$$

$$- \frac{4\lambda}{k^3} \sin(k\lambda) + \frac{4}{k^4} \cos(kR_{21}) - \frac{4}{k^4} \cos(k\lambda)$$

$$I_{25} = -\frac{2R_{21}^4}{k^2} \cos(kR_{21}) + \frac{2\lambda^4}{k^2} \cos(k\lambda) + \frac{8R_{21}^3}{k^3} \sin(kR_{21})$$

$$- \frac{8\lambda^3}{k^3} \sin(k\lambda) + \frac{24R_{21}^2}{k^4} \cos(kR_{21}) - \frac{24\lambda^2}{k^4} \cos(k\lambda)$$

$$- \frac{48R_{21}}{k^5} \sin(kR_{21}) \tag{A14}$$

$$+ \frac{48\lambda}{k^5} \sin(k\lambda) - \frac{48}{k^6} \cos(kR_{21}) + \frac{48}{k^6} \cos(k\lambda)$$

$$R_{21} = \tfrac{1}{2}(R_1 + R_2) \tag{A15}$$

$$b = -6R_{21}g_{21}[\eta_1 R_1 g_{11} + \eta_2 R_2 g_{22}] \tag{A16}$$

$$g_{11} = \frac{1}{(1 - \xi)^2} \left[1 + \frac{1}{2}\xi + \frac{3}{2}\eta_2 R_2^2(R_1 - R_2) \right] \tag{A17}$$

$$g_{22} = \frac{1}{(1 - \xi)^2} \left[1 + \frac{1}{2}\xi + \frac{3}{2}\eta_1 R_1^2(R_2 - R_1) \right] \tag{A18}$$

$$g_{21} = \frac{1}{2R_{21}}[R_2 g_{11} + R_1 g_{22}] \tag{A19}$$

$$d = \frac{1}{2}[\eta_1 a_1 + \eta_2 a_2] \tag{A20}$$

$$a_2 = A \left[B_2 + \frac{\pi R_2^3}{6}(\rho_1 + \rho_2)(1 + 2\xi) - 2\pi\lambda^2\rho_1\{R_1 \right.$$

$$\left. + R_2 + \frac{\pi R_1 R_2}{6}(\rho_1 R_1^2 + 2\rho_2 R_2^2)\} \right] \tag{A21}$$

$$+ (B - C)\frac{\pi}{2}R_2^3(1 - \xi)^{-4}$$

R_1 and R_2 are the solvent and solute hard sphere diameters, respectively, and ρ_1 and ρ_2 are the solvent and solute number densities, respectively.

REFERENCES

1. Owrutsky JC, Raftery D, Hochstrasser RM. Ann Rev Phys Chem 45:519, 1994.
2. Oxtoby DW. Adv Chem Phys 47:487, 1981.
3. Oxtoby DW. Ann Rev Phys Chem 32:77, 1981.
4. Elsaesser T, Kaiser W. Ann Rev Phys Chem 42:83, 1991.
5. Chesnoy J, Gale GM. Ann Phys Fr 9:893, 1984.
6. Zwanzig R. Chem J Phys 34:1931, 1961.
7. Bader JS, Berne BJ. J Chem Phys 100:8359, 1994.
8. Adelman SA, Stote RH, Muralidhar R. J Chem Phys 99:1320, 1993.

9. Goodyear G, Stratt RM. J Chem Phys 107:3098, 1997.
10. Egorov SA, Skinner JL. J Chem Phys 105:7047, 1996.
11. Herzfeld KF, Litovitz TA. Absorption and Dispersion of Ultrasonic Waves. New York: Academic, 1959.
12. Delalande C, Gale GM. J Chem Phys 71:4804, 1979.
13. Madigosky WM, Litovitz TA. J Chem Phys 34:489, 1961.
14. Andrew JJ, Harriss AP, McDermott DC, Williams HT, Madden PA, Simpson CJSM. Chem Phys 139:369, 1989.
15. Harris CB, Smith DE, Russell DJ. Chem Rev 90:481, 1990.
16. Tokmakoff A, Sauter B, Fayer MD. J Chem Phys 100:9035, 1994.
17. Myers DJ, Chen S, Shigeiwa M, Cherayil BJ, Fayer MD. J Chem Phys 109:5971, 1998.
18. Woutersen S, Emmerichs U, Nienhuys H-K, Bakker HJ. Phys Rev Lett 81:1106, 1998.
19. Bright FV, McNally MEP, eds. Supercritical Fluid Technology. Vol. 488. Washington, DC: American Chemical Society, 1992.
20. Johnston KP, Penninger JML, cds. Supercritical Fluid Science and Technology. Vol. 406. Washington, DC: American Chemical Society, 1989.
21. Kiran E, Levelt Sengers JMH, eds. Supercritical Fluids: Fundamentals for Application. Vol. 273. Boston: Kluwer, 1994.
22. Kiran E, Brennecke JF, eds. Supercritical Fluid Engineering Science. Vol. 514. Washington, DC: American Chemical Society, 1993.
23. Kendall JL, Canelas DA, Young JL, DeSimone JM. Chem Rev 99:543, 1999.
24. Darr JA, Poliakoff M. Chem Rev 99:495, 1999.
25. Dean JR, ed. Applications of Supercritical Fluids in Industrial Analysis. Boca Raton, FL: CRC Press, Inc., FL: 1993.
26. Shaw RW, Brill TB, Clifford AA, Eckert CA, Franck EU. Chem Eng News 69:26, 1991.
27. Eckert CA, Ziger DH, Johnston KP, Ellison TK. Fluid Phase Equilibria 14:167, 1983.
28. Eckert CA, Ziger DH, Johnston KP, Kim S. J Phys Chem 90:2738, 1986.
29. Economou IG, Donohue MD. AIChE J 36:1920, 1990.
30. Wheeler JC. Ber Bunsenges Phys Chem 76:308, 1972.
31. Sun Y-P, Fox MA, Johnston KP. J Am Chem Soc 114:1187, 1992.
32. Maiwald M, Schneider GM. Ber Bunsenges Phys Chem 102:960, 1998.
33. Sun Y-P, Bunker CE. Ber Bunsenges Phys Chem 99:976, 1995.
34. Tucker SC. Chem Rev 99:391, 1999.
35. Brennecke JF, Tomasko DL, Eckert CA. J Phys Chem 94:7692, 1990.
36. Zagrobelny J, Betts TA, Bright FV. J Am Chem Soc 114:5249, 1992.
37. Zagrobelny J, Bright FV. J Am Chem Soc 114:7821, 1992.
38. Anderton RM, Kauffman JF. J Phys Chem 99:13759, 1995.
39. Heitz MP, Bright FV. J Phys Chem 100:6889, 1996.
40. Heitz MP, Maroncelli M. J Phys Chem A 101:5852, 1997.

41. Carlier C, Randolph TW. AIChE J 39:876, 1993.
42. Randolph TW, Carlier C. J Phys Chem 96:5146, 1992.
43. Kajimoto O. Chem Rev 99:355, 1999.
44. Ishii R, Okazaki S, Okada I, Furusaka M, Watanabe N, Misawa M, Fukunaga T. Chem Phys Lett 240:84, 1995.
45. Nishikawa K, Tanaka I, Amemiya Y. J Phys Chem 100:418, 1996.
46. Sengers JV, Levelt Sengers JMH. In: Croxton CA, ed. Progress in Liquid Physics New York: Wiley, 1978.
47. Poliakoff M, Howdle SM, Kazarian SG. Angew Chem Int Ed Engl 34:1275, 1995.
48. Ben D-Amotz, LaPlant F, Shea D, Gardecki J, List D. In: Supercritical Fluid Technology. Washington, DC: American Chemical Society, 1992, pp 19–30.
49. Blitz JP, Yonker CR, Smith RD. J Phys Chem 93:6661, 1989.
50. Pan X, McDonald JC, MacPhail RA. J Chem Phys 110:1677, 1999.
51. Wood KA, Strauss HL. J Chem Phys 78:3455, 1983.
52. Howdle SM, Bagratashvili VN. Chem Phys Lett 214:215, 1993.
53. Echargui MA, Marsault-Herail F. Chem Phys Lett 179:317, 1991.
54. Echargui MA, Marsault F-Herail. Mol Phys 60:605, 1987.
55. Clouter MJ, Kiefte H, Ali N. Phys Rev Lett 40:1170, 1978.
56. Marsault-Herail F, Salmoun F, Dubessy J, Garrabos Y. J Mol Liq 62:251, 1994.
57. Salmoun F, Dubessy J, Garrabos Y, Marsault-Herail F. J Raman Spect 25:281, 1994.
58. Clouter MJ, Kiefte H. Phys Rev Lett 52:763, 1984.
59. Mukamel S, Stern PS, Ronis D. Phys Rev Lett 50:590, 1983.
60. Strauss HL, Mukamel S. J Chem Phys 80:6328, 1984.
61. Schwarzer D, Troe J, Votsmeier M, Zerezke M. J Chem Phys 105:3121, 1996.
62. Schwarzer D, Troe J, Zerezke M. J Chem Phys 107:8380, 1997.
63. Schwarzer D, Troe J, Votsmeier M, Zerezke M. Ber Bunsenges Phys Chem 101:595, 1997.
64. Schwarzer D, Troe J, Zerezke M. Phys J Chem A 102:4207, 1998.
65. Kim S, Johnston KP. AIChE J 33:1603, 1987.
66. Kim S, Johnston KP. Ind Eng Chem Res 26:1206, 1987.
67. Goodyear G, Tucker SC. J Chem Phys 110:3643, 1999.
68. Urdahl RS, Myers DJ, Rector KD, Davis PH, Cherayil BJ, Fayer MD. J Chem Phys 107:3747, 1997.
69. Myers DJ, Shigeiwa M, Cherayil BJ, Fayer MD. Chem Phys Lett 313:592, 1999.
70. Myers DJ, Shigeiwa M, Cherayil BJ, Fayer MD. J Phys Chem B 104:2402, 2000.
71. Myers DJ, Shigeiwa M, Silbey RJ, Fayer MD. Chem Phys Lett 312:399, 1999.
72. Cherayil BJ, Fayer MD. J Chem Phys 107:7642, 1997.

73. Tokmakoff A, Marshall CD, Fayer MD. JOSA B 10:1785, 1993.

74. N.I.o. Standards, NIST Database ST 14, 9.08 ed. Boulder, Co: U.S. Department of Commerce, 1992.

75. Younglove BA, Ely JF. J Phys Chem Ref Data 16:543, 1987.

76. Rubio RG, Zollweg JA, Streett WB. Ber Bunsenges Phys Chem 93:791, 1989.

77. Jones LH. Spec Acta 19:329, 1963.

78. Jones LH, McDowell RS, Goldblatt M. Inorg Chem 8:2349, 1969.

79. Amster RL, Hannan RB, Tobin MC. Spec Acta 19:1489, 1963.

80. Astholz DC, Troe J, Wieters W. J Chem Phys 70:5107, 1979.

81. Tokmakoff A, Urdahl RS, Zimdars D, Francis RS, Kwok AS, Fayer MD. J Chem Phys 102:3919, 1995.

82. Egorov SA, Everitt KF, Skinner JL. J Phys Chem A 103:9494, 1999.

83. Lebowitz J, Percus JK. J Math Phys 4:116, 1963.

84. Bagchi B. J Chem Phys 100:6658, 1994.

85. Evans R. Adv Phys 28:143, 1979.

86. Lebowitz JL. Phys Rev 133:895, 1964.

87. Stanley HE. Introduction to Phase Transitions and Critical Phenomena. New York: Oxford, 1971.

88. Kawasaki K. Ann Phys 61:1, 1970.

89. Postol TA, Pelizzari CA. Phys Rev A 18:2321, 1978.

90. de Schepper IM, Verkerk P, van AA Well, de Graaf LA. Phys Rev Lett 50:974, 1983.

91. Bafile U, Verkerk P, Barocchi F, de Graaf LA, Suck J-B, Mutka H. Phys Rev Lett 65:2394, 1990.

92. de Schepper IM, Cohen EGD. Phys Rev A 22:287, 1980.

93. Alley WE, Alder BJ. Phys Rev A 27:3158, 1983.

94. Alley WE, Alder BJ, Yip S. Phys Rev A 27:3174, 1983.

95. de Schepper IM, van Rijs JC, van Well AA, Verkerk P, de Graaf LA. Phys Rev A 29:1602, 1984.

96. Egelstaff PA. Adv Phys 11:203, 1962.

97. Everitt KF, Egorov SA, Skinner JL. Chem Phys 235:115, 1998.

98. Adelman SA, Stote RH. J Chem Phys 88:4397, 1988.

99. Friend DG, Ingham H, Ely JF. J Phys Chem Ref Data 20:275, 1991.

100. Altunin VV. Thermophysical Properties of Freons: Methane Series. Washington, DC: Hemisphere Publishing Company, 1987.

101. Platzer B, Polt A, Maurer G, eds. Thermophysical Properties of Refrigerants. Vol. VII. Berlin: Springer-Verlag, 1990.

102. Kenkre VM, Tokmakoff A, Fayer MD. J Chem Phys 101:10618, 1994.

103. Heinemann F, Schmidt H, Peters K, Thiery D. Kristallogr Z 198:123, 1992.

104. Ben-Amotz D, Herschbach DR. J Phys Chem 94:1038, 1990.

105. Torrie BH, Binbrek OS, Powell BM. Mol Phys 87:1007, 1996.

106. Moore P, Tokmakoff A, Keyes T, Fayer MD. J Chem Phys 103:3325, 1995.

107. Moore PB, Ji X, Ahlborn H, Space B. Chem Phys Lett 296:259, 1998.

108. Vijayadamodar GV, Nitzan A. J Chem Phys 103:2169, 1995.
109. Ishii R, Okazaki S, Okada I, Furusaka M, Watanabe N, Misawa M, Fukunaga T. Chem J Phys 105:7011, 1996.
110. Nitzan A, Mukamel S, Jortner J. J Chem Phys 60:3929, 1974.
111. Myers DJ, Urdahl RS, Cherayil BJ, Fayer MD. J Chem Phys 107:9741, 1997.
112. Debenedetti PG, Kumar SK. AIChE J 34:645, 1988.
113. Debenedetti PG, Petsche IB, Mohamed RS. Fluid Phase Equilibria 52:347, 1989.
114. Egoroy SA, Skinner JL. J Chem Phys 112:275, 2000.

16
Vibrational Energy Relaxation in Liquids and Supercritical Fluids

James L. Skinner and Karl F. Everitt
University of Wisconsin–Madison, Madison, Wisconsin

Sergei A. Egorov
University of Virginia, Charlottesville, Virginia

I. INTRODUCTION

This chapter is concerned with how energy deposited into a specific vibrational mode of a solute is dissipated into other modes of the solute-solvent system, and particularly with how to calculate the rates of such processes. For a polyatomic solute in a polyatomic solvent, there are many pathways for vibrational energy relaxation (VER), including intramolecular vibrational redistribution (IVR), where the energy flows solely into other vibrational modes of the solute, and those involving solvent-assisted processes, where the energy flows into vibrational, rotational, and/or translational modes of both the solute and the solvent.

The basic theoretical framework for understanding the rates of these processes is Fermi's golden rule. The solute-solvent Hamiltonian is partitioned into three terms: one for selected vibrational modes of the solute, including the vibrational mode that is initially excited, one for all other degrees of freedom (the bath), and one for the interaction between these two sets of variables. One then calculates rate constants for transitions between eigenstates of the first term, taking the interaction term to lowest order in perturbation theory. The rate constants are related to Fourier transforms of quantum time-correlation functions of bath variables. The most common implementation of this scheme involves performing a classical molecular dynamics simulation for the solute-solvent system, calculating the required time-correlation functions classically, and then possibly multiplying by a quantum correction factor.

In this chapter we first review the general theoretical framework of VER, including a discussion of various different quantum correction factors. We then consider three specific systems, extending and developing the basic framework as needed, and in each case then make detailed comparison with experiment.

The first system we consider is the solute iodine in liquid and super-critical xenon (1). In this case there is clearly no IVR, and presumably the predominant pathway involves transfer of energy from the excited iodine vibration to translations of both the solute and solvent. We intro-duce a breathing sphere model of the solute, and with this model calculate the required classical time-correlation function analytically (2). Information about solute-solvent structure is obtained from integral equation theories. In this case the issue of the quantum correction factor is not really important because the iodine vibrational frequency is comparable to thermal energies and so the system is nearly classical.

The second system is neat liquid oxygen. While one VER pathway involves resonant intermolecular vibration-vibration energy transfer, this is not measured in ensemble-averaged VER experiments (3), and so we do not include this pathway in our theory. The remaining pathways involve vibration-rotation and vibration-translation energy transfer. Our approach to this problem is entirely conventional, performing a classical molec-ular dynamics simulation of a reasonably realistic model of two-site rigid oxygen (4). In this case, however, because of the high vibrational frequency for oxygen, one is nowhere near the classical limit, and so the issue of the quantum correction factor becomes extremely important (5,6). For reasons discussed elsewhere (4,6) we use the Egelstaff approximation scheme (7) for this problem.

Finally, we consider VER of a specific CO stretch mode of $W(CO)_6$ in supercritical ethane (8–11). In principle, all of the pathways described in the first paragraph are operative in this system, and one could proceed with an all-atom classical simulation. Instead, we have tried to simplify the problem, explicitly considering only the solute's vibrations and the translations of both solvent and solute molecules, and again treating the solute as a breathing sphere (12). We again pursue an analytical approach to this problem, making certain arguments concerning the relative time scale of intramolecular vibrational and intermolecular translational dynamics. Since some of the thermodynamic state points considered experimentally involve a near-critical solvent, we had to abandon our integral equation theory for the solute-solvent structure and instead use Monte Carlo simulations.

II. GENERAL THEORY OF VIBRATIONAL ENERGY RELAXATION

We begin this section with a quite general quantum mechanical discussion of the energy relaxation of an oscillator coupled to a bath. Thus the Hamiltonian is

$$H = H_q + H_b + V \tag{1}$$

where H_q is the Hamiltonian for a one-dimensional oscillator with coordinate q, H_b is the Hamiltonian associated with all other (bath) coordinates of the system, and V, the coupling of the bath to the oscillator, involves both q and the bath coordinates.

The oscillator Hamiltonian, which is not necessarily harmonic, has a set of eigenvalues and orthonormal eigenstates defined by

$$H_q|n\rangle = E_n|n\rangle \tag{2}$$

where $n = 0, 1, 2, \cdots$, and the eigenvalues are labeled in order of increasing energy.

If the oscillator is weakly coupled to the bath, in canonical thermal equilibrium the probability of finding the oscillator in the nth state is of course $P_n^{eq} = e^{-\beta E_n}/Z_q$, where $\beta = 1/kT$ and the oscillator's canonical partition function is $Z_q = \sum_n e^{-\beta E_n}$. In addition, the oscillator's off-diagonal (in this energy representation) density matrix elements are zero. The average oscillator energy (in thermal equilibrium) is $E_{eq} = \sum_n E_n P_n^{eq}$.

The general question is, if the oscillator is prepared at time $t = 0$ in a nonequilibrium state with some initial density matrix $\rho(0)$, how does the oscillator's energy relax back to equilibrium? Or to be more precise, defining the nonequilibrium average energy by $\overline{E}(t) = \sum_n E_n P_n(t)$, where the oscillator's nonequilibrium occupation probabilities are $P_n(t)$, what is the time dependence of $\overline{E}(t)$?

In order to answer this question, one needs to know the nonequilibrium probabilities $P_n(t)$. Under certain circumstances it is well understood that these probabilities satisfy the master equation (13):

$$P_n(t) = -\sum_m k_{n \to m} P_n(t) + \sum_m k_{m \to n} P_m(t) \tag{3}$$

where $k_{n \to m}$ is the time-independent rate constant for making a bath-induced transition from oscillator state $|n\rangle$ to state $|m\rangle$. While a detailed discussion of the validity of this equation is not warranted here, suffice it to say that if (1) the oscillator-bath coupling is sufficiently weak (which means that one can use lowest-order perturbation theory to calculate the rate constants and that the "bath correlation time" is much shorter than the inverse of these rate constants), and (2) the equations of motion for the diagonal and off-diagonal density matrix elements are uncoupled (either because of special symmetry properties of the problem or because the off-diagonal elements decay to zero sufficiently rapidly), then the master equation is valid. We will simply assume this to be the case in what follows.

The rate constants in the master equation can be derived from Fermi's golden rule, with the result that (14)

$$k_{n \to m} = \frac{1}{\hbar^2} \int_{-\infty}^{\infty} dt \ e^{i\omega_{nm}t} \langle V_{nm}(t)V_{mn}(0)\rangle \tag{4}$$

where $\omega_{nm} = (E_n - E_m)/\hbar$, $\langle \cdots \rangle = \mathrm{Tr}_b[e^{-\beta H_b}\cdots]/\mathrm{Tr}_b[e^{-\beta H_b}]$, $V_{nm} = \langle n|V|m\rangle$, and for any bath operator O, $O(t) = e^{iH_b t/\hbar} O e^{-iH_b t/\hbar}$. It is easy to show that these rate constants so defined satisfy the all-important (without it the system would not reach equilibrium) property of detailed balance, such that (15)

$$k_{m \to n} = e^{-\beta\hbar\omega_{nm}} k_{n \to m} \tag{5}$$

In general the relaxation to equilibrium of $\bar{E}(t)$ is nonexponential, since the rate matrix in the master equation has an infinite number of (in principle) nondegenerate eigenvalues if there are an infinite number of states $|n\rangle$. There are, however, two instances where the relaxation is approximately exponential. In the first instance one assumes that the initial nonequilibrium state has appreciable population only in the first two oscillator eigenstates, and further that $k_{1 \to 0} \gg k_{1 \to m}$ and $k_{0 \to 1} \gg k_{0 \to m}$ for $m \geq 2$. If one neglects terms involving these small rate constants, the master equation reduces to a pair of coupled rate equations for a two-level system:

$$\dot{P}_0(t) = -k_{0 \to 1}P_0(t) + k_{1 \to 0}P_1(t) \tag{6}$$
$$\dot{P}_1(t) = k_{0 \to 1}P_0(t) - k_{1 \to 0}P_1(t) \tag{7}$$

It is well known that for this pair of equations both populations $P_0(t)$ and $P_1(t)$ relax to equilibrium exponentially, with the same relaxation time, T_1, where $1/T_1 = k_{0 \to 1} + k_{1 \to 0}$. It follows then that the nonequilibrium energy obeys

$$\dot{\overline{E}}(t) = -\frac{1}{T_1}[\overline{E}(t) - E_{eq}] \tag{8}$$

that is, it also relaxes to equilibrium with the same time constant. Note that using detailed balance one can write

$$\frac{1}{T_1} = [1 + e^{-\beta\hbar\omega_{10}}]k_{1\to 0} \tag{9}$$

Since the assumptions leading to the truncation of the state space to two states also typically imply that $e^{-\beta\hbar\omega_{10}} \ll 1$, to an excellent approximation one can then write $1/T_1 \simeq k_{1\to 0}$.

The second instance leading to exponential decay of $\overline{E}(t)$ follows from a more detailed specification of the model. The oscillator-bath coupling term V is a function of oscillator and bath coordinates and can be expanded in powers of q. Matrix elements between oscillator states $|n\rangle$ and $|m\rangle$ of the zeroth-order term in this expansion will vanish, and so this term does not contribute to the rate constants. The leading order term, then, is first order in q. Defining the bath operator F by $F = -(\partial V/\partial q)|_{q=0}$, the rate constants are approximately given by

$$k_{n\to m} = \frac{|q_{nm}|^2}{\hbar^2} \int_{-\infty}^{\infty} dt \ e^{i\omega_{nm}t}\langle F(t)F(0)\rangle \tag{10}$$

We next assume that the oscillator is harmonic, so that $E_n = \hbar\omega_0(n + \frac{1}{2})$. It is also well-known that $q_{nm} \propto \delta_{m,n\pm 1}$, and that $q_{n,n-1} = \sqrt{n\hbar/2\mu\omega_0}$, where μ is the oscillator mass. Therefore, all rate constants vanish except between neighboring pairs of levels, and $k_{n\to n-1} = nk_0$, where k_0 is given by

$$k_0 = \frac{1}{2\mu\hbar\omega_0} \int_{-\infty}^{\infty} dt \ e^{i\omega_0 t}\langle F(t)F(0)\rangle \tag{11}$$

The rate constants $k_{n\to n+1}$ are determined by detailed balance. In this special case one can show (16), perhaps surprisingly, that $\overline{E}(t)$ relaxes to equilibrium exponentially, with the single relaxation time T_1 given by

$$\frac{1}{T_1} = [1 - e^{-\beta\hbar\omega_0}]k_0 \tag{12}$$

which is distinctly different from Equation (9). Of course, if the oscillator is harmonic and the oscillator-bath coupling is expanded to first order in q (in which case $\omega_{10} = \omega_0$ and $k_{1\to 0} = k_0$), then these two expressions must agree in the low-temperature ($\beta\hbar\omega_0 \gg 1$) limit, as is the case.

It is now interesting to consider the classical ($\hbar \to 0$) limits of these two expressions. Actually, in the first instance it is inappropriate to take this limit, since just the opposite limit ($\beta\hbar\omega_{10} \gg 1$) is invoked in truncating the state space to two levels. However, in the second instance, one can smoothly take the classical limit, and one finds that in this case T_1 is given by the usual classical Landau-Teller result:

$$\frac{1}{T_1} = \frac{1}{kT\mu} \int_0^\infty dt \cos(\omega_0 t)\langle F(t)F(0)\rangle_{cl} \tag{13}$$

where the subscript cl on the angular brackets indicates that this is now a classical time correlation function. We have used the fact that in the classical limit any time autocorrelation function is real and even.

There are also situations when one is not in the classical limit, and so Equation (13) would not seem applicable, and instead one would like to approximate one of the quantum mechanical expressions for T_1 by relating the relevant quantum time-correlation function to its classical analog. For the sake of definiteness, let us consider the case where the oscillator is harmonic and the oscillator-bath coupling is linear in q, as discussed above. In this case $k_{1\to0}$ can be written as

$$k_{1\to0} = k_0 = \frac{\hat{G}(\omega_0)}{2\mu\hbar\omega_0} \tag{14}$$

where

$$\hat{G}(\omega) = \int_{-\infty}^\infty dt\, e^{i\omega t}G(t) \tag{15}$$

with $G(t) = \langle F(t)F(0)\rangle$. That is, $\hat{G}(\omega)$ is the Fourier transform of the quantum force-force time-correlation function. We (5,6) and others (7,14,16–18) have discussed at some length various approximate schemes for relating $\hat{G}(\omega)$ to its classical analog

$$\hat{G}_{cl}(\omega) = \int_{-\infty}^\infty dt\, e^{i\omega t}G_{cl}(t) \tag{16}$$

with $G_{cl}(t) = \langle F(t)F(0)\rangle_{cl}$. Here we list three such schemes:

1. The "standard" scheme, where

$$\hat{G}(\omega) = \frac{2}{1 + e^{-\beta\hbar\omega}}\hat{G}_{cl}(\omega) \tag{17}$$

2. The "harmonic" scheme, where

$$\hat{G}(\omega) = \frac{\beta\hbar\omega}{1 - e^{-\beta\hbar\omega}}\hat{G}_{cl}(\omega) \tag{18}$$

3. The Egelstaff scheme (7), where

$$\hat{G}(\omega) = e^{\beta\hbar\omega/2} \int_{-\infty}^{\infty} dt\, e^{i\omega t} G_{cl}(\sqrt{t^2 + (\beta\hbar/2)^2}) \tag{19}$$

Each of these schemes satisfies the important property of detailed balance:

$$\hat{G}(-\omega) = e^{-\beta\hbar\omega}\hat{G}(\omega) \tag{20}$$

The standard scheme has been the most popular one for the last 20 years or so. The harmonic scheme gets its name because if the bath is harmonic (more precisely, if F can be represented as a linear combination of harmonic coordinates), then this scheme is exact (16,19). By comparing these schemes to results from exactly solvable models we have concluded that the standard scheme is really not very accurate at high frequencies and that the harmonic and Egelstaff schemes are more promising (5,6). A fourth scheme satisfying detailed balance,

$$\hat{G}(\omega) = e^{\beta\hbar\omega/4} \left(\frac{\beta\hbar\omega}{1 - e^{-\beta\hbar\omega}} \right)^{1/2} \hat{G}_{cl}(\omega) \tag{21}$$

may also have some merit (6).

Here is a fascinating result: if the harmonic scheme of Equation (18) is applied to Equation (14), using Equation (12) for T_1 recovers exactly the classical Landau-Teller result, as recently shown by Bader and Berne (16). Thus, if the oscillator and bath are *both* harmonic and are bilinearly coupled, the exact quantum result and the exact classical result are identical! This provides some justification for using the purely classical result of Equation (13) even in situations where one clearly is not in the classical limit.

III. I_2 IN LIQUID AND SUPERCRITICAL XENON

With the goal of describing some VER experiments on the solute iodine in Xe solvent (1), in this section we specialize to the case of a diatomic solute in an atomic solvent. In fact, we consider a simplified model where the diatomic solute is replaced with a "breathing" sphere (2). We take the oscillator Hamiltonian for the one-dimensional breathing coordinate to be harmonic:

$$H_q = \frac{p^2}{2\mu} + \frac{\mu \omega_0^2 q^2}{2} \tag{22}$$

The bath consists of the translational motions of the solute and all the solvent atoms. Since all potentials are spherically symmetrical, we write

$$H_b = T + \sum_i \phi(r_i) + \sum_{i<j} \phi_s(r_{ij}) \tag{23}$$

where T is the translational kinetic energy of the solute molecule and the solvent atoms, i and j label solvent atoms, r_i is the distance between the solute and solvent atom i, r_{ij} is the distance between solvent atoms i and j, $\phi(r)$ is the solute-solvent pair potential, and $\phi_s(r)$ is the solvent-solvent pair potential. Assuming all interactions to be of the Lennard-Jones form, we have

$$\phi(r) = 4\varepsilon \left[\left(\frac{\sigma}{r} \right)^{12} - \left(\frac{\sigma}{r} \right)^6 \right] \tag{24}$$

$$\phi_s(r) = 4\varepsilon_s \left[\left(\frac{\sigma_s}{r} \right)^{12} - \left(\frac{\sigma_s}{r} \right)^6 \right] \tag{25}$$

To obtain the oscillator-bath interaction term, we argue that the solute's instantaneous size depends linearly on the breathing coordinate q multiplied by a dimensionless coefficient α. The latter is treated as the single adjustable parameter in the theory, which should on physical grounds be less than but on the order of unity. This leads to (2)

$$V \simeq -qF \tag{26}$$

with

$$F = \sum_i f(r_i) \tag{27}$$

and

$$f(r) = -\frac{12\varepsilon\alpha}{\sigma} \left[2 \left(\frac{\sigma}{r} \right)^{12} - \left(\frac{\sigma}{r} \right)^6 \right] \tag{28}$$

In the iodine experiments the solute is first excited electronically and then enters the ground electronic state with many quanta of vibrational excitation (1). The subsequent vibrational relaxation therefore involves the whole manifold of vibrational states. Since in our model H_q is harmonic, we use Equations (12) and (14), together with the "standard" semiclassical approximation scheme from Equation (17), to calculate the (single-exponential) VER rate. [Note that we did this calculation (2) before we discovered (5,6) how poor the standard approximation scheme generally is, but this is not a serious issue in this case since (as we will see) the system is nearly classical ($\hbar\omega_0 \simeq kT$), and the frequency-domain approximation schemes of Equations (17), (18), and (21) all give essentially the same result that $\hat{G}(\omega_0) \simeq \hat{G}_{cl}(\omega_0)$.] This gives

$$\frac{1}{T_1} = \frac{\tanh(\beta\hbar\omega_0/2)\hat{G}_{cl}(\omega_0)}{\hbar\omega_0\mu} \tag{29}$$

Finally, again using the facts that $G_{cl}(t)$ is even in time and that one can subtract any constant from $G_{cl}(t)$ within the (finite-frequency) Fourier transform without affecting the result, we obtain

$$\frac{1}{T_1} = \frac{2\tanh(\beta\hbar\omega_0/2)}{\hbar\omega_0\mu} \int_0^\infty dt\, \cos(\omega_0 t) C(t) \tag{30}$$

where

$$C(t) = \langle F(t) F(0)\rangle_{cl} - \langle F\rangle_{cl}^2 \tag{31}$$

Note that for $\beta\hbar\omega_0 \gg 1$ this becomes the Landau-Teller result of Equation (13).

To proceed further we now consider analytic approximations to $C(t)$. We first expand $C(t)$ in powers of t:

$$C(t) = C(0)[1 - At^2 + Bt^4 \ldots] \tag{32}$$

and derive (2) approximate but accurate analytic expressions for $C(0)$, A, and B that depend on the previously defined $f(r)$, $\phi(r)$, and $\phi_s(r)$, and on $g(r)$ and $g_s(r)$, the solute-solvent and solvent-solvent radial distribution functions, respectively. To obtain these distribution functions we use the HMSA integral equation approach (20,21). We next approximate the actual time correlation function by the ansatz (22)

$$C(t) = C(0) \cos(bt)/\cosh(at) \tag{33}$$

and choose the parameters a and b so as to reproduce exactly the short-time expansion of Equation (32) through order t^4. This expression for $C(t)$ can be Fourier transformed analytically, leading to an analytic expression for T_1 (2).

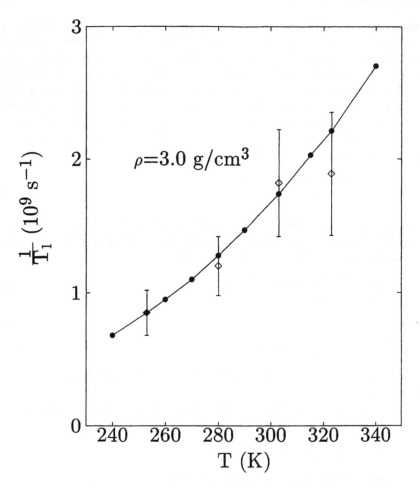

Figure 1 $1/T_1$ vs. T for I_2 in xenon at a density of $\rho = 3.0$ g/cm^3. The open diamonds are the experimental points, and the solid circles (with connecting lines) are from theory.

We then use this theory to analyze the experimental data of Paige and Harris (1), who studied the VER rate of I_2 in liquid Xe at 280 K for a variety of solvent densities from 1.8 to 3.4 g/cm^3 and at a density of 3.0 g/cm^3 for several temperatures from 253 to 323 K (at the higher temperatures in this range Xe is supercritical). We take the following Lennard-Jones parameters: (2) $\varepsilon/k = 349$ K, $\sigma = 4.456$ Å, $\varepsilon_s/k = 221.7$ K, and $\sigma_s = 3.930$ Å. The vibrational frequency of iodine is $\omega_0/2\pi c = 214.6$ cm^{-1}, which in temperature units is 309 K, showing that for this system $\hbar\omega_0 \simeq kT$. For each density and temperature pair we calculate T_1 as described above, and then make a global comparison to all of the experimental data using α as a fit parameter, obtaining $\alpha = 0.7$. In Figs. 1 and 2 we compare theory and experiment at constant density and temperature, respectively. The agreement between theory and experiment at constant density is excellent. The agreement at constant temperature is satisfactory, although the theoretical density dependence is clearly steeper than that observed experimentally.

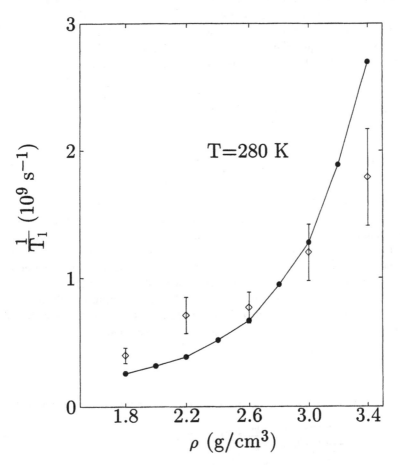

Figure 2 $1/T_1$ vs. ρ for I_2 in xenon at T $= 280$ K. The open diamonds are the experimental points, and the solid circles (with connecting lines) are from theory.

IV. NEAT LIQUID O_2

We next consider a more challenging problem, that of VER in neat liquid oxygen (where now the solute and solvent molecules are both oxygen). This is more challenging for two reasons: (1) Whereas in the last problem VER was dominated by the vibration-translation mechanism, in this case since the solvent is molecular, and one can anticipate that VER will have contributions from both vibration-translation and vibration-rotation channels. This means that rotations of the solvent will have to be considered explicitly, and, for example, we will not be able to use the analytical results discussed above. (2) Whereas the last problem was nearly classical ($kT \simeq \hbar\omega_0$), in this case, because of the much higher vibrational frequency of oxygen, one is deeply in the quantum regime ($\hbar\omega_0 \gg kT$), and so the issue of the most appropriate semiclassical approximation scheme becomes very important.

The Hamiltonian for the vibration of a "tagged" (solute) oxygen molecule is taken to be harmonic, as in Equation (22). The bath Hamiltonian involves the translations and rotations of all (solute and solvent) oxygen molecules (4):

$$H_b = \sum_i \left(\frac{P_i^2}{2M} + \frac{L_i^2}{2I} \right) + \sum_{i<j} \sum_{\alpha,\beta} \phi(|\vec{r}_{i\alpha} - \vec{r}_{j\beta}|) \tag{34}$$

where i and j index the molecules ($i = 0$ is the solute and $i = 1, 2, \ldots$ are solvent molecules), $\vec{P}_i$ is the momentum of the ith molecule, M is the molecular mass, $\vec{L}_i$ is the angular momentum of the ith molecule, and I is the moment of inertia. $\alpha, \beta = 1, 2$ index the two sites (atoms) of each molecule, and $\vec{r}_{i\alpha}$ is the position of site α on atom i. Thus the potential energy involves the site-site pairwise potential, $\phi(r)$, which is taken to be of the Lennard-Jones form in Equation (24). The oscillator-bath interaction has the form $V = -qF$, and F has two contributions, from the centrifugal and potential forces on the oscillator, respectively (4):

$$F = \frac{L_0^2}{I r_e} + \frac{1}{2} \sum_\beta \vec{F}_\beta \cdot \hat{r}_\beta \tag{35}$$

where r_e is the separation of the sites in a molecule (the bond length),

$$\vec{F}_\beta = \sum_\alpha \sum_i \frac{\vec{r}_{i\alpha} - \vec{r}_{0\beta}}{|\vec{r}_{i\alpha} - \vec{r}_{0\beta}|} \phi'(|\vec{r}_{i\alpha} - \vec{r}_{0\beta}|) \tag{36}$$

is the force on site β of the tagged diatomic, and $\hat{r}_\beta$ is the unit vector pointing from the center of mass of the tagged diatomic to site β.

In the oxygen VER experiments (3) the $n = 1$ vibrational state of a given oxygen molecule is prepared with a laser, and the population of that state, probed at some later time, decays exponentially. Since in this case $\hbar\omega_0 \gg kT$, we are in the limit where the state space can be truncated to two levels, and $1/T_1 \simeq k_{1\to0}$. Thus the rate constant $k_{1\to0}$ is measured directly in these experiments. Our starting point for the theoretical discussion is then Equation (14). For reasons discussed in some detail elsewhere (6), for this problem we use the Egelstaff scheme in Equation (19) to relate the Fourier transform of the quantum force-force time-correlation function to the classical time-correlation function, which we then calculate from a classical molecular dynamics computer simulation. The details of the simulation are reported elsewhere (4); here we simply list the site-site potential parameters used therein: $\varepsilon/k = 38.003$ K, and $\sigma = 3.210$ Å, and the distance between sites is $r_e = 0.7063$ Å.

Because of numerical problems emanating from the inherent statistical noise in $G_{cl}(t)$, it is impossible to evaluate the Fourier transform in Equation (19) at the required oxygen frequency of $\omega_0/2\pi c = 1552.5$ cm^{-1}. We have used several methods to extrapolate the numerical results to high frequency, including the use of the ansatz in Equation (33) (4,23). Here, however, we simply assume that the rate would follow an exponential energy gap law (2), $k_{1\to0} \propto e^{-\gamma\omega_0}$, and therefore we perform a linear extrapolation on a log plot (6).

The experimental rate at 70 K is $k_{1\to0} = 360$ s^{-1} (3). Our theoretical rate, obtained from the classical force-force time-correlation function from the molecular dynamics simulation together with the Egelstaff approximation and the extrapolation described above gives $k_{1\to0} = 270$ s^{-1}, in fine agreement with experiment (considering the extremely slow time scale). It is important to understand, however, that the theoretical error bar on this number is probably in the neighborhood of one order of magnitude, due to uncertainties in the potential, in the extrapolation, and in the semiclassical approximation scheme. One might note that the popular "standard" approximation scheme of Equation (17) produces a number that differs from experiment by over five orders of magnitude (6).

V. W(CO)$_6$ IN SUPERCRITICAL ETHANE

Our last specific system involves the solute W(CO)$_6$ in supercritical ethane. In the experiments (8–11) a particular vibrational mode of the solute is excited to $n = 1$, and the population of this level is subsequently probed as a function of time. VER in this system in principle embodies all the theoretical complications of a polyatomic solute in a polyatomic solvent: intramolecular vibration-vibration energy transfer, as well as intermolecular vibration-vibration, vibration-rotation, and vibration-translation relaxation pathways. We have not attempted to provide a general solution for this problem. In fact, in what follows we implicitly assume that in this system the dominant pathways involve only other intramolecular vibrations on the solute and translations. And even so, we pursue a simplified model where the solute-solvent and solvent-solvent potentials are both isotropic (12).

The Hamiltonian again has the form of Equation (1), and in this case we will not assume that H_q (the Hamiltonian for the mode that is excited) is harmonic, but only that Equation (2) applies. The bath involves the other intramolecular coordinates of the solute, $\vec{Q}$, as well as all translational degrees of freedom:

$$H_b = H_{\vec{Q}} + T + \sum_i \phi(r_i) + \sum_{i<j} \phi_s(r_{ij}) \tag{37}$$

where $H_{\vec{Q}}$ is the Hamiltonian associated with $\vec{Q}$, and, as in Section III, T is the translational kinetic energy of the solute and solvent molecules, $\phi(r)$ is the solute-solvent pair potential, given by Equation (24), and $\phi_s(r)$ is the solvent-solvent pair potential, given by Equation (25).

To define the oscillator-bath interaction term, we write the full solute-solvent pair potential as $\phi(r, q, \vec{Q})$, which we take to be

$$\phi(r, q, \vec{Q}) = 4\varepsilon(q, \vec{Q}) \left[\left(\frac{\sigma(q, \vec{Q})}{r} \right)^{12} - \left(\frac{\sigma(q, \vec{Q})}{r} \right)^6 \right] \tag{38}$$

Thus we again assume a Lennard-Jones form, where now the well depth and range parameters depend on the solute's internal vibrational coordinates. Without loss of generality we can define these coordinates so that $q = \vec{Q} = 0$ corresponds to the minimum in the intramolecular potential. The solute-solvent potential in H_b above is actually then $\phi(r) \equiv \phi(r, 0, 0)$, where clearly $\varepsilon \equiv \varepsilon(0, 0)$ and $\sigma \equiv \sigma(0, 0)$. The oscillator-bath interaction term is

$$V = \sum_i [\phi(r_i, q, \vec{Q}) - \phi(r_i)] \tag{39}$$

which can be written as

$$V = \sum_\alpha A_\alpha F_\alpha - \sum_i \phi(r_i) \tag{40}$$

where $\alpha = 1, 2$, and

$$A_1(q, \vec{Q}) = 4\varepsilon(q, \vec{Q}) \left(\frac{\sigma(q, \vec{Q})}{\sigma} \right)^{12} \tag{41}$$

$$A_2(q, \vec{Q}) = 4\varepsilon(q, \vec{Q}) \left(\frac{\sigma(q, \vec{Q})}{\sigma} \right)^{6} \tag{42}$$

$$F_\alpha = \sum_i f_\alpha(r_i) \tag{43}$$

$$f_1(r) = \left(\frac{\sigma}{r} \right)^{12} \tag{44}$$

$$f_2(r) = - \left(\frac{\sigma}{r} \right)^{6} \tag{45}$$

To calculate the VER rate from state $|1\rangle$ to state $|0\rangle$ for the mode with coordinate q, we use Equation (4):

$$k_{1 \to 0} = \frac{1}{\hbar^2} \int_{-\infty}^{\infty} dt \, e^{i\omega_{10}t} \langle V_{10}(t) V_{01}(0) \rangle \tag{46}$$

In this case, because the bath includes intramolecular coordinates, the evaluation of this formula, which involves products of translational and vibrational time-correlation functions, is quite complicated (12). For a polyatomic solute with a large enough number of vibrational modes, we argue that the time-correlation functions for translations decay on the time scale of the inverse of the characteristic frequencies of the translational bath, which is much slower than the decay of time-correlation functions for the intramolecular vibrations. Therefore we can replace the translational time-correlation functions by their initial values, which we evaluate classically. The upshot is that the rate constant can be written as (12)

$$k_{1 \to 0} = \gamma_1 \langle F_1^2 \rangle + \gamma_2 \langle F_2^2 \rangle + \gamma_{12} \langle F_1 F_2 \rangle \tag{47}$$

where

$$\langle F_\alpha F_\beta \rangle = \rho \int d\vec{r} \, f_\alpha(r) f_\beta(r) g(r)$$
$$+ \rho^2 \int d\vec{r}_1 d\vec{r}_2 f_\alpha(r_1) f_\beta(r_2) g(r_1) g(r_2) g_s(r_{12}) \tag{48}$$

ρ is the solvent number density, $f_1(r)$ and $f_2(r)$ are given in Equations (44) and (45), and $g(r)$ and $g_s(r)$ are the solute-solvent and solvent-solvent radial distribution functions, as discussed earlier. The (temperature-dependent) constants γ_1, γ_2, and γ_{12} are related to matrix elements involving $A_1(q, \vec{Q})$ and $A_2(q, \vec{Q})$ [see Equations (41) and (42)], but since we do not have explicit models for those, we treat these constants as adjustable parameters.

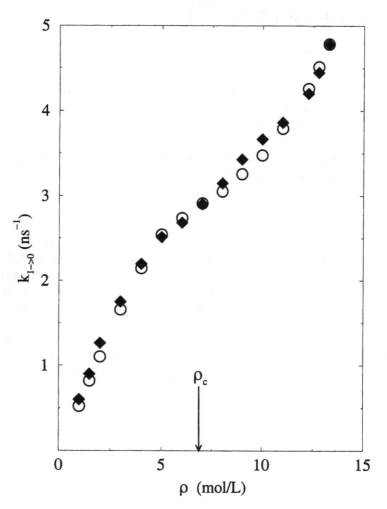

Figure 3 Solvent-induced VER rate for the asymmetrical CO stretch mode of $W(CO)_6$ in supercritical ethane at 307.15 K (critical temperature = 305.33 K) as a function of density. The solid diamonds are the experimental points, and the theory is given by the open circles.

The experiments by Fayer and coworkers measure VER rates of the asymmetrical (T_{1u}) CO stretching mode of $W(CO)_6$ over a wide range of densities and temperatures in supercritical ethane (8–11) (as well as in other solvents that we do not consider herein). We use the following interaction parameters for this system (12): $\varepsilon_s/k = 233$ K, $\sigma_s = 4.24$ Å,

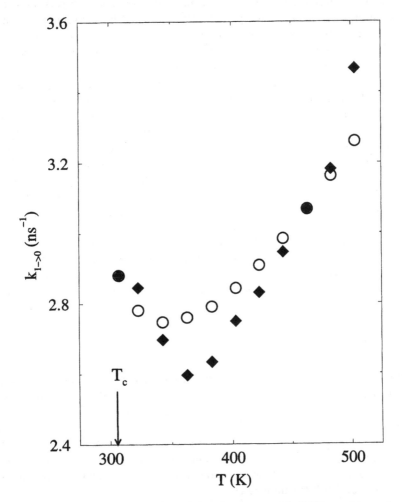

Figure 4 Solvent-induced VER rate for the asymmetrical CO stretch mode of $W(CO)_6$ in supercritical ethane at the critical density (6.87 mol/L) as a function of temperature. The solid diamonds are the experimental points, and the theory is given by the open circles.

$\varepsilon/k = 500$ K, and $\sigma = 5.50$ Å. The other quantities we need before we can undertake a fit of the data are solute-solvent and solvent-solvent radial distribution functions. In Section III we obtained these distribution functions using the HMSA integral equation theory. Unfortunately, however, this and other popular integral equation closures either do not converge or are very inaccurate for infinitely dilute mixtures near the solvent's critical point (24). Since we are particularly interested in this region, we have been forced instead to calculate the radial distribution functions from Monte Carlo simulations for each density-temperature pair studied experimentally.

There are two sets of experimental data. The first is a study of the VER rate constant as a function of density at a constant temperature of 307.15 K (a little less than two degrees above the critical temperature) (8,10,11). The data for the solvent-induced contribution to the rate constant (the gas-phase IVR contribution has been subtracted) are shown in Fig. 3. In Fig. 4 are shown experimental data for the solvent-induced VER rate constant at the critical density as a function of temperature (9). Our theoretical fits to the experimental data are also shown in Figs. 3 and 4. As discussed, for each density-temperature pair, the relevant radial distribution functions were obtained from simulation and the averages $\langle F_1^2 \rangle$, $\langle F_2^2 \rangle$, and $\langle F_1 F_2 \rangle$ were calculated. A global least-squares fit to both sets of data gives $\gamma_1 = 0.0104$ ns^{-1}, $\gamma_2 = 0.00181$ ns^{-1}, and $\gamma_{12} = 0.00862$ ns^{-1}. As seen in Fig. 3, the fit to the density-dependent data is very good. In Fig. 4 one sees that the theory reproduces the interesting trend that the rate decreases and then increases with increasing temperature, but that theory and experiment are not in quantitative agreement.

VI. CONCLUSION

In this chapter we have reviewed the general theory of vibrational energy relaxation for a single oscillator coupled to a bath, and we have discussed the application of these results to three specific systems: iodine in xenon, neat liquid oxygen, and $W(CO)_6$ in ethane. In the first case the bath is the translations of the solute and solvent molecules, in the second case it is the translations and rotations of solute and solvent molecules, and in the third case it is the solute's other intramolecular vibrations and the translations of solute and solvent molecules.

As discussed above, the conventional approach to VER in liquids involves a classical molecular dynamics simulation of the solute (with one or more vibrational modes constrained to be rigid) in the solvent. The required time-correlation functions are computed classically and then multiplied by a quantum correction. The major limitation in this approach is uncertainty in the appropriate quantum correction factor. We have shown that different choices can lead to wildly different results for high-frequency oscillators (6). Although we have made some suggestions about the best way to proceed, at this point this remains an unsolved problem.

One aspect of the last set of experiments on $W(CO)_6$ in supercritical ethane that we have not yet discussed involves the possible role of "local density enhancements" in VER and other experimental observables for near-critical mixtures. The term local density enhancement refers to the anomalously high solvent coordination number around a solute in "attractive" (where the solute-solvent attraction is stronger than that for the solvent with itself) near-critical mixtures (24,25). Although Fayer and coworkers can fit their data with a theory that does not contain these local density enhancements (10,11) (since in their theory the solute-solvent interaction has no attraction), based on our theory, which is quite sensitive to short-range solute-solvent structure and which does properly include local density enhancements if present, we conclude that local density enhancements do play an important play in VER and other spectroscopic observables (26) in near-critical attractive mixtures.

ACKNOWLEDGMENT

The authors are grateful for support from the National Science Foundation (Grant Nos. CHE-9816235 and CHE-9522057).

REFERENCES

1. Paige ME, Harris CB. Chem Phys 149:37, 1990.
2. Egorov SA, Skinner JL. Chem J Phys 105:7047, 1996.
3. Faltermeier B, Protz R, Maier M. Chem Phys 62:377, 1981.
4. Everitt KF, Egorov SA, Skinner JL. Chem Phys 235:115, 1998.
5. Egorov SA, Skinner JL. Chem Phys Lett 293:469, 1998.
6. Egorov SA, Everitt KF, Skinner JL. J Phys Chem A 103:9494, 1999.
7. Egelstaff PA. Adv Phys 11:203, 1962.
8. Urdahl RS, Myers DJ, Rector KD, Davis PH, Cherayil BJ, Fayer MD. J Chem Phys 107:3747, 1997.
9. Myers DJ, Chen S, Shigeiwa M, Cherayil BJ, Fayer MD. J Chem Phys 109:5971, 1998.
10. Myers DJ, Shigeiwa M, Fayer MD, Cherayil BJ. Chem Phys Lett 313:592, 1999.

11. Myers DJ, Shigeiwa M, Fayer MD, Cherayil BJ. J Phys Chem B 104:2402, 2000.
12. Egorov SA, Skinner JL. J Chem Phys 112:275, 2000.
13. Oppenheim I, Shuler KE, Weiss GH. Stochastic Processes in Chemical Physics: The Master Equation. Cambridge, MA: MIT Press, 1977.
14. Oxtoby DW. Adv Chem Phys 47 (Part 2):487, 1981.
15. Berne BJ, Harp GD. Adv Chem Phys 17:63, 1970.
16. Bader JS, Berne BJ. J Chem Phys 100:8359, 1994.
17. Frommhold L. Collision-Induced Absorption in Gases. Vol. 2. Cambridge: Cambridge University Press, 1993.
18. Egorov SA, Berne BJ. J Chem Phys 107:6050, 1997.
19. Skinner JL. J Chem Phys 107:8717, 1997.
20. Zerah G, Hansen J-P. J Chem Phys 84:2336, 1986.
21. Egorov SA, Stephens MD, Yethiraj A, Skinner JL. Mol Phys 88:477, 1996.
22. Douglass DC. J Chem Phys 35:81, 1961.
23. Everitt KF, Skinner JL. J Chem Phys 110:4467, 1999.
24. Egorov SA, Skinner JL. Chem Phys Lett 317:558, 2000.
25. Tucker SC. Chem Rev 99:391, 1999.
26. Egorov SA, Skinner JL. J Phys Chem A 104:483, 2000.

Index

Milton Keynes UK
Ingram Content Group UK Ltd.
UKHW020002071024
449327UK00031B/2621